Laser and Coherence Spectroscopy

Laser and Coherence Spectroscopy

Edited by

Jeffrey I. Steinfeld
Massachusetts Institute of Technology

PLENUM PRESS · NEW YORK AND LONDON

PHYSICS

6 229- 3/38

Library of Congress Cataloging in Publication Data

Main entry under title:

Laser and coherence spectroscopy.

Includes bibliographies and index.
1. Laser spectroscopy. 2. Coherence (Optics) I. Steinfeld, Jeffrey I.   II.
Title: Coherence spectroscopy.
QC454.L3L33   535.5'8   77-11119
ISBN 0-306-31027-9

© 1978 Plenum Press, New York
A Division of Plenum Publishing Corporation
227 West 17th Street, New York, N.Y. 10011

Printed in the United States of America

Contributors

William G. Breiland • Department of Chemistry, University of Illinois, Urbana, Illinois

W. H. Flygare • Noyes Chemical Laboratory, University of Illinois, Urbana, Illinois

J. M. Friedman • Department of Chemistry and Laboratory for Research on the Structure of Matter, The University of Pennsylvania, Philadelphia, Pennsylvania. Present address: Bell Telephone Laboratories, Murray Hill, New Jersey

Charles B. Harris • Department of Chemistry, and Materials and Molecular Research Division of Lawrence Berkeley Laboratory, University of California, Berkeley, California

R. M. Hochstrasser • Department of Chemistry and Laboratory for Research on the Structure of Matter, The University of Pennsylvania, Philadelphia, Pennsylvania

Paul L. Houston • Department of Chemistry, Cornell University, Ithaca, New York

F. A. Novak • Department of Chemistry and Laboratory for Research on the Structure of Matter, The University of Pennsylvania, Philadelphia, Pennsylvania

T. G. Schmalz • Noyes Chemical Laboratory, University of Illinois, Urbana, Illinois

R. L. Shoemaker • Optical Sciences Center, The University of Arizona, Tucson, Arizona

Jeffrey I. Steinfeld • Department of Chemistry, Massachusetts Institute of Technology, Cambridge, Massachusetts

v

Preface

The impact which has been made on spectroscopy by lasers, and by this route on major segments of physics and chemistry, has received ample documentation in the past several years. Two principal themes emerge from examination of the numerous books and monographs now available on this subject: first, an increase in spectral resolution to levels previously undreamed of; and, second, the generation of nonlinear phenomena as a result of the intense radiation fields available from laser devices. There is one additional property of laser radiation which, although used extensively in experiments, does not appear to have been as thoroughly reviewed as the foregoing aspects. This is the spatial and temporal coherence of the radiation field produced by the laser, which makes possible the coherent excitation of molecular energy states. This feature is the subject of the present volume.

While the use of coherence methods in spectroscopy has been paced by lasers, it is by no means restricted to this technology. In the second and fourth chapters, microwave sources are discussed as generators of coherent radiation fields and are used to probe both rotational energy levels and spin states of electronically excited molecules. The phenomena discussed in this book, such as nutation, free induction decay, radiative echoes, rapid passage, and so forth, are really the same in different regions of the spectrum, and themselves echo from one chapter to the next.

The first chapter deals primarily with the response of molecular systems to two or more applied radiation fields, one of which is sufficiently intense to drive the system into a saturation or coherent-excitation regime. These methods are collectively termed "double resonance." In addition, a brief review of spectroscopic notation and experimental devices is given, to serve as an introduction to the remaining four chapters. The chapter is coauthored by Jeffrey I. Steinfeld, of the Massachusetts Institute of Technology, who is also the editor of this volume, and Paul L. Houston, of Cornell University. Houston was formerly a graduate student in Steinfeld's group at M.I.T.

The second chapter deals with coherent transient spectroscopy in the microwave region. While these experiments do not directly involve the use of laser technology, interest in them has greatly increased following the observation of similar phenomena in laser spectroscopy. We are also given a glimpse into the exciting prospect of carrying out Fourier transform spectroscopy using rapid-transient techniques. This chapter was written by Willis Flygare and T. G. Schmalz, of the University of Illinois at Urbana.

The subject of the following chapter is the extension of these techniques to molecular vibrational transitions in the infrared spectrum. Following a thorough development of the theory by means of density matrices, a number of individual phenomena are discussed in detail, with particular attention to superradiance. Richard L. Shoemaker, the author of this chapter, received his graduate training with Flygare at the University of Illinois.

The fourth chapter deals with the use of microwave fields with optical excitation and detection to probe coherent phenomena in electronically excited states of aromatic hydrocarbons. Since the fine structure of the energy levels of these molecules is due to electron spin, we see in this chapter the closest connection with the magnetic resonance methods which have served as models for developing many of the experiments discussed in the preceding chapters. This chapter is by Charles B. Harris, of the University of California at Berkeley, and William G. Breiland, now at the University of Illinois.

The final chapter deals once more with a number of the ideas raised in the earlier part of the book, but from a different point of view. The emphasis here is on scattering of light from coherent radiation sources, when time dependence is of interest. The radiation field is treated quantum-mechanically, in contrast to the semiclassical analysis employed in the first four chapters. A Green's operator technique is used to analyze resonance fluorescence, Raman scattering, two-photon absorption, and other phenomena. The authors—Robin Hochstrasser, J. M. Friedman, and F. A. Novak—are at the University of Pennsylvania.

It is our hope that by collecting into a single volume these treatments of essentially similar phenomena in diverse physical systems, we will have provided the reader with new insights into this exciting area of spectroscopy and may perhaps stimulate the development of new experimental approaches. If suitable areas are identified, additional volumes of this type may be produced in the future.

The authors would like collectively to thank their colleagues, secretaries, and family members who aided in the production of their individual contributions. A special acknowledgment is due Thomas Lanigan, Chemistry Editor at Plenum Publishing Corporation, for nursing this project through to completion.

Jeffrey Steinfeld

Cambridge, Massachusetts

Contents

2 Coherent Transient Microwave Spectroscopy and Fourier Transform Methods

T. G. Schmalz and W. H. Flygare

3 Coherent Transient Infrared Spectroscopy

R. L. Shoemaker

4 Coherent Spectroscopy in Electronically Excited States

Charles B. Harris and William G. Breiland

5   Resonant Scattering of Light by Molecules: Time-Dependent and Coherent Effects

F. A. Novak, J. M. Friedman, and R. M. Hochstrasser

Laser and Coherence Spectroscopy

Double-Resonance Spectroscopy

Jeffrey I. Steinfeld and Paul L. Houston

1.1. Introduction to Double-Resonance Methods

1.1.1. Introduction

The domain of spectroscopy is a large and richly varied one, constituting one of the basic methodologies of physical science. And yet, the way in which spectroscopy is usually carried out is severely limited: typically, one considers the interaction of a resonant medium with a single light wave, and derives absorption and emission properties separately for each frequency of radiation. In this chapter we shall consider those phenomena resulting from the simultaneous application of two radiation fields to a molecular system. With conventional, incoherent optical fields, no new physical features would be introduced by the simultaneous use of two fields; the intensities of each field are sufficiently low that the absorbing system would respond to each field independently, and the overall effect would simply be the sum of the two interactions. But the ever-widening use of intense, coherent, monochromatic laser radiation sources has led to the possibility of using one field to alter the internal state distribution of an absorbing system, thereby influencing its response to a second radiation field and leading to new physical information. For convenience, we shall term such use of two radiation fields to interrogate a sample "double-resonance spectroscopy."

Jeffrey I. Steinfeld • Department of Chemistry, Massachusetts Institute of Technology, Cambridge, Massachusetts   *Paul L. Houston* • Department of Chemistry, Cornell University, Ithaca, New York

In this introductory section we shall review some of the basic principles of the interaction of radiation and matter that are particularly important in double-resonance methods. In Section 1.1.2 we consider electric-dipole interactions between molecular quantum states, which are the source of most strongly allowed spectroscopic transitions. We shall also need to develop expressions for resonance line shapes under various conditions, and for the strength of the coupling of the transition to the electromagnetic field. Section 1.1.3 presents a brief summary of spectroscopic term-value formulas pertinent to the systems that have been extensively studied, and a guide to further reading in this area. In Section 1.1.4 we distinguish between several basic types of double-resonance schemes. Section 1.1.5 is a brief historical survey of the development of these methods from nuclear magnetic resonance, through optical pumping of atoms, to laser techniques in which a wide range of spectroscopic transitions can be studied. The remaining sections of this chapter will discuss the response of a system to saturating fields; experimental double-resonance procedures; results of microwave, infrared, and optical double-resonance experiments; and finally some of the molecular information that is gained from these experiments.

1.1.2. Dynamics of the Interaction of Radiation and Matter

Electromagnetic fields interact with atoms and molecules by way of their time-varying charge distributions. The leading term in the expansion of a field and a charge distribution is the electric-dipole interaction, followed with diminishing strength by electric quadrupole, octupole, and so forth, and various magnetic-multipole interactions. All of these possible interactions may contribute to the absorption and emission properties of a molecule, but in the present treatment we shall be concerned with only the strongest of these, the electric dipole. For weak fields, conventional perturbation-theory treatment gives the rate of induced transitions as

$$R_{k \to m} = \frac{2\pi}{c\hbar^2} I(\nu_0) |\langle m | \boldsymbol{\mu} | k \rangle|^2 \tag{1-1}$$

where $I(\nu_0)$ is the intensity of light at the resonant frequency ν_0, and $\boldsymbol{\mu}$ is the dipole-moment operator between states m and k. When the field strength is sufficiently high, the system no longer responds linearly to the exciting radiation, and instead is driven into an optical pumping regime. This situation is considered further in Section 1.2.

The interaction of a molecule with the radiation field does not occur at the single sharp frequency ν_0, but is instead broadened by a variety of mechanisms to give a resonance line shape with some finite width $\Delta\nu$.

Homogeneous or Lorentzian broadening gives rise to a line shape of the form

$$k_{abs}(\nu) = k_0 \frac{\Gamma}{(\nu - \nu_0)^2 + \Gamma^2/4} \tag{1-2}$$

where $\Gamma = \Delta\nu$ is the half-width of the line at its half-height. This type of broadening arises from radiative decay ("natural" broadening), collisions, or power-broadening resulting from saturation by high-intensity radiation. Inhomogeneous broadening, arising from a distribution of resonant frequencies in an ensemble of molecules, generally has a Gaussian line shape:

$$k_{abs}(\nu) = k_0 \frac{2\sqrt{\ln 2}}{\pi \, \Delta\nu} \exp\left[-4\ln 2 \frac{(\nu - \nu_0)^2}{(\Delta\nu)^2} \right] \tag{1-3}$$

For gas-phase molecules, the dominant mechanism leading to this type of broadening is the Doppler effect arising from the velocity distribution, for which the half-width is

$$\Delta\nu_0 = 2\sqrt{R \ln 2} \, \nu_0 \sqrt{\frac{T}{M}} \approx 7.2 \times 10^{-7} \sqrt{\frac{T}{M}} \tag{1-4}$$

Here T is the absolute translation temperature of the molecules and M the mass in atomic mass units (H atom = 1 amu).

The Lorentzian and Doppler line shapes are compared in Fig. 1-1. Each line has been normalized to unit total area. It is easy to see that the Doppler

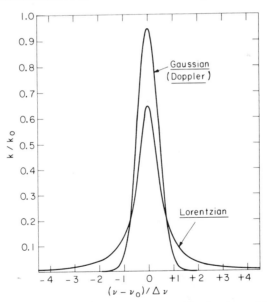

Figure 1-1. Comparison of Gaussian (Doppler) and Lorentzian line shapes. The line shapes are normalized such that $\int k \, d\nu = 1$.

line is more sharply peaked, with very little intensity further than two linewidths away from the center. The Lorentzian line, on the other hand, possesses wings which extend quite far out, to higher and lower frequencies. An important difference between these two broadening mechanisms is in their response to high-intensity radiation. A sufficiently monochromatic radiation source can saturate only a particular velocity group of molecules, leaving the rest unaffected; it is thus possible to "burn a hole" into a Doppler line profile. A homogeneous, or Lorentzian line, on the other hand, is uniformly affected by pumping at any point along the line shape. In this case, the effect of monochromatic pumping is to "bleach" the entire line rather than to "hole-burn." This behavior will be discussed in greater detail in subsequent sections.

It is also useful to get some qualitative feeling for the magnitudes involved in spectroscopic transition strength. A fully allowed electric-dipole transition would have a moment μ of the order of 1 Debye (D) or 10^{-18} esu-cm. The Einstein absorption coefficient is given by

$$B_{12} = \frac{8\pi^3}{3h^2 c} \mu_{12}^2 \qquad (1\text{-}5)$$

or $B_{12} = 2 \times 10^8 \, \text{s g}^{-1}$, for $\mu_{12} = 1$ D. The rate at which molecules are transferred between states 1 and 2 by a field of intensity I is approximately $B_{12}I/\Delta\nu$. Thus, for $I = 1 \, \text{W/cm}^2$ ($10^7 \, \text{ergs s}^{-1} \, \text{cm}^{-2}$), and a transition linewidth $\Delta\nu = 100 \, \text{MHz}$, this rate is of the order of $10^7 \, \text{s}^{-1}$.

It is also possible to express this quantity as an absorption cross section σ,

$$B_{12} = \frac{1}{h\nu_0} \int \sigma \, d\nu \qquad (1\text{-}6)$$

For a transition moment of 1 D and a linewidth of 100 MHz, the cross section is of the order of $10^{-12} \, \text{cm}^2$, which is several thousand times the gas-kinetic collision cross section for a molecule. Such a cross section would be equivalent to a peak absorption coefficient of $5 \times 10^4 \, \text{Torr}^{-1} \, \text{cm}^{-1}$. These values are characteristic of fully allowed atomic transitions, such as the D lines of sodium. For a molecule, in which the transition strength is distributed over a large number of vibrational and rotational components (equivalent to a large effective linewidth), more typical values would be absorption cross sections comparable to molecular dimensions ($1 \, \text{Å}^2 = 10^{-16} \, \text{cm}^2$) and absorption coefficients of the order of a few $\text{Torr}^{-1} \, \text{cm}^{-1}$.

1.1.3. Summary of Molecular Spectroscopy

In this section we shall review some of the expressions for the frequencies of dipole-allowed transitions between energy levels of molecules. The transition frequency is given by

$$\Omega_0 = 2\pi\nu_0 = \frac{E_a - E_b}{\hbar} \tag{1-7}$$

where E_a and E_b are the energies (term values) of the upper and lower states, respectively, involved in the transition. In the experiments to be discussed in this and the succeeding chapters, these term values will refer primarily to rotational, vibrational, and electronic levels in molecules.

The subject of molecular spectroscopy is, of course, far more extensive than can be adequately discussed in this summary. Our intention is just to provide a brief compendium of useful expressions to be used in conjunction with the rest of this book. For further discussion the reader should consult any of a number of textbooks in this area, such as those by Herzberg (1945, 1950, 1966), King (1964), and Steinfeld (1974a).

1.1.3.1. Rotational Energy Levels

Experiments involving microwave radiation generally probe molecular rotational levels. The expressions given here are for a rigid rotor, neglecting centrifugal distortion and vibration–rotation interaction. For further details, the reader is referred to Townes and Schawlow (1955).

Linear molecules, i.e., diatomics or polyatomics such as OCS or HCN, have a particularly simple rotational spectrum. The energy levels are given by

$$E_{\text{rot}} = BJ(J+1) \tag{1-8}$$

with the inertial constant $B = \hbar^2/2I$, I the moment of inertia, and J taking on integer values $0, 1, 2, \ldots$. As a consequence of the selection rule for dipole-allowed transitions, the absorption frequencies form an equally spaced array,

$$\nu_{\text{rot}} = 2B(J+1) \tag{1-9}$$

Nonlinear polyatomic molecules having sufficiently high symmetry are *symmetric rotors*, in which two of the moments of inertia are equal to each other and different from the third. Ammonia, NH_3, and methyl fluoride, CH_3F, are examples of symmetric rotors. For such molecules,

$$E_{\text{rot}} = BJ(J+1) + (C-B)K^2 \tag{1-10}$$

Here $B = \hbar^2/2I_b$ and $C = \hbar^2/2I_c$, where I_c is the moment of inertia about the unique axis and K is the projection of the rotational angular momentum along that axis. Since the transitions usually observed are $\Delta J = \pm 1$ and $\Delta K = 0$, the rotational frequencies still have the form of Eq. (1-9).

Asymmetric rotors are polyatomic molecules in which all three moments of inertia are unequal, such as formaldehyde, H_2CO, and ethylene oxide, $(CH_2)_2O$. The rotational spectrum of such molecules is, in general, exceedingly complex. Approximate closed-form expressions exist for the rotational energies of low-J levels, but in general the terms must be computed numerically. Levels are designated by $J_{K_oK_p}$, where K_o and K_p label the limiting values of K for the prolate and oblate symmetric tops most similar to the actual molecule. Each $\Delta J = 1$ line in the spectrum is split into $2J + 1$ components.

In addition to the overall rotational-energy-level structure, there are several types of fine structure commonly observed in microwave spectroscopy. Two of these in particular should be mentioned, since extensive use has been made of such transitions in double-resonance experiments.

The phenomenon of l doubling occurs when a degenerate vibration (e.g., the bending mode v_2 in a linear molecule) interacts with the rotation. The complete expression for vibrational and rotational energy in this case is

$$E_{vr} = \hbar\omega_2(v_2+1)+B_v[J(J+1)-l^2] \tag{1-11}$$

where the quantum number l can take on the values $v, v-2, \ldots, -v+2, -v$. The rotational line frequencies are given by

$$\nu_{rot} = \left[2B_v \pm \frac{q_l}{2}(v_2+1)J \right](J+1) \tag{1-12}$$

It is also possible to have $\Delta J = 0$ transitions with $\Delta l = 2$, in which case the transition frequencies are

$$\Delta l = \frac{q_l}{2}(v_2+1)J(J+1) \tag{1-13}$$

The splitting constant q_l is approximately equal to $2.6B_e^2/\hbar\omega_2$.

In the $v_2 = 1$ level of the HCN molecule, for example, a series of lines is found corresponding to pure Δl transitions. The constant $q_l = 224.2$ MHz, and some representative transitions are

$$6_1 \rightarrow 6_{-1} \qquad 9,243.3 \text{ MHz}$$

$$8_1 \rightarrow 8_{-1} \qquad 16,147.8 \text{ MHz}$$

$$10_1 \rightarrow 10_{-1} \qquad 24,660.4 \text{ MHz}$$

Such transitions often provide a convenient probe of high-J states in the X- and K-band microwave regions, when the pure rotational lines lie at inconveniently high frequencies.

Inversion doubling of rotational levels occurs when a double minimum in the potential surface permits a molecule to exist in two configurations, as for example the ground state of the ammonia molecule. In this case each

(J,K) level is split into two sublevels of opposite parity, and dipole-allowed transitions exist between the sublevels with $\Delta J = \Delta K = 0$. The magnitude of the splitting depends on the details of the potential barrier between the local minima; generally, an expansion in J and K is possible, as is the case for ammonia:

$$\nu_{\text{inv}} = \nu_0 - aJ(J+1) + bK^2 + cJ^2(J+1)^2 + dJ(J+1)K^2 + eK^4 \quad (1\text{-}14)$$

For ammonia, this results in a series of lines between 16 and 40 GHz for $(J,K) = 1\text{-}16$, which have proved to be very useful in monitoring these rotational levels.

1.1.3.2. Vibrational Energy Levels

Molecular absorption in the infrared is due primarily to the excitation of vibrational motions. Diatomic molecules possess a single vibrational fundamental frequency; nonlinear polyatomic molecules containing N atoms have $3N - 6$ different normal modes of vibration, many if not all of which are infrared-active (Herzberg, 1945). Most of the experiments to be discussed in this book involve excitation of the fundamental vibration in each mode ($v = 1 \leftarrow v = 0$), although some experiments involving *overtones* ($\Delta v > 1$) and *hot bands* (originating on $v > 0$) will also be discussed.

In laser excitation of vibrational modes, it is generally necessary to specify rotational fine structure as well. In the rigid-rotor harmonic-oscillator approximation, rotation is fully separated from vibrational motion and the energies of the two are additive. We shall consider briefly the vibration–rotation spectra of linear or diatomic molecules, and of polyatomic molecules.

In linear molecules, the rotational energy is $E_{\text{rot}} = BJ(J+1)$, and electric-dipole transitions permit $\Delta J = \pm 1$ in addition to $\Delta v = 1$. This gives rise to a band with two branches: the P branch, for J' (upper) $= J''$ (lower) $- 1$, and the R branch for $J' = J'' + 1$. The frequencies of the lines in each branch are given by the following expressions:

$$\nu_P(J) = \nu_0 - (B' + B'')J + (B' - B'')J^2$$
$$\nu_R(J) = \nu_0 + (B' + B'')(J+1) + (B' - B'')(J+1)^2 \quad (1\text{-}15)$$

In the absence of vibration–rotation interaction, $B' = B''$, and each branch consists of equally spaced lines. Each branch goes through a maximum at that value of J corresponding to the most highly populated state in the Boltzmann distribution of rotational energies.

In polyatomic molecules, both rotational quantum numbers J and K can change in a vibrational transition. In addition, if the vibrational symmetry of the upper and lower states is different, there is the possibility of ΔJ

or $\Delta K = 0$, giving rise to a Q branch. This gives rise to a complex series of subbands associated with each vibrational transition; for details, the reader is referred to the treatments cited at the beginning of this section. We will simply explain the notation that will be encountered in ensuing discussions of specific systems. For example, the transition in methyl fluoride that is pumped by the CO_2 laser is assigned as the $^Q R_3(4)$ line of the ν_3 band. The superscript Q refers to the change in the K quantum number, and means that $K'' = 3$ goes to $K' = 3$ in this particular transition; the R(4) has the same significance as in linear molecules, i.e., $J'' = 4$ going to $J' = 5$. One would then expect to see double-resonance effects on those rotational transitions sharing a level with the infrared transition, such as the $(J,K = 4,3 \leftarrow 3,3)$ and $(J,K = 5,3 \leftarrow 4,3)$ lines; and this is indeed the case.

A good example of complexity in infrared spectra and how to deal with it may be had in the ν_3 band of SF_6, which has received an enormous amount of attention because of its strong interaction with CO_2 laser radiation. Figure 1-2a shows the appearance of this band under conventional low resolution and at room temperature. No rotational fine structure can be discerned, and there is a complex series of peaks that cannot be analyzed in terms of any simple rotational structure. This is because the band is extensively overlapped by hot bands originating on thermally populated low-lying vibrational levels of the molecule. When the sample is cooled to about 140°K, as in Figure 1-2b, the structure reduces to that of a simple P,Q,R form expected for this molecule; but individual rotational lines still cannot be resolved (Houston and Steinfeld, 1975). It is only when conventional spectrometry is abandoned and the cooled sample is probed with an ultra-high-resolution tunable diode laser that the rotational structure is resolved (Figure 1-2c; Aldridge *et al.*, 1975). Here we can see the P(J) and R(J) lines equally spaced on either side of the central Q branch. This is the structure expected for a spherical top molecule such as SF_6, in which the central sulfur atom is octahedrally coordinated to the six fluorine atoms, and the rotational energy levels follow a simple $BJ(J + 1)$ form. At higher J levels, each level is split into a number of components by a vibration–rotation interaction known as Coriolis coupling; but we will not go into those details here.

1.1.3.3. Electronic Energy Levels

1.1.3.3a. Atoms. The subject of atomic spectroscopy has been dealt with extensively by Herzberg (1944) and Condon and Shortley (1963), among many others. Here we shall only review the notation necessary for the understanding of experiments to be described in this chapter, particularly in Section 1.6.

Electronic states of atoms are described by the term symbol $^{(2S+1)}(L)_{(J)}$, where S is the total electron spin and L is the total electron orbital angular

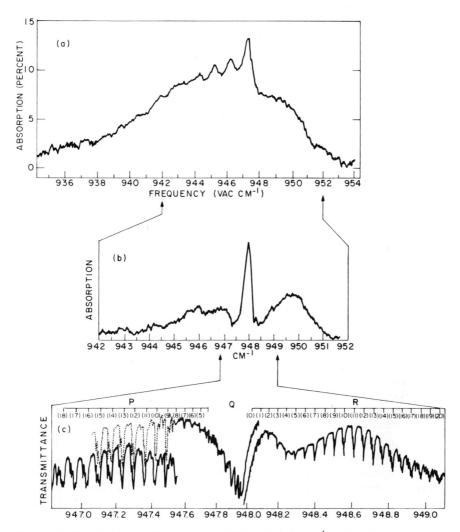

Figure 1-2. Vibrational absorption spectrum of SF_6 near 948 cm^{-1} in the infrared; resolution *increases* and temperature *decreases* from top to bottom. (a) Conventional low-resolution spectrum of SF_6 at room temperature. (b) Grating spectrum of SF_6 at −150°C. Note that the subsidiary Q-branch peaks due to hot bands have disappeared, and that a simple PQR structure is revealed (from Houston and Steinfeld, 1975). (c) Spectrum of SF_6 at −150°C taken with a tunable diode laser having a resolution of 0.001 cm^{-1}. The P and R branches are seen to consist of a number of equally spaced lines, which are further split by Coriolis interactions at higher *J*. The spectrum is a composite of several laser modes which have been "patched" together (from Aldridge *et al.*, 1975).

momentum. States are named S, P, D, F, ..., according to whether $L = 0, 1, 2, 3, \ldots$. The quantum number J is obtained by vector addition of L and S and can take on the values of $|L + S|, \ldots, |L - S|$. Thus, in a ^3S atom, in which $S = 1$ and $L = 0$, we can have only $J = 1$; but in a ^3P atom, such as an excited state of mercury, we can have $^3P_2$, $^3P_1$, and $^3P_0$ states. In the presence of a magnetic field, each J state can be further split into $(2J + 1)$ magnetic sublevels. The energy levels for most of the experimentally known states of the atoms are tabulated in the C. E. Moore tables (1958). In general, the term values must be calculated by self-consistent many-electron methods; there are no simple closed form expressions for the energies in terms of the quantum numbers except for hydrogen-like atoms with only one electron.

Nuclear spin I can couple with total electronic angular momentum J to form a new quantum number $F = I + J$. Again, as a result of vector addition, the allowed values of F are unit steps between $|I + J|$ and $|I - J|$, so that a $^3P_2$ atomic state with $I = \frac{1}{2}$ can have F values of $\frac{5}{2}$ and $\frac{3}{2}$. The hyperfine splitting energies are given by

$$E_{\text{hfs}} = \frac{a}{2}[F(F+1) - J(J+1) - I(I+1)] \tag{1-16}$$

where a is the hyperfine constant of the atom.

1.1.3.3b. Molecules. Molecules are generally treated in the Born–Oppenheimer approximation, in which the total energy is separated into electronic and rotational–vibrational parts. The latter energy expressions were treated in the two preceding subsections.

Electronic states of diatomic molecules are labeled analogously to those of atoms, i.e., by total electron spin Σ, total electron angular momentum Λ, and their sum $\Omega = \Lambda + \Sigma$. For $\Lambda = 0, 1, 2, \ldots$, we have Σ, Π, Δ, ... states. There are additional symmetry quantum numbers $+$ or $-$ for Σ states and u or g for homonuclear molecules. Molecular electronic states are designated by letters of the Roman alphabet, with X always reserved for the ground state. Polyatomic molecules are labeled by the symmetry-group characters of their electronic wavefunctions, e.g., the ^2A and ^2B states of NO_2 (A and B are irreducible representations of the C_{2v} point group to which NO_2 belongs).

There are several additional interactions which must be considered in order to describe the systems with which we shall be dealing later in this chapter.

1. Lambda doubling arises as a result of a coupling between molecular rotation and electronic orbital angular momentum in states with $\Lambda \geq 1$, such as the CN $A^2\Pi$ and CS $A^1\Pi$ states. Each rotational level is split into two components with energy levels

$$F_{\pm}(J) = B_v J(J+1) \pm qJ(J+1) \tag{1-17}$$

where q is the lambda-doubling parameter.

2. Spin doubling occurs in $^2\Sigma$ states such as the CN B($^2\Sigma^+$) state discussed in Section 1.6.1.2, as a result of interaction between molecular rotation N and electron spin S, with total angular momentum $J = N + S$. The two components of each rotational level are

and
$$F_1(N) = B_v N(N+1) + \tfrac{1}{2}\gamma N$$
$$F_2(N) = B_v N(N+1) - \tfrac{1}{2}\gamma(N+1)$$
$$(1\text{-}18)$$

where γ is the spin–rotation interaction constant.

3. Nuclear hyperfine splitting can also occur, as in the case of atoms, by the coupling of nuclear spin I to total molecular angular momentum J to give $F = I + J$.

1.1.4. Definition of Double-Resonance Spectroscopy

In order to make clear the types of experiments we will be discussing in this chapter, we will define double-resonance spectroscopy as the *use of two resonant one-photon interactions in a single molecule* to probe molecular structure and relaxation properties. In this way we distinguish the phenomenon from two-photon spectroscopy and other techniques, which will not be covered in this treatment.

The several types of double-resonance experiments may be distinguished with the aid of energy-level diagrams, shown in Figure 1-3. The most fundamental type is that of *three-level double resonance*, in which fields at two different frequencies, ν_1 and ν_2, couple a given molecular energy level to two other levels. The common level may be either lower in energy than the other two, as in Figure 1-3a, or higher than the other two, as in Figure 1-3b. Another possible variation is to have the common level intermediate in energy, Figure 1-3c. In such a case it is common for detection of the resonance signal to be carried out by observing the spontaneous fluorescence from level a to some other final states of the system. The theory of the three-level system coupled by radiation was first treated by Javan (1957).

A second type is *four-level double-resonance spectroscopy*, in which the two radiation fields probe pairs of levels not having a level in common. For the double-resonance effect to occur, at least one level in each of the two pairs must be coupled to the other by collisional relaxation processes. This can occur by relaxation in the excited state (Figure 1-3d), in the ground state (Figure 1-3e), or both. Another variant is shown in Figure 1-3f, in which relaxation occurs to an intermediate level d, which is then further excited to a higher level c by the field at ν_2, which can then be detected by its fluorescence.

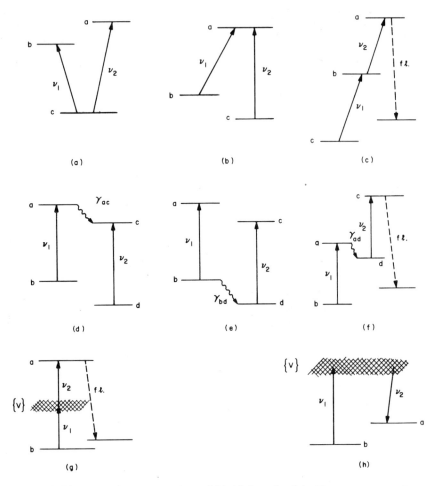

Figure 1-3. Diagrammatic representations of (a)–(c) three-level double-resonance spectroscopy, (d)–(f) four-level double-resonance spectroscopy, (g) two-photon fluorescence, and (h) Raman scattering. The latter two processes are shown as taking place by way of a set of virtual levels in the molecule, $\{v\}$, which is one way of representing the probability of the process in perturbation theory.

The various double-resonance configurations enumerated here are addressed to rather different purposes. The three-level experiments (Figures 1-3a–c) are generally used to achieve *spectrum simplification*. If one of the radiation fields, say ν_1, is sufficiently intense to saturate its transition (see Section 1.2), then all other transition frequencies in the system which possess a level in common with ν_1 will undergo an intensity change—either enhancement or diminution, depending on whether the population in the common level is increased or reduced. Detection at other frequencies, when referenced to the pumping field ν_1, will show a signal at a greatly reduced number of discrete frequencies, compared with the normal absorption or emission spectrum. To a first approximation, all other spectroscopic transitions in the system will be unaffected, and thus not observed. As collisions transfer population from directly pumped levels to others in the system, however (Figures 1-3d–f), additional double-resonance transitions will become active. The magnitude and pressure dependence of these latter signals may then be used to study *molecular relaxation processes* in the system.

We will not consider here such processes as two-photon spectroscopy (Figure 1-3g) or Raman spectroscopy (Figure 1-3h), in which neither ν_1 nor ν_2 are resonant with real transitions in the molecule. (Of course, Figure 1-3c can be considered as a limiting case of Figure 1-3g in which a real molecular state lies at an energy $E_b + h\nu_1$). Lamb dip spectroscopy is a special case of Figure 1-3a in which states a and b are identical, so that $\nu_1 = \nu_2$, but the fields are counterpropagating, so $k_1 = -k_2$. This type of experiment has been rather thoroughly reviewed recently (Letokhov and Chebotayev, 1975; Chebotayev and Letokhov, 1975) and will not be considered in detail here. Furthermore, we will omit experiments in which one field is resonantly absorbed by the molecule and a second, nonresonant laser field is used to probe refractive index changes or other thermal effects. Experiments of this class, such as thermal lensing (Siebert *et al.*, 1974), can indeed be very useful in elucidating relaxation and energy transfer processes, and indeed one must often be careful in distinguishing such thermal effects from true double-resonance phenomena; but they lie somewhat outside the scope of the present discussion.

1.1.5. Historical Survey

The first true double-resonance experiment was carried out at M.I.T. by Brossel and Bitter in 1952. They used a mercury resonance lamp to excite mercury atoms which were simultaneously subject to a radio-frequency (rf) field. Atomic fluorescence, observed at right angles, was used to monitor the changes of populations in Zeeman-split hyperfine levels. The actual rf resonances were observed at constant applied frequency by varying the strength of an applied dc magnetic field. The atomic g factor and resonance linewidths were measured in that experiment.

For the following 10 or so years, double-resonance methods were exclusively the province of spin spectroscopy (nmr and esr). The reason for this was simply that radiation sources of the required intensity and monochromaticity were available only in the rf and microwave regions, using conventional electronic technology. Manufacturers provided double-resonance accessories on their commercial nmr spectrometers, and techniques involving both rf irradiation of nuclear spins and microwave irradiation of electron spins (ENDOR) were also developed. We shall not deal any further with this type of spin spectroscopy; the field is regularly and extensively reviewed elsewhere (see, for example, Dalton and Dalton, 1973).

Double resonance as a general technique in molecular spectroscopy was first realized in microwave spectroscopy of rotational levels, by E. B. Wilson's group at Harvard University and T. Oka at the National Research Council of Canada. Developments were again paced by the technology available for producing high-intensity, narrow-bandwidth radiation fields. With the introduction of the laser, it became possible to extend these techniques to the infrared and optical regions of the spectrum. Each of these regions will be discussed in detail in Sections 1.4–1.6; first, however, let us review in the following two sections, some general considerations about the theory and the experimental procedures involved in double-resonance methods.

1.2. Response of a System to Pumping and Analyzing Radiation Fields

In this section we shall analyze the effects on absorption of radiation of a second radiation field present in the same molecule. Related theory is presented elsewhere in this volume and will not be repeated here; specifically, a density matrix description of the interaction of molecules with radiation fields (Section 3.2) and the Feynman–Vernon–Hellwarth vector representation of a two-level system, leading to the optical Bloch equations (Section 3.4). Instead, we shall confine ourselves to describing the saturation of an absorption line by high-intensity radiation, in Section 1.2.1; double resonance in a three-level system, in Section 1.2.2; and an equivalent rate-constant description, in Section 1.2.3.

1.2.1. Saturation of Molecular Absorption Lines

When a molecular absorption line is subjected to high-intensity radiation, it is often found that the fraction of light absorbed by the sample becomes less than that for low-intensity light; i.e., the absorption coefficient appears to saturate. Let us consider how this comes about, as discussed in the text by Allen and Eberly (1975).

Consider a two-level system, with states a and b coupled by electric-dipole radiation:

$$H = H_0 - \mu_{ab}\varepsilon \cos \omega t \qquad (1\text{-}19)$$

Let the quantum-mechanical amplitude of the upper state, a, be a_2, and that of the lower state, b, be a_1. Then

$$\frac{da_1}{dt} = \frac{i}{2}\kappa\varepsilon a_2[e^{i(\omega-\omega_0)t} + e^{-i(\omega+\omega_0)t}]$$

$$\frac{da_2}{dt} = \frac{i}{2}\kappa\varepsilon a_1[e^{-i(\omega-\omega_0)t} + e^{i(\omega+\omega_0)t}] \qquad (1\text{-}20)$$

where $\kappa = 2\mu_{ab}/\hbar$ and $\omega_0 = (E_a - E_b)/\hbar$. As long as the Rabi frequency $\omega_1 = \kappa\varepsilon \ll \omega_0$, we may neglect the high-frequency terms to give the rotating-wave approximation,

$$\frac{d^2a_2}{dt^2} + i(\omega - \omega_0)\frac{da_2}{dt} + \frac{(\kappa\varepsilon)^2}{4}a_2 = 0 \qquad (1\text{-}21)$$

The solution to Eq. (1-21) is

$$a_2(t) = e^{-i\Delta t/2}(Ae^{i\Omega t/2} + Be^{-i\Omega t/2})$$

and

$$a_1(t) = -\frac{1}{\kappa\varepsilon}e^{i\Delta t/2}[(\Delta-\Omega)Ae^{i\Omega t/2} + (\Delta+\Omega)Be^{-i\Omega t/2}] \qquad (1\text{-}22)$$

with $\Delta = \omega - \omega_0$ and $\Omega = [\Delta^2 + (\kappa\varepsilon)^2]^{1/2}$.

Let us assume the initial conditions $a_1(t_0) = e^{i\theta}$ (i.e., one multiplied by an arbitrary phase factor) and $a_2(t_0) = 0$. This gives the coefficients

$$a_1(t) = e^{[i\theta + i\Delta(t-t_0)/2]}\left[\cos\frac{\Omega}{2}(t-t_0) - i\frac{\Delta}{\Omega}\sin\frac{\Omega}{2}(t-t_0)\right]$$

and

$$a_2(t) = i\frac{\kappa\varepsilon}{\Omega}e^{[i\theta - i\Delta(t-t_0)/2]}\sin\frac{\Omega}{2}(t-t_0) \qquad (1\text{-}23)$$

The corresponding expressions for the populations are

$$n_1(t) = |a_1(t)|^2 = \frac{\Delta^2}{\Omega^2} + \frac{(\kappa\varepsilon)^2}{\Omega^2}\cos^2\frac{\Omega}{2}(t-t_0)$$

and

$$n_2(t) = |a_2(t)|^2 = \frac{\kappa\varepsilon}{\Omega^2}\sin^2\frac{\Omega}{2}(t-t_0) \qquad (1\text{-}24)$$

(See Figure 1.4.)

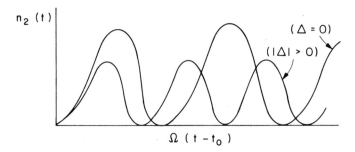

Figure 1-4. Time evolution of the population of the excited state of a two-level system subject to a coherent dipole perturbation. Note that on-resonance pumping ($\Delta = 0$) results in the slowest oscillations having the greatest amplitude.

The effect of collisions can be incorporated in this treatment by averaging Eq. (1-24) over a Poisson distribution of dephasing collisions with characteristic time T_2',

$$dn(t) = \frac{1}{T_2'} e^{-(t-t_0)/T_2'} dt \tag{1-25}$$

to give

$$\langle |a_2|^2 \rangle_{av} = \frac{1}{T_2'} \int_{-\infty}^{t} |a_2(t, t_0)|^2 e^{-(t-t_0)/T_2'} dt_0$$

$$= \frac{1}{2} \frac{(\kappa\varepsilon)^2}{(\omega - \omega_0)^2 + (1/T_2')^2 + (\kappa\varepsilon)^2} \tag{1-26}$$

If this expression is compared with Eq. (1-2), it will be seen that the line shape is that of a Lorentzian, modified by power broadening proportional to ε^2, that is, to the intensity of the radiation field. Note that the limit of $\langle |a_2|^2 \rangle_{av}$, as $\varepsilon \to \infty$, is 0.5. This means that a very intense field will eventually equalize populations between upper and lower levels of a transition.

The power absorbed, which is the observable in this system, can be obtained as

$$\Delta P = \frac{dW}{dt} = \frac{N_1 - N_2}{2T_2'} \frac{\hbar\omega(\kappa\varepsilon)^2}{(\omega - \omega_0)^2 + (1/T_2')^2 + (\kappa\varepsilon)^2} \tag{1-27}$$

As $\varepsilon \to \infty$, ΔP becomes a constant. Thus, the absorption coefficient

$$\alpha = \frac{\Delta P}{P} = \frac{\Delta P}{c|\varepsilon|^2/8\pi}$$

$$= \frac{(N_1 - N_2)}{T_2'} \frac{4\pi\mu_{12}^2\omega}{c\hbar[(\omega - \omega_0)^2 + (1/T_2')^2 + (\kappa\varepsilon)^2]} \to 0 \tag{1-28}$$

and the medium saturates. This can be rewritten as the phenomenological expression

$$\alpha = \frac{\alpha_0}{1 - I/I_{sat}} \tag{1-29}$$

where all the appropriate factors are incorporated into α_0 and I_{sat}. The radiation intensity reaches the order of I_{sat} when $\mu_{12}\varepsilon T'_2 \gtrsim \hbar$, so the saturating field intensity is inversely proportional to the transition moment strength (or the absorption cross section) and the relaxation time of the system. Since the radiation field intensity is proportional to the square of the field strength, this means that I_{sat} is inversely proportional to the square of the product of the transition moment times the relaxation time. Rotational transitions in low-density gases may often be saturated with a few milliwatts of microwave power. It typically requires several watts of monochromatic infrared power to saturate corresponding vibrational transitions, which accounts for the absence of saturated infrared absorption spectroscopy until the advent of suitable lasers. Kilowatts or megawatts of power at optical wavelengths may be required to saturate the broad electronic transitions of molecules in solution; but narrow, intense electronic transitions in atomic vapors may be saturated at relatively low power levels, 50–100 mW.

Saturation of a Doppler-broadened absorption line has been considered by Shimoda and Shimizu (1972) and by others.

1.2.2. Double Resonance in a Three-Level System

The semiclassical theory of double resonance in a three-level system has been treated by Shimoda and Shimizu (1972) and Shimoda (1974, 1976). We summarize their results here.

Consider a three-level system of the type shown in Figure 1-3b, and let a_1, a_2, and a_3 be the quantum-mechanical amplitudes of the states c, b, and a, respectively. Suppose that a strong field ε_p at frequency ω_p is applied close to the transition frequency $\omega_{31} = \omega_{ac}$, and a weaker field ε is applied at frequency ω close to $\omega_{32} = \omega_{ab}$. The equations of motion for the amplitudes are then

$$\frac{da_1}{dt} = \frac{i}{2}\kappa_p \varepsilon_p a_3 e^{i(\omega_p - \omega_{31})t}$$

$$\frac{da_2}{dt} = \frac{i}{2}\kappa \varepsilon a_3 e^{i(\omega - \omega_{32})t} \tag{1-30}$$

$$\frac{da_3}{dt} = \frac{i}{2}[\kappa_p \varepsilon_p a_1 e^{-i(\omega_p - \omega_{31})t} + \kappa a_2 e^{-i(\omega - \omega_{32})t}]$$

with $\kappa_p = 2\mu_{13}/\hbar$ and $\kappa = 2\mu_{23}/\hbar$. Let us impose the initial conditions $a_1(t_0) = e^{i\theta}$ [so that $|a_1(t_0)|^2 = 1$], $a_2(t_0) = a_3(t_0) = 0$. The general steady-state solution for ε_p and ε being constant may be found by setting $da_i/dt = 0$, yielding

$$a_1(t) = A_1 e^{i(\omega_p - \omega_{31} - \lambda)t}$$
$$a_2(t) = A_2 e^{i(\omega - \omega_{32} - \lambda)t} \tag{1-31}$$
$$a_3(t) = A_3 e^{-i\lambda t}$$

where

$$\lambda^3 - (\omega_p - \omega_{31} + \omega - \omega_{32})\lambda^2$$
$$+ [(\omega_p - \omega_{31})(\omega - \omega_{32}) - (\kappa_p \varepsilon_p)^2 - (\kappa \varepsilon)^2]\lambda$$
$$+ (\kappa_p \varepsilon_p)(\omega - \omega_{32}) + (\kappa \varepsilon)(\omega_p - \omega_{31}) = 0 \tag{1-32}$$

This complicated cubic expression for λ can be solved approximately by recognizing that $\varepsilon_p \gg \varepsilon$ in typical experiments. We then find that

$$a_{13}(t) = i\frac{\kappa_p \varepsilon_p}{\Omega_p} e^{i\theta - i(\omega_p - \omega_{31})(t + t_0)/2} \sin\frac{\Omega_p(t - t_0)}{2} \tag{1-33}$$

where $\Omega_p = [(\omega_p - \omega_{31})^2 + (\kappa_p \varepsilon_p)^2]^{1/2}$ and

$$a_{12}(t) = \frac{\kappa \varepsilon \kappa_p \varepsilon_p}{2\Omega_p} \exp\left\{i\theta\left[\frac{e^{i(\Omega_p + \delta)(t - t_0)} - 1}{\Omega_p + \delta} + \frac{e^{-i(\Omega_p - \delta)(t - t_0)} - 1}{\Omega_p - \delta}\right]\right\} \tag{1-34}$$

where $\delta = 2(\omega - \omega_{32}) - (\omega_p - \omega_{31})$.

The net transition probabilities are P_{12} between states c and b and P_{32} between states a and b. These may be found as in Eqs. (1-25) and (1-26) by assuming a Poisson distribution of dephasing collisions and averaging the square of the appropriate probability amplitudes over this distribution. The quantity actually measured in the double-resonance experiment is $\Delta P(\omega)$, the amount of power absorbed from the probing field at frequency ω. This is given by

$$\Delta P(\omega) = \frac{\hbar\omega}{T_2'}[-(N_1 - N_2)\langle P_{12}\rangle_{av} + (N_2 - N_3)\langle P_{32}\rangle_{av}] \tag{1-35}$$

Explicit expressions for the various terms in Eq. (1-35) are, in general, fairly lengthy, but we can calculate a result here for the particular case in which $\omega_p = \omega_{31}$; i.e., the pumping field is exactly on resonance. We then obtain

$$\langle P_{12}\rangle_{av} = \frac{\kappa^2 \varepsilon^2 (T_2')^2}{8}\left[\frac{1}{1 + \omega_+^2(T_2')^2} + \frac{1}{1 + \omega_-^2(T_2')^2}\right.$$
$$\left. - \frac{2 + [2 + \kappa_p^2 \varepsilon_p^2(T_2')^2]\omega_+ \omega_-(T_2')^2}{[1 + \kappa_p^2 \varepsilon_p^2(T_2')^2][1 + \omega_+^2(T_2')^2][1 + \omega_-^2(T_2')^2]}\right] \tag{1-36}$$

with $\omega_\pm = \omega - \omega_{32} \pm (\kappa_p \varepsilon_p/2)$.

The expression for $\langle P_{32} \rangle_{av}$ is the same as that for $\langle P_{12} \rangle_{av}$ [Eq. (1-36)], with the minus sign replaced by a plus sign. Combining these in Eq. (1-35), we have the result

$$\Delta P(\omega) = \hbar\omega \frac{\kappa^2 \varepsilon^2 T_2'}{8} \left[(2N_2 - N_1 - N_3)\left(\frac{1}{1 + \omega_+^2 (T_2')^2} + \frac{1}{1 + \omega_-^2 (T_2')^2} \right) \right.$$

$$\left. + (N_1 - N_3) \frac{2 + [2 + \kappa_p^2 \varepsilon_p^2 (T_2')^2]\omega_+\omega_- (T_2')^2}{[1 + \kappa_p^2 \varepsilon_p^2 (T_2')^2][1 + \omega_+^2 (T_2')^2][1 + \omega_-^2 (T_2')^2]} \right] \quad (1\text{-}37)$$

One interesting point about this result is that the resonance denominator does not peak at the observed molecular frequency ω_{32} but at $\omega = \omega_{32} \pm \frac{1}{2}\kappa_p \varepsilon_p$. This gives rise to a line shape such as that shown in Figure 1-5. The effect is known as resonant modulation splitting of the absorption line. If it is observed in a particular system, it permits measurement of μ_{13}.

Additional results, and incorporation of Doppler broadening into the line shape, are given in the references cited at the beginning of this section.

1.2.3. Rate-Equation Analysis of Double Resonance

The treatment in the preceding section has explicitly considered the coherent interaction between the radiation field and the molecular states; i.e., phase information is retained in the description of both the field- and the molecular-state amplitudes. In many cases a rate-equation description will be adequate. This is especially so for four-level double-resonance experiments involving collisional energy transfer, since collisions which alter the internal state of a molecule almost always lead to dephasing as well. In such cases it is sufficient to use a kinetic analysis in which only the populations in

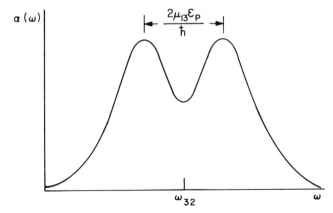

Figure 1-5. Resonant modulation splitting in a three-level double-resonance system.

the various energy levels are considered (Gordon, 1967; Gordon and Steinfeld, 1976).

The evolution of n_f, the fractional population in state f, is given by

$$\frac{dn_f}{dt} = -\sum_i \Pi_{fi} n_i$$

or in matrix form,

$$\frac{d\mathbf{n}}{dt} = -\mathbf{\Pi} \cdot \mathbf{n} \tag{1-38}$$

The matrix $\mathbf{\Pi}$ is the array of inelastic transition rates coupling the levels of the system, and has the property

$$\Pi_{ii} = -\sum_{i \neq j} \Pi_{ij} \tag{1-39}$$

The sums indicated in Eqs. (1-38) and (1-39) go from 1 to N distinct energy states of the system.

The steady-state solution of Eq. (1-38) is obtained by setting

$$\mathbf{\Pi} \cdot \mathbf{n}_{eq} = 0 \tag{1-40}$$

When a continuous-wave (cw) pumping radiation field is applied to the system, the equilibrium populations $\mathbf{n}_{eq}$ are shifted to new steady-state populations $\mathbf{n}_p$, satisfying the equation

$$(\mathbf{\Pi} + \delta\mathbf{\Pi}) \cdot \mathbf{n}_p = 0 \tag{1-41}$$

Intensity changes monitored in absorption or emission will be proportional to $(n_p - n_{eq})_i$, where the transition originates on the ith state. In principle, it is possible to use these data to infer elements of the $\mathbf{\Pi}$ matrix connecting this state with others in the system, but in practice it is often impossible to disentangle competing sequential pathways between different states. This problem can be alleviated by introducing modulated rather than cw pumping of the system. In such a case, the level populations are given by

$$\mathbf{n}(t) = \exp[-(\mathbf{\Pi} + \delta\mathbf{\Pi})t] \cdot \mathbf{n}(0) \tag{1-42}$$

where $\delta\mathbf{\Pi}(t)$ may be square-wave-modulated, for example. This introduces both a ΔI and a phase shift between $\mathbf{n}(t)$ and $\delta\mathbf{\Pi}(t)$, from which individual energy transfer rates may be found.

In the case of pulsed excitation, $\mathbf{n}(t)$ displays exponential relaxation in which the time constants are the eigenvalues of the $\mathbf{\Pi}$ matrix. Because of the stochastic property expressed in Eq. (1-39), and its corollary $\sum_{i=1}^{n} n_i = 1$, one eigenvalue is identically zero. To avoid the computational difficulties introduced by this zero eigenvalue, it is useful to reduce $\mathbf{n}$ to a matrix of order $(N-1)$, which obeys the equation of motion

$$\frac{d\mathbf{n}'}{dt} = \mathbf{A} \cdot \mathbf{n}' + \mathbf{c} \tag{1-43}$$

where $\mathbf{n}'$ is $\mathbf{n}$ reduced by its last element (n_N),

$$\mathbf{c} = -(\Pi_{1N}, \Pi_{2N}, \ldots, \Pi_{N-1,N})$$

and $\mathbf{A} = -(\mathbf{\Pi} + \mathbf{c})$. Diagonalizing this $(N-1) \times (N-1)$ problem, we have

$$\mathbf{A}\mathbf{R}_i = \lambda_i \mathbf{R}_i \qquad (i = 1, 2, \ldots, N-1)$$

where the $\mathbf{R}_i$ are the normalized eigenvectors of the matrix. The solutions to the time-dependent fractional populations are then

$$\mathbf{n}'(t) = \sum_{i=1}^{N-1} \alpha_i \mathbf{R}_i e^{\lambda_i t} - (\mathbf{A})^{-1} \cdot \mathbf{c} \tag{1-44}$$

The coefficients α_i may be found by solving the set of inhomogeneous equations

$$\mathbf{n}'(0) = \sum_i \alpha_i \mathbf{R}_i - (\mathbf{A})^{-1} \cdot \mathbf{c} \tag{1-45}$$

where $\mathbf{n}'(0)$ is the set of populations before the saturating pulse is applied to the system. Inverting this equation gives

$$\alpha_i = \mathbf{R}_i^{-1}[\mathbf{n}'(0) + (\mathbf{A})^{-1} \cdot \mathbf{c}] \tag{1-46}$$

If we let $t \to \infty$ in Eq. (1-44), and note that the λ_i are all negative, we find that

$$\mathbf{n}'_{eq} = -(\mathbf{A})^{-1} \cdot \mathbf{c} \tag{1-47}$$

This is equivalent to the solution of the steady-state problem expressed by Eq. (1-40).

The modulated double-resonance method has most frequently been applied to the microwave experiments discussed in Section 1.4.1. Pumping with pulses short compared to the relaxation time is often used in infrared double-resonance experiments (Sections 1.4.2 and 1.5.2). If observations are made at times very much longer than the slowest relaxation time of the system, the specific population changes due to the pumping radiation are no longer to be seen; instead, one has an overall thermal heating of the system. Several of the initial attempts at double-resonance and fluorescence spectroscopy, using pulsed infrared lasers to excite the system, were complicated by such heating effects. These effects can be reduced or eliminated by going to short observation times, low absorber pressures, and/or addition of an inert buffer gas to increase the heat capacity of the system.

1.3. Experimental Considerations

The rate of progress in any experimental science can only be as fast as the rate at which its experimental tools advance. This section will attempt to outline some of the advances that have made double-resonance methods in spectroscopy possible. Double resonance, as defined in Section 1.4.4, requires two or more radiation sources. It is not surprising, therefore, that many of the significant experimental advances in this field are linked to the introduction of new and more powerful sources of radiation. At the same time, double-resonance experiments are becoming increasingly dependent on sophisticated methods to detect and enhance the signals produced. Advances in these two areas, radiation sources and detection methods, will be considered next.

1.3.1. Radiation Sources

1.3.1.1. Klystrons

For frequencies in the microwave region of the spectrum, the first source available for saturating molecular transitions was the klystron tube (Hamilton *et al.*, 1947). The basic function of this device can be understood by consideration of an electron beam accelerated by a potential, V_0, and made to cross two cavities. An input cavity, called the buncher, applies an oscillating field, $V \cos \omega t$, to the electrons through two grids. This field modulates the electron velocity from an original value, v_0, so that some electrons are accelerated while others are decelerated. Eventually, after a distance z, the electrons become bunched together. This bunching gives rise to a current modulation (Slater, 1950)

$$I = I_0[1 + 2J_1(x) \sin \omega (t - z/v_0)] \tag{1-48}$$

where I_0 is the current in the absence of modulation, J_1 is the first Bessel function, and x, called the bunching parameter, is

$$x = \frac{1}{2} \frac{z}{v_0} \frac{V}{V_0} \omega \tag{1-49}$$

The maximum modulation of current occurs when $J_1(x)$ is a maximum, or when $x = 1.84$. At this point $I = 1.16 I_0$.

If we place the second cavity, called the catcher, at a position z along the electron beam axis, some of the beam power can be captured. For low values of V/V_0 and small grid spacings in the cavity, the power is given by

$$P = I_0^2 J_1(x)^2 R_s \tag{1-50}$$

where R_s is the shunt resistance of the second cavity and is related to its Q.

In the present configuration, the klystron acts as an amplifier. For use as a microwave source, we need to feed back some of the output power to the input cavity. Under these conditions, stable operation at constant power implies a constant magnitude of the bunching parameter, x. It can be seen from formula (1-49) that the frequency ω will be related to the dimensions of the klystron, z.

Actual klystrons dispense with the second cavity and, instead, use a reflector to cause the electron beam to retraverse the original cavity. However, the frequency is still dependent on z and may be tuned slightly by changing the spacing between the reflector and the cavity. Several excellent klystron tubes are now commercially available at frequencies of 3–60 GHz. Some of these have been summarized by Townes and Schawlow (1955), who also give a review of other microwave circuit elements and techniques.

1.3.1.2. Fixed-Frequency Lasers

Many advances in double-resonance techniques are directly related to the development of new laser systems. While the detailed understanding of the mechanisms of these lasers and the phenomena related to them would require an extensive application of quantum electronics, the following sections will attempt only to summarize these properties which make specific lasers useful to the experimentalist. The reader who is unfamiliar with laser physics can be referred to excellent texts on the subject by Yariv (1967), Röss (1969), Levine and DeMaria (1971), Lengyel (1966), and Pantell and Puthoff (1969).

A number of double-resonance applications have relied on the fortuitous coincidence between fixed-frequency lasers and molecular transitions. Table 1-1 lists some of the available fixed-frequency lasers with their output wavelengths and powers. In double-resonance applications to date, the argon ion, CO_2, N_2O, and HF lasers have been the most commonly used. These will be described briefly below.

1.3.1.2a. Argon Ion. Laser oscillation has now been achieved in at least 29 atomic ions [Pressley (1971) and Davis and King (1975)], the most important of which is the argon-ion laser [Bridges (1964)]. While the Ar^+ ground state is the $(3s)^2(3p)^5\,^2P_{3/2}$ state lying 15.75 eV above the ground state for neutral Ar, lasing takes place between two excited states. In each of these excited states a $3p$ electron has been promoted to a higher level. The $(3s)^2(3p)^4 4p$ combination gives rise to the upper laser states, $^2D_{5/2}$, $^2D_{3/2}$, and $^4D_{5/2}$, among others, while the $(3s)^2(3p)^4 4s$ combination gives rise to the lower $^2P_{3/2}$ and $^2P_{1/2}$ laser states. Electron collisions excite the upper D states preferentially so that an inversion is created between the D states and the lower P states. Combinations allowed by the $\Delta J = \pm 1$ selection rule give

Table 1-1   *Common Fixed-Frequency Lasers*

Material	Wavelength	Power Pulsed	Power Continuous
CO_2	9.2–11.0 μm	20 MW	100 W
N_2O	10.3–11.0 μm	1 MW	2 W
HF	2.6–3.0 μm	350 kW	5 W
HCl	3.7–4.0 μm	2 kW	
HBr	4.0–4.6 μm	2 kW	
CO	4.7–6.6 μm	10 kW	20 W
Ho:YLF	2.08 μm	100 kW	100 W
Nd:YAG	1.06 μm	50 MW	100 W
Ruby	694.3 nm	100 MW	
Ar ion	351.1–514.5 nm		10 W
Kr ion	350.7–799.3 nm		1 W
He:Ne	632.8 nm		50 mW
	3.391		10 mW
He:Cd	325–441 nm		50 mW
N_2	337.1 nm	1 MW	
KrF	250 nm	3 MW	
XeF	350 nm	2 MW	
I*	1.315 μm	10 kW	

rise to the dominant 487.99 nm ($^2D_{4/2}$–$^2P_{3/2}$), 496.51 nm ($^2D_{3/2}$–$^2P_{1/2}$), and 514.53 nm ($^4D_{5/2}$–$^3P_{3/2}$) lines. A number of other transitions is also observed in Ar^+, of which the most common are at 501.7, 476.5, 472.7, 465.8, and 457.9 nm. In addition to these Ar^+ transitions, laser action in Ar^{2+} has been observed at wavelengths of 351.1 and 363.5 nm, among others. The large variety of wavelengths available and the relatively simple operation of argon-ion lasers have made them an extremely useful laboratory tool.

Commercial argon-ion lasers use the combination of changeable dielectric mirrors and an intracavity prism for selection of the output lines. The output linewidth may be narrowed to roughly 10 MHz by the use of an intracavity etalon.

A recent development has allowed the modification of a continuously operating argon-ion laser to produce pulses of 600-ps duration every 10 ns (Johnson, 1973). In this configuration, the index of refraction of the prism is modulated at a frequency of $c/2L$, where c is the speed of light and L is the cavity length. For each longitudinal laser mode, this modulation produces sidebands in the frequency domain spaced by $c/2L$. Since this is also the spacing *between* the longitudinal modes, the sidebands interfere with the fundamental of the adjacent mode. The modes then become locked together so that they all have a definite phase and amplitude relationship. Since the

Fourier transform of this frequency-domain pattern is a series of pulses in the time domain, the cavity contains pulses of radiation propagating at a rate of $c/2L$. By deflecting the beam with an acousto-optically driven cavity dumper, output may be extracted at roughly half the rate, $c/4L$.

1.3.1.2b. Carbon Dioxide and Nitrous Oxide. Since its discovery by Patel in 1964, the CO_2 laser has grown to be one of the most powerful and useful radiation sources available. Commercial lasers routinely operate continuously at 30 W or with a pulsed output of 1 MW in 10^{-7} s. An energy-level diagram for this important laser is shown in Figure 1-6.

While early work on the CO_2 laser was done with direct electrical excitation of the CO_2 (Patel, 1964a), it was soon found possible to make use of the near-resonance between $N_2(v = 1)$ and $CO_2(00°1)$ to create inverted CO_2 by specific energy transfer (Legay and Legay-Sommaire, 1964; Patel, 1964b). Lasing action takes place at 10.6 and 9.6 μm from the $CO_2(00°1)$ to either the 10°0 or 02°0 levels, respectively. Since each of these vibrational transitions has a number of rotational levels associated with it, a total of nearly 80 laser lines is available between 9.2 and 11 μm. Additional lines may be obtained with the use of isotopically substituted CO_2.

Continuously excited CO_2 lasers may be Q-switched by either a saturable absorber (Wood and Schwarz, 1967) or by mechanical methods, such as rotation of one of the mirrors. Peak powers of about 10–100 kW at 200 Hz are obtainable by such methods. Alternatively, the discharge itself may be pulsed. In the most popular version of the pulsed discharge CO_2 laser, electrical excitation occurs transverse to the optical axis, and pressures in

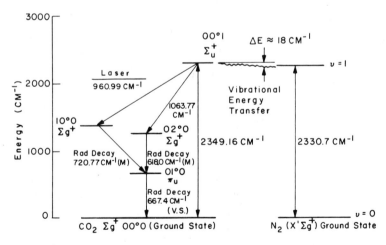

Figure 1-6. Energy-level diagram showing pertinent vibrational levels of CO_2 and N_2. Rotational levels for the vibrational levels have not been shown for simplicity (from Patel, 1964b).

excess of 1 atm are attainable. Such TEA (transversely excited atmospheric) lasers (Beaulieu, 1970; Fortin, 1971) routinely now produce 1–2 J of radiation in 50- to 500-ns pulses at 2 Hz.

The N_2O molecular laser is analogous to the CO_2 laser in design and operation. Important differences are (1) the wavelength is shifted to longer values, (2) the available output powers are lower, and (3) roughly twice as many lines are available. The last difference stems from the fact that alternate rotational states in CO_2 have zero degeneracy, owing to the symmetric nature of the molecule and the nuclear spin of zero on the oxygen atom. Since N_2O is asymmetric, all levels are allowed.

The output wavelengths of the CO_2 and N_2O lasers make them ideal sources for saturating vibrational transitions. Several double-resonance applications of these lasers will be mentioned in Sections 1.4.2, 1.5.2, and 1.6.2.

1.3.1.2c. Hydrogen Fluoride. The first chemical laser, discovered by Kasper and Pimentel in 1965, was based on the reaction $H + Cl_2 \rightarrow HCl^* + Cl$. Since that time, many other chemical reactions have been found to produce lasing action in the hydrogen and deuterium halides. One of the more commonly used of these lasers is the HF chemical laser.

The $F + H_2 \rightarrow HF^* + H$ reaction responsible for inversion in most HF lasers may be initiated by several methods. Kompa and Pimentel (1967) and Kompa *et al.* (1968) have used flash photolysis of hydrogen mixtures with fluorine-containing compounds. One of the more convenient methods for initiation is pulsed electrical excitation of a mixture of SF_6 and H_2 (Wood *et al.*, 1971). Continuous operation of the HF chemical laser is also possible (Spencer *et al.*, 1969). Under most of these conditions, at least five P-branch lines on each of the $3 \rightarrow 2$, $2 \rightarrow 1$, and $1 \rightarrow 0$ vibrational bands are available. These lines cover a spectral range of 2.6–3.0 μm. By substituting D_2 for H_2 in the gas mixture, a DF laser may be produced in the 3.8- to 4.0-μm region.

1.3.1.3. Tunable Lasers

To an increasingly large extent, new double-resonance techniques are becoming dependent on the availability of laser sources which can be tuned to molecular absorptions of interest. Of the many methods for producing tunable radiation which have been reviewed recently (Leone and Moore, 1974; Burdett and Poliakoff, 1974; Yardley, 1975), this section will consider four: dye lasers, diode lasers, lasers based on nonlinear techniques, and the spin-flip Raman laser.

1.3.1.3a. Dye Lasers. Lasers which employ organic dye solutions have provided tunable coverage from the near-ir to the near-uv region of the spectrum (Bass *et al.*, 1971; Schäfer, 1963; Magyov, 1974). The method by which inversion is created in these large molecules is illustrated in Figure

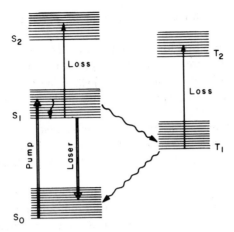

Figure 1-7. Energy-level scheme for an organic dye laser. Radiative processes are indicated by straight arrows and radiationless processes by wavy arrows (from Leone and Moore, 1974).

1-7. A pumping source promotes the dye molecule to its first excited singlet state, S_1. Rapid radiationless relaxation in the vibrational manifold of S_1 then causes population of the lowest vibrational level, which becomes inverted relative to high vibrational levels of the ground state singlet, S_0. Because the line-shape functions in solution are so broad, a nearly continuous spectrum of S_1–S_0 laser emission is obtained. However, in order for lasing to be possible, the gain from the inversion must overcome three types of optical-loss mechanisms. The first is loss of inversion from spontaneous emission. Since the lifetime of S_1 is on the order of 5–10 ns, inversion requires a high pumping level before stimulated emission can compete favorably with spontaneous radiation. A second loss may occur due to S_1–S_2 absorption of the laser frequency. Finally, intersystem crossing to the triplet state T_1 not only depletes the upper lasing level but also makes possible a loss due to triplet absorption at the lasing frequency. By using suitable triplet quenchers, this last absorptive loss may be minimized.

Dye lasers differ from one another in two main features: the structural and spectral properties of the dye and the type of pump source used for excitation. Drexhage (1973) has reviewed the requirements and availability of laser dyes prior to 1973. The recent discovery of new dyes for the near-infrared region of the spectrum (Webb *et al.*, 1974) has greatly extended the available spectral range of dye lasers.

A variety of pumping sources is available for dye excitation. The least costly pump source is the flash lamp, first used by Sorokin *et al.* (1968). Commercial systems based on flash lamp pumping typically provide energies of 6 mJ per pulse at 30 Hz with a pulse width of about 1 μs. Technical considerations for this type of excitation include the speed of the flash lamp, especially for shorter wavelengths, and the problems associated with thermal instability.

The simplest method for pumping a dye laser is to use another laser. Two pulsed systems are commonly available. Figure 1-8 shows a typical configuration for a N_2-laser pumped dye laser (Hänsch, 1972). Commercial systems based on this design produce 10-ns pulses of 15-kW peak power at a repetition rate of 50 Hz and a linewidth of 0.03 cm^{-1}. A second common laser for pumping dyes is the frequency-doubled Nd:YAG laser at 530 nm. This system has the disadvantage of not being able to pump dyes whose lasing wavelength is shorter than 530 nm unless the pump laser is redoubled to 265 nm.

Continuously operating dye lasers were first demonstrated by Peterson *et al.* (1970). A focused Ar$^+$ laser was used to pump a flowing dye solution contained in a cell. The use of a windowed dye cell has been almost totally supplanted now by the use of a free-flowing jet stream from a nozzle (Runge, 1972). This improvement eliminates the possibility of window burning by the pump laser. Commercially available argon-ion pumped dye lasers provide 1–4 W of power of the 440- to 780-nm spectral range. While linewidths of 10 MHz are commercially available, with extreme care it is possible to obtain resolution approaching 500 kHz (Wu *et al.*, 1974).

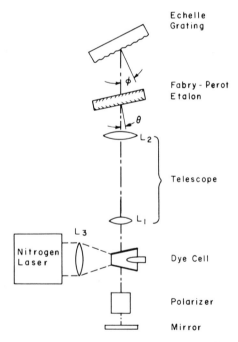

Figure 1-8. Basic components of a narrow-band tunable dye laser (from Hänsch, 1972).

With its broad tunability, high peak and continuous powers, and narrow linewidth, the dye laser is the most versatile of all the available tunable lasers.

1.3.1.3b. Diode Lasers. The band gap of a semiconductor diode may be pumped by electron injection to produce an excess of holes in the n region and of electrons in the p region. As these holes and electrons recombine with free charge carriers, radiation is emitted at a frequency corresponding to the band-gap energy difference. If the faces of the semiconductor crystals are polished to form a cavity, and if pumping can overcome the optical losses, stimulated emission will build up at those frequencies allowed by the relationship $v_p = p(c/2Ln)$, where p is an integer, L is the cavity length, c is the speed of light, and n is the index of refraction of the semiconductor material. Since the bandwidth of the spontaneous recombination radiation may be 5–50 cm^{-1} wide, whereas the modes are separated by some 1–2 cm^{-1}, several modes will normally lase at the same time. These can easily be separated, however, by passing the output through a monochromator.

Coarse tuning of the laser is accomplished by doping the semiconductor to select the bandgap, while fine tuning is performed by varying either the length of the crystal or its index of refraction. The length may be altered by application of external pressure, while the index of refraction is most easily altered by changing the current and, thus, the temperature. The band gap may also be altered by application of an external magnetic field (Butler, 1969).

Disadvantages of diode lasers include the lower power (10–300 μW) and small tuning range (10–50 cm^{-1}) for a particular crystal. Advantages include the narrow bandwidth (50 kHz) and the relative ease of operation. Lead salt tunable diode lasers have recently been made available commercially (Carson Alexiou, Laser Analytics). These lasers may be purchased to operate at a frequency range of >50 cm^{-1} centered between 300 and 3400 cm^{-1} with about 250 μW of cw power. Although this low power level precludes use of the tunable diode laser as a general saturating source, the laser's narrow linewidth and stability make it an excellent probe source for double-resonance applications.

1.3.1.3c. Nonlinear Devices. If the macroscopic polarization of a medium is expanded as a function of the applied electric field, we obtain

$$P = \chi^{(1)}E + \chi^{(2)}E^2 + \cdots \tag{1-51}$$

where $\chi^{(i)}$ is the ith-order susceptibility. Under normal conditions, the electric field E is small, and only the first term of the expansion is important. However, the intense fields generated by lasers induce polarizations for which the $\chi^{(2)}E^2$ term is a significant contribution. One manifestation of this component is frequency doubling. If a strong field at frequency v_p passes through a crystal of the proper symmetry, a polarization may be developed at $2\nu_p$ due to the $\chi^{(2)}E^2$ term in the expansion. A more subtle manifestation of the $\chi^{(2)}E^2$ term is the phenomenon known as parametric oscillation. In this case the pump beam at ν_p is split into two beams, the idler at ν_i and the signal at ν_s. Conservation of energy requires that $\nu_i + \nu_s = \nu_p$, while conservation of momentum requires that the three beams travel through the medium in phase with one another. The latter condition is most easily met in a temperature-controlled birefringent crystal such as $LiNbO_3$. In practice, an optical cavity encloses the crystal and allows either ν_s or ν_i to oscillate. The output may be tuned by selecting phase matching conditions for the desired combination of idler, signal, and pump. In most cases this control is derived from the temperature or angle of the crystal, or from the frequency of the pumping radiation. One commercially available unit (Chromatix) pumps a temperature-controlled $LiNbO_3$ crystal with either the frequency-doubled Nd:YAG laser or a tunable dye laser. The output wavelength is tunable from about 0.55 to 3.5 μm. A more advanced system demonstrated by Herbst *et al.* (1974) uses a Nd:YAG laser at 1.06 μm to pump an angle-tuned crystal. The output from this system was tunable from 1.5 to 4.5 μm and was powerful enough that second nonlinear processes were possible.

1.3.1.3d. Spin-Flip Raman Laser. The InSb spin-flip Raman laser has shown great potential as a tunable source in the 5.2- to 6.0-μm and 9.0- to 12.8 μm regions of the spectrum (Patel and Shaw, 1971; Dennis *et al.*, 1972; Häfele, 1974; Butcher *et al.*, 1975). To date, however, the high cost and inconvenient operation of this laser have limited its usefulness in double-resonance applications. Future advances both in new materials and in extension of operation to higher temperatures or lower fields may make it a more prevalent experimental tool.

Like the parametric oscillator, the spin-flip Raman laser is based on a nonlinear mixing rather than on a population inversion. In the normal Raman effect, Stokes scattering is observed at a frequency $\nu_s = \nu_p - \nu_t$, where ν_p is the pumping frequency and ν_t is the frequency corresponding to an allowed Raman transition. In the InSb crystal, the transition of interest arises because the conduction electrons may align themselves either parallel or antiparallel to an applied magnetic field. The frequency separation between these two states is given by $\nu_t = g^* \mu_B B / h$ where g^* is the effective g factor, μ_B is the Bohr magneton, and B is a variable external magnetic field. If the pump frequency ν_p is intense enough, terms in the polarization of the type $\chi^{(3)}E^3$ cause coupling of the pump radiation and the Stokes scattered radiation and produce stimulated Raman emission at $\nu_s = \nu_p - \nu_t$. Since ν_t is proportional to the magnetic field B, the output is tunable, usually by about 1.6 cm^{-1} per kG of field.

Patel and Shaw (1970) were the first to obtain spin-flip Raman lasing in InSb. In their experiments a CO_2 laser was used as a pump source.

Mooradian *et al.* (1970) soon realized the advantage of the CO laser as the pumping source. At roughly 5.3 μm the stimulated Raman effect is enhanced by resonance between the band-gap energy and the pump energy. Continuous output powers of roughly 200 mW have been obtained from such systems. By pumping with a frequency-doubled pulsed CO_2 source, it is also possible to obtain pulses of kilowatt peak power in the 5-μm region (Wood *et al.*, 1973).

While spin-flip Raman lasers are not commercially available, they show promise as both pump and probe sources for double-resonance applications.

1.3.2. Signal Detection and Enhancement

Although the development of laser sources was the single most important experimental advance needed to extend double-resonance spectroscopy, many double-resonance experiments could not have been performed if advances in detection and signal processing had not occurred simultaneously. This section will summarize the methods available for detection (Section 1.3.2.1) and averaging with lock-in amplifiers (Section 1.3.2.2), boxcar averagers (Section 1.3.2.3), and transient recorders (Section 1.3.2.4).

1.3.2.1. Detectors

While the methods for detection of radiation are by now fairly standard, especially in the microwave region (Section 2.2), steady advances during the last decade have increased both the efficiency and frequency response of most common detection systems. In the visible and ultraviolet regions of the spectrum, photomultiplier tubes are now available with quantum efficiencies in excess of 10% for the spectral range 120–850 nm, and in excess of 1% between 850 and 1000 nm. The time response for these tubes can typically be made less than 3 ns, while the dark current can be kept, in most cases, to below a few nanoamps. A useful guide to photomultiplier tubes is published by RCA (1970).

The recent development of the silicon vidicon has extended detection from the single spatial element to the multichannel regime. These instruments (Olson, 1972) have a broad spectral range and a high quantum efficiency. Coupled with a suitable spectrometer, they speed data acquisition by recording the entire spectrum rather than a single spectral feature.

For the infrared region of the spectrum, although semiconductor detectors are now made which approach ideal performance, these are frequently limited by the presence of background radiation from the room temperature surroundings. Methods of reducing this background by using cooled interference filters and by limiting the field of view of the detector to include only the area from which the signal emanates have been discussed in two excellent texts by Hudson (1969) and Hudson and Hudson (1975). Infrared detectors are characterized by a parameter D^* which compensates for the variation of detectivity on area and bandwidth,

$$D^* = \frac{(A\,\Delta F)^{1/2}}{\text{NEP}} \qquad (1\text{-}52)$$

where A is the area of the detector, ΔF is the electronic bandwidth, and NEP is the equivalent radiant power developed by the noise. Typical values of D^* for a variety of detectors as a function of wavelength are shown in Figure 1-9.

1.3.2.2. Lock-In Amplifier

For the detection of signals which are either dc or slowly varying in time, a substantial reduction in noise may be obtained by decreasing the bandwidth of the detection electronics with a simple low-pass filter. Both shot noise and Johnson noise, which are proportional to the square root of

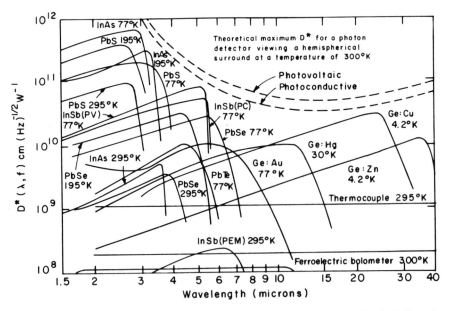

Figure 1-9. Comparison of the D^* of various infrared detectors when operated at the indicated temperature. Chopping frequency is 1800 Hz for all detectors except InSb (PEM), which is 100 Hz; ferroelectric bolometer, 100 Hz; thermocouple, 10 Hz; and thermistor bolometer, 10 Hz. Each detector is assumed to view a hemispherical surround at a temperature of 300°K (from Hudson, 1969).

the frequency bandwidth, will then be dramatically reduced. This method is effective, however, only until the shot and Johnson noise become less significant than the flicker noise, whose amplitude increases as $1/f$ and is nonnegligible below about 100 Hz.

A convenient method for overcoming this problem is to modulate the dc signal at some frequency f_c which is well above the region where flicker noise is significant. A narrow-band tuned amplifier centered at f_c will then effectively reduce the shot and Johnson noise and not be limited by flicker noise. In practice, this technique of shifting the measurement frequency from dc to f_c is itself limited unless the modulation frequency f_c and detection electronics can be locked together to prevent f_c from drifting relative to the bandpass of the narrow filter.

The lock-in amplifier overcomes the drift problem while taking advantage of a simple correlation effect. Consider a reference square wave at f_c which modulates a signal of interest. The cross-correlation between this reference signal R and any other function of time S is given by

$$C_{RS}(\tau) = \lim_{T \to \infty} \frac{1}{2T} \int_{-T}^{+T} R(t)S(t-\tau)\, dt \qquad (1\text{-}53)$$

where τ is a delay time. In particular, if S is random noise, which by definition is uncorrelated with R, the correlation function C_{RS} will tend to zero as the measurement time increases. However, the cross-correlation of a signal at f_c with the reference signal which causes it will yield a definite constant whose value will depend on the phase delay, τ, between the signal and the reference. If both signal and noise are cross-correlated with the reference, the result will be an output proportional to the signal with relatively little noise. Signal-to-noise improvements of 100–1000 (40–60 db) are typical.

The conceptual form of the lock-in amplifier is illustrated in Figure 1-10. The reference signal activates a switch which couples alternate 180° phase components of the signal and noise to a capacitor. The noise, which

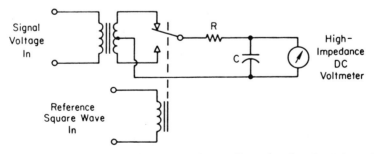

Figure 1-10. Synchronous detector in conceptual form to illustrate action of lock-in amplifier (from *P.A.R. Tech. Publ. T-198A*).

has no correlation with the reference, will be averaged to zero by the *RC* filter. The square-wave-modulated signal, however, will be rectified by the switching system and will produce a finite dc component.

Figure 1-11 illustrates the block diagram of a commercial lock-in amplifier. A phase shifter is used to align the reference and signal so that the maximum output is obtained. In some applications, for example the measurement of vibrational energy transfer in methane excited by a modulated cw source (Yardley and Moore, 1963), it is this phase information which is of interest. In other applications, once the phase has been adjusted, the amplitude of the dc output is measured as a function of external variables such as wavelength, pressure, or magnetic field.

1.3.2.3.   *Boxcar Averager*

The application of cross-correlation signal processing need not be limited to square-wave-modulated dc signals. Any repetitive waveform related in time to a reference signal or trigger may be enhanced with respect to random noise by a device called a boxcar integrator. This instrument uses an aperture pulse of variable delay, τ, with respect to the trigger to interrogate the signal waveform for a small duration of time. When the aperture is open, the signal voltage is switched to charge a capacitive memory. Random noise, which is completely uncorrelated with the trigger signal, averages to zero after a large number of repetitions. The signal waveform, on the other hand, always has the same voltage at the delay τ with respect to the trigger. After a suitable number of repetitions the capacitively stored voltage will bear a fixed relation to the waveform voltage at the chosen delay. As this delay is slowly swept through the time interval of interest, an averaged waveform is presented at the output of the capacitive memory.

The signal-to-noise improvement ratio in such a system depends on which of two types of averaging is used. For linear averaging, in which the output voltage increases by roughly the same amount with each repetition, the signal-to-noise improvement ratio is simply the square root of the number of repetitions at each value of τ. For exponential averaging, the output voltage asymptotically approaches a final value proportional to the input signal. In this mode of operation, the signal-to-noise ratio varies as $(2TC/AD)^{1/2}$, where TC is the *RC* time constant for charging the capacitive memory and AD is the aperture duration.

Commercial boxcar integrators are available with aperture duration as short as 10 ns. However, by coupling the integrator to a specially designed sampling head, it is possible to attain 25-ps aperture durations and system response times of 100 ps. Signal-to-noise improvements of 10–100 are typical.

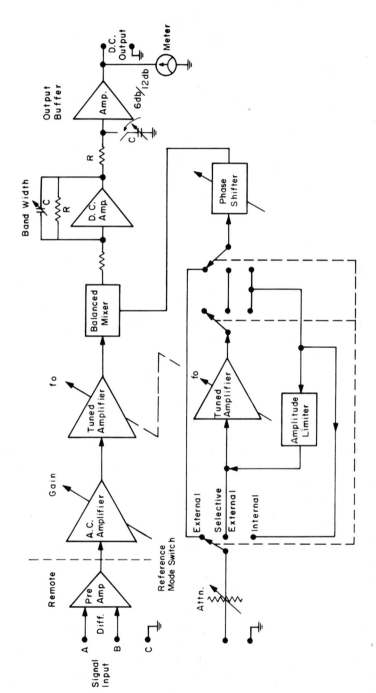

Figure 1-11. Block diagram of a typical lock-in amplifier. With the exception of the preamplifier, all the elements of the unit are usually contained in a relay rack chassis (from *P.A.R. Techn. Publ. T–198A*).

There are two main difficulties with the boxcar averager as described above. The first is that the averaging is inefficient. If 100 points of a waveform are desired with a signal-to-noise improvement of 10, the number of repetitions sampled must be at least $100 \times (10)^2 = 10^4$. The Waveform Eductor improves this situation by providing 100 separate apertures, each of which samples and stores a different part of the waveform. The second disadvantage of the boxcar averager is that, for low repetition rates, capacitor leakage limits the accuracy of the capacitive memory. While this problem has been overcome by sequential analog-to-digital and digital-to-analog conversion to refresh the memory, the resulting system is somewhat clumsy. Both of these problems, lack of efficiency and imperfect storage, may be solved by the use of a transient recorder coupled to a computer or hard-wired averager, as described below.

1.3.2.4. Transient Recorder

The block diagram of a typical transient recorder is shown in Figure 1-12. The time interval containing the signal of interest is divided into 2048 segments which are timed with respect to a trigger input. During each of these segments or channels, the input signal is sampled and held long enough for a fast analog-to-digital converter to digitize the value of the signal voltage and store it in a memory. After all 2048 channels have been converted, the transient recorder presents the output either as sequential digital words or as a reconstructed analog signal.

So far, no averaging has occurred. However, if the digital output is stored by a computer or hard-wired digital averager, the transient recorder can be cleared and made ready to analyze a second signal. Following acquisition of the second signal, the computer or averager must add the digital output to its memory and again free the recorder to acquire a third signal waveform. After this sequence has been repeated n times, the computer will contain a sum in each location of its memory which is simply n times the average signal voltage at the time interval corresponding to that location. Any random noise occurring with the signal will be uncorrelated with respect to the trigger signal. Consequently, its value will decrease to zero in proportion to $n^{-1/2}$.

Since on the order of 1000–2000 points are sampled during the time interval of interest, the transient recorder/computer combination acquires information from the entire signal on each repetition. The system is quite efficient as long as the digital transfer to the computer occurs in less than the time between repetitive waveforms. Current speeds for digitally adding 2048 channels to a computer memory limit the efficiency of the transient recorder/computer system for repetition rates above about 50 Hz. For repetition rates of 2 kHz or more, the boxcar averager is more efficient.

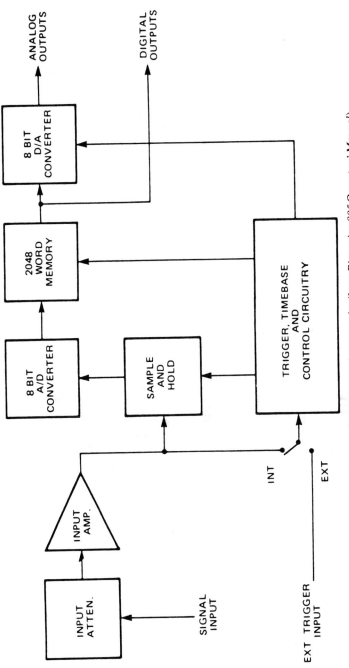

Figure 1-12. Block diagram of a typical transient recorder (from *Biomation 805 Operators' Manual*).

Because the storage of information in the transient recorder is digital rather than capacitive, it does not suffer from the leakage problem found in most boxcar averagers at low repetition rates.

Typical transient recorders have word memories between 512 and 2048 locations and maximum sampling rates of one word per 10 ns. A recent technique which sweeps an electron beam across a diode array has allowed a sampling rate of one word per 10 ps to be attained. This system has the disadvantage that $512 \times 512 = 2.6 \times 10^5$ diodes must be examined by the computer to obtain 512 points of the waveform.

Besides the efficiency in averaging, there are several more subtle advantages to the transient recorder/computer system. For example, the transient recorder can continuously sample the input and stop at any time relative to the trigger signal. In this "pretrigger" mode the recorder actually acquires signals that precede the trigger pulse. This mode of operation is particularly useful for obtaining base-line information. A second advantage is that the time base may be switched for some fraction of the sampling scan. For example, the first 1000 points may be recorded at one word per microsecond while the remainder are recorded at one word per millisecond. This mode of operation is useful if the signal contains two time components of interest.

The biggest advantage to the transient recorder/computer system is that the data are available in digital form for computer analysis. In practice, this property makes the results of one experiment immediately available as a guide for the next experiment. The primary disadvantage of this type of averaging system is the high cost. Typical recorder/computer systems are priced in the $15,000–60,000 range, depending on the time resolution and number of memory words desired.

1.3.3. Detection by Fluorescence versus Absorption Techniques

It frequently happens that a choice must be made between the detection of induced population changes by either fluorescence or absorption of radiation. A few general guidelines will serve to explain the experimental choices made in investigations described later in the chapter.

Suppose that we have a modulated pumping source which produces N_p photons per pulse. These pump photons are absorbed by the molecule under investigation to give an excited-state density of

$$\rho_1 = \frac{N_1}{V} \tag{1-54}$$

where N_1 is the number of excited molecules and V is the volume that they occupy. For a pump beam area of A_p and an absorption length of L,

$$V = A_p L \tag{1-55}$$

The number of excited molecules in the absence of saturation is simply given by Beer's law,

$$N_1 = N_0[1 - \exp(-\alpha_0 L P_0)] \tag{1-56}$$

where N_0 is the number of ground-state molecules in the volume V, α_0 is their absorption coefficient, and P_0 is their pressure. Assuming that the degree of excitation is small, we can expand the exponential and retain only the first two terms. The result is that

$$N_1 = N_0 \alpha_0 L P_0 \tag{1-57}$$

and

$$\rho_1 = \frac{N_0 \alpha_0 P_0}{A_p} \tag{1-58}$$

We will now turn to consideration of detection of this excited population, first by its fluorescence and then by its absorption of radiation. The fluorescence signal is given by

$$S = \frac{N_1 F}{\tau_{\text{rad}}} \frac{hc}{\lambda_f} \tag{1-59}$$

where F is the fraction of the fluorescence collected, τ_{rad} is the radiative lifetime for the upper state, and λ_f is the fluorescence wavelength. The noise is generally limited by the noise equivalent power of the detector, NEP. Taking Eq. (1-57) into account, the signal-to-noise ratio for fluorescence detection is, therefore,

$$(S/N)_{\text{fluor}} = \frac{N_0 \alpha_0 L_f P_0 F h c}{\tau_{\text{rad}} \lambda_f \text{NEP}} \tag{1-60}$$

where L_f is the length over which the fluorescence is observed.

Detection by absorption depends on the density of excited states ρ_1, given by Eq. (1-58). The presence of this density of excited states causes an absorption given by

$$\frac{I}{I_0} = \exp(-\sigma_1 L_a \rho_1) \tag{1-61}$$

where σ_1 is the cross section for absorption by an excited molecule and L_a is the absorption length over which the excited molecules are observed. The signal is actually proportional to the change in absorption,

$$S = \Delta I = I_0 - I = I_0[1 - \exp(-\sigma_1 L_a \rho_1)] \tag{1-62}$$

For small signals, using Eq. (1-58), we obtain

$$S = \frac{I_0 \sigma_1 L_a N_0 \alpha_0 P_0}{A_p} \tag{1-63}$$

In most cases, the noise in the absorption measurement will be caused by fluctuations of magnitude δI_0 occurring on the signal carrier I_0. The signal-to-noise ratio is given by

$$(S/N)_{abs} = \frac{I_0}{\delta I_0} \frac{\sigma_1 L_a N_0 \alpha_0 P_0}{A_p} \tag{1-64}$$

We can compare the $(S/N)_{abs}$ and the $(S/N)_{fluor}$ by making the observation that τ_{rad} is related to σ_0, the absorption cross section for ground-state molecules. For Doppler broadening with $T = 300°K$ and molecules of a mass of 28 amu, the relation between σ_0 and τ_{rad} is given by

$$\sigma_0 = C\lambda_f^3/\tau_{rad} \tag{1-65}$$

where $C = 5.32 \times 10^{-7}$ s/cm. Combining Eq. (1-65) with (1-60), we obtain

$$\frac{(S/N)_{abs}}{(S/N)_{fluor}} = NEP \ \lambda^4 \frac{C}{hc} \frac{\sigma_1}{\sigma_0} \frac{L_a}{L_f} \frac{1}{A_p F} \frac{I_0}{\delta I_0} \tag{1-66}$$

For infrared detection, the NEP of a good detector is on the order of 10^{-8} W. If we use the reasonable values of $\sigma_1/\sigma_0 = 1$, $L_a/L_f = 10$, $1/A_p = 1/0.1$ cm $= 10$ cm^{-1}, $F = 0.01$, and $I_0/\delta I_0 = 100$, the ratio $(S/N)_{abs}/(S/N)_{fluor}$ becomes greater than 1 for values of λ_f greater than about 2.4 μm. Consequently, throughout most of the infrared region of the spectrum it is more advantageous to observe excited populations by absorption than by emission. As λ_f decreases into the visible from 2.4 μm, fluorescence detection is favored. In regions where photomultiplier tubes can be used, the NEP drops below 10^{-8} W and fluorescence detection becomes even more favorable. Consequently, most optically detected double-resonance methods use fluorescence detection, as we shall see shortly. In the microwave region, the NEP rises rapidly, owing to the large thermal background. Absorption methods are exclusively used for this region of the spectrum.

1.3.4. Experimental Configurations

The remainder of this chapter will deal with various experimental arrangements for observing double resonance and with the information that they provide. Sections 1.4–1.6 are organized according to the pumping and probing radiation employed, as shown in Table 1-2. Section 1.7 will then briefly summarize the general molecular properties which may be measured by these techniques.

Table 1-2   Section Organization Based on Pumping and Probing Radiation

Pumping radiation	Probing radiation		
	Microwave	Infrared	Visible/uv
Microwave	1.4.1	1.5.1	1.6.1
Infrared	1.4.2	1.5.2	1.6.2
Visible/uv	1.4.3	1.5.3	1.6.3

1.4. Microwave-Detected Double Resonance

The first extension of double-resonance techniques beyond the radio-frequency regime of spin spectroscopy was to microwave investigations of rotational energy levels. This particular direction was dictated by the availability of radiation sources of sufficient power and monochromaticity to satisfy the criterion for saturation of the pumped transition, discussed in Section 1.2, in the form of microwave klystrons. Thus, microwave–microwave double-resonance experiments (Section 1.4.1) were the first to be carried out; infrared microwave techniques (Section 1.4.2) followed the development of infrared lasers.

1.4.1. Microwave Pumping

In the microwave–microwave double-resonance (MMDR) experiment, it is necessary to subject the molecules under study to the simultaneous influence of two microwave fields. One field, the pumping field, must be intense enough to saturate the rotational transition with which it is resonant; while this generally requires pumping power of the order of a few milliwatts, it is customary to use pumping klystrons delivering 1 W or more of power. The second, probing field is much weaker in intensity, and the art of the experiment lies in preventing the strong pumping radiation from leaking back into the signal channel.

A typical MMDR apparatus is shown in Figure 1-13. This particular configuration employs the cw pumping method extensively utilized by Oka and his coworkers (Oka, 1973), in which both pump and probe fields are held at constant amplitude. The microwave line is scanned by means of Stark modulation, and the change in its intensity between pumping field on and pumping field off indicates the double resonance. In the modulated MMDR method introduced by Wilson and his coworkers (Cox *et al.*, 1965), the pumping field is periodically turned on and off, and the in-phase component

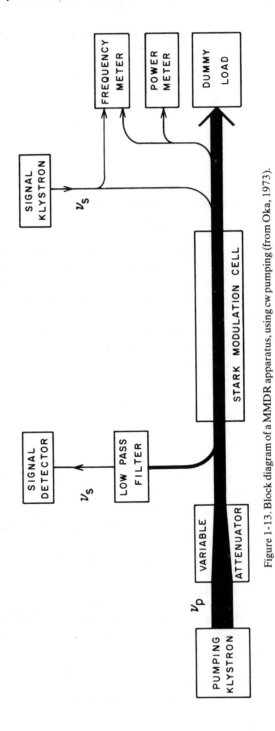

Figure 1-13. Block diagram of a MMDR apparatus, using cw pumping (from Oka, 1973).

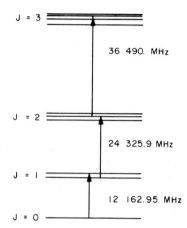

Figure 1-14. Energy-level diagram for the first four rotational levels of OCS. Stark splitting patterns are indicated, but hyperfine structure has been omitted.

of the probe signal is detected using lock-in amplification (Section 1.3.2.2). In this mode, Stark modulation of the transition frequency may be dispensed with. It is also possible to pulse either or both of the microwave fields, and observe transient response in the time domain. However, the information that can be obtained from this last approach is more easily acquired by the use of the Stark-switching methods described by Flygare in Chapter 2 of this volume.

1.4.1.1. Carbon Oxysulfide

Since the rotational transitions in OCS are generally used as calibration standards by microwave spectroscopists, it is not surprising that MMDR was first observed in this particular system (Battaglia *et al.*, 1959). In this experiment, the $2 \leftarrow 1$ transition at 24,325 MHz was pumped, and double-resonance signals were observed at the $3 \leftarrow 2$ transition at 36,490 MHz (Figure 1-14). The observations served to confirm the theory of the MMDR effect put forth by DiGiacomo (1959).

These transitions were also the first to be studied by Wilson and his group at Harvard (Cox *et al.*, 1965), and again furnished a test of the theory discussed in Section 1.2. New molecular information was obtained with the use of modulated double resonance (Unland and Flygare, 1966), with which rotational relaxation times for OCS were measured with itself and with Ar, He, and O_2 as collision partners. These results are discussed in more detail in Section 1.7.

When the pumping microwave field is pulsed in a time short compared with the relaxation time of the system, coherent transient effects will often appear (Brown, 1974; Glorieux and Macke, 1974). From the decay time of these transients, Brown measured a coherence relaxation time constant of

$(2\pi p\tau)^{-1} = 5.66\,\mathrm{MHz/Torr}$. The OCS system has been extensively investigated by means of transient Stark-switching methods, and the reader should refer to Section 2.7 for additional information on relaxation behavior in this system.

1.4.1.2. Ammonia

The double-resonance experiments described in the previous section were all of the three-level type, i.e., with one energy level in common between the two applied fields. Four-level experiments in OCS have thus far yielded negative results (Oka, 1973; Cox *et al.*, 1965), implying the absence of strong selection rules for the transfer of rotational energy between OCS molecules. In the ammonia (NH_3) molecule, however, the situation is quite different.

In NH_3, populations of various rotational levels up to $J = 10$ can be very conveniently monitored by making use of the inversion splitting of each rotational level, discussed in Section 1.3.1. Each (J,K) level is split into two levels of opposite parity. A dipole-allowed transition exists between the upper and lower inversion doublet states, with frequencies lying in the K-band (16–25 GHz) region. (The rotation spectrum of ammonia may be seen in Figure 1-16.) In an extensive series of experiments, Oka and his coworkers (Oka, 1967a; Oka, 1968a, b, c; Daly and Oka, 1970; Fabris and Oka, 1972) systematically saturated one after another of these doublets with an intense pumping field, and examined intensity changes in the other doublets. From the existence and magnitudes of these intensity changes, one can infer the preferred pathways for rotational energy transfer, and the relative rates of the inelastic processes involved (see Section 1.4.2.2 for a fuller discussion of the point). The steady-state type of experiment employed by Oka cannot, however, yield absolute magnitudes for the rotational relaxation rates; these must be obtained from time-resolved measurements, such as the pulsed double-resonance experiments described below.

The principal result of the ammonia double-resonance study is that dipole-like selection rules appear to be obeyed in most sorts of collisions. That is, $\Delta J = 0$ or 1 tends to be favored, and parity changes sign ($+$ to $-$, and vice versa). The $\Delta J = 0$ (intradoublet) rate is an order of magnitude faster than the $\Delta J = 1$ (rotationally inelastic) rate, and $\Delta J > 1$ rates are much smaller still. Also, $\Delta K = 0$ is preferred. These selection rules are most pronounced in collisions between two ammonia molecules, and are also obeyed in collisions with H_2, HD, D_2, O_2, N_2, CH_4, SF_6, and other gases. In NH_3–rare gas (He, Ar, Xe) collisions, these dipole-type rules tend to be relaxed, but there remains a strong symmetry restriction on rotational states which may couple to each other, dictated by nuclear-spin statistics. This rule is $A \leftrightarrow A$ and $E \leftrightarrow E$, but $A \nleftrightarrow E$, with the result that $\Delta K = \pm 3n$ (n an integer) becomes possible (Fabris and Oka, 1972).

These results are also borne out by a unique "triple-resonance" study on ammonia (Oka, 1968c). In this experiment, the ($J = 4, K = 1$) inversion doublet is pumped at 21,134.37 MHz, and the ($J = 3, K = 1$) doublet is "clamped" at constant population by a second strong field at 22,234.51 MHz. The ($J = 2, K = 1$) doublet at 23,098.78 MHz is probed for population changes accruing from pumping ($J = 4, K = 1$). In this way, direct $\Delta J = 2$ processes can be distinguished from those occurring via two successive $\Delta J = 1$ steps. It was found that $\Delta J = 2$ processes do indeed occur in NH_3–He collisions but not in pure ammonia. This is in accord with the collision "selection rules" deduced from the other experiments carried out on this system.

1.4.1.3. Formaldehyde

Another fairly simple system that has received extensive study is formaldehyde (H_2CO) and its deuterated analogs. Formaldehyde is a near-prolate symmetric rotor (see Section 1.3.1), with $A = 8.7517 \text{ cm}^{-1}$, $B = 1.1245 \text{ cm}^{-1}$, and $C = 1.0123 \text{ cm}^{-1}$. Many $\Delta J = 0$ transitions lie in a convenient microwave region. An early experiment was carried out by Yajima and Shimoda (1960) on HDCO, using a three-level scheme with the $13_{3,10}$ level in common. A series of four-level double-resonance experiments has been carried out by Oka (1967b) and his coworkers (Chu *et al.*, 1973), in which a $\Delta K_p = 1$ transition in one J manifold is pumped, and the corresponding transition in the ($J - 1$) or ($J - 2$) manifold is probed. Once again, dipole-like selection rules were found for rotationally inelastic collisions; i.e., $J = 0$ or ± 1 tends to be favored.

1.4.1.4. Ethylene Oxide

The first system in which collision-coupled four-level MMDR was observed was ethylene oxide, $(CH_2)_2O$ (Oka, 1966a). The rotational energy levels involved are shown in Figure 1-15. When the $2_{21}-2_{12}$ transition is pumped at 34,157.1 MHz, intensity changes are observed in the $3_{21}-3_{12}$ and $3_{30}-3_{21}$ transitions, but not in the nearby $2_{11}-2_{02}$ transition. This indicates that rotational energy transfer proceeds with simultaneous changes of ± 1 in the J-type and both K-type quantum numbers. This again corresponds to dipole-like selection rules for the collisions. An extensive series of pairs of levels in this system has been examined by modulated MMDR (Ronn and Wilson, 1967), with similar results. The analogous behavior was also found in trimethylene oxide, $(CH_2)_3$, by Cohen and Wilson (1973a). In the latter work, double resonance on the resolved Stark components of the ($3_{21}-3_{12}$)

and $(2_{21}-2_{12})$ transitions confirmed that collision-induced transitions between different M states followed the dipole-like selection rules $\Delta M = 0$ or ± 1 as well.

1.4.1.5. Hydrogen Cyanide

Transitions between different rotational levels of HCN and DCN lie in the far infrared. However, a situation arises, analogous to that in ammonia, making it convenient to study these levels in the microwave region. HCN molecules possessing one quantum of the low-lying (ν_2) bending vibration have their rotational levels split into states of odd and even parity by l doubling (see Section 1.1.3.1). This provides a series of transitions associated with each J level in the 10 to 40-GHz region, which have been used for MMDR experiments (Oka, 1967b; Gordon *et al.*, 1969; Cohen and Wilson, 1973b). In these experiments, one l doublet was saturated, and double-resonance signals were detected in l doublets corresponding to rotational states differing by one or more quantum numbers. Unlike the systems considered thus far, in which dipole-like collision selection rules obtained, $\Delta J > 1$ is possible in HCN. This was confirmed by a triple-resonance experiment (Oka, 1968c) similar to that carried out in ammonia. In this case the $J = 11$ l doublet at 29,585.12 MHz was pumped and the $J = 10$ l doublet at 24,660.40 MHz was "clamped" at constant population. Double-resonance signals were then observed in the $J = 9$ l doublet at 20.181.39 MHz, confirming the existence of $\Delta J = 2$ processes in HCN. In

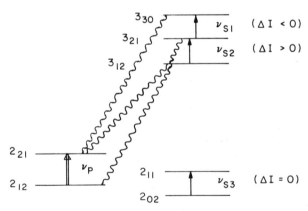

Figure 1-15. Four-level double resonance in ethylene oxide. When the $2_{21}-2_{12}$ transition is pumped, double resonance is observed on the $3_{30}-3_{21}$ and $3_{21}-3_{12}$ transitions, but not on the $2_{11}-2_{02}$ transition. This indicates the existence of the pathways for collision-induced rotational energy changes shown by the wavy lines, which follow dipole-like selection rules (adapted from Fig. 13 of Oka, 1973).

ammonia, collision-induced transitions of this type were observed only in the presence of added He.

1.4.1.6. Other Systems

MMDR has been applied to a considerable number of systems. These, along with the specific systems discussed in the preceding subsections, are listed in Table 1-3. Generally, there are three ways in which MMDR has been a useful tool.

1. Three-level experiments have served to confirm rotational assignments of lines in complex microwave spectra. A good example of this application is the recent MMDR study of isoxazole, C_3H_3NO, by Stiefvater (1975), in which numerous assignments were made in isotopic varieties of the molecule.

2. In addition, the intensity of transitions which are weak due to either low populations or unfavorable selection rules can be enhanced.

3. Four-level experiments have elucidated the detailed pathways by which rotational energy is transferred in molecular collisions. In most instances, the rotational levels which are coupled by collisions are those which would be coupled by an electric-dipole transition moment operator. The implications of this result for our understanding of molecular interactions will be discussed in more detail in Section 1.7.

1.4.2. Infrared Pumping

There is one major limitation inherent in the microwave double-resonance experiments described in the previous section. Any given pair of rotational levels will have very nearly equal populations in the upper and lower states at normal temperatures ($\Delta E/kT = 0.005$ for a 30-GHz transition at 300°K). Thus, the effect of a saturating microwave field is to induce only a small net change in molecular state populations, leading to a corresponding small $\Delta I/I$ to be measured in the experiment. Clearly, if one could begin with a pair of levels having very different equilibrium populations, much larger double-resonance signals might be anticipated. It is this consideration which led to the use of infrared pumping of vibration–rotation levels to replace the microwave pumping field. In order for such an experiment to be practical, infrared fields of the order of several watts/cm^2, in a bandwidth narrow compared to the molecular linewidth, are required to saturate the vibrational transition. Thus, it was only after infrared laser sources became available that infrared–microwave double-resonance experiments could be carried out.

Table 1-3 Microwave–Microwave Double Resonance

Reference	Molecule	Transition pumped		Transition probed		Results
Battaglia et al. (1959)	OCS	$2 \leftarrow 1$	24,325 MHz	$3 \leftarrow 2$	36,490 MHz	Observation; verified theory
Oka (1966a)	C_2H_5I	$3_{03}-0_{00}(\tfrac{7}{2} \leftarrow \tfrac{5}{2})$	34,513.2	$4_{04}-3_{03}(\tfrac{9}{2} \leftarrow \tfrac{7}{2})$	23,078.3	Enhancement of weakly allowed $\Delta J = 3$ transition
Oka (1966b)	$(CH_2)_2O$	$3_{03}-0_{00}(\tfrac{5}{2} \leftarrow \tfrac{5}{2})$	34,649.6	$4_{04}-3_{03}(\tfrac{5}{2} \leftarrow \tfrac{5}{2})$	23,203.5	Collisional "selection rules" for $\Delta J = 1$
		$2_{21}-2_{12}$	34,157.1	$3_{21}-3_{12}$	23,611.5	
Oka (1967b)	H_2CO	$10_{28}-10_{29}$	34,100.32	$3_{30}-3_{21}$	23,135.5	Dipole-like collisional "selection rules" ($\Delta J = 0$ or ± 1; parity $+ \leftrightarrow -$) found for H_2CO, HDCO
				$9_{27}-9_{28}$	22,965.71	
				$8_{26}-8_{27}$	14,726.74	
		$18_{3,15}-18_{3,16}$	33,270.80	$17_{3,14}-17_{3,15}$	24,068.31	
				$16_{3,13}-16_{3,14}$	17,027.60	Quadrupole-like collisional "selection rules" ($\Delta J = 2$) found for HCN, DCN, H_2CCO
	HDCO	$15_{3,12}-15_{3,13}$	33,274.70	$14_{3,11}-14_{3,12}$	22,624.65	
				$13_{3,10}-13_{3,11}$	14,873.02	
		$22_{4,18}-22_{4,19}$	33,303.12	$21_{4,17}-21_{4,19}$	23,578.92	
				$20_{4,16}-20_{4,17}$	16,343.41	
	HCN ($v_2=1$)	$J=12$ l doublet	34,953.5	$J=8, 9, 10, 11$ l doublets	16–29 GHz	
	DCN ($v_2=1$)	$J=13$ l doublet	33,814.3	$J=9, 10, 11, 12$ l doublets	16–29 GHz	
	$H_2C\!=\!C\!=\!O$	$13_{1,12}-13_{1,13}$	34,333.14	$12_{1,11}-12_{1,12}$	29,430.02	
				$11_{1,10}-11_{1,11}$	24,903.53	
				$10_{1,9}-10_{1,10}$	20,753.90	
				$9_{1,8}-9_{1,9}$	16,980.97	
Oka (1968c)	HCN	$\left.\begin{array}{l} J=11 \\ \text{l doublet} \\ J=10 \\ \text{l doublet} \end{array}\right\}$	29,585.12 24,660.40	$J=9$ l doublet	20,181.39	By "clamping" $\Delta J = 1$ transitions, it was possible to observe directly $\Delta J = 2$ transitions in HCN–HCN, NH_3–He, and

	Molecule					Observations
	NH₃	$(J=4, K=1)$ i doublet / $(J=3, K=1)$ i doublet	21,134.37 / 22,234.51	$(J=2, K=1)$ i doublet	23,098.78	CH₃OH–He, but *not* in NH₃–NH₃ or CH₃OH–CH₃OH
	CH₃OH	$9,2$–$9,1$ / $10,2$–$10,1$	25,541.43 / 25,878.18	$11,2$–$11,1$	26,313.11	
Oka (1967a) Oka (1968a) Oka (1968b) Daly and Oka (1970) Fabris and Oka (1972)	NH₃	*Inversion doublets* $J=2, K=1$ $J=3, K=2$ $J=4, K=3$ $J=5, K=4$ $J=6, K=5$ $J=7, K=6$ $J=3, K=1$ $J=4, K=2$ $J=5, K=3$ $J=6, K=4$ $J=7, K=5$ $J=8, K=6$ etc.	23,098.78 22,834.10 22,688.24 22,653.00 22,732.45 22,924.91 22,234.51 21,703.34 21,285.30 20,994.62 20,804.80 20,719.20	*Inversion doublets* $J=1, K=1$ $J=2, K=2$ $J=3, K=3$ $J=4, K=4$ $J=5, K=5$ $J=6, K=6$ $J=2, K=1$ $J=3, K=2$ $J=4, K=3$ $J=5, K=4$ $J=6, K=5$ $J=7, K=6$ etc.	23,694.48 23,722.61 23,870.11 24,139.39 24,532.94 25,056.04 23,098.78 22,834.10 22,688.24 22,653.00 22,732.45 22,924.91	Dipole-like collisional selection rules ($\Delta J = 0$ or ± 1; parity $+ \leftrightarrow -$) Rate for $\Delta J > 1$ much less than for ($\Delta J = 1$) Rate for $\Delta K \neq 0$ much less than for $\Delta K = 0$ In rare-gas collisions (He, Ar, Xe): $A \leftrightarrow A, E \leftrightarrow E$ "allowed" $A \leftrightarrow E$ "forbidden" $\Delta K = 3n$ "Dipole-like" selection rules also found for collisions with H₂, HD, D₂, O₂, N₂, CH₄, SF₆
Seibt (1972)	C₂H₅OH CH₂DCH₂OH	Many transitions from 3_{12}–2_{21} to $21_{7,15}$–$20_{8,12}$ 21–24 GHz		Many transitions from 2_{12}–3_{03} to $20_{7,14}$–$19_{8,11}$ 27–41 GHz		Measured rotational relaxation pathways in self- and He-collisions
Cox *et al.* (1965)	OCS	$1 \leftarrow 0$ $2 \leftarrow 1$ 2_{02}–1_{01}	12,162.97 24,325.92	$2 \leftarrow 1$ $1 \leftarrow 0$ 1_{10}–1_{01}	24,325.92 12,162.97	Observation; verified theory and spectrum assignments
	(CH₂)₂O					

continued

Table 1–3   Microwave–Microwave Double Resonance (continued)

Reference	Molecule	Transition pumped	Transition probed	Results
Ronn and Wilson (1967)	$(CH_2)_2O$	$2_{11}-2_{02}$ $3_{30}-3_{21}$ 23,134.2 $3_{21}-3_{12}$ 23,610.4 $4_{31}-4_{22}$ 24,834.26	$1_{11}-0_{00}$ 39,581.8 $2_{21}-2_{12}$ 34,157.1 $2_{21}-2_{12}$ 34,157.1 $3_{21}-3_{22}$	Use of *modulated* double resonance and comparison with theory of Gordon (1967)
	$H_2C=CF_2$	$4_{22}-4_{31}$ 23,134.2 $3_{30}-3_{21}$ 26,627.6 $4_{22}-4_{23}$ 21,745.2 $8_{62}-8_{63}$ 24,771.0 $6_{42}-6_{43}$	$5_{23}-5_{32}$ 37,781.0 $4_{40}-4_{31}$ 34,148.3 $5_{23}-5_{24}$ $9_{63}-9_{64}$ $7_{53}-7_{34}$	Collisional energy transfer observed to follow dipole-like selection rules
	$(CH_2)_2S$	$6_{24}-6_{25}$ 22,976.4	$7_{25}-7_{26}$ 35,515.3	
	HCN	$J = 10$ 24,660.40 l doublet	$J = 11$ 29,585.12 l doublet $J = 12$ 34,953.5 l doublet	
	$FHC=CF_2$	$6_{06}-6_{15}$ 24,903.51 $7_{53}-7_{52}$	$7_{17}-7_{26}$ $6_{52}-6_{33}$	
	HNO_3	$8_{62}-8_{63}$ 22,147.00	$9_{63}-9_{64}$	
	$(CH_3)_3CCHO$ $CH_3CH{-}O{-}CH_2$	$4_{04}-4_{03}$ 23,970.0 $1_{11}-0_{00}$	$6_{06}-6_{05}$ $2_{20}-2_{11}$ 34,100.0	
Gordon et al. (1969)	HCN	$J = 10$ 24,660.40 l doublet	$J = 11$ 29,585.12 l doublet	Relative rates of $\Delta J = 1$, $\Delta J = 2$, and $\Delta J = 3$ collisions follow dipole-like selection rules; also with He, Ar, Xe
Cohen and Wilson (1973b)		$J = 12$ l doublet	$J = 12$ 34,953.5 l doublet $J = 13$ 40,766.0 l doublet	
Unland et al. (1965)	$CH_3CCl=CH_2$	$1_{01}-0_{00}$ 8,288.24	$2_{02}-1_{01}$ 16,171.80	Confirmed spectroscopic assignment of hyperfine levels

Reference	Molecule					Comments
Unland and Flygare (1966)	OCS	$1 \leftarrow 0$	12,162.97	$2 \leftarrow 1$	24,325.92	Measured relaxation times with self, Ar, He, O_2
Cohen and Wilson (1973a)	$(CH_2)_3O$	$7_{53}-7_{53}$	24,592.0	$8_{53}-8_{54}$	35,825.0	Further confirmation of dipole-like collisional selection rules
				$6_{33}-6_{34}$	36,022.6	
				$9_{73}-9_{54}$	35,994.0	
				$10_{73}-10_{74}$	35,330.0	No double-resonance signal observed in cyclobutanone
		$7_{52}-7_{53}\ (v=1)$	24,240.0	$8_{53}-8_{54}\ (v=1)$	35,501.5	
		$6_{52}-6_{33}$	26,081.0	$7_{53}-7_{54}$	36,032.7	
		$5_{42}-5_{23}$	25,928.0	$7_{43}-7_{44}$	35,948.0	
		$7_{62}-7_{43}$	26,331.0	$8_{63}-8_{44}$	36,010.5	
		$7_{25}-7_{16}$	23,724.0	$6_{15}-6_{06}$	28,821.6	
		$3_{13}-2_{12}$	25,172.3	$3_{12}-3_{11}$	29,283.4	
	HFC=CFH	$13_{3,11}-12_{48}$	24,852.2	$15_{5,11}-14_{68}$	11,893.0	
		$13_{3,11}-12_{48}\ (v=1)$	23,913.0	$14_{4,10}-13_{59}\ (v=1)$	11,716.6	
	$(CH_2)_2O$	$3_{21}-3_{12}$	23,610.4	$2_{21}-2_{12}$	34,157.1	$(\Delta M = 0, \pm 1$ observed in separated M levels)
	$CF_3C \equiv CH$	$4 \leftarrow 3$	23,023.0	$6 \leftarrow 5$	23,039.3	
	ICN	$4 \leftarrow 3$	25,837.6	$6 \leftarrow 5$	38,800.0	$(\Delta F = 0, \pm 1$ observed in separated hyperfine levels)
Brown (1974) Glorieux and Macke (1974)	OCS	$1 \leftarrow 0$	12,162.95	$2 \leftarrow 1$	24,325.92	Pulsed double resonance, coherent transients observed and compared with theory; $1/2\pi p\tau = 5.66$ MHz/Torr
Chu *et al.* (1973)	H_2CO	$22_{3,19}-23_{1,22}$	59,352.0	$23_{3,19}-22_{3,20}$	100,511.0	Observed $\Delta J = 1$, $\Delta K_a = 2$, $\Delta K_c = 3$ transitions; confirmed assignments; improved spectroscopic constants
	D_2CO	$14_{1,13}-13_{3,10}$	27,522.0	$13_{3,10}-13_{3,11}$	44,113.0	
Yajima and Shimoda (1960)	HDCO	$3_{03}-2_{12}$	34,916.4	$2_{11}-2_{12}$	16,038.06	Observed weak transitions, confirmed assignments

continued

Table 1-3   Microwave–Microwave Double Resonance (continued)

Reference	Molecule	Transition pumped		Transition probed		Results
Yajima (1961)	HCOOH	$2_{11}-1_{10}$	46,581.3	$2_{12}-2_{11}$	4,916.35	Comparison with theory
Stiefvater (1975)	Isoxazole, C_3H_3NO	Numerous		Numerous		Assigned lines and increased sensitivity on weak isotopic satellites for structure determination
Ford (1976)	$ClCH_2CHO$	$3_{13}-4_{04}$	9,590.4	$4_{04}-3_{03}$	20,966.9	Assignment of lines in cis, trans, gauche configurations and in vibrationally excited states; structural constants
				$4_{14}-3_{13}$	20,683.7	
		$2_{12}-3_{03}$ $(v=3)$	10,965.6	$4_{04}-3_{03}$ $(v=3)$	21,507.3	
		$2_{12}-3_{03}$ $(v=4)$	10,075.8	$4_{04}-3_{03}$ $(v=4)$	21,655.3	
		$2_{12}-3_{03}$ $(v=5)$	9,275.0	$4_{04}-3_{03}$ $(v=5)$	21,798.0	
		$2_{12}-3_{03}$ $(v=6)$	8,567.1	$4_{04}-3_{03}$ $(v=6)$	21,934.1	

1.4.2.1. Methyl Halides

The first experiment of this type was reported by Ronn and Lide in 1967. They pumped methyl bromide in a waveguide with the P(20) line of the CO_2 laser, and found intensity changes on several low-J rotational transitions. Similar results were obtained by Lemaire *et al.* (1969). However, the effect appears to be greatest at pressures of a few Torr, at which collisions dominate, and it was later found that the laser line misses the methyl bromide absorption line by several hundred MHz (Winnewisser *et al.*, 1972). Thus, these experiments must be interpreted as a combination of pumping and thermalization effects rather than a double resonance. More recently, three- and four-level experiments were carried out on $^{13}CH_3F$ (Jetter *et al.*, 1973), CH_3Cl (Frenkel *et al.*, 1971), and CF_3I (Jones and Kohler, 1975). In the former two systems, detection was by means of submillimeter wave spectroscopy ($\nu > 100\,GHz$), since relatively high-J methyl halide transitions were involved. Rotational energy transfer was observed in both of the latter two investigations, and it was concluded that $\Delta J = 1$ transitions were once again dominant.

1.4.2.2. Ammonia

The first true infrared–microwave three-level double resonance was discovered by Shimizu and Oka in 1970. This involves the $^{O}Q_{as}(8, 7)$ line of the ν_2 band of the ammonia molecule near $11\,\mu m$. As shown in the energy-level diagram in Figure 1-16, this transition couples the upper member of the ($J = 8$, $K = 7$) inversion doublet to a vibrationally excited level. This transition is very nearly coincident with the P(13) line of the N_2O laser at $927.739\,cm^{-1}$, and pumping with this line removes a large portion of the population in the upper level of the inversion doublet. This produces a large increase in the absorption intensity of that doublet line at $23{,}232.24\,MHz$.

A number of investigations have been carried out on this resonance (see Table 1-4). Several techniques have been used to increase the number of observable laser-ammonia coincidences. These include use of isotopically substituted molecules (Shimizu and Oka, 1970a, b), shifting the energy levels with a high-voltage Stark field (Fourrier and Redon, 1972, 1974; Redon *et al.*, 1975), and introduction of a second, off-resonant microwave field to induce two-photon transitions (Freund and Oka, 1972; see also the results of Freund *et al.*, 1973, discussed in Section 1.5.2.1). Four-level double-resonance experiments have confirmed the dipolar selection rules for NH_3–NH_3 collisions found in microwave double-resonance experiments. The way in which these are manifested may be seen from Figure 1-16. The laser line [N_2O P(13), in this instance] removes a large fraction

Table 1-4 Infrared–Microwave Double Resonance

Reference	Molecule	Transition	Pumped by	at	Transition probed		Results
Ronn and Lide (1967)	CH_3Br	ν_6 band	CO_2 laser	940 cm^{-1}	2,1←1,1	~38,300 MHz	First experimental observation; later shown to be thermal in origin
Lemaire et al. (1969)	CH_3Br	$\nu_6[P_1(9)]$	CO_2 laser P(20)	944.195	3,1←2,1 2,1←1,1	57,450 38,300	"Confirmed" Ronn and Lide result, but with $(\Delta I/I)_{max}$ at approx. 1–2 Torr
Jetter et al. (1973)	$^{13}CH_3F$	$\nu_3[^O R_3(4)]$	CO_2 001–020 P(32)	1035.47	4,3←3,3 5,3←4,3	199,087 248,850	No coherence effects; measured $(2\pi pT)^{-1} = (15.2 \pm 0.8)$ MHz/Torr
Frenkel et al. (1971)	CH_3Cl	$\nu_6[^R Q_3(6)]$	CO_2 001–100 P(26)	938.69	5,0←4,0 5,3←4,3 6,0←5,0 6,3←5,3 5,4←4,4 6,4←5,4 7,4←6,4	132,919 132,900 159,498 159,480 132,420 158,890 185,370	Observed and assigned rotational levels; observed $\Delta K = 0$ and 3, $\Delta J = 1$ in collisions
Jones and Kohler (1975)	CF_3I	$\nu_1[^O R_{21}(7),$ $F=\frac{21}{2}\leftarrow\frac{19}{2}]$	CO_2 001–020 R(16)	1076.0	$J=5\leftarrow4$ through 13←12, K = 2	15–39 GHz	$\Delta J = \pm1$, $\Delta K = 0$ in relaxation; also measured ΔF transitions and constants
Shimizu and Oka (1970a,b)	$^{14}NH_3$	$\nu_2[^O Q_{as}(8,7)]$	N_2O laser P(13)	927.739	(J = 8, K = 7) i doublet	23,232.24	First authenticated infrared–microwave double-resonance experiment
	$^{15}NH_3$	$\nu_2[^O Q_{as}(4,4)]$	N_2O laser P(15)	925.979	(J = 4, K = 4) i doublet	23,046.10	
Kreiner and Jones (1974)	NH_3	$\nu_2[^O Q_{as}(8,7)]$	N_2O laser P(13)	927.739	(J = 8, K = 7) i doublet	23,232.24	Observed energy transfer among hyperfine levels
Fourrier and Redon (1972)	NH_3	$\nu_2[^O Q_{sa}(9,6)]$	N_2O R(30)	962.99	(J = 9, K = 6) i doublet	18,499.28	Used high Stark fields to pull lines into coincidence with laser frequencies (n.b.: tabulated doublet frequencies *not* corrected for Stark shifts)
	NH_3	$\nu_2[^O Q_{as}(3,2)]$	N_2O P(8)	932.02	(J = 3, K = 2) i doublet	22,834.17	

Reference	Molecule	Assignment	Laser line		Transition		Comments
		$\nu_2[^OQ_{as}(11,9)]$	N_2O P(14)	926.80	$(J=11, K=9)$ i doublet	21,070.70	
		$\nu_2[^OR_{ss}(5,3)]$	CO_2 001–020 P(12)	1053.92	$(J=5, K=3)$ i doublet	21,285.27	
		$\nu_2[^OQ_{aa}(4,2)]$	CO_2 001–100 R(6)	966.25	$(J=4, K=2)$ i doublet	21,703.26	
		$\nu_2[^OQ_{as}(5,3)]$	CO_2 001–100 R(32)	932.96	$(J=5, K=3)$ i doublet	21,285.27	
Lemaire et al. (1974)	NH_3	$\nu_2[^OQ_{as}(8,7)]$	N_2O P(13)	927.739	(1,1) through (6,5) i doublets	19–24 GHz	Measured rotational energy transfer for various ΔJ, ΔK
Levy et al. (1972)	NH_3	$\nu_2[^OQ_{as}(8,7)]$	N_2O P(13)	927.739	$(J=8, K=7)$ i doublet	23,232.24	Observed transient nutation in pulsed double-resonance experiment
Levy et al. (1973)	NH_3	$\nu_2[^OQ_{as}(8,7)]$	N_2O P(13)	927.739	$(J=9, K=7)$ i doublet; $(J=7, K=4)$ i doublet	20,735.44; 19,218.36	Measured rotational relaxation time constant in pulsed experiment
Fourrier and Redon (1974)	NH_3	$\nu_2[^OQ_{sa}(9,6)]$	N_2O R(30)	962.99	$(J=9, K=6)$ i doublet, 21,080 at $E_0 = 11.4$ kV/cm		Produced inversion on (9,6) doublet, observed emission
Kano et al. (1974)	NH_3	$\nu_2[^OQ_{as}(8,7)]$	N_2O P(13)	927.739	(2,1) through (10,10) i doublets		Apparent "nondipolar" energy transfer found, but may be thermal or cascading effects
Freund and Oka (1972)	$^{14}NH_3$	$\nu_2[^OQ_{as}(4,4)]$	N_2O P(15)	925.982	$(J=4, K=4)$ i doublet	23,360	Two-photon Lamb dip; second off-resonant microwave field used to compensate frequency difference
	$^{15}NH_3$	$\nu_2[^OQ_{sa}(5,4)]$	CO_2 R(6)	966.251	$(J=5, K=4)$ i doublet	22,095	
Dobbs et al. (1975)	NH_3	$\nu_2[^OQ_{as}(8,7)]$	N_2O P(13)	927.739	$(J=8, K=7)$ i doublet	23,232.24	Used two-pulse technique to measure $1/2\pi T_1 = 1/2\pi T_2 = (25 \pm 2)$ MHz/Torr

continued

Table 1-4 Infrared–Microwave Double Resonance (continued)

Reference	Molecule	Transition	Pumped by	at	Transition probed		Results
Redon et al. (1975)	NH_3	$\nu_2[^oQ_{as}(5,3)]$	CO_2 001–100 P(32)	932.9	$(J=5, K=3)$ i doublet	21,287.27	Used high (7–15 kV/cm) dc fields to bring transitions into coincidence; measured resolved M levels; dipolar selection rules found
		$\nu_2[^oQ_{as}(3,2)]$	N_2O P(8)	932.0	$(J=3, K=2)$ i doublet	22,834.17	
Kano et al. (1976)	NH_3	$\nu_2[^oQ_{as}(8,7)]$	N_2O P(13)	927.739	(2,1) through (10,9) i doublets		Distinguished direct from cascade rotational energy transfer; observed transfer between $^{14}NH_3$ and $^{15}NH_3$
Dobbs et al. (1976)	OCS	$02^o0 \leftarrow 00^o0$ P(5)	CO_2 001–020 P(22)	1076.0	$J=4 \leftarrow 3, v_2=2$	48,801.1	Confirmed assignment of laser-pumped level

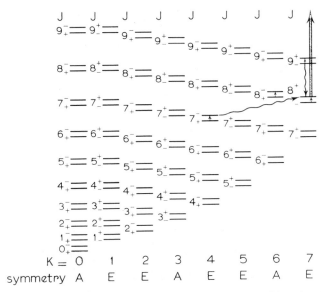

Figure 1-16. Energy-level diagram for rotational states of ammonia, with parity of inversion levels shown (splittings exaggerated). The heavy stippled arrow indicates the pumping of the $(J = 8, K = 7, P^+)$ level by the N_2O laser. Allowed rotational energy transfer pathways are indicated by the wavy arrows (from Dobbs *et al.*, 1975).

of the population from the even($+$)-parity upper level of the $(J = 8, K = 7)$ inversion transition. By dipolar selection rules, collisions return population to this state from the odd($-$)-parity levels of the (9,7) and (7,7) doublets, corresponding to $\Delta J = \pm 1$, $K = 0$. This manifests itself as a *decrease* in the intensity of these doublet transitions, since population is removed from the lower, rather than the upper, member of the doublet. The $(J = 8, K = 6)$ inversion doublet is close in energy to the (8, 7) doublet, but no signal is seen in this doublet at low pressures. This is because $\Delta K = 1$ transition would involve intercombination between A and E symmetry species, dictated by hydrogen spin statistics; such a transition is dipole-forbidden. The E-symmetry (8,7) doublet can, however, be coupled to the E-symmetry (7,4) doublet, and a double-resonance signal is seen in that transition. Since the upper level is of ($-$) parity in the $K = 4$ stack, it is this that couples to the ($+$) parity level that is pumped by the laser. This results in an increased absorption in the $\Delta K = 3$ transition, unlike the decrease seen in the $\Delta K = 0$ transition.

The experiments described thus far made use of cw or slowly modulated pumping radiation, and thus were able to determine the pathways of energy transfer but not the absolute rate. This was measured in pulsed double-resonance experiments using a Q-switched laser (Levy *et al.*, 1972, 1973). A new phenomenon appeared in these experiments—the coherently excited

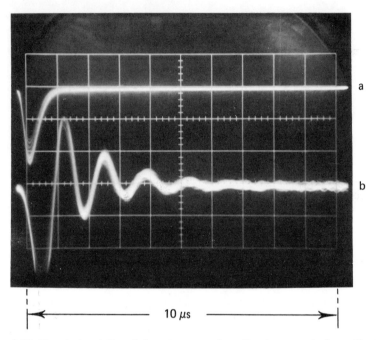

a

b

|← —————————— 10 μs —————————— →|

Figure 1-17. Transient nutation of the microwave absorption in ammonia (trace *b*) when pumped by an infrared laser pulse (trace *a*) (from Steinfeld, 1974b).

ammonia molecules displayed transient nutation; i.e., the coherence of the laser field was transferred to the microwave transition, resulting in a damped oscillating absorption signal. A typical infrared–microwave transient nutation in ammonia is shown in Figure 1-17. (For a more detailed discussion of transient nutation and related phenomena, see Chapters 2 and 3.) From this effect it was possible to measure the coherence dephasing time, T_2, in the inversion $J = 8$, $K = 7$ doublet. The value of $\frac{1}{2}\pi p T_2 = (24 \pm 1)$ MHz/Torr found in this experiment was in good agreement with the results of other measurements on this system. A modification of this method, using a laser pulse followed by a delayed electric-field shift, was able to measure T_1 for this transition as well (Dobbs *et al.*, 1975); the apparatus for carrying out this experiment is shown in Figure 1-18. For the $J = 8$, $K = 7$ doublet transition, $T_1 = T_2$ was found. Since pure microwave measurements on lower (J, K) inversion doublets of ammonia have yielded the result that $T_1 < T_2$ (Hoke *et al.*, 1976; see also Section 2.7), the intriguing mechanism accounting for the difference in relaxation behavior with rotational state remains in question.

To date, relatively few other species have been studied by infrared–microwave double resonance. This situation will surely change as tunable, high-powered pulsed infrared lasers become increasingly available.

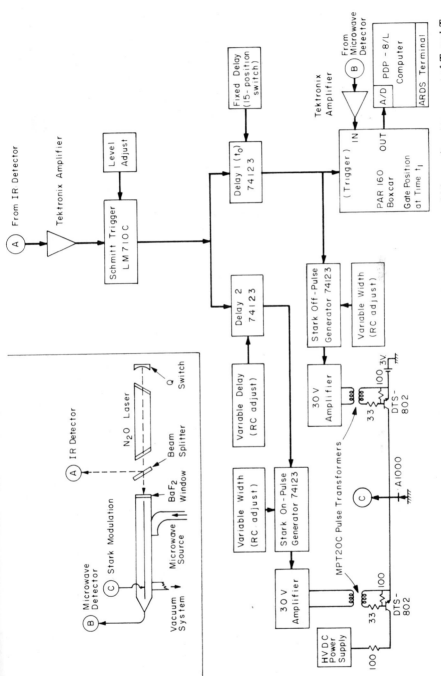

Figure 1-18. Apparatus for Stark-modulated, infrared-laser-enhanced transient nutation spectroscopy, which permits measurement of T_1 and T_2 for the same transition (from Dobbs *et al.*, 1975).

Table 1.5   *Microwave–Infrared Double Resonance*

Reference	Molecule	Transition pumped	at	Transition probed by	at	Results
Takami and Shimoda (1972)	H_2CO	$5_{14} \leftarrow 5_{15}$	72,409 MHz	$\nu_5(6_{06} \leftarrow 5_{14})$	Xe laser	2850.63 cm^{-1}	Effect observed and interpreted
Takami and Shimoda (1973a)	H_2CO	$5_{14} \leftarrow 5_{15}$ $4_{13} \leftarrow 4_{14}$	72,409 48,285	$\nu_5(6_{06} \leftarrow 5_{14})$	Xe laser	2850.63	Energy-transferring collisions observed
Takami and Shimoda (1973b)	HDCO	$2_{11} \leftarrow 2_{12}$ $3_{12} \rightarrow 3_{13}(\nu_2 = 1)$	16,038 33,245	$\nu_1(3_{12} \leftarrow 2_{11})$	Xe laser	2850.59	Microwave transitions observed in vibrationally excited state
Takami et al. (1973); Curl and Oka, (1973)	CH_4	$J_L = 6_7, F_2^{(1)} \leftarrow F_1^{(2)}$ $(\nu_3 = 1)$ $J_L = 6_7, F_1^{(2)} \rightarrow F_2^{(2)}$ $(\nu_3 = 1)$	15,601.85 6,895.20	$\nu_3 P(7)$	Ne laser	2947.91	Rotational transitions observed in vibrationally excited state
Curl (1973); Curl et al. (1973)	CH_4	$J = 7, F_2^{(2)} \leftarrow F_1^{(2)}$ $J = 7, F_2^{(2)} \rightarrow F_1^{(1)}$	423.02 1,246.55	$\nu_3 P(7)$	Ne laser	2947.91	Ground-state hyperfine splittings observed, measured D_t, H_{4t}, H_{6t}

Takami and Shimoda (1974)	HCOOH	$13_{2,11} \to 13_{2,12}$ ($v_1 = 1$)	37,337.68	$\nu_1(13_{2,11} \leftarrow 13_{1,12})$	Ne laser	2947.91	Measured rotational constants in vibrationally excited state
		$13_{2,11} \to 14_{1,14}$ ($v_1 = 1$)	12,740.85				
		$12_{3,10} \leftarrow 13_{2,11}$ ($v_1 = 1$)	10,567.33				
		$14_{2,12} \to 14_{2,13}$ ($v_1 = 1$)	48,198.15	$\nu_1(14_{2,12} \leftarrow 14_{1,13})$	Ne laser	2947.91	
		$14_{2,12} \to 13_{3,11}$ ($v_1 = 1$)	19,516.66				
		and others					
Takami and Shimoda (1976)	HC≡CCHO	$3_{13} \to 4_{04}$ ($v_2 = 1$)	25,016.03	$\nu_2(3_{13} \leftarrow 4_{22})$	Xe laser	2850.61	Assigned infrared absorption lines, measured rotational constants of $v_2 = 1$ state
		$3_{13} \to 2_{12}$ ($v_2 = 1$)	27,473.74	$\nu_2(3_{12} \leftarrow 4_{23})$	Xe laser	2947.91	
		$3_{12} \to 3_{03}$ ($v_2 = 1$)	64,266.34				
		$4_{13} \leftarrow 3_{12}$ ($v_2 = 1$)	37,943.21				
		and others					

1.4.3. Optical Pumping

As discussed in earlier sections, electronic transitions are the most difficult to saturate, requiring the highest intensity of radiation. This fact, coupled with the relatively weak absorption at microwave frequencies, makes optically pumped, microwave-detected double-resonance experiments the least favorable of all the configurations one can contemplate. We have not been able to find any examples of such experiments in the literature.

1.5. Infrared-Detected Double Resonance

With the advent of infrared lasers, it became possible to carry out absorption measurements on vibrational transitions with the same sensitivity and resolution as microwave klystrons had permitted for rotational spectroscopy. Double-resonance experiments on vibrational transitions, using infrared lasers as probing sources, soon followed. This section follows the same outline as Sections 1.4 and 1.6; we consider first microwave pumping (Section 1.5.1), then infrared (Section 1.5.2) and optical pumping (Section 1.5.3).

Virtually all experiments done to date in this area have used single-frequency lasers as probing sources, notably rare-gas atom lasers (Ne, Xe), CO_2 CO, or HF. The further development of tunable infrared laser sources, such as optical parametric oscillators, semiconductor diodes, and spin-flip lasers (see Section 1.3.1.3), will make possible a much wider variety of investigations than was possible previously. Some results using these devices have already begun to appear (Jensen and Steinfeld, 1976).

1.5.1. Microwave Pumping

If a sample placed within a laser cavity has an absorption transition coinciding with the laser line frequency, the resulting loss will affect the laser's output intensity. If the population of the absorbing levels is then altered in some way, as by a second ratiation field coupling with one of those levels, a corresponding change will occur in the laser output. In this way one can use an infrared laser to monitor rotational-energy-level population changes induced by microwave pump fields. Experiments of this type have been carried out on a relatively small number of molecules, which are listed in Table 1-5.

One type of system, extensively studied by Takami, Shimoda, and coworkers in Tokyo (1972, 1973a,b, 1974, 1976), includes molecules such

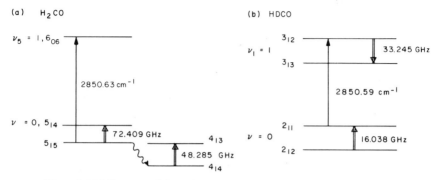

Figure 1-19. Microwave–infrared double resonance in (a) H_2CO and (b) HDCO.

as formaldehyde (Figure 1-19), formic acid, and propynal. The C–H stretching modes in these molecules are pumped by either the He–Xe laser at 2850.63 cm^{-1} or the He–Ne laser at 2947.91 cm^{-1} (3.39-μm transition). In either case the laser frequency can be tuned over several GHz by an axial magnetic field, in order to coincide with a number of rotation–vibration transitions. The objective of most of these transitions was measurement of rotational constants in the vibrationally excited state connected to the laser transition.

The P(7) line of the ν_3 transition in methane, pumped by the 3.39-μm line of the He–Ne laser, has also been investigated. The transitions studied in this system are shown in Figure 1-20. Spectroscopy was carried out on both the ground (Curl, 1973; Curl *et al.*, 1973) and vibrationally excited (Takami *et al.*, 1973; Curl and Oka, 1973) states of the infrared transition. Since the $\nu_3 = 1$ level, lying 3000 cm^{-1} above the ground state, is essentially unpopulated at room temperature, the use of the double-resonance method had an obvious function in producing population in that level. In the case of the ground-state transitions, which were between hyperfine components within the $J'' = 7$ level, the energy splittings were so small (400 and 1200 MHz) that the transitions were too weak to be observed in conventional spectroscopy. Double resonance was used to amplify these very weak transitions, using the infrared laser as a nonlinear circuit element.

1.5.2. Infrared Pumping

Infrared detection of double resonance caused by infrared pumping has proved to be a useful method for monitoring energy transfer processes. As we have seen in Section 1.3.3, detection by absorption techniques generally becomes more favorable than detection by fluorescence techniques for

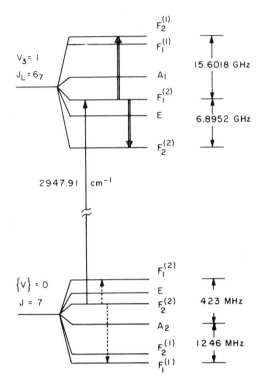

Figure 1-20. Energy-level diagram for the hyperfine components of the $\nu_2 P(7)$ line in methane, showing microwave pumping transitions and infrared probing transition (from Curl and Oka, 1973).

wavelengths above 2.4 μm. It is perhaps surprising, then, to find that most studies in the infrared have used fluorescence detection techniques. To a large extent, this preference stems from the lack of available tunable lasers to probe absorption changes. All applications of infrared–infrared double resonance to date either have relied on a fortuitous coincidence between an existing laser and a molecular absorption (e.g., the CO_2 laser with SF_6) or have probed the molecule directly responsible for the laser itself (e.g., the CO_2 laser with CO_2). As newer tunable sources, particularly diode lasers (Section 1.3.1.3), become more available, we can expect to see an increase in the use of infrared–infrared double resonance (IRIRDR).

Experiments using IRIRDR will be discussed in the context of their application. Sections 1.5.2.1 and 1.5.3.2 will discuss applications to vibrational and rotational energy transfer problems, respectively, while Section 1.5.2.3 will cover the less well-developed applications to dephasing, momentum transfer, and molecular alignment. Table 1-6 gives a complete summary of these experimental results.

Table 1-6 Infrared–Infrared Double Resonance

Reference	Molecule	Transitions pumped, source	Transitions monitored, source	Results
Rhodes et al. (1968)	CO_2	$02^\circ 0$–$00^\circ 1$; 9.6-μm CO_2 laser	$10^\circ 0$–$00^\circ 1$; 10.6 μm CO_2 laser	Collisional relaxation of $10^\circ 0$ is $k = 4\times 10^5$ s^{-1} Torr^{-1}; $k = 3\times 10^5$ s^{-1} Torr^{-1}
Jacobs et al. (1975)	CO_2	$10^\circ 0$–$00^\circ 1$; 10.6 μm CO_2 laser	$02^\circ 0$–$00^\circ 1$; 9.6-μm CO_2 laser	
Steinfeld et al. (1970)	SF_6	$\nu_3 = 0$–1; P(20) line of 10.6 μm CO_2 laser	$\Delta\nu_3 = +1$; P(16), P(18), P(20), P(22), P(24), P(26), P(28), and P(32) lines of 10.6 μm CO_2 laser	$k_{VT} = 8.2\times 10^3$ s^{-1} Torr^{-1}; mapped excited-state spectrum
Frankel (1976a,b)	SF_6	$\nu_3 = 0$–1; P(20) line of 10.6 μm CO_2 laser	$\Delta\nu_3 = +1$; P(22), P(24), P(26), P(28), P(30), P(32), P(34), and P(36) lines of 10.6 μm CO_2 laser	Evidence for collisionless energy transfer
Steinfeld and Jensen (1976)	SF_6		$\Delta\nu_3 = 0$–1; PbSnSe diode laser	Measurement of width of spectral "hole"
Houston et al. (1973)	BCl_3	$^{11}BCl_3$ $\nu_3 = 0$–1; P(12) line of 10.6-μm CO_2 laser	$\Delta\nu_3 = 1$ in $^{11}BCl_3$; P(12)–P(38) lines of 10.6 μm CO_2 laser $\Delta\nu_3 = 1$ in $^{10}BCl_3$; R(18) and R(20) lines of 10.6 μm CO_2 laser	$k_{VT} = 1.73\times 10^5$ s^{-1} Torr^{-1} $k_{VV} \geq 2\times 10^6$ s^{-1} Torr^{-1}
Frankel et al. (1974)	BCl_3	$^{11}BCl_3$ $\nu_3 = 0$–1; P(12) line of 10.6-μm CO_2 laser	$\Delta\nu_3 = 1$ in $^{11}BCl_3$; P(28) line of 10.6-μm CO_2 laser	BCl_3–HCl relax. rate as function of temperature
Yuan et al. (1973)	C_2H_4	$\nu_7 = 0$–1; P(26) line of 10.6-μm CO_2 laser	$\nu_7 = 0$–1; P(14) of 10.6-μm CO_2 laser	$k_{VV} = 3.49\times 10^5$ s^{-1}; $k_{VT} = 7.14\times 10^3$ s^{-1} Torr^{-1}
Nachson and Coleman (1974)	NO	$(v, J) = (0, \frac{17}{2})$–$(2, \frac{15}{2})$; $P_1(6)$ HF laser line	(see below)	Rate measured for NO(1)+NO(0) = 2NO(0) is $k = 2.6\times 10^3$ s^{-1} Torr^{-1}

E	v'	J'	v''	J''	CO line
$^2\Pi_{1/2}$	1	$\frac{11}{2}$	2	$\frac{9}{2}$	$P_{11-10}(14)$
$^2\Pi_{1/2}$	2	$\frac{9}{2}$	3	$\frac{11}{2}$	$P_{11-10}(12)$
$^2\Pi_{3/2}$	1	$\frac{13}{2}$	2	$\frac{15}{2}$	$P_{9-8}(16)$
$^2\Pi_{3/2}$	2	$\frac{15}{2}$	3	$\frac{17}{2}$	$P_{10-9}(16)$

continued

Table 1.6   Infrared–Infrared Double Resonance (continued)

Reference	Molecule	Transitions pumped, source	Transitions monitored, source	Results
Brechignac et al. (1975)	CO	$(v,J) = (8,9)-(9,8)$, $P_{9-8}(9)$ CO laser	$P_6(11) \equiv (v,J) = (5,11)-(6,10)$; $P_7(10)$, $P_8(9)$, $P_8(10)$, $P_8(12)$, $P_9(8)$, $P_9(10)$, $P_{10}(7)$, $P_{10}(8)$, $P_{10}(9)$, $P_{11}(7)$, $P_{11}(8)$, $P_{11}(9)$, $P_{12}(8)$, $P_{12}(9)$; all with cw CO operating on same transition	Measured rates for $CO(v)+CO(v') = CO(v-1)+CO(v'+1)$
Lee and Faust (1971)	H^-/CaF_2	$n = 0-1$; $R(4)$ 10.6-μm CO_2	$n = 2-3$; $P(38)$, $P(12)$, and $R(24)$ lines of 10.6-μm CO_2 laser	$T_1 = 7$ ps
Abouaf-Marguin et al. (1973)	NH_3/matrix	$\nu_2 = 0-1$; $R(10)$, $R(12)$ of 10.6-μm CO_2 laser	$\Delta\nu_2 = +1$; $R(10)$, $R(12)$ of 10.6-μm CO_2 laser	Relaxation of ν_2 is $1.5-3$ μs
Carroll and Marcus (1968)	CO_2	10^00^01; R branch of 10.6-μm CO_2 laser	10^00^01; P branch of 10.6-μm CO_2 laser	$k_{rot} = 2\times 10^{-10}$ cm^3 $molec^{-1}$ s^{-1}
Cheo and Abrams (1969)	CO_2	10^00, $J = 20-00^01$, $J = 19$; $P(20)$ of 10.6-μm CO_2 laser	02^00, $J = 20-00^01$, $J = 19$; $P(20)$ of 9.6-μm CO_2 laser	$k_{rot} = 3.4\times 10^{-10}$ cm^3 $molec^{-1}$ s^{-1}
Jacobs et al. (1974a,b)	CO_2	10^00, $J = 20-00^01$, $J = 19$; $P(20)$ of 10.6-μm CO_2 laser	02^00, $J = 20-00^01$, $J = 19$; $P(20)$ of 9.6-μm CO_2 laser	$k_{rot} = 4\times 10^{-10}$ cm^3 $molec^{-1}$ s^{-1}
Peterson et al. (1974)	HF	$(v,J) = (0,5)-(1,4)$, $(0,6)-(1,5)$ and $P_1(5)$ and $P_1(6)$ of HF laser	$(v,J) = (0,5)-(1,4)$; $P_1(5)$ $(0,6)-(1,5)$; $P_1(6)$	$k_{rot} = 4.9-7.8\times 10^7$ s^{-1} $Torr^{-1}$
Hinchen (1975);Hinchen and Hobbs (1976)	HF	$(v,J) = (0,4)-(1,3)$; $P_1(4)$ of HF laser	$(v,J) = (0,5)-(1,4)$; $P_1(5)$ $(0,6)-(1,5)$; $P_1(6)$ $(0,7)-(1,6)$; $P_1(7)$ $(0,8)-(1,7)$; $P_1(8)$	$k_{rot} = 1.2-4.0\times 10^7$ s^{-1} $Torr^{-1}$
Alimpiev and Karlov (1974)	SF_6 BCl_3	$\nu_3 = 0-1$; $P(12)$, $P(16)$, $P(18)$, or $P(20)$ of 10.6-μm CO_2 laser	$\nu_3 = 0-1$; $P(12)$, $P(16)$, $P(18)$, or $P(20)$ of 10.6 μm CO_2 laser	Observation of nutation

Shoemaker et al. (1974)	CH_3F	$(\nu_3, J, K) = (0,4,3)-(1,5,3)$; P(32) of 9.6-$\mu$m CO_2 laser	ΔM transitions caused by Stark tuning	Velocity changes little on molecular reorientation
Meyer and Rhodes (1974)	CO_2	$10^\circ 0-00^\circ 1$, P(18); P(18) of 10.6-μm CO_2 laser	$10^\circ 0-00^\circ 1$ P(20); using P(20) 10.6-μm CO_2 laser	Velocity changes little in rotationally inelastic collisions
Meyer et al. (1974)	CO_2	$10^\circ 0-00^\circ 1$, P(20); P(20) of 10.6-μm CO_2 laser	$02^\circ 0-00^\circ 1$, P(18); P(18) of 9.6-μm CO_2 laser	Momentum transfer for $00^\circ 1$ state
Frankel and Steinfeld (1975)	SF_6 BCl_3	$\nu_3 = 0-1$; 10.6-μm of CO_2: P(12) P(12) P(18)	$\Delta\nu_3 = +1$; 10.6-μm CO_2 laser: P(28) P(12) P(28)	Molecular orientation changes little in vibrationally inelastic collisions
Freund et al. (1973)	NH_3	$\nu_2 = 0 \leftarrow 1$: $^OQ_{aa}(8,7)$; N_2O P(13) +klystron $^OQ_{ss}(7,7)$; N_2O R(32) +klystron $^OQ_{ss}(4,4)$; N_2O P(15) +klystron $^OQ_{aa}(4,4)$; N_2O R(28) +klystron	Coupled three- and four-level transitions	Two-photon (ir+microwave) transitions used for *both* pump and probe; collisionally transferred Lamb dip found in four-level experiments
Preses and Flynn (1977)	CH_3F	$^{12}CH_3F$: $\nu_3 = 1 \leftarrow 0$; P(20) of 9.6-μm CO_2 laser	$^{13}CH_3F$: $\nu_3 = 1 \leftarrow 0$; P(32) of 9.6-μm CO_2 laser	One quantum of ν_2 transferred between $^{12}CH_3F$ and $^{13}CH_3F$ in six collisions

1.5.2.1. Vibrational Energy Transfer

1.5.2.1a. Carbon Dioxide. It is not surprising that the first molecule to be investigated by double resonance with CO_2 lasers was CO_2 itself. A diagram of energy levels important to the CO_2 laser is shown in Figure 1-6. Lasing action takes place from the $00°1$ state, populated by energy transfer from $N_2(v=1)$, to either the $10°0$ state ($10.4\ \mu m$) or to the $02°0$ state ($9.4\ \mu m$). Since lasing depends both on the net rate of production of the upper-state population and on the rate of relaxation of the lower state, many efforts to clarify the CO_2 laser mechanism attempted to measure the vibrational relaxation of the lower $10°0$ and $02°0$ states. A double-resonance experiment for this purpose was devised by Rhodes *et al.* (1968), in which a Q-switched laser operating at $9.6\ \mu m$ saturated a $23°C$ sample of CO_2 on the $02°0–00°1$ transition. A second CO_2 laser operating continuously at $10.6\ \mu m$ probed the population difference between the $10°0$ and $00°1$ states. Immediately following the Q-switched pulse, the absorption of the cw laser decreased, owing to the increase of molecules in the upper $(00°1)$ level. The absorption then relaxed toward its equilibrium value with a rate of $(4\pm1)\times10^5\ s^{-1}\ Torr^{-1}$. Since the relaxation of the upper level was known to be much slower than this rate, the authors concluded that the observed effect must be due to a decrease in $CO_2(10°0)$ population. They attributed the decrease to a near-resonant process such as

$$10°0+00°0\rightarrow01^10+01^10$$

However, Jacobs *et al.* (1975) have recently performed an analogous experiment by saturating the $10.4\ \mu m$ transition and monitoring the 9.4-μm absorption. Their results suggest that the process observed by Rhodes *et al.* is actually

$$10°0+00°0\rightarrow02°0+00°0+102.8.cm^{-1}$$

Under this assumption, the rates obtained by these two investigations are in complete agreement.

1.5.2.1b. Sulfur Hexafluoride. The interaction of CO_2 laser photons with the SF_6 molecule has been extensively studied since the observation by Wood and Schwarz (1967) that SF_6 could be used to passively Q-switch the CO_2 laser. Attempts to understand the Q-switching mechanism quickly focused on the saturation and relaxation behavior of this molecule. Wood *et al.* (1969) and Burak *et al.* (1969a) developed kinetic models to explain the decrease in SF_6 absorption as the intensity of the pumping CO_2 laser increased. Although both models predicted the existence of a substantial excited-state population, they differed radically on the assumed rate constant for vibrational deactivation of this population. To resolve the controversy over this rate, the M.I.T. group (Burak *et al.*, 1969b; Steinfeld *et al.*,

1970) probed the relaxation of the excited state by an infrared double-resonance experiment.

The experimental arrangement used in this work is shown in Figure 1-21. A pulsed CO_2 laser operating near the P(20) line at 10.6 μm was used to saturate a part of the ν_3 transition in SF_6. A weaker probe laser, collinear with the pumping laser, monitored changes of absorption caused by the excited-state population. Three types of changes were observed, depending on the frequency of the probe laser:

1. When the probe laser operated on the same CO_2 transition as the pump laser, a concurrent decrease or "hole" in the absorption spectrum was observed corresponding to the saturation of those levels in resonance with the pump field.

2. At probe laser lines between P(22) and P(32), the authors observed a transient increase in absorption whose decay provided a measure of the rate at which vibrationally excited SF_6 molecules were deactivated. Measurement of this rate as a function of pressure and composition of the gas mixture yielded the vibrational deactivation rates listed in Table 1-7.

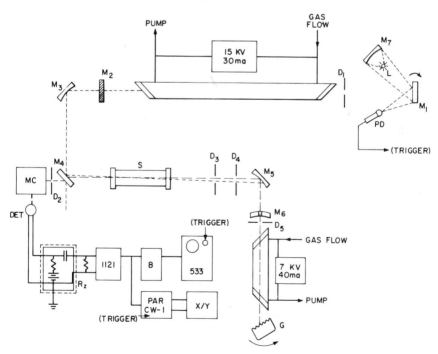

Figure 1-21. Block diagram of infrared double-resonance apparatus (from Steinfeld *et al.*, 1970).

Table 1-7   Relaxation Times for SF_6

Collision partner	$p\tau$ (μs-Torr)
SF_6	122
He	41
Ne	194
Ar	470
Kr	3040
C_2H_6	8.6
$(CH_3)_2O$	5.3
CH_3Br	14
CHF_2Cl	19
H_2	9.3
N_2	103
Cl_2	80
H_2O	11.3
CH_4	20

3. When the probe laser was tuned to the P(16) or P(14) CO_2 laser lines, a transient decrease in the absorption was observed which relaxed to its equilibrium value with the same rate as measured by type 2 above.

These effects were interpreted with the aid of the following model, illustrated in Figure 1-22. The pump laser couples ground-state molecules to a specific rotational level of the $\nu_3 = 1$ state. Vibration–vibration energy transfer processes rapidly redistribute the excited molecules to a manifold of states represented by n_3 in the figure. Quasi-equilibrium among these states is reached within a few microseconds at 1 Torr of SF_6. Finally, the excited vibrational manifold relaxes via a vibration-to-translation energy transfer process whose rate-limiting step is the deactivation of the ν_6 level at 363 cm^{-1}.

Because of anharmonicity, the transient absorption spectrum of the n_3 distribution will be shifted to longer wavelengths than that of the ground-state SF_6. This fact accounts for the variation in the sign of the change in absorption as the probe laser is tuned to different CO_2 lines. At longer wavelengths, the excited-state absorption exceeds the ground-state absorption, while at shorter wavelengths the opposite is true.

The infrared–infrared double-resonance experiment on SF_6 settled the controversy over the relaxation rate which had arisen during the saturation experiments. The rate of vibrational relaxation had now been measured for collisions of SF_6 with a variety of partners (Table 1–7). Since these rates were in good agreement with those subsequently found by Bates *et al.* (1970, 1971) using fluorescence detection, they could reliably predict the saturation and Q-switching behavior. In fact, Burak *et al.* (1971) successfully described experimental observations of Q-switching using a model based on

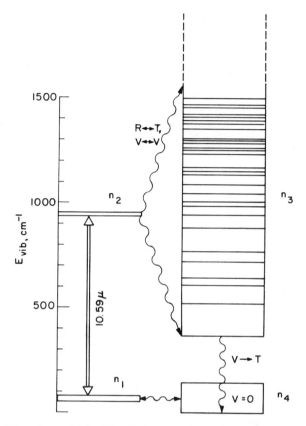

Figure 1-22. Relaxation model for SF_6. The levels on the left-hand represent those rotational components of the $\{v_1\} = 0$ and $v_3 = 1$ that can be directly coupled by CO_2 laser radiation. The stack of vibrational levels in the upper right-hand part of the diagram are assumed to be in rapid equilibrium; specific levels only up to 1500 cm^{-1} are shown (from Steinfeld *et al.*, 1970).

the rates obtained from the double-resonance experiment. On the other hand, the double-resonance experiment pointed to certain deficiencies in the saturation models. Since the double-resonance experiment had demonstrated an excited-state absorption, it was then imperative to include this effect in any new saturation model. More recent studies of the saturation behavior by Oppenheim and Melman (1971) and by Armstrong and Gaddy (1972) have included this important effect.

Frankel (1976a, b) has extended the application of double-resonance techniques in SF_6 to much shorter time scales. By using a single pulse from a mode-locked, transverse discharge CO_2 laser as his pump source, he was able to measure relaxation times on the order of 4 ns. The results of his measurements suggest that energy transfer between vibrational modes of SF_6 occurs rapidly even under collisionless conditions. Such collisionless energy transfer following high-intensity radiation may have important consequences for multiphoton laser isotope separation.

Steinfeld and Jensen (1976) have suggested that the coupling between vibrational levels observed by Frankel could be induced by power broadening. If this were the case, the width of the "hole" bleached in the SF_6 absorption spectrum should be given by the Rabi frequency

$$\Delta \nu = \frac{\mu |\varepsilon|}{h} \tag{1-67}$$

where μ is the dipole moment and

$$|\varepsilon| = \left(\frac{2\pi I}{c}\right)^{1/2} \tag{1-68}$$

where I is the laser intensity. Consequently, if this mechanism is correct, the width of the bleached absorption hole should be proportional to $I^{1/2}$. Currently, both Steinfeld and Jensen (1976) and P. L. Kelley and P. Moulton (private communication, 1976) are attempting to measure the width of this hole by double-resonance schemes. In these studies a tunable diode laser will probe the frequency width of the SF_6 absorption hole produced by a high-power pump pulse.

1.5.2.1c. Boron Trichloride. A double-resonance experiment analogous to that in SF_6 has been used by Houston et al. (1973) to examine vibrational relaxation in BCl_3. A diagram of the ν_3 absorption in naturally occurring BCl_3 is shown in Figure 1-23. Note that this absorption band consists of two distinct peaks, corresponding to $^{11}BCl_3$ and $^{10}BCl_3$. Using the P(12) CO_2 laser line as the pump source, double-resonance signals could be observed when the probe laser was tuned to any of the CO_2 transitions between P(12) and P(38), or R(18) and R(20). From the sign and amplitude of the double-resonance signal, the excited-state absorption spectrum could be deduced.

By pumping $^{11}BCl_3$ with the P(12) CO_2 laser line and monitoring the rise of increase in $^{10}BCl_3$ absorption at the R(20) line, Houston et al. were able to show that the process

$$^{11}BCl_3(\nu_3 = 1) + {}^{10}BCl_3(\nu_3 = 0) \rightarrow {}^{11}BCl_3(\nu_3 = 0) + {}^{10}BCl_3(\nu_3 = 1)$$

takes place in fewer than 13 collisions. This fact supports a relaxation model similar to that for SF_6 in which the initial ν_3 vibrational energy is rapidly transferred by collisions to a distribution of molecular levels. This distribution subsequently relaxes via a vibration-to-translation energy transfer process whose rate-determining step is the deactivation of the ν_4 level at $243\ cm^{-1}$.

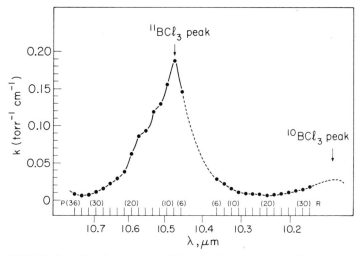

Figure 1-23. CO_2 laser line absorption coefficients for boron trichloride (Green and Steinfeld, 1975, unpublished results).

Probabilities for deactivation of BCl_3 by various collision partners are summarized in Figure 1-24. Particular attention should be drawn to the fact that HCl relaxes BCl_3 very efficiently. Since the rotational transitions $(J = 4 \rightarrow 6)$ and $(J = 3 \rightarrow 5)$ for HCl are in near resonance with the vibrational transition in $BCl_3(\nu_4 = 1 \rightarrow 0)$, vibration-to-rotation energy transfer might account for the large magnitude of the deactivation rate. A calculation (Poulsen *et al.*, 1973) was successful in accounting for the HCl–BCl_3 and DCl–BCl_3 rates at 300°K. However, this theory, based on short-range forces, predicted an increase in deactivation probability with increasing temperature proportional to $\exp(T^{-1/3})$. By contrast, subsequent measurements by Frankel *et al.* (1974) using the double-resonance scheme showed that the probability for deactivation decreased with increasing temperature. Such behavior is much more typical of collisions in which long-range forces play the dominant role.

1.5.2.1d. Ethylene. Vibrational relaxation in ethylene has been investigated using a double-resonance technique by Yuan *et al.* (1973). These authors used a pulsed CO_2 laser operating on the P(26) line at 10.6 μm to pump the ν_7 mode of C_2H_4. A continuous CO_2 laser operating on the P(14) line monitored the population of the same ν_7 mode, but in a different part of the rotational manifold than that pumped by the pulsed laser. Population in this mode exhibited a double exponential decay due to bimolecular collisions. The faster rate of 349 ms^{-1} Torr^{-1} was thought to correspond to the process

$$C_2H_4(\nu_7) + C_2H_4(0) \rightarrow C_2H_4(\nu_{10}) + C_2H_4(0) + 123 \text{ cm}^{-1}$$

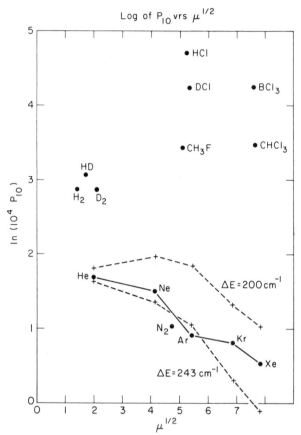

Figure 1-24. Vibrational deactivation probabilities for the ν_4 mode of BCl_3 with various collision partners. The dashed lines represent the predictions of the Tanczos–Stretton breathing-sphere model for the monatomic gases, with potential parameters $\alpha = 4\ \text{Å}^{-1}$ and $\Delta E = 200$ or $243\ \text{cm}^{-1}$ (from Houston et al., 1973).

The slower rate of $7.14\ \text{ms}^{-1}\ \text{Torr}^{-1}$ was in good agreement with the $V–T$ rate measured by fluorescence decay techniques (Yuan and Flynn, 1972, 1973).

1.5.2.1e. Nitric Oxide. The first gas-phase IRIRDR scheme using lasers other than CO_2 lasers was performed by Nachshon and Coleman (1974). In this work, a pulsed HF laser operating on the $P_1(6)$ line (P(6) of the $1 \to 0$ transition) was used to pump the P(17/2) line of the $2 \leftarrow 0$ transition in ground-state $(^2\Pi_{1/2})$NO. Immediately following the pump pulse, a cw electrically excited CO laser probed the $v = 2 \leftarrow 1$ and $v = 3 \leftarrow 2$ transitions of both ground-state $(^2\Pi_{1/2})$ and excited-state $(^2\Pi_{3/2})$ NO. A diagram showing the appropriate energy levels is given in Figure 1-25. Since the match between the NO absorption lines and the CO laser transitions is not exact, argon at a pressure of about 700 Torr was required to broaden the NO absorption lines.

The authors were able to show that spin exchange processes such as

$$NO(^{2}\Pi_{1/2}, v = 1) + M \rightarrow NO(^{2}\Pi_{3/2}, v = 1) + M$$

and

$$NO(^{2}\Pi_{1/2}, v = 2) + M \rightarrow NO(^{2}\Pi_{3/2}, v = 2) + M$$

occur on a time scale of less than 0.05 μs when M is 400 Torr of argon with 2.5 Torr of NO. By monitoring the decay of the NO($v = 1 \rightarrow 2$) absorption signal, the authors obtained a rate of 2.6×10^3 s^{-1} Torr^{-1} for the process

$$NO(v = 1) + NO(v = 0) \rightarrow 2NO(v = 0)$$

This rate was found to be in good agreement with that measured by Stephenson (1973) using fluorescence techniques.

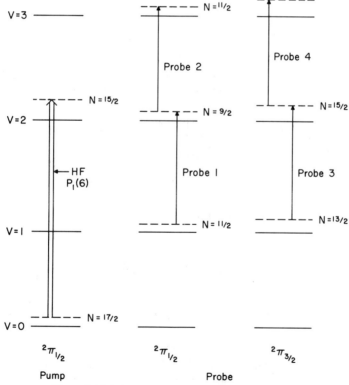

Figure 1-25. Infrared–infrared double resonance in NO.

1.5.2.1f. Carbon Monoxide. Vibrational relaxation in CO has been investigated by Brechignac *et al.* (1975) using a Q-switched CO laser as the pump source and a cw CO laser as the probe. In order to study energy transfer among high-lying vibrational levels in the CO gas, the sample was excited by an electric discharge.

When the pumping laser directly connected the $v = 8$ and $v = 9$ levels, double-resonance signals were observed in levels ranging from $v = 5$ to $v = 11$ (see Table 1-6). The time dependence of the population change in each case was well matched by a computer calculation which assumed that rates for such processes as

$$CO(v) + CO(v') \rightarrow CO(v-1) + CO(v'+1) \pm \Delta E$$

followed the theory of Sharma and Brau (1969). Small discrepancies seemed to indicate the presence of $\Delta v = 2$ collisional transitions. The rates for the $\Delta v = 1$ processes were found to be on the order of 100 times the known rates for

$$CO(v) + CO(0) \rightarrow (v-1) + CO(1)$$

1.5.2.1g. Hydrogen Ion/Calcium Fluoride. The first application of IRIRDR to the solid state was performed by Lee and Faust (1971) in their investigation of the vibrational spectrum of an H^- ion on an F^- site in CaF_2. Two pulsed CO_2 lasers were used. The first (QS) was Q-switched by rotation of the grating which formed the tuning element, while the second (NQS) was pulsed by modulating the discharge. Under normal operation, the QS laser produced by 20-kW pulses of 400-ns duration and the NQS laser produced 10- to 30-W pulses of much longer duration. Two types of experiments were performed.

In the first, the QS laser alone was used to investigate the saturation of the $n = 0 \rightarrow 1$ vibrational transition for the CaF_2/H^- system. The decrease in absorption as a function of increased intensity of the R(4) 964.7-cm^{-1} line was matched by the formula

$$\frac{1}{I} \frac{dI}{dx} = -\alpha_L - \frac{\alpha_H}{1 + WI} \tag{1-69}$$

where I is the laser intensity, α_L and α_H are the absorption coefficients of the lattice and ion, respectively, and W is a saturation parameter found to be $W = 6 \times 10^{-16}$ cm^2 erg^{-1} s. From the value of W a relaxation time T_1 was calculated to be 7 ps by use of the equation $T_1 = \frac{1}{2}\hbar\omega_{01} W/\sigma$, where ω is the cross section for absorption of photons of frequency ω_{01}.

In a second experiment, stepwise absorption to the $n = 2$ level was monitored by observing changes in the NQS absorption concurrent with the

QS pulse. The three $n = 2$ sublevels found by this double-resonance technique agreed with those found previously by Elliot *et al.* (1965).

1.5.2.1h. Ammonia. In a second solid-state application, infrared–infrared double resonance was used by Abouaf-Marguin *et al.* (1973) to study the vibrational relaxation of matrix-isolated NH_3 at 8°K. A pulsed CO_2 laser operating on the R(10) or R(12) line of the 00°1–10°0 transition was employed for excitation of the ν_2 inversion mode of NH_3. A continuous CO_2 laser probed changes in absorption either at the pumping frequency or at the adjacent R(10) or R(12) transition. A variety of effects was observed.

In a nitrogen matrix with both lasers operating on the same transition, a rapid increase in transmission occurred immediately following the excitation pulse. The transmission relaxed exponentially on a short time scale with τ_1 of a few microseconds and then returned to its equilibrium value on a time scale of a few hundred microseconds. As the concentration was increased from $M/R = 2000$ to $M/R = 20$, the fast time constant decreased from about 3 μs to 1.5 μs. If the pump and probe lasers were not tuned to the same transition, or if CH_4 or Ar was used for the matrix, the rapid increase in transmission was absent and the only change in transmission occurred on the long time scale.

The short exponential decay observed in the N_2 matrix was interpeted to be the relaxation of the ν_2 mode of NH_3. As expected, when interactions between ammonia molecules were increased by increasing the concentration, the time constant for this relaxation decreased. The change in transmission observed on the slow time scale was then interpreted as resulting from the heating of the matrix as the vibrational energy was transferred to the lattice. The absence of the fast component when the pump and probe lasers were not tuned to the same frequency indicated that the lasers interacted with two different sets of molecules, while the absence of the fast component in Ar and CH_4 matrices was taken to mean that the vibrational relaxation in these matrices was faster than the detection response time, about 400 ns.

The relatively short vibrational relaxation of the ν_2 mode in the gas phase (about 2 ns) has been interpreted as being due either to vibration–rotation energy transfer or to the high inversion frequency of the NH_3 molecule (Jones *et al.*, 1969; Bass and Winter, 1972). The fact that the relaxation is much longer in an N_2 matrix where neither inversion nor rotation is dominant supports this conclusion.

A rather different type of infrared double resonance on ammonia was carried out by Freund *et al.* (1973). They employed two N_2O lasers and two microwave klystrons to induce a pair of infrared–microwave two-photon transitions on a number of lines in the ν_2 band of NH_3, without the use of external Stark fields or other tuning methods. In four-level experiments involving states coupled by inelastic collisions between the two inversion doublet levels, they observed a collisionally transferred "Lamb dip," or hole burned into the velocity distribution, on the double-resonance signal. This showed that, in this system, the cross section for momentum transfer was less than that for relaxation of the inversion states. No such collisionally transferred hole was observed on four-level experiments involving rotational inelasticity, such as $\Delta J = 1$ and $\Delta K = 0$.

1.5.2.2. Rotational Energy Transfer

1.5.2.2a. Carbon Dioxide. Rotational relaxation in CO_2 has been extensively studied by IRIRDR due to the availability of resonant laser sources. In the first application of this technique, Carroll and Marcus (1968) used a pulsed pumping laser to deplete a rotational state in the 00°1 level of CO_2 contained in an active gain medium. A second cw probe laser monitored changes in population either of the depleted rotational level or of adjacent rotational levels. From the time dependence of the probe laser absorption, they deduced a rotational relaxation rate of roughly 2×10^{-10} cm^2 $molec^{-1}$ s^{-1} for pure CO_2 at 298°K. From the variation of the relaxation rate with ΔJ, they further concluded that $|\Delta J| = 2$ transitions were most effective in the relaxation. ($|\Delta J| = 1$ transitions cannot occur, owing to the absence of alternate rotational levels in CO_2.)

Cheo and Abrams (1969) subsequently improved on this double-resonance technique by using a GaAs electro-optic Q switch to reduce the CO_2 pump laser pulse width to 20 ns. A rotational relaxation rate for CO_2–CO_2 collisions of 3.4×10^{-10} cm^3 $molec^{-1}$ s^{-1} was obtained from the rate of change in the probe laser absorption following the Q-switched pulse. Although this rate differed slightly from that measured by Carroll and Marcus, a more serious contradiction developed over the $|\Delta J| = 2$ selection rule. Cheo and Abrams found that the observed probe signals did not depend strongly on the difference between the probed J-level and the pumped one. Their results suggest, therefore, that rotational relaxation in CO_2 may not depend strongly on ΔJ or on the amount of energy transferred.

The CO_2 relaxation problem has more recently been investigated by Jacobs et al. (1974a,b) who employed the same double-resonance scheme with yet another increase in time resolution. These investigators used a mode-locked CO_2 laser in conjunction with two amplifiers to reduce their pumping pulse width to about 2 ns. Results of their work (Jacobs et al., 1974a) placed the CO_2–CO_2 rotational relaxation rate at $(4.0 \pm 0.6) \times 10^{-10}$ cm^3 $molec^{-1}$ s^{-1}, in good agreement with that of Cheo and Abrams (1969). Decay times for various J levels were found to increase for increasing or decreasing ΔJ centered around the pumped level (Jacobs et al., 1974b). Qualitative agreement of these results with a computer model based on $|\Delta J| = 2$ collisions was obtained. However, the discrepancies indicated that $|\Delta J| > 2$ collisions could not be ignored.

1.5.2.2b. Hydrogen Fluoride. The rotational relaxation of HF has also been investigated by IRIRDR techniques. Peterson *et al.* (1974) have used a single HF laser to pump and probe the rotational levels connected by the $P_1(5)$ and $P_1(6)$ laser lines. Pulses from the HF laser were terminated abruptly by a CdTe Pockels cell. Residual radiation which leaked through the cross polarizer was used to probe the rotational relaxation. Bleaching of absorption on the $P_1(6)$ and $P_1(6)$ lines relaxed exponentially with rates of 4.9×10^7 and $7.8 \times 10^7 \, s^{-1} \, Torr^{-1}$, respectively.

Hinchen (1975) and Hinchen and Hobbs (1976) have improved on this technique by using two lasers. The absorption of a stable cw laser was used to determine the loss or arrival of population in a particular J or J' level following population of the J level with a second pulsed laser. Specifically, when the pulsed HF laser pumped HF to the $(v, J) = (1, 3)$ level, population subsequently increased in levels with $J = 4, 5, 6$, and 7. The rates associated with this change in population decreased from $4.0 \times 10^7 \, s^{-1} \, Torr^{-1}$ to $1.2 \times 10^{-7} \, s^{-1} \, Torr^{-1}$ as ΔJ increased from 1 to 4. Collisional selection rules of $|\Delta J| = 1$, therefore, seem to be obeyed in HF–HF collisions.

1.5.2.3. Dephasing, Momentum Transfer, and Molecular Alignment

An increasing number of infrared double-resonance applications are taking advantage of the coherence, polarization, and spectral narrowness of the laser source. Although these applications are less prevalent than those to vibrational and rotational energy transfer, they offer the future promise of information hitherto obtained only from molecular beam studies or less direct methods. Some typical instances follow.

1.5.2.3a. Dephasing. The optical coherence of the laser wavefront may be used to create an ensemble of molecules whose quantum-mechanical wavefunctions all have the same phase. Collisions among prepared molecules destroy this coherence and give rise to a "dephasing" relaxation which is related to the off-diagonal elements of the density matrix (see Chapter 3). As described in Section 1.4.2.2, one manifestation of this dephasing is the appearance of nutation on infrared–microwave double-resonance signals. A similar nutation effect has been observed in infrared–infrared double resonance by Alimpiev and Karlov (1974).

In their experiment a Q-switched and a cw CO_2 laser were propagated collinearly through low-pressure samples of SF_6 or BCl_3. When the lasers were tuned to the P(12), P(16), P(18), or P(20) lines, an oscillation was observed in the recovery of the cw laser absorption following bleaching of the sample by the pulsed pump laser. The qualitative features of this effect were related to the strength of the laser field, the dipole moment of the transition, and the number of molecules participating in the transition.

Experimental results were found to be in good agreement with a theory which extended the nutation effect to the infrared region of the spectrum.

1.5.2.3b. Momentum Transfer. The Doppler effect has long been known to spectroscopists as an annoying consequence of molecular velocities which broadens spectral lines and obscures the intrinsic spectrum (see Section 1.1.2). Recently, however, spectroscopists have started to make use of the Doppler effect to study collisions in which velocity or momentum is transferred. Although a complete examination of velocity-selective spectroscopy is beyond the scope of this chapter, a few examples will be cited which demonstrate double-resonance applications to the study of velocity-changing collisions.

Shoemaker *et al.* (1974) have shown that collisions of methyl fluoride molecules which tip the angular momentum vector relative to a fixed axis do so while nearly preserving the original velocity. In their experiments two lasers operating on the same CO_2 laser transition were frequency locked to one another to maintain a constant difference of 30.008 MHz. The two lasers were propagated collinearly through a sample of CH_3F contained in a Stark cell. As the Stark field caused interaction of M sublevels of the $(\nu_3, J, K) = (0,4,3) \rightarrow (1,5,3)$ methyl fluoride transition with the lasers, sharp resonances were observed which corresponded to level combinations diagrammed in Figure 1-26a. Resonance occurs when molecules from, say, the right-hand lower level can be pumped by Ω_1 to the upper level and simultaneously interact with field Ω_2 to finish in the lower left-hand level. Molecules can exhibit such resonance only if the Doppler shift is the same for both transitions. Consequently, the width of the resonance corresponds to the homogeneous (Lorentz) width rather than to the Doppler width.

In addition to the sharp resonances corresponding to level arrangements such as in Figure 1-26a, satellite resonances were also observed corresponding to the level arrangement of Figure 1-26b. These satellite

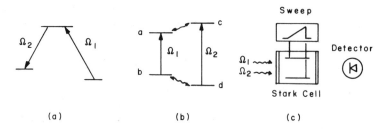

(a)   (b)   (c)

Figure 1-26. (a) Tranditional three-level double-resonance configuration (see Figure 1-3); (b) collision-induced double-resonance level configuration, with wavy lines indicating collision-induced transitions; (c) experimental arrangement for monitoring optical double-resonance signals (from Shoemaker *et al.*, 1974).

resonances are due to collisional coupling between two pairs of M sublevels of the same (J,K) methyl fluoride line. From the width of the resonance, which is proportional to the velocity smearing on collision, the authors were able to show that collisions which change the orientation of CH_3F molecules nearly preserve the original molecular velocity.

The experiment outlined above measured the change in velocity associated with collisional change in molecular orientation. Meyer and Rhodes (1974) and Meyer *et al.* (1974) used a similar technique to measure the change in velocity associated with collisions which change the *rotational* state of a molecule. Two probe CO_2 lasers were each frequency-locked to a stable CO_2 laser and propagated collinearly through a sample cell containing CO_2 and H_2. The total 4.3-μm fluorescence from the $00°1$ level of CO_2 was detected as a function of the difference in frequency between the two lasers. When the lasers both operated on the $P(20)$ 10.6 μm CO_2 line, a resonant decrease in fluorescence was observed at zero frequency difference, because of the fact that the lasers each interacted with molecules in the same velocity group.

The lasers were then tuned to different lines. One laser was locked at a frequency somewhat below its transition center, while the other was tuned. A broad resonance was observed at a frequency of the tuned laser which again corresponded to having both lasers interact with the same molecular velocity group. Since the lasers monitored adjacent rotational levels, the width of the resonance could be used to measure the change in velocity associated with a collisional change in rotational state. Their results indicated that the average velocity change was $3 \pm 2 \times 10^3$ cm/s for collisions between H_2 and CO_2 which took CO_2 from $J = 20$ to $J = 18$. This change in velocity is much less than the mean thermal speed.

1.5.2.3c. Molecular Alignment. Frankel and Steinfeld (1975) have investigated molecular reorientation following vibrationally inelastic collisions in BCl_3 and SF_6. Two CO_2 lasers were used as in the experiments of Section 1.5.2.1. In the present case, however, the amplitude of the double-resonance signal was monitored as a function of the relative polarization of the two lasers. The signals showed a minimum when the two beams were perpendicularly polarized, even though the signals monitored levels connected by V–V equilibration requiring 5–10 collisions. They concluded from their observations that the molecular alignment must remain partially unchanged during these vibrationally inelastic collisions.

1.5.3. Optical Pumping

As was the case for microwave-detected double resonance (Section 1.4.3), no instances have been found in which infrared radiation has been used to monitor changes in molecular-level populations brought about by pumping an electronic transition.

1.6. Optically Detected Double Resonance

In the preceding sections we have outlined the development of double-resonance techniques in which transient responses to radiation were monitored in the microwave and infrared spectral regions. Because of the greater efficiency in detection of visible photons, it is not surprising to find that double-resonance methods using optical monitoring actually predated those using infrared and microwave detection. Optical detection of transients produced by microwave pumping will be the subject of the following subsection, while later sections will focus on optical detection of both infrared and optical pumping.

1.6.1. Microwave–Optical Double Resonance

In the first application of the microwave–optical double-resonance (MODR) technique, Brossel and Bitter (1952) excited mercury atoms to the $m_J = 0$ sublevel of the $^3P_1$ state using π-polarized light from a quartz-mercury arc lamp at 253.7 nm. In the absence of collisions or external fields, fluorescence observed at right angles to the excitation consisted solely of the π component of polarization. However, upon the application of a radio-frequency field at $\omega = \gamma H_z$, where H_z is a constant magnetic field strength and γ is the gyromagnetic ratio for the upper state, atoms were transferred from the $m_J = 0$ level to levels with $m_J = \pm 1$. This transfer was observed as a decrease in polarization of the fluorescent emission. From the resonance linewidth, a lifetime of $1 \cdot 55 \times 10^{-7}$ s was obtained for the Hg $^3P_1$ state. The Landé g factor was found to be $g = 1.4838$.

The experiment outlined above illustrates an important difference between double-resonance techniques involving optical detection and those involving infrared or microwave detection. For most optical double-resonance schemes, three photons are involved. One of these, at a radio frequency in the example above, serves to pump population density from one state to another. A second photon, the optical fluorescence in the example above, serves as a monitor of those population changes. The function of the third photon is simply to create a state which can fluoresce. As we shall see below, even if the pump photon interacts with molecules in the ground electronic state, many optical double-resonance techniques will employ a three-photon scheme. The advantage of the three-photon technique stems from the fact, elaborated in Section 1.3.3, that it is much easier

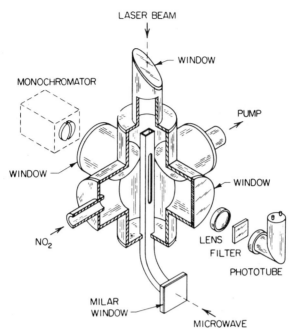

Figure 1-27. Experimental arrangement for observing MODR (from Tanaka *et al.*, 1974).

to detect fluorescence than absorption for $\lambda < 2\ \mu$m. Except in cases where the excited electronic states may be created by other means, e.g. by a chemical reaction as in CN (see below), the function of one of three photons will be to create a fluorescing excited state.

A typical experimental arrangement for microwave–optical double resonance is shown in Figure 1-27. Microwaves and laser photons are crossed in an interaction zone containing the molecules under study. Fluorescence is viewed at right angles by a photomultiplier, using either a filter or a monochromator to eliminate scattered laser light. A summary of the various systems studied by the MODR technique is given below and in Table 1-8.

1.6.1.1. Microwave–Optical Double Resonance in Atoms

Since the first application of MODR to $Hg(^3P_1)$ by Brossel and Bitter (1952), excited states of a large number of atoms have been investigated. Much of this earlier work has been reviewed by Series (1959) and by Burdick (1967) and will not be elaborated here. For a more recent application of this technique, the reader is referred to the study of ^{4}He $7\ ^1D_2$–$7\ ^1F_3$ and $7\ ^1D_2 - 7\ ^3F_3$ fine-structure intervals by Wing and Lamb (1972).

Table 1-8   Microwave–Optical Double Resonance

Reference	Molecule	Transition(s) pumped	Transition monitored	Results
Brossel and Bitter (1952)	Hg	$\Delta m_F = \pm 1$; $\omega = 50\text{–}150$ MHz in $^3P_1$ state; $H_z = 0\text{–}71$ G; $\text{pump}_2 = 253.7$ Hg lamp $^1S_0\text{–}{}^3P_1$	Polarization changes in $^3P_1\text{–}{}^1S_0$ fluorescence at 253.7 nm	Landé g factor for $^3P_1$ and radiative lifetime determined
Wing and Lamb (1972)	^{4}He	$1s7d\ {}^1D_2\text{–}1s7f\,{}^1F_3$ 31.55826 GHz $1s7d\ {}^1D_2\text{–}1s7f\,{}^3F_3$ 31.41207 GHz	$7\,{}^1D\text{–}2\,{}^1P$ fluorescence at 400.9 nm	Decrease in fluorescence on microwave resonance
Barger et al. (1962); Evenson et al., (1964)	CN	$E(v)'' = A^2\Pi_{3/2}(10);\ E(v)' = B^2\Sigma^+(0)$	Intensity of $B^2\Sigma^+(v=0)\rightarrow X^2\Sigma^+(v=0)$ near 370.0 nm	Determined 9 of the 12 hyperfine energy levels in the $A^2\Pi\text{-}B^2\Sigma$ perturbation complex

J''	P''	F''	J'	P'	F'	$\omega(\text{MHz})$
$\frac{7}{2}$	$-$	$\frac{5}{2}$	$\frac{7}{2}$	$+$	$\frac{7}{2}$	8971.0
				$+$	$\frac{7}{2}$	9354.9
				$+$	$\frac{5}{2}$	9599.3
$\frac{7}{2}$	$-$	$\frac{7}{2}$	$\frac{9}{2}$	$+$	$\frac{7}{2}$	8838.7
				$+$	$\frac{5}{2}$	8952.7
				$+$	$\frac{7}{2}$	9336.5
				$+$	$\frac{9}{2}$	9580.3
				$+$	$\frac{7}{2}$	9741.6
$\frac{7}{2}$	$-$	$\frac{9}{2}$	$\frac{9}{2}$	$+$	$\frac{7}{2}$	8927.1
				$+$	$\frac{9}{2}$	8813.2
				$+$	$\frac{9}{2}$	9554.0
				$+$	$\frac{9}{2}$	9715.5
				$+$	$\frac{11}{2}$	9854.9

Reference	Molecule	Transition(s) pumped	Transition monitored	Results
Evenson (1969)	CN	$E(v)'' = A^2\Pi_{3/2}(10);\ E(v)' = B^2\Sigma^+(0)$	$B^2\Sigma^+(v=0)\rightarrow X^2\Sigma^+(v=0)$ fluorescence at 387.5 nm	Determination of the 12 hyperfine energy levels in the $A^2\Pi\text{-}B^2\Sigma$ perturbation complex

J''	P''	F''	J'	P'	F'	$\omega\ (\text{MHz})$
$\frac{7}{2}$	$+$	$\frac{9}{2}$	$\frac{7}{2}$	$+$	$\frac{9}{2}$	10,462.6
			$\frac{7}{2}$	$+$	$\frac{7}{2}$	10,570.0
			$\frac{9}{2}$	$+$	$\frac{7}{2}$	11,201.0
			$\frac{9}{2}$	$+$	$\frac{9}{2}$	11,365.3
			$\frac{9}{2}$	$+$	$\frac{11}{2}$	11,503.5

CN — Pratt and Broida (1969); Evenson (1968)

$B^2\Sigma^+(v=0)\to X^2\Sigma^+(v=0)$ fluorescence at 387.5 nm

Cross section for $A\to B$ energy transfer was found to be larger than that for rotational relaxation

						ω (MHz)
						11,352.0
						11,187.0
						10,544.0
						10,935.0
						11,176.0

$E(v)'' = E(v)' = A^2\Pi_{3/2}(10)$

J''	P''	F''	J'	P'	F'	ω (MHz)
$\tfrac{7}{2}$	+	$\tfrac{9}{2}$	$\tfrac{7}{2}$	—	$\tfrac{9}{2}$	1649.1
						1622.0
$\tfrac{7}{2}$	+	$\tfrac{7}{2}$	$\tfrac{7}{2}$	—	$\tfrac{7}{2}$	1636.5
						1610.5
$\tfrac{7}{2}$	+	$\tfrac{5}{2}$	$\tfrac{5}{2}$	—	$\tfrac{5}{2}$	1601.0
						1579.5

OH — German and Zare (1969a,b)

$\Delta m_F = \pm 1$ in $A^2\Sigma^+(v=0, J=\tfrac{3}{2},$ $K=2, F=1,2)$, $\omega = 1\text{–}11$ MHz, $H_z = 0\text{–}26$ G; pump$_2$ = 307.206 nm Zn I

Polarization changes in $A\to X$ fluorescence at 308.0 nm

$g(F=2)=0.301$; $g(F=1)=0.492$; $\tau = 7.7\times 10^{-7}$ s

OD — German and Zare (1970)

$m_F = \pm 1$ in $A^2\Sigma^+(v=0, J=\tfrac{3}{2},$ $K=1, F=\tfrac{5}{2})$; pump$_2$ = 307.16 nm Ba I line

As above

$\tau = 6.3\times 10^{-7}$ s

CS — Silvers et al. (1970)

$A^1\Pi(v=0)$, $J=8 + \to -$ at 1099.4 MHz; pump$_2$ = 257.61 nm Mn II line

$A^1\Pi(v=0, J=8)\to X^1\Sigma^+$ $(v=0, J=7,8,9)$ fluorescence near 257.6 nm

Lambda-doublet splitting with Stark-level crossing gives $\mu_8 = 0.68$D

CS — Field and Bergeman (1971)

Pump$_2$ = CS molecular lamp near 257 nm; pump$_1$ = $A^1\Pi(v=0,J)+ \to -$

J	ω (MHz)
1	13.6
2	42.1
3	89.5
4	161.83
5	271.33
6	436.38
7	691.29
8	1099.37
9	1787.19

As above

Stark effect and lambda splitting gives dipole moment:

J	$\mu_J(D)$
3	0.6703
4	0.6699
5	0.6713
6	0.6718
7	0.6733
8	0.6752
9	0.6786

continued

Table 1-8   Microwave–Optical Double Resonance (continued)

Reference	Molecule	Transition(s) pumped	Transition monitored	Results
Field *et al.* (1972a)	BaO	Pump$_1$: X$^1\Sigma$ ($v=0$), $J=0\to1$, $J=1\to2$; $\omega = 37.4039$ GHz; pump$_2$: Ar$^+$ laser 496.51 nm	A$^1\Sigma \to$ X$^1\Sigma$(7,1) band fluorescence at 513.53 nm	Observed increase in fluorescence on microwave resonance
Field *et al.* (1972b)	BaO	Pump$_1$: A$^1\Sigma$($v=7$); $J=2\to3$, $\omega=44{,}891.4$ MHz; $J=1\to2$, $\omega=29{,}927.6$ MHz; pump$_2$: Ar$^+$ laser 496.51 nm	Change in polarization of rotationally unresolved A–X (7,1) band fluorescence; change in intensity of rotationally resolved fluorescence at 513.5 nm	Rotational constant for A$^1\Sigma$($v=7$) is 7482.01 MHz
Field *et al.* (1973)	BaO	(see sub-table below)	Rotationally resolved A–X (v',1) band fluorescence at 513 nm	Perturbations analyzed in A$^1\Sigma$ state; rotational constants obtained for A$^1\Sigma$

Transition(s) pumped for Field *et al.* (1973):

Pump 1					Pump 2: cw dye		
E	v	J''	J'	ω(MHz)	$A(v)$	$X(v)$	λ (nm)
X$^1\Sigma$	0	0	1	18,702.1	1	0	580.8
	0	1	2	37,404.1	1	0	580.8
	0	2	3	56,106.1	1	0	580.8
	1	2	3	55,855.2	3	1	570.3
A$^1\Sigma$	0	2	3	46,376.2	0	0	598.1
	1	1	2	30,760.5	1	0	580.8
	1	2	3	46,142.1	1	0	580.8
	2	2	3	45,986.1	2	1	586.7
	3	1	2	30,493.2	3	1	570.3
	3	3	4	45,740.1	3	1	570.3
	3	4	5	60,984.5	3	1	570.3
	3	2	3	76,226.8	3	1	570.3
	4	2	3	45,551.1	4	2	576.1
	5	2	3	45,397.3	5	3	581.9

Reference	Molecule	Pump and transition data	Experiment	Result
Solarz and Levy (1973, 1974); Solarz et al. (1974)	NO₂	Pump₂: Ar⁺ 488 nm $^2A_1$(000) $N=4$, $K=3 \to {}^2B_2$(?) $N=5$, $K=2$; pump₁ in $^2B_2$(?) $N''\,K''\,J''$ | $N'\,K'\,J'$ | ω (GHz) 5 2 ? | 5 ? ? | 9.2685	Resolved fluorescence near 510.5 nm from $^2B_2$ electronic state, B_2 vibrational symmetry, $N=5$, $K=2$ to $N=4,5,6$ and $K=2$ of $^2A_1$(000)	Assignment of 488-nm Ar⁺ line, measurement of linewidth of upper state yields lifetime $\tau = 3.39 \pm 0.36\ \mu s$
Tanaka et al. (1974a)	NO₂	Pump₂: Ar⁺ 457.9 nm line; pump₁: in $^2A_1$(000) $N''\,K''\,J''$ | $N'\,K'\,J'$ | ω (GHz) 9 1 $\frac{19}{2}$ | 10 0 $\frac{21}{2}$ | 40.7 9 1 $\frac{17}{2}$ | 10 0 $\frac{21}{2}$ | 40.4	Resolved fluorescence near 474.2 nm from $^2B_2$ electronic state, A_1 vibrational symmetry, $N,K = 9,0$ to $N,K = 10,0$ and $8,0$ in $^2A_1$(010), (020), (100)	Assignment of 457.9 nm line to $J = \frac{21}{2} \to \frac{19}{2}$ of $10_{0,10} \to 9_{0,9}$ transition from $^2A_1$(000) to $^2B_2$ with A_1 vibrational symmetry
Tanaka et al. (1973)	NO₂	Pump₂: tunable dye laser near 593.6 nm; pump₁: in $^2A_1$(000) $N''\,K''\,J''$ | $N'\,K'\,J'$ | ω (GHz) 9 1 $\frac{19}{2}$ | 10 0 $\frac{21}{2}$ | 40.7 9 1 $\frac{17}{2}$ | 10 0 $\frac{21}{2}$ | 40.4 9 1 $\frac{17}{2}$ | 10 0 $\frac{19}{2}$ | 40.9	Unresolved fluorescence near 620 nm from $^2B_2$ with $N=8,9$ to $^2A_1$(010) $N=(9,7)$, $(10,8)$	Used tunable dye laser to assign lines near 593.6 nm
Tanaka et al. (1974b)	NO₂	Pump₂: tunable dye laser near 593.6 nm; pump₁: in $^2B_2$ $N''\,K''\,J''$ | $N'\,K'\,J'$ | ω (GHz) 8 1 $\frac{17}{2}$ | 9 0 $\frac{19}{2}$ | 24.4 8 1 $\frac{15}{2}$ | 9 0 $\frac{17}{2}$ | 33.1 8 1 $\frac{17}{2}$ | 9 0 $\frac{17}{2}$ | 38.2	As above	Obtained hyperfine structure for $^2B_2\ 8_{1,8} \to 9_{0,9}$ in an unknown vibrational state
Tanaka and Harris (1976)	NO₂	As above $N''\,K''\,J''$ | $N'\,K'\,J'$ | ω (GHz) 6 1 $\frac{13}{2}$ | 7 0 $\frac{15}{2}$ | 42.4 6 1 $\frac{13}{2}$ | 7 0 $\frac{13}{2}$ | 31.5 6 1 $\frac{11}{2}$ | 7 0 $\frac{13}{2}$ | 25.2	As above	Obtained hyperfine structure for $^2B_2\ 6_{1,6} \to 7_{0,7}$

continued

Table 1-8 *Microwave-Optical Double Resonance (continued)*

Reference	Molecule	Transition(s) pumped	Transition monitored	Results
Tanaka et al. (1975)	NO_2	Pump$_2$: tunable dye laser near 593.6 nm; pump$_1$: in $^2A_1$ (000)	Resolved fluorescence from $K'_a =$ 0,1,2,3,4 in $^2B_2$ electronic state	Assignment of nearly 80 lines of the 593-nm band. Obtained rotational constants and spin splittings for $^2B_2$ (for a more complete summary of assigned transitions, see reference)
Cook et al. (1976)	NH_2	Pump$_2$: tunable dye laser near 576 nm on X $^2B_1 \rightarrow$ A $^2A_1$ transition; pump$_1$: microwaves in vibrationless ground state; five HF components observed for $6_{16}(J = \tfrac{13}{2}) \rightarrow 5_{23}(J = \tfrac{11}{2})$	Total A–X fluorescence	Hyperfine assignments obtained
Hills et al. (1976)	NH_2	Pumps 1 and 2 as above	As above	Hyperfine assignments obtained

Tanaka et al. sub-table:

N''	K''	J''	N'	K'	J'	ω (GHz)
5	1	$\tfrac{9}{2}$	6	0	$\tfrac{11}{2}$	69.8
5	1	$\tfrac{11}{2}$	6	0	$\tfrac{13}{2}$	70.6
7	1	$\tfrac{13}{2}$	8	0	$\tfrac{15}{2}$	15.0
7	1	$\tfrac{15}{2}$	8	0	$\tfrac{17}{2}$	15.5
9	1	$\tfrac{17}{2}$	10	0	$\tfrac{19}{2}$	40.9
9	1	$\tfrac{19}{2}$	10	0	$\tfrac{21}{2}$	40.7

Hills et al. sub-table:

N''	K''	J''	F''	N'	K'	J'	F'	ω (MHz)
5	1	$\tfrac{11}{2}$	$\tfrac{13}{2}$	5	1	$\tfrac{9}{2}$	$\tfrac{11}{2}$	7351.21
5	1	$\tfrac{11}{2}$	$\tfrac{11}{2}$	5	1	$\tfrac{9}{2}$	$\tfrac{9}{2}$	7389.63
5	1	$\tfrac{11}{2}$	$\tfrac{9}{2}$	5	1	$\tfrac{9}{2}$	$\tfrac{7}{2}$	7422.73
1	1	$\tfrac{3}{2}$	$\tfrac{3}{2}$	1	1	$\tfrac{1}{2}$	$\tfrac{1}{2}$	7910.53
1	1	$\tfrac{3}{2}$	$\tfrac{1}{2}$	1	1	$\tfrac{1}{2}$	$\tfrac{1}{2}$	7916.63
1	1	$\tfrac{3}{2}$	$\tfrac{5}{2}$	1	1	$\tfrac{1}{2}$	$\tfrac{3}{2}$	7975.56
1	1	$\tfrac{3}{2}$	$\tfrac{3}{2}$	1	1	$\tfrac{1}{2}$	$\tfrac{3}{2}$	7983.72

N''	K''	J''	F''	N'	K'	J'	F'	ω (MHz)
6	2	$\frac{13}{2}$	$\frac{15}{2}$	6	2	$\frac{11}{2}$	$\frac{13}{2}$	9,645.66
6	2	$\frac{13}{2}$	$\frac{13}{2}$	6	2	$\frac{11}{2}$	$\frac{11}{2}$	9,704.72
6	2	$\frac{13}{2}$	$\frac{11}{2}$	6	2	$\frac{11}{2}$	$\frac{9}{2}$	9,754.59
5	2	$\frac{11}{2}$	$\frac{13}{2}$	5	2	$\frac{9}{2}$	$\frac{11}{2}$	10,553.66
5	2	$\frac{11}{2}$	$\frac{11}{2}$	5	2	$\frac{9}{2}$	$\frac{9}{2}$	10,562.33
5	2	$\frac{11}{2}$	$\frac{9}{2}$	5	2	$\frac{9}{2}$	$\frac{7}{2}$	10,570.95
4	2	$\frac{9}{2}$	$\frac{11}{2}$	4	2	$\frac{7}{2}$	$\frac{9}{2}$	10,698.10
4	2	$\frac{9}{2}$	$\frac{9}{2}$	4	2	$\frac{7}{2}$	$\frac{7}{2}$	10,741.97
4	2	$\frac{9}{2}$	$\frac{7}{2}$	4	2	$\frac{7}{2}$	$\frac{5}{2}$	10,777.55
7	2	$\frac{15}{2}$	$\frac{17}{2}$	7	2	$\frac{13}{2}$	$\frac{15}{2}$	11,537.14
7	2	$\frac{15}{2}$	$\frac{15}{2}$	7	2	$\frac{13}{2}$	$\frac{13}{2}$	11,558.85
7	2	$\frac{15}{2}$	$\frac{13}{2}$	7	2	$\frac{13}{2}$	$\frac{11}{2}$	11,578.69

Kim *et al.* (1976)

BO_2

Pump$_1$: $^2\Pi_u$ (010) $\rightarrow$ $^2\Pi_g$ (110) R(7) line with 514.5-nm Ar$^+$ laser; pump$_2$: microwaves in Zeeman sublevels of $J = \frac{15}{2}$ in ground state

Fluorescence from $^2\Pi_g$

Zeeman assignments obtained

1.6.1.2. Microwave–Optical Double Resonance in CN

The first application of the microwave–optical double-resonance technique to the study of a diatomic molecule, CN, was reported by Barger *et al.* (1962). Subsequent studies of the CN radical by Evenson *et al.* (1964), Evenson (1968, 1969), and Pratt and Broida (1969) further elucidated this system. CN was produced in the $A^2\Pi$ metastable state by the chemical reaction of active nitrogen with methylene chloride. Microwave pumping from the $A^2\Pi$ state of nearby $B^2\Sigma^+$ levels was monitored by measuring an increase in intensity of the $B^2\Sigma^+ \rightarrow X^2\Sigma^+(0,0)$ violet band of CN near 387.5 nm. The relevant energy levels, which are shown in Figure 1-28, arise

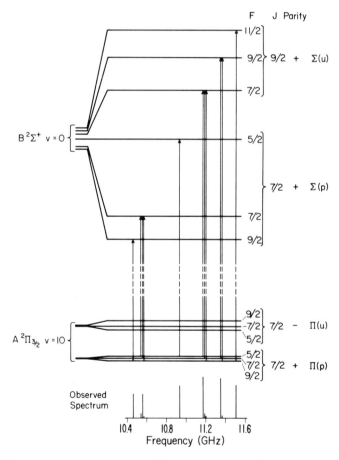

Figure 1-28. Energy-level diagram of the $B^2\Sigma^+(v = 0, K' = 4)$ and $A^2\Pi_{3/2}(v = 10)$ states of CN. The observed microwave magnetic-dipole spectrum is shown (from Evenson, 1969).

from the perturbation of one component of the spin doublet of the $B^2\Sigma^+$ state by one component of the Λ doublet of the $A^2\Pi$ state. If we denote by $\Pi(p)$ and $\Pi(u)$ the perturbed and unperturbed Λ doublets of the $A^2\Pi$ state and by $\Sigma(p)$ and $\Sigma(u)$ the perturbed and unperturbed spin components of the $B^2\Sigma^+$ state, we can see that five possible transitions may be observed. Transitions from $\Pi(p)$ to $\Sigma(p)$ and $\Sigma(u)$ shown in Figure 1-28 will be magnetic-dipole-allowed and result in an increase in $B \rightarrow X$ fluorescence. Transitions from $\Pi(u)$ to $\Sigma(p)$ and $\Sigma(u)$ will be electric-dipole-allowed and will also result in an increase in $B \rightarrow X$ fluorescence. Finally, the $\Pi(u) \rightarrow \Pi(p)$ electric-dipole transitions can be expected to cause an increase in $B \rightarrow X$ fluorescence because $\Pi(p)$ loses its population to $B^2\Sigma^+$ through mixing with $\Sigma(p)$. Nearly all of these transitions have been observed by the application of microwave radiation at the frequencies listed in Table 1-8. In addition to providing a detailed spectroscopic study of the perturbation complex, these studies also provide some dynamic information. From the pressure dependence of the microwave linewidth, it was deduced that the cross section for collisional energy transfer from the $A^2\Pi$ state to the $B^2\Sigma^+$ state is larger than that for rotational relaxation.

1.6.1.3. Microwave–Optical Double Resonance in OH and OD

A second example of the application of optical double-resonance techniques to the study of diatomic molecules was performed by German and Zare (1969a), who used the technique to obtain the radiative lifetime of $OH(^2\Sigma^+)$. The OH radical was produced by the reaction of $H + NO_2$ and excited to $v' = 0$, $K' = 2$, $J' = \frac{3}{2}$ of the $A^2\Sigma^+$ state by the coincidence with the Zn I 307.206-nm line. The upper state has two hyperfine levels: $F = J \pm I = \frac{3}{2} \pm \frac{1}{2} = 1,2$. Each of these hyperfine levels is further split into $2F + 1$ components by the application of a magnetic field. Transitions between these Zeeman sublevels with $\Delta m_F = \pm 1$ were induced by a radio-frequency field applied perpendicular to a static magnetic field. Resonance was detected by a change in the polarization of $A \rightarrow X$ fluorescence following excitation with π-polarized light. The experimental arrangement is consequently very similar to that used by Brossel and Bitter (1952) in their study of Hg. From the position of the double-resonance signals, German and Zare were able to determine the g values for the two hyperfine components: $g(F = 2) = 0.301 \pm 0.0015$, and $g(F = 1) = 0.492 \pm 0.0045$. These values confirmed that the state obeys Hund's case (b) coupling. By combining the g values with a value for $g\tau$ obtained previously by measurement of the Hanle effect (German and Zare, 1969b) a radiative lifetime of $\tau = (7.7 \pm 0.8) \times 10^{-7}$ s was obtained. A similar study for OD yielded $\tau = (6.3 \pm 0.7) \times 10^{-7}$ s (German and Zare, 1970). More recently, German (1975) and Brophy *et al.* (1974)

have reinvestigated the lifetimes of OH and OD($A^2\Sigma^+$) by the more direct technique of laser-induced fluorescence. The values obtained are in fair agreement with those obtained by combining the double resonance and Hanle effect results; however, the lifetime of OH($A^2\Sigma^+$) appears to be dependent on the particular K', J' level which is excited.

1.6.1.4. Microwave–Optical Double Resonance in CS

The $A^1\Pi$ state of CS has also been examined by microwave–optical double-resonance techniques. Silvers *et al.* (1970) have used a polarized Mn II atomic lamp at 257.61 nm to excite CS to the upper (+) lambda doublet of the $A^1\Pi(v' = 0$, $J' = 8)$ level. When an applied radio-frequency field was tuned to the difference between the lambda doublets (1099.4 MHz), a change in the polarization of the fluorescent $A \to X$ emission was observed. By measuring the position of the double-resonance signal as a function of an applied Stark field, the dipole moment for CS $A^1\Pi(v = 0, J = 8)$ was found to be $\mu_8 = 0.68 \pm 0.01$. Field and Bergeman (1971) have extended this technique by using a molecular CS lamp in place of the Mn II atomic source. Lambda-doubling intervals between 13 and 1787 MHz were found for $A^1\Pi(v = 0$, $J = 1$–9$)$, which yielded dipole moments between 0.6699 and 0.6786 D (see Table 1-8). These measurements were combined with information from the CS optical spectrum to extend the analysis of perturbations in the $A^1\Pi(v = 0)$ level.

1.6.1.5. Microwave–Optical Double Resonance in BaO

The first application of laser techniques to MODR was performed by Field *et al.* (1972a) in their study of BaO. This molecule has also recently been examined using optical–optical double resonance as discussed in Section 1.6.3.2. In the MODR experiment an argon-ion laser operating at 496.51 nm was used to pump molecules from $v'' = 0$, $J'' = 1$ of the ground electronic state ($X^1\Sigma$) to $v' = 7$, $J' = 2$ of the $A^1\Sigma$ excited state. The microwave transition between $J = 2$ and $J = 1$ in the ground state was then detected as a change in intensity of fluorescence emitted from the $A^1\Sigma$ state. Subsequent work (Field *et al.*, 1972b) also detected the two $\Delta J = \pm 1$ microwave transitions in the upper electronic state that shared the common $J' = 2$ level. The scope of these investigations was significantly expanded by the use of a tunable cw dye laser in place of the fixed-frequency argon-ion laser (Field *et al.*, 1973). Fourteen microwave rotational transitions in the $X^1\Sigma(v = 0, 1)$ and $A^1\Sigma(v = 0$–5$)$ states of $^{138}Ba^{16}O$ and one transition in the $A^1\Sigma(v = 1)$ state of $^{137}Ba^{16}O$ were observed. A summary of these transitions is given in Table 1-8. Analysis of the rotational constants for $v = 0$–5 in the $A^1\Sigma$ state demonstrated that the A state is significantly perturbed.

1.6.1.6. Microwave–Optical Double Resonance in NO_2

Microwave–optical double resonance as a spectroscopic tool has shown the most promise in applications to molecules for which the visible absorption spectrum is much too complicated for analysis by conventional means. One such molecule is NO_2. Low-lying electronic states and high vibrational levels of the ground electronic state of NO_2 mutually perturb one another to yield an extremely complicated visible absorption spectrum. Although many aspects of this spectrum are still poorly understood, the application of MODR to the study of NO_2 has proved to be a considerable aid in assigning specific absorption lines.

Solarz and Levy (1973) were the first to examine NO_2 by this technique. In their study, the 488-nm Ar^+ line was used to excite NO_2 to a level of the $^2B_2$ electronic state. The absorption in this upper electronic state of 100% modulated microwave radiation at 9.27 GHz was monitored by observing a small modulated component of the visible fluorescence. Measurement of the linewidth of this microwave transition (Solarz and Levy, 1974) indicated that the upper state had a lifetime of $\tau = 3.39 \pm 0.36$ μs. Subsequent analysis of the resolved fluorescence (Solarz *et al.*, 1974) identified the transition pumped by the 488-nm Ar^+ line as being $^2A_1(N=4, K=3) \to {}^2B_2(N=5, K=2)$.

A similar method was used by Tanaka *et al.* (1974a) to identify the transition pumped by the 457.9-nm Ar^+ line. In this case, however, the microwave transitions occurred in the ground electronic state between spin components of the $N=9$, $K=1$ level and the $N=10$, $K=0$ level. By rotationally resolving the visible fluorescence, the authors were able to show that the vibrationless $^2A_1(N=10, J=\frac{21}{2}, K=0)$ level was responsible for absorption of the 457.9 nm line and that absorption took place to a $^2B_2(N=9, K=0)$ level of A_1 vibrational symmetry.

By coincidence, Tanaka *et al.* (1973) had previously observed MODR signals between the same two ground-state levels for optical pumping by a tunable cw dye laser near 593.6 nm. The observed transitions from these two ground levels ($10_{0,10}$ and $9_{1,9}$) are summarized in Figure 1-29. The authors assigned the first four optical transitions on the left of this figure by rotationally resolving the photoluminescence from the $^2B_2$ state when a cw dye laser was tuned to transitions which shared a common level with the microwave pump transition. The assignment of the fifth transition in Figure 1-29 (Ar^+, 457.9 nm) by Tanaka *et al.* (1974a) was discussed above. The remaining optical transitions were recently assigned by Tanaka *et al.* (1975) using similar techniques. In this more recent work the assignment of nearly 80 lines with $K'_a = 0, 1, 2, 3,$ and 4 allowed the derivation of the following rotational constants for the $^2B_2$ electronic state: $A = 8.52$ cm^{-1}, $B = 0.458$ cm^{-1}, and $C = 0.388$ cm^{-1}.

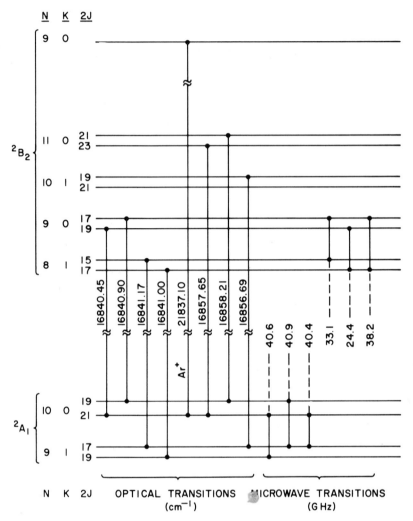

Figure 1-29. MODR transitions in NO_2 involving the $10_{0,10}$ and 9_{19} levels. Hyperfine structure has been omitted.

Tanaka *et al.* (1974b) have also observed MODR signals for microwave transitions in the upper, $^2B_2$, electronic state. Three of these transitions between components of the $N = 8$, $K = 1$ and $N = 9$, $K = 0$ levels are shown in Figure 1-29. A more recent study (Tanaka and Harris, 1976) has also uncovered three transitions between the $N = 6$, $K = 1$ and $N = 7$, $K = 0$ levels.

In summary, the application of the MODR technique to the complex optical spectrum of NO_2 has allowed the assignment of a large number of specific absorption lines. A complete summary of these double-resonance combinations is presented in Table 1-8. A tentative structure for the $^2B_2$ electronic state has been assigned (Tanaka *et al.*, 1975) and a lifetime for this state of $3.39 \pm 0.36 \, \mu s$ has been obtained (Solarz and Levy, 1974). Perturbations of $^2B_2$ levels by levels with a longer radiative lifetime have been identified. Although the exact nature of the perturbing state(s) is still unknown, substantial progress has been made in characterizing the perturbations in limited spectral regions.

1.6.1.7. Microwave–Optical Double Resonance in NH₂

Microwave–optical double resonance has recently been used to assign the microwave transitions in the vibrationless X^2B_1 state of NH_2 (Cook *et al.*, 1976). NH_2 was prepared by the reaction of microwave-discharged water with hydrazine in a flow reactor. A modulated source induced changes in the population of ground-state levels. These changes were then detected by looking for a modulated component of the total fluorescence following excitation of NH_2 to the $A\,^2A_1$ state using a cw dye laser at about 576 nm. Both electric-dipole transitions (Cook *et al.*, 1976) and magnetic-dipole transitions (Hills *et al.*, 1976) were detected in the microwave region.

1.6.1.8. Microwave–Optical Double Resonance in BO₂

Kim *et al.* (1976) have recently reported the observation of microwave–optical double resonance in the BO_2 radical. A single-mode argon-ion laser was used to pump the $R(7)$ line of the $^2\Pi_u(010) \rightarrow {}^2\Pi_g(110)$ band at 514.5 nm. Microwave transitions between Zeeman components of the $J = \frac{15}{2}$ level in the ground state were observed by monitoring fluorescence from the $^2\Pi_g$ excited state.

1.6.2. Infrared–Optical Double Resonance

Although the microwave–optical double-resonance techniques described in the preceding section have received fairly wide usage, the extension to infrared pumping has received little attention to date. Undoubtedly, the reason for this situation is that sufficiently powerful and tunable infrared laser sources for perturbing Boltzmann population distributions have not been readily available. With the development of new nonlinear mixing techniques as described in Section 1.3.1.3, the area of infrared–optical double resonance should undergo significant growth in the future.

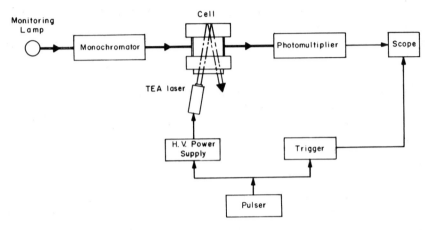

Figure 1-30. Experimental apparatus for infrared–visible double-resonance measurements (from Yogev and Haas, 1973).

A typical apparatus for observation of infrared–optical double resonance is shown in Figure 1-30. A powerful pulsed infrared laser is used to vibrationally excite the molecules under study. Optical absorption by these excited molecules is then monitored using a conventional monochromator/lamp arrangement with photomultiplier detection. Improvements in the apparatus might be made by replacing the monochromator/lamp with a narrow-band cw tunable dye laser and by propagating the infrared and dye laser beams collinearly. Additionally, one might use a three-photon technique, as discussed above for MODR. A summary of infrared–optical double-resonance applications is given below and in Table 1-9.

1.6.2.1. Infrared–Optical Double Resonance in NH_3

In the first application of the infrared–optical double-resonance technique, Ambartzumian *et al.* (1972) used the P(32) line of the 10.6 μm CO_2 laser to excite the Q(5,3),ν_2 transition in NH_3. The degree of vibrational excitation was monitored by observing an increase in optical absorption on the hot band $X(\nu_2 = 1) \to A(\nu_2 = 0)$ near 225 nm. The increase in optical absorption occurs because of two effects: (1) the initial vibrational excitation, and (2) subsequent bulk heating of the gas following vibration–translation energy transfer. From observation of the optical absorption, Ambartzumian *et al* determined that roughly 40% of the NH_2 molecules had been excited to $\nu_2 = 1$ by the CO_2 laser. This value corresponds to a vibrational temperature of 900°K.

Table 1-9   Infrared–Optical Double Resonance

Reference	Molecule	Transition pumped	Transition monitored	Results
Ambartzumian et al. (1972)	NH_3	$Q(5,3)$ of $\nu_2 = 0 \rightarrow 1$ pumped by $P(32)$ line of 10.6-μm CO_2 laser	NH_3 absorption near 200–225 nm on $X(\nu_2 = 1)$ $\rightarrow A(\nu_2 = 0)$	Measurement of the extent of excited-state population (40%)
Ambartzumian et al. (1975a)	OsO_4	$\nu_3 = 0$–1 pumped by $P(2)$ or $R(2)$ of 10.6 μm CO_2 laser	Absorption on the X–A band 270–320 nm	Observation of excited vibrational levels with $v \geq 10$
Yogev and Haas (1973)	Biacetyl, $CH_3COCOCH_3$	C–$CH_3(?)$ stretch of ground electronic state pumped by CO_2 laser on $P(20,22)$, 10.6 μm	Carbonyl n–π^* transitions near $21,700$ cm^{-1}	Decrease in absorption near 0–0 band, new absorption $21,700$–$19,600$ cm^{-1}
Kugel et al. (1975)	F^{8+}	$2\,^2S_{1/2}$–$2\,^2P_{3/2}$ by $P_2(5)$ line of HBr chemical laser at 2382.5 cm^{-1}	$2\,^2P_{3/2}$–$1\,^2S_{1/2}$ emission at 826 eV	Lamb shift for $2\,^2S_{1/2}$–$2\,^2P_{1/2}$
Laubereau et al. (1975)	Coumarin-6	Pump$_1$: CH_3 and CH_2 symmetrical and assymetrical vibrations at $\nu = 2970$ cm^{-1} by Nd:glass pumped OPO; pump$_2$: S_0–S_1 transition at $18,910$ cm^{-1} by doubled Nd:glass	S_1–S_0 fluorescence	Vibrational relaxation time of 1.3 ± 0.3 ps for 3×10^{-5} M coumarin-6 in CCl_4 at room temperature
Orr (1976)	Biacetyl, $CH_3COCOCH_3$	C–CH_3 stretch pumped by CO_2 laser at 10.6 μm	Excited at $22,645$ cm^{-1}, luminescence observed at 460 and 510 nm	Used He–Cd 4416-Å laser as second field

1.6.2.2. Infrared–Optical Double Resonance in OsO₄

Ambartzumian *et al.* (1975a) have also used ultraviolet probe absorption to measure the nonequilibrium distribution produced by a 10-MW, 90-ns CO_2 laser pulse in OsO_4. Totally symmetric (ν_1) hot bands of the $A \rightarrow X$ transition (320–270 nm) were monitored immediately following excitation of the ν_3 vibrational mode. A finite delay was observed between the laser pulse and the change in ultraviolet absorption, which corresponded to the vibrational equilibrium between the ν_3 and ν_1 modes. From the wavelength dependence of the change in ultraviolet absorption, the authors determined that levels with at least 10 quanta of ν_1 vibration had been populated following the laser pulse. By focusing the laser and using a radical scavenger, they were also able to demonstrate selective dissociation of either $^{192}OsO_4$ or $^{187}OsO_4$ with the R(2) or P(2) 10.6 μm CO_2 laser lines, respectively.

1.6.2.3. Infrared–Optical Double Resonance in Biacetyl

Yogev and Haas (1973) have used the infrared–optical double-resonance method to study vibrational relaxation of biacetyl, $CH_3COCOCH_3$. Population changes in the C–CH₃ stretching mode were produced with a TEA CO_2 laser, and vibrational relaxation was monitored by observing the decay of increased hot-band optical absorption near $21,600 \text{ cm}^{-1}$. Decay times on the order of 1–10 ms were found for samples of 0.5–40 Torr. In addition, a finite delay between the excitation pulse and the onset of optical absorption was observed. This delay was interpreted to be due to the fact that, while the laser excited a C–CH₃ stretch, the $n-\pi^*$ transition responsible for the optical absorption is probably promoted by an out-of-plane bending mode. Consequently, vibration–vibration energy transfer must occur before the onset of the optical absorption.

1.6.2.4. Infrared–Optical Double Resonance in F^{8+}

While the previous examples of infrared–optical double resonance have employed optical absorption as the monitoring technique, we have seen in Section 1.3.3 that the detection of optical fluorescence is intrinsically more sensitive. For most cases, however, the price paid for the increased sensitivity is that a third photon must be used to raise the molecule to a fluorescing excited state. On the other hand, it is sometimes possible, as in the MODR case of CN, for example, to obtain the fluorescing state by different means, e.g., by a chemical reaction. The case of F^{8+} is similar.

Kugel *et al.* (1975) have produced the hydrogen-like F^{8+} ion by stripping and then adding electrons to ions in the high-energy beam of an

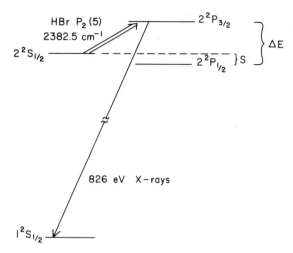

Figure 1-31. Infrared–optical double resonance in F^{8+}.

electrostatic accelerator. A small fraction of the resultant F^{8+} ions are found to be in the metastable $2\,^2S_{1/2}$ state. As shown in Figure 1-31, infrared–optical double resonance (here extended to include x-rays) was performed on these $2\,^2S_{1/2}$ ions to obtain a value for the Lamb shift or splitting between the $2\,^2S_{1/2}$ and $2\,^2P_{1/2}$.

A pulsed HBr chemical laser operating at 2382.5 cm^{-1} [the $P_2(5)$ line] was "scanned" over the frequency $\Delta E - S$ by changing the angle of intersection between the ion beam and the laser radiation. Changing this angle "tunes" the molecules in and out of resonance due to the Doppler shift for the line absorption. Absorption was monitored by counting x-rays emitted from the $2\,^2P_{3/2}$ state to the ground $1\,^2S_{1/2}$ state. From the angle at which the resonance occurred, a value of $\Delta E - S = 68853.9 \pm 35$ GHz was obtained. When combined with the accurately calculated value of 72,192.96 GHz for ΔE, this measurement provides a value for the Lamb shift of $S = 339 \pm 35$ GHz.

1.6.2.5. Infrared–Optical Double Resonance in Coumarin-6

A three-photon infrared–optical double resonance scheme has been used by Laubereau *et al.* (1975) to measure ultrashort relaxation times for coumarin-6 dissolved in CCl_4. A 6-ps pulse selected from the mode-locked train of a Nd:glass laser was used to produce a 3-ps infrared pulse of $\nu = 2970$ cm^{-1} by parametric generation in a $LiNbO_3$ crystal (see Section 1.3.1.3). Part of the pumping beam was separated with a beam splitter and doubled in a KDP crystal to produce $\nu = 18,910$ cm^{-1} visible radiation. The visible pulse was delayed by 0–10 ps with respect to the infrared pulse by the use of a variable-distance optical delay line.

The sequence of events was as follows. The infrared pulse was absorbed by the CH_3 and CH_2 symmetric and asymmetric vibrational modes of the coumarin-6. After a short delay, some of the excited molecules relaxed, leaving a diminished population in the vibrationally excited state. The visible pulse then probed the remaining excited-state population by further pumping excited molecules to the first excited singlet state. Fluorescence from this state was detected as a function of the delay between the infrared and ultraviolet pulses to give a relaxation time of 1.3 ± 0.3 ps for $3 \times 10^{-5} M$ coumarin-6 in CCl_4 at room temperature.

1.6.3. Optical–Optical Double Resonance

The technique of optical detection of nonequilibrium populations produced by optical pumping is one of the more recent additions to the spectroscopist's set of tools. To a large extent its late development has awaited advances in laser technology. Previous to these advances, experiments using the optical–optical double-resonance (OODR) technique were limited to the study of atoms for which strong atomic lamp resonances were available. Zito and Schraeder (1963) and Kibble and Pancharatnam (1965), for example, have used Hg I arc lamps to excite mercury in a two-step mechanism. The latter pair of authors was able to observe the Hanle effect for the $7\,^3S_1$ excited state. Similarly, Smith and Eck (1970) have observed level-crossing signals in $3\,^2D$ lithium atoms excited by a two-step mechanism using a Li I spectral lamp. The advent of the laser has allowed a significant generalization of these techniques. Normally, optically accessible excited electronic states have short radiative lifetimes. In order to build up a large enough concentration of the intermediate level in an optical double-resonance scheme, therefore, the rate of pumping must be large enough to overcome the radiative losses. Pulsed tunable dye lasers are providing high-enough pumping rates so that OODR techniques may now be applied in a more general manner to both atomic and polyatomic species.

Before reviewing some of these applications, it would be advisable to recall from Section 1.1.4 the difference between stepwise optical excitation, which proceeds through real intermediate states, and true multiphoton absorption, which proceeds through virtual intermediate states. While the former process easily falls into the category of double resonance, the latter process, although it frequently yields similar results, will not be the subject of our consideration below. For a general review of the multiphoton case the reader is referred to the publications of Worlock (1972) and McClain (1974) as well as to Section 5.7 of this volume.

1.6.3.1. Optical–Optical Double Resonance in Atoms

The application of OODR to atomic species is still the most prevalent use of this technique. Recently, there has been a great deal of interest in highly excited Rydberg atomic states (Scott, 1975). As we shall see below, optical–optical double resonance provides a simple procedure for producing and studying such states.

1.6.3.1a. Rb and Mg. It is noteworthy that some of the earliest applications of OODR occurred in laboratories which were simultaneously interested in dye laser development. For example, Bradley *et al.* (1970) have used a ruby-pumped dye laser at 780 nm to excited Rb ($5\,^2S_{1/2}$) to the $5\,^2P_{3/2}$ level. The absorption of the $5\,^2P_{3/2}$–$5\,^2D_{5/2}$ line at 775.8 nm was then observed using a second broad-band dye laser as a continuum source. Subsequently, Bradley and his coworkers used similar techniques to investigate Mg (Bradley *et al.*, 1973a) and Ba (Bradley *et al.*, 1973b). For the study of magnesium, a doubled Nd:glass laser at 530 nm was used to pump a tunable dye laser oscillating near 570 nm, as shown in Figure 1-32. The dye laser was doubled in an ADP crystal to give output near 285.2 nm which pumped Mg atoms from the ground state ($3s^2,{}^1S_0$) to the $3s3p,{}^1P_1$ excited state. About 20% of the 530 nm Nd:glass pump laser was split out of the original beam, redoubled to the ultraviolet at 265 nm, and used to pump an

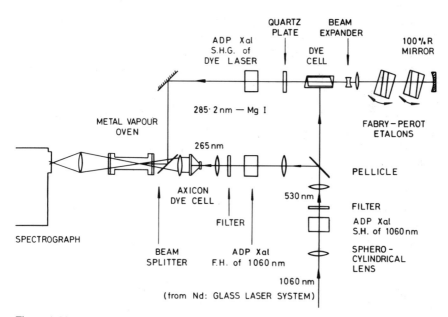

Figure 1-32. General experimental arrangement for observing optical–optical double resonance in metal vapors (from Bradley *et al.*, 1974).

Axicon dye cell, which provided an absorption continuum from 400 to 290 nm. Absorption by the Mg $3s3p, {}^1P_1$ atoms was recorded photographically and revealed a series of $3s3p, {}^1P_1 \rightarrow 3snd, {}^1D_2$ lines with $n = 12$ to 24. Strong absorption from the $3s3p, {}^1P_1$ state to the $3p^2, {}^1S_0$ autoionization state was also observed at 300.9 nm.

1.6.3.1b. Ba. For the study of barium (Bradley *et al.*, 1973b) the experimental arrangement was the same as shown in Figure 1-32 with the exception that the dye laser was tuned to 553.5 nm and not frequency-doubled. In this experiment Ba $(6s^2, {}^1S_0)$ was excited by the dye laser to the $6s6p, {}^1P_1$ state, whose absorption at 418–455 nm was then recorded photographically, again using the Axicon dye cell as a continuum source. Absorption to the $6snd, {}^1D_2$ levels with $n = 9$–41 was observed by using a variety of dyes in the continuum cell. In addition, a $6s6p, {}^1P_1 \rightarrow 5dns, {}^1D_2$ series with a limit at 29,649.7 cm^{-1} was observed.

1.6.3.1c. Ca. McIlrath (1969) and McIlrath and Carlsten (1973) have used a somewhat different version of the OODR technique to study excited-state absorption in calcium. A Nd:glass laser was doubled to 530 nm and used to pump a dye laser oscillating at 657.28 nm. The dye laser output excited Ca from the $4s^2, {}^1S_0$ ground state to the $4s4p, {}^3P_1$ lowest triplet level. Subsequent absorption by the ${}^3P \rightarrow 4sns, {}^3S_1$ and ${}^3P \rightarrow 4sns, {}^3D$ series was recorded photographically using a flash-lamp continuum background. In addition to these transitions, the six possible lines of the $4s4p, {}^3P \rightarrow 4p^2, {}^3P$ transitions were observed near 430 nm, and transitions to the continuum states $3d4d, {}^3D$ and $3d4d, {}^3S$ (autoionizing) and $3d4d, {}^3P$ and $3d5d, {}^3P$ (nonautoionizing) were observed near 275 nm. These transitions are summarized in Table 1-10. It is interesting to note that, although the dye laser populates the $4s4p, {}^3P_1$ level, absorption from all the triplet levels ${}^3P_{0,1,2}$ is observed. This is because the relaxation among the fine-structure levels is fast compared to the laser pumping time. In a related experiment, the authors were able to show that the rate for this relaxation was faster than 0.25×10^8 s^{-1} for Ca in 10 Torr of helium buffer gas. Finally, it should be noted that on the order of 10^{16} Ca atoms were produced in the excited ${}^3P$ levels by absorption of the pulsed dye laser.

1.6.3.1d. Cs and Rb. All the examples above of OODR in atoms used optical absorption to monitor population changes caused by the optical pump source. It has already been noted (Section 1.3.3) that, in the optical spectral region, fluorescence techniques are often more sensitive. In the examples above, the decreased sensitivity of photographic detection was overcome by production of large densities of excited states using laser absorption. Svanberg *et al.* (1973) and Svanberg and Belin (1974) have utilized the increased sensitivity of fluorescence detection to allow OODR using an atomic lamp as the primary pumping source. As we have seen before, however, fluorescence detection often requires a second excitation

Table 1-10 Optical–Optical Double Resonance in Atoms

Reference	Atom	Pump 1 Initial	Pump 1 Final	Pump 1 λ, nm	Pump 2 Initial	Pump 2 Final	Pump 2 λ, nm	Monitor Initial	Monitor Final	Monitor λ, nm
Zito and Schraeder (1963)	Hg	$6\,^1S_0$	$6\,^3P_1$	253.652	$6\,^3P_0$	$7\,^3S_1$	404.666	$7\,^3S_1$	$6\,^3P_2$	546.074
Kibble and Pancharatnam (1965)	Hg	$6\,^1S_0$	$6\,^3P_1$	253.652	$6\,^3P_1$	$7\,^3S_1$	435.834	$7\,^3S_1$ $7\,^3S_1$	$6\,^3P_2$ $6\,^3P_0$	546.074 404.666
Smith and Eck (1970)	Li	$2\,^2S_{1/2}$	$2\,^2P$	670.8	$2\,^2P$	$3\,^2D$	610.4	$3\,^2D$	$2\,^2P$	610.4
Bradley et al. (1970)	Rb	$5\,^2S_{1/2}$	$5\,^2P_{3/2}$	780.0				$5\,^2P_{3/2}$	$5\,^2D_{5/2}$	775.8
Bradley et al. (1973a)	Mg	$3s^2\,^1S_0$	$3s3p\,^1P_1^0$	285.2				$3s3p\,^1P_1^0$ $3s3p\,^1P_1^0$	$3snd\,^1D_2$ $3p^2\,^1S_0$	$n=12-24$ 378–398 300.9
Bradley et al. (1973b)	Ba	$6s^2\,^1S_0$	$6s6p\,^1P_1^0$	553.5				$6s6p\,^1P_1^0$ $6s6p\,^1P_1^0$	$6snd\,^1D_2$ $5dns\,^1D_2$	$n=9-41$ 418–455 $n=8-11$ 367–451
McIlrath (1969); McIlrath and Carlsten 1973	Ca	$4s^2\,^1S_0$	$4s4p\,^3P_1$	657.28				$4s4p\,^3P_0$ $^3P_1$ $^3P_2$ $4s4p\,^3P_0$ $^3P_1$ $^3P_2$ $4s4p\,^3P$ $4s4p\,^3P_0$ $^3P_1$ $^3P_1$ $^3P_1$ $^3P_2$ $^3P_2$	$4s5s\,^3S_1$ $^3S_1$ $^3S_1$ $4s6s\,^3S_1$ $^3S_1$ $^3S_1$ $4sns\,^3S_1$ $4p^2\,^3P_1$ $^3P_0$ $^3P_1$ $^3P_2$ $^3P_1$ $^3P_2$	610.27 612.22 616.22 394.89 395.70 397.37 $n=7-14$ 428.94 430.77 429.90 428.30 431.87 430.25

continued

Table 1-10   Optical–Optical Double Resonance in Atoms (continued)

Reference	Atom	Pump 1 Initial	Pump 1 Final	Pump 1 λ, nm	Pump 2 Initial	Pump 2 Final	Pump 2 λ, nm	Monitor Initial	Monitor Final	Monitor λ, nm
McIlrath and Carlsten (continued)	Ca	$4s^2\,^1S_0$	$4s4p\,^3P_1$	657.28				$4s4p\,^3P_0$	$4s4d\,^3D_1$	442.54
								$^3P_1$	$^3D_1$	443.57
								$^3P_1$	$^3D_2$	443.50
								$^3P_2$	$^3D_2$	445.48
								$^3P_2$	$^3D_3$	445.54
								$4s4p\,^3P_1$	$4s5d\,^3D_2$	363.08
								$^3P_2$	$^3D_3$	364.44
								$4s4p\,^3P_0$	$4s6d\,^3D_1$	334.5
								$^3P_1$	$^3D_{1,2}$	335.0
								$^3P_2$	$^3D_{1,2,3}$	336.2
								$4s4p\,^3P_0$	$4s7d\,^3D_1$	321.0
								$^3P_1$	$^3D_{1,2}$	321.5
								$^3P_2$	$^3D_{1,2,3}$	322.6
								$4s4p\,^3P_0$	$4s8d\,^3D_1$	313.6
								$^3P_1$	$^3D_{1,2}$	314.1
								$^3P_2$	$^3D_{1,2,3}$	315.1
								$4s4p\,^3P$	$4snd\,^3D$	$n = 9\text{–}22$
Svanberg et al. (1973); Svanberg and Belin (1974)	Cs	$6^2S_{1/2}$	$6^2P_{3/2}$	852.1	$6^2P_{3/2}$	$8^2D_{3/2}$	621.8	$8^2D_{3/2}$	$6^2P_{1/2}$	601.0
					$6^2P_{3/2}$	$9^2D_{3/2}$	584.8	$9^2D_{3/2}$	$6^2P_{1/2}$	566.4
					$6^2P_{3/2}$	$10^2D_{3/2}$	—	$10^2D_{3/2}$	$6^2P_{1/2}$	—
					$6^2P_{3/2}$	$11^2D_{3/2}$	550.4	$11^2D_{3/2}$	$6^2P_{1/2}$	534.1
					$6^2P_{3/2}$	$12^2D_{3/2}$	541.4	$12^2D_{3/2}$	$6^2P_{1/2}$	525.7
					$6^2P_{3/2}$	$13^2D_{3/2}$	535.1	$13^2D_{3/2}$	$6^2P_{1/2}$	519.7
					$6^2P_{3/2}$	$14^2D_{3/2}$	530.4	$14^2D_{3/2}$	$6^2P_{1/2}$	515.3
					$6^2P_{3/2}$	$n^2S_{1/2}$	$n = 9\text{–}11$	$n^2S_{1/2}$	$6^2P_{1/2}$	—

Reference	Atom	Lower state	λ (nm)	Intermediate state	λ (nm)	Upper state	Fluorescence / lower state	λ (nm) / n
	Rb	$5^2S_{1/2}$	794.8	$5^2P_{1/2}$	620.6	$6^2D_{3/2}$	$5^2P_{3/2}$	629.9
		$5^2S_{1/2}$	780.0	$5^2P_{3/2}$	572.4	$7^2D_{3/2}$	$5^2P_{1/2}$	564.8
		$5^2S_{1/2}$	780.0	$5^2P_{3/2}$	543.2	$8^2D_{3/2}$	$5^2P_{1/2}$	536.3
		$5^2S_{1/2}$	780.0	$5^2P_{3/2}$	543.1	$8^2D_{5/2}$	$5^2S_{1/2}$	420.2
		$5^2S_{1/2}$	780.0	$5^2P_{3/2}$	$n = 8,9$	$n^2S_{1/2}$	$5^2P_{1/2}$	—
Haroche et al. (1974)		$3^2S_{1/2}$	589.0	$3^2P_{3/2}$	$n^2D_{3/2,5/2}$	$n^2D_{3/2,5/2}$	$3^2P_{3/2}$	$n = 7$–10
Duong et al. (1974)	Na	$3^2S_{1/2}$	589.6	$3^2P_{1/2}$	615.4	$5^2S_{1/2}$		
Takagi et al. (1975)	Na	$3^2S_{1/2}$ $F=2$	589.0	$3^2P_{3/2}$ $F=1$	$F=0$	$3^2S_{1/2}$ $F=0$	$3^2P_{1/2}$ $F=1$	589.0
Gornik et al. (1973)	Na	$3^2S_{1/2}$	589.0	3^2P	568.0	4^2D	3^2P	568.0
Gallagher et al. (1975a,b)	Na	$3^2S_{1/2}$	589.0	$3^2P_{3/2}$		$n^2S_{1/2}$	3^2P	$n = 7$–13
				$3^2P_{3/2}$		n^2D	3^2P	$n = 5$–13
Ducas et al. (1975)	Na	$3^2S_{1/2}$	589.0	$3^2P_{3/2}$		$n^2S_{1/2}$	3^2P	$n = 15$–30
				$3^2P_{3/2}$		n^2D	3^2P	$n = 15$–30

source to create the fluorescing state. In this case the second excitation source was a tunable cw dye laser. The authors succeeded in measuring the hyperfine structure of excited D and F levels in Cs and Rb. In both cases, the atomic lamp excited the alkali vapor from the $n\ ^2S_{1/2}$ ground state to either the $n\ ^2P_{1/2}$ or the $n\ ^2P_{3/2}$ excited states where $n = 5$ for Rb and $n = 6$ for Cs. Subsequent dye laser absorption from $n\ ^2P_{1/2}$ or $n\ ^2P_{3/2}$ to $m\ ^2D$ or $m\ ^2S_{1/2}$ states with $m > n$ was monitored by observing one component of the $^2D-^2P$ or $^2S-^2P$ fluorescence. A complete summary of these transitions is given in Table 1-10. By observing the level crossings in an applied magnetic field, the authors were able to measure a number of dipole and quadrupole interaction constants for the excited D states in Cs and Rb.

    1.6.3.1e. Na. By far, the atom which has been most closely examined using the OODR technique is sodium. Haroche *et al.* (1974) have used pulsed dye lasers to excite sodium in a stepwise fashion from the $3\ ^2S_{1/2}$ ground state through the $3\ ^2P_{3/2}$ state to a superposition of $n\ ^2D_{5/2,3/2}$ fine-structure components. For $n = 9$ and 19, these authors were able to deduce the $n\ ^2D_{5/2}-n\ ^2D_{3/2}$ splitting from the frequency of the quantum beats superimposed on the $n\ ^2D-3\ ^2P$ fluorescence. Duong *et al.* (1974) have measured the hyperfine structure of the $5\ ^2S_{1/2}$ state produced by similar excitation through the $3\ ^2P_{1/2}$ intermediate. By directing the excitation photons in opposite directions, these authors were able to partially eliminate the effect of Doppler broadening. The increased resolution thus obtained made possible tht spectral separation of the $F = 1$ and 2 hyperfine levels of the $5\ ^2S_{1/2}$ state, which differ by only 159 MHz.

    Partial elimination of the Doppler effect in optical–optical double resonance has also been obtained by Takagi *et al.* (1975). Their experiment was analogous to that performed by Shoemaker *et al.* (1974) as described in Section 1.5.2.3, with the exception that optical frequencies were used to excite Na from the $^2S_{1/2}$ ground state to the $^2P_{1/2}$ excited state. Resonance was observed for transitions between Zeeman sublevels of the hyperfine components.

    One of the most important applications of OODR is to the study of highly excited Rydberg states of atoms. These states are created by the promotion of an electron into a large, loosely bound orbit. In sodium, a prototypical example for such studies, the $11\ ^2D$ state, for example, is characterized by a radius of nearly 100 Å and a binding energy of only 0.12 eV. Rydberg atoms are, therefore, expected to have high polarizability and large collisional cross sections.

    A number of workers have used the sodium $^2P_{1/2,3/2}$ states as intermediates in a two-step laser excitation mechanism to create high Rydberg states. Optical–optical double resonance can then provide information about these states through measurement of the lifetime. For example, Gornik *et al.* (1973) have used OODR to study the lifetime of the $4\ ^2D$ state.

A flash-lamp pumped dye laser provided the initial excitation to the $2\,^2P$ state, while a nitrogen laser pumped dye laser was used to excite the $2\,^2P$–$4\,^2D$ transition. The lifetime of the $4\,^2D$–$2\,^2P$ fluorescence decay was found to be 57 ns. Gallagher *et al.* (1975a) have extended this technique to obtain lifetimes of the $n\,^2S_{1/2}$ states ($n = 7$–13) and $n\,^2D$ states ($n = 5$–13). The lifetimes increased approximately as the cube of the effective principal quantum number, in good agreement with a Coulomb approximation theory. These same authors (Gallagher *et al.*, 1975b) subsequently studied the collisional processes affecting the $n\,^2D$ Rydberg states. It was found that collisions with rare gas atoms actually lengthened the apparent $n\,^2D$ lifetime by collisionally coupling the $n\,^2D$ states with states for which $l \geq 2$. The cross section for the collisional process appeared to increase as the geometrical cross section of the excited atom. Finally, two groups have used the same technique to obtain $n\,^2S_{1/2}$ and $n\,^2D$ states with $n = 15$–30 (Ducas *et al.*, 1975) or $n = 13$–18 (Ambartzumian *et al.*, 1975b). The former group observed, $n\,^2S, n\,^2D \rightarrow n\,^2P$ fluorescence from $n \geq 23$ and also detected the excited atoms by direct ionization in an applied field. The threshold for ionization varied as $E_n^{\,0} = 16[(n^*)^4]^{-1}$, where n^* is the effective quantum number.

1.6.3.2. Optical–Optical Double Resonance in Diatomic Molecules

1.6.3.2a. ICl. The technique of optical–optical double resonance has also recently been applied to the study of diatomic molecules. Barnes *et al.* (1974) have excited $E \rightarrow A$ state fluorescence in ICl by the two-step excitation scheme shown in Figure 1-33. An optical parametric oscillator (OPO) was used for the $A \leftarrow X$ state pump source, while a nitrogen laser-pumped dye laser fired 3 ns after the OPO pulse was employed for the $E \leftarrow A$ excitation source. Fluorescence from $E \rightarrow A$ was monitored at right angles to the collinear laser beams through a spectrometer. While the $E \leftarrow A$ excitation laser was fixed at 416.9 nm, $E \rightarrow A$ fluorescence at the same wavelength was recorded as a function of the $A \leftarrow X$ OPO excitation wavelength. An increase in $E \rightarrow A$ fluorescence was observed when the two pump lasers shared a common level in the A state. The spectral features obtained are listed in Table 1-11 and compare favorably with the wavelengths reported by Clyne and Coxon (1967) for the A–X system and Haranath and Rao (1957) for the E–A system.

1.6.3.2b. I_2. A similar optical–optical double-resonance scheme has been applied to I_2 by Rousseau and Williams (1974). In this experiment a tunable dye laser was used to excite a particular $X(^1\Sigma_{0^+g}) \rightarrow B(^3\Pi_{0^+u})$ transition while a fixed-frequency (28,514 cm^{-1}) cw krypton-ion laser was used to excite $B \rightarrow E$ transitions. Fluorescence from E to B was recorded at right

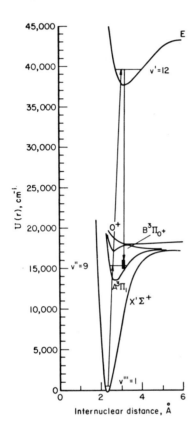

Figure 1-33. Potential-energy diagram for ICl showing the two-step excitation of fluorescence (from Barnes *et al.*, 1974).

angles to the collinear pump beams and resolved using a double monochromator. For given $X \to B$ and $B \to E$ transitions, the high-energy $E \to B$ fluorescence consisted of a series of sharp lines corresponding to transitions to bound levels of the B state, while the low-energy fluorescence consisted of much broader features which result from transitions between a bound level in the E state and a continuum level of the B state. The intensity variations of the continuum emission arise because of the phase relationships between the initial and final wave functions as predicted by Condon (1928). Because the continuum emission originated from a single excited level, this experiment forms one of the first unambiguous observations of the predicted effect, which Condon called "internal diffraction." A partial list of the stronger of the observed transitions is given in Table 1-11.

In a subsequent experiment, Rousseau (1975) modulated the ultraviolet laser acousto-optically to produce pulses which decayed in less than 10 ns. The pulsed fluorescence on the $E \to B$ transition was found to have a zero-pressure lifetime of 27 ± 2 ns.

Table 1-11 Optical–Optical Double Resonance in Diatomic Molecules

Reference	Molecule	Pump 1				Pump 2				Monitor
Barnes *et al.* (1974)	ICl	$X^1\Sigma^+(v'') \to A^3\Pi_i(v')$				$A^3\Pi_i(v') \to E(v^*)$				$E(v^*) \to A^3\Pi_i$ fluorescence
			v''	v'	λ (nm)		v'	v^*	λ (nm)	
			1	9	669		9	12	416.9	
			1	8	677		8	11	416.9	
			2	10	679		10	13	416.9	
			2	9	687		9	12	416.9	
			2	8	695		8	11	416.9	
			3	9	705		9	12	416.9	
Rousseau and Williams (1974)	I_2	$X^1\Sigma_0^+{}_g \to B^3\Pi_{0^+u}(v,J)$ $(v,J) = (11,38)$ $(16,15)$ $(17,87)$				$B^3\Pi_{0u} \to E(v^*)$ $v^* = 53$				$E(v^*) \to B^3\Pi_{0^+u}$ fluorescence
Field *et al.* (1975)	BaO	$X^1\Sigma(v''=0, J''=48) \to$ $A^1\Sigma(v'=8, J'=49)$; 488 nm				$A^1\Sigma(v'=8, J'=49) \to$ $Z^1\Sigma(v^*=?, J^*=48, 50)$; 568.3–624.6 nm				11 pairs of pump 2 transitions observed by Z–X or A–X fluorescence
		$X^1\Sigma(v''=0, J''=6) \to$ $A^1\Sigma(v'=7, J'=7)$; 496.5 nm				$A^1\Sigma(v'=7, J'=7) \to$ $Z^1\Sigma(v^*=?, J^*=6, 8)$; 573.9–623.7 nm				15 pairs of pump 2 transitions observed by Z–X or A–X fluorescence
Wong *et al.* (1975)	^{24}MgO	$X^1\Sigma(v'', J'') \to B^1\Sigma(v', J')$				$B^1\Sigma \to ?^1\Pi$ 566–625 nm and $B \to E^1\Sigma^+$				$?^1\Pi \to X^1\Sigma$ fluorescence $E^1\Sigma^+ \to X^1\Sigma$ fluorescence
		v''	J''	v'	J'	λ (nm)				
		4	26	5	27	476.5				
		3	49	4	50	476.5				
		2	70	3	71	476.5				
		5	10	4	9	514.5				
		2	70	1	71	514.5				
	^{25}MgO	3	50	2	51	514.5				
	^{26}MgO	2	36	2	37	496.5				
		4	25	3	26	514.5				
Allamandola and Nibler (1974)	C_2^-	$v=0 \to v=1,2$ in ground electronic state via optical pump at 21,050 cm^{-1} to higher electronic state and subsequent fluorescence				$v=1 \to v=0$ electronic band absorption at 17,339 cm^{-1}				Fluorescence from upper electronic state

1.6.3.2c. BaO. Field *et al.* (1975) have recently extended previous microwave optical double-resonance spectroscopy in BaO (Section 1.6.1.5) by the use of optical–optical double-resonance techniques. For convenience in the following discussion, we will use the notation presented in their original paper. Three energy levels were involved in the OODR scheme. Molecules in the $X^1\Sigma(v'',J'')$ ground state were excited to the $A^1\Sigma(v',J')$ intermediate state by a cw argon-ion laser (488.0 nm: $v'' = 0$, $J'' = 48$, $v' = 8$, $J' = 49$; 496.5 nm: $v'' = 0$, $J'' = 1, 6$, $v' = 7$, $J' = 2, 7$). Concurrently, a tunable dye laser excited the intermediate state to the v^*, J^* levels of a higher-lying $^1\Sigma$ electronic state which we will denote as Z. Two types of experiments were performed, which the authors called OODR excitation spectroscopy and OODR photoluminescence spectroscopy.

OODR excitation spectroscopy records the absorption spectrum of $A^1\Sigma(v',J')$ molecules by observing either a *decrease* in the A–X fluorescence or an *increase* in the Z–X fluorescence as the dye laser is tuned through the A → Z absorption region. By using this technique, Field *et al.* were able to locate 19 vibrational levels of the upper Z state. Because of the high density of vibrational states $(1/100\ \text{cm}^{-1})$ and because of the large perturbations among them, it was impossible to make specific vibrational assignments. In fact, Z is actually at least two electronic states, each of $^1\Sigma$ symmetry.

In the OODR photoluminescence experiment, both lasers were held at fixed frequencies and the $Z^1\Sigma(v^*, J^* = 50) \to X^1\Sigma(v'', J'' = 49, 51)$ fluorescence was resolved using a monochromator. The observation of transitions to $X^1\Sigma$ with $v'' = 0$–34 yielded improved rotational and vibrational constants for BaO $X^1\Sigma$ from which an accurate RKR potential was derived.

The importance of these techniques is twofold. First, since only specific v, J levels are excited, the resulting OODR excitation or photoluminescence spectra are relatively easy to interpret. Second, because only molecules within a narrow range of velocities relative to the laser propagation direction are excited, the resulting spectra may be obtained with a resolving power of almost two orders of magnitude beyond the Doppler limit. This feature may prove useful for optical Stark, Zeeman, and hyperfine spectroscopy.

1.6.3.2d. MgO. Wong *et al.* (1975) have briefly reported the observation of OODR in MgO. A variety of cw argon lines was used to excite specific $X^1\Sigma^+ \to B^1\Sigma^+$ transitions (Table 1-11). A tunable dye laser (566–625 nm) then raised the intermediate B state to higher electronic states. Four pairs of (P,R) transitions were found to the $E^1\Sigma^+$ state, while six groups of (P,Q,R) lines were located for transitions to a $^1\Pi$ state. OODR excitation spectra were recorded by observing the $^1\Pi \to X$ or $E^1\Sigma^+ \to X$ fluorescence which occurs in the ultraviolet region of the spectrum.

1.6.3.2e. C_2^-. Finally, OODR has recently been coupled with the technique of matrix isolation to yield vibrational lifetimes of C_2^- in N_2 or Ar matrices at 14–24°K (Allamandola and Nibler, 1974). The C_2^- molecular

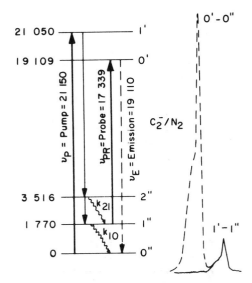

Figure 1-34. Energy levels and part of the fluorescence spectrum for C_2^- in N_2 matrix at 16°K. Solid line, probe laser off; dashed line, probe laser on (from Allamandola and Nibler, 1974).

ion was prepared by 121.6-nm hydrogen resonance photolysis of an $N_2(Ar)/C_2H_2$ mixture and deposited onto the cold tip of a closed-cycle refrigerator at 16°K. As shown in Figure 1-34, ground-state C_2^- ions were pumped to $v' = 1$ of an excited electronic state by absorption of a chopped argon-ion laser. Subsequent fluorescent emission populated the $v'' = 1$ and 2 levels of the ground state. After a variable time delay, the population of the $v'' = 1$ level was probed by monitoring the 0–0 electronic band fluorescence following excitation of the $1 \to 0$ band with a tunable dye laser. Lifetimes were independent of temperature (14–24°K) or matrix composition (Ar or N_2) and showed the unusual property that the lifetime of $v'' = 2$ (about 1.2 ms) was longer than that of $v'' = 1$ (about 0.2 ms).

1.6.3.3. Optical–Optical Double Resonance in Polyatomic Molecules

Although it is clear that OODR is a versatile and useful spectroscopic technique, to date its application has been limited to the study of atomic and diatomic species. There seems to be no inherent reason for this limitation, however. In fact, it appears likely that the major future application of OODR will be to the study of polyatomic species whose spectra are too complicated for analysis by conventional means. One proposed example for such a study will be discussed below.

The NO_2 molecule, despite the simplicity of its structure, has long been the subject of spectroscopic debate. In the visible region, at least two excited states contribute to its absorption. The $^2B_1$ excited state is formed as the upper Renner–Teller components of the $^2A_1$ ground state. It is degenerate with the ground state in the linear configuration and correlates to $NO(^2\Pi)$ and $O(^3P)$. In addition, a $^2B_2$ state correlating with $NO(^2\Pi)$ and $O(^1D)$ is thought to have its spectroscopic origin below that of the $^2B_1$ state (Brand *et al.*, 1973). As a result of the perturbations among the $^2A_1$, $^2B_1$, $^2B_2$, and possibly other electronic states, the visible absorption spectrum of NO_2 is extremely dense and complex. Only small regions have been analyzed with any success, despite much effort in this field (e.g., Douglas and Huber, 1965; Stevens *et al.*, 1973; Abe *et al.*, 1974; Solarz and Levy, 1973; Tanaka *et al.*, 1973, 1975; Brand *et al.*, 1975; Stevens and Zare, 1975; Smalley *et al.*, 1975). Optical–optical double resonance may provide a method for partially unraveling this spectrum.

The basic idea behind this approach is relatively straightforward. One optical monitor laser would select a particular transition and thereby "tag" one $[v,N,K,J]$ level of the ground electronic state. Recalling the assignments made by MODR and referring to Figure 1-29, we might, for example, choose the Ar^+ laser line at 21,837.1 cm^{-1} as our monitoring laser. In this case, From Figure 1-29, the $[v,N,K,J]$ level would then be $[(0,0,0),10,0,\frac{21}{2}]$. Simultaneously, a second optical laser, the pump source, would be scanned through the absorption region of interest. This second laser might be a tunable pulsed (or modulated cw) dye laser. When the pump and monitor lasers share the common $[v,N,K,J]$ level of the ground state, a decrease in the fluorescence excited by the monitor laser should be observed. For example, referring again to Figure 1-29, if the pump laser were to be tuned to 16,840.45 or 16,857.65 cm^{-1}, it would deplete the $[(0,0,0)10,0,\frac{21}{2}]$ level by exciting it to the $^2B_2$ electronic state. Consequently, fewer molecules would be available to absorb the argon-ion line. The absorption of the monitoring laser line as a function of the frequency of the pumping laser would thus map the absorption spectrum for molecules which are restricted to the selected $[v,N,K,J]$ state. By using a cw tunable dye laser as the monitor rather than the Ar^+ line, absorption spectra could be obtained as if the absorbing species were in any selected $[v,N,K,J]$ level. Such a simplification should greatly improve our understanding of the NO_2 spectral features and the perturbations which cause them.

1.6.4. Optically Detected Double Resonance in Large Molecules

A rather different type of experiment makes use of fluorescence or phosphorescence, excited by uv radiation, to monitor changes in level populations or phase relationships induced by saturating microwave fields. This type of experiment is most often carried out on aromatic hydrocarbons at very low temperatures. This field is thoroughly discussed by Harris and Breiland in Chapter 4.

1.7. Molecular Information from Double-Resonance Experiments

While Sections 1.4–1.6 have outlined the field of double-resonance spectroscopy from the point of view of the experimental configurations employed, this section will review the double-resonance results from the point of view of the molecular information these techniques provide. In Sections 1.7.1 and 1.7.2 we will briefly discuss the two broad categories, spectroscopic and kinetic, into which this information falls. We will then conclude the chapter with a discussion of some of the future directions which double-resonance spectroscopy might be expected to follow.

1.7.1. Spectroscopic Information

One of the major advantages offered by double-resonance techniques is the simplification of spectral assignments. If the sample pressure is sufficiently small so that collisional coupling between levels is negligible, then only three-level double resonances may be observed (see Section 1.1.4 and Figure 1-3). Under these conditions, the observation of a double-resonance signal implies that the two molecular transitions share a common level. If one of the transitions is known, the second may often be deduced from the frequency of the resonance and knowledge of the common level.

The variety of spectral regions in which this technique may be used has been demonstrated in Sections 1.4–1.6. Highlights include the use of microwave–microwave double resonance to assign the rotational spectra of HDCO and H_2CO (Section 1.4.1.3), and of isoxazole and chloroacetaldehyde (Section 1.4.1.6). Infrared–microwave methods were used to assign rotational–vibrational levels in CH_3Cl (Frenkel *et al.*, 1971) and OCS (Dobbs *et al.*, 1976). In optically detected double resonance, microwave pumping in the ground state has been extensively used to assign both rotational and electronic transitions. Examples of this application include the spectral assignments in BaO (Section 1.6.1.5), NO_2 (Section 1.6.1.6), NH_2 (Section 1.6.1.7), and BO_2 (Section 1.6.1.8). Finally, a potentially useful optical–optical double-resonance scheme for assigning the visible absorption spectrum of NO_2 has been presented in Section 1.6.3.3.

The second major advantage of double-resonance-method acquisition of spectroscopic information is that it allows investigation of states not normally populated at room temperature. In this type of application, an infrared or optical source is used to produce an excited state whose absorption is then probed in the microwave, infrared, or optical regions of the spectrum. For example, microwave–infrared double resonance has been used to obtain rotational spectra of vibrationally excited levels in HDCO, CH_4, HCOOH, and $HC\equiv CCHO$ (Section 1.5.1). Infrared–infrared double resonance has been used to measure the excited-state absorption profiles in SF_6 and BCl_3 (Section 1.5.2.1). However, it is in optically detected double resonance that these methods have been most widely used for the investigation of excited states. Microwave–optical double resonance has helped to assign the perturbation complex in electronically excited CN (Section 1.6.1.2), to obtain the lambda splittings and hence dipole moments in excited CS (Section 1.6.1.4), and to probe the rotational structure in excited BaO (Section 1.6.1.5) and NO_3 (Section 1.6.1.6). Optical–optical double resonance has allowed the spectral observation both of many high-lying atomic states (Section 1.6.3.1), as well as of a few new excited diatomic states (ICl, I_2, BaO, and MgO; Section 1.6.3.2). Several of these states cannot be observed by single-photon techniques, because they have the same parity as the ground state and thus do not possess a dipole-allowed one-photon transition from that state.

The double-resonance technique of tagging a molecular transition with one radiation source while monitoring transitions which share common levels with a second radiation source should receive wider usage as tunable lasers become more readily available.

1.7.2. Energy Transfer and Interaction Potentials

As we have seen in previous sections, information on molecular inelastic processes can be gained from the density dependence of double-resonance signal strengths and decay times. This information can then lead to an improved understanding of interaction potentials and collision dynamics.

The dephasing time analogous to T_2 in nmr experiments can be obtained from transient coherent experiments (see Sections 1.4.2.2 and 1.5.2.3 and Chapters 2 and 3). On a phenomenological level, this parameter may be compared with the pressure-broadened linewidth of the transition. While in most instances it is found that $1/2\pi pT_2 = \Delta\nu/p$, the linewidth data are often of insufficient accuracy to permit a valid comparison to be made. An analysis of these relaxation times in terms of collision dynamics has been made by Marcus and Liu (1975) for the OCS and NH_3 systems.

The microwave–microwave double-resonance experiments (Section 1.4.1) have been particularly valuable in revealing details of rotational energy transfer. In most cases the rotational states which can be easily coupled with one another are those between which there exist nonvanishing electric-dipole matrix elements. This is to be expected if the leading term in the interaction potential is the dipole–dipole r^{-3} term (Oka, 1973; Sharma and Brau, 1969). Since this potential is long-range, we would also expect that the cross sections for relaxation should be quite large. This is indeed the case in ammonia, for example, in which both the dephasing time (T_2) and the relaxation time between the inversion doublet levels (T_1) are 25 MHz/Torr, which corresponds to a collision rate 10 times gas-kinetic. The time for deactivation of the widely spaced rotational levels is much slower, however, with $Z_{rot} \simeq 5$. A similar result was found for methyl fluoride by Jetter *et al.* (1973), who found $1/2\pi pT = 15$ MHz/Torr for the decay of infrared–microwave double-resonance signals. This corresponds to $(p\tau_{rot})^{-1} = 9 \times 10^7 \, s^{-1} \, Torr^{-1}$, while the gas-kinetic collision rate is $2 \times 10^7 \, s^{-1} \, Torr^{-1}$. Recent CO_2 laser Q-switching experiments using CH_3F also point to a rotational relaxation cross section which is larger than gas-kinetic (A. Devir, private communication).

Several of the double-resonance experiments described in this chapter have been particularly revealing of the relative efficiencies of competing molecular relaxation processes. For example, in a number of highly polar systems, it has been found that the cross sections for rotational energy transfer are considerably larger than those for linear momentum transfer, i.e., velocity-changing collisions. This is the case, for example, in HF–HF collisions (Hinchen, 1975; Hinchen and Hobbs, 1976). This is entirely consistent with having a rotational relaxation cross section larger than gas-kinetic. A similar result was found by Freund *et al.* (1973) in ammonia, in which a hole burned into the velocity distribution was collisionally transferred between members of an inversion doublet. This velocity group did not survive a rotationally inelastic collision, however.

In addition, Shoemaker *et al.* (1974) found in their double-resonance experiments in CH_3F that the cross section for angular momentum reorientation exceeded that for velocity change. In SF_6, however, which is a nonpolar, nearly spherical molecule, Frankel and Steinfeld (1975) found that the orientation of the optically pumped molecules survived even $V \rightarrow V$ transferring collisions.

In the case of vibrational relaxation, as studied by infrared double resonance (Section 1.5.2), there exists a number of other, more traditional techniques with which to compare the results, such as shock-tube or ultrasonic relaxation methods. Such a comparison has been made by Breshears (1973) and Breshears and Blair (1973) for the particular case of SF_6. Double-resonance methods are uniquely suited to the study of the important problem of vibrational relaxation in low-temperature solids, to which the traditional techniques are not applicable. Examples of this application are the studies of ammonia (Section 1.5.2.1) and C_2^- (Section 1.6.3.2).

Finally, double-resonance methods can be very useful in studying collision properties of exotic species, such as the "giant" Rydberg atoms, which are not to be found under normal equilibrium conditions. Experiments with $n\,^2D$ states of sodium having $n \geq 5$ have elucidated electronic quenching cross sections many times larger than gas kinetic (Section 1.6.3.1).

1.7.3. Future Directions

We will certainly see a wider variety of experiments, on more species in additional regions of the spectrum, as tunable laser sources become more available from the far infrared to the far ultraviolet. In addition, new types of information will be forthcoming from this research. One specific area may be mentioned here. By carrying out double-resonance spectroscopy with narrow-bandwidth sources within the Doppler profile of an absorption line, it will be possible to measure in detail the velocity dependence of cross sections for a wide variety of molecular processes. Some experiments of this type have already been carried out, and are discussed in a report by Robinson (1976). Velocity selection has thus far been the exclusive domain of the molecular beam experimenter; the combination of velocity and state selection made possible by double-resonance spectroscopic methods using ultra-high-resolution light sources represents a completely new and powerful capability, and should yield a wealth of new and surprising insights into molecular collision phenomena.

NOTE ADDED IN PROOF

Shortly after the completion of the manuscript for this chapter, several additional descriptions of double-resonance experiments appeared in the literature which are worthy of special notice. Takami (1976) has presented a summary of the theory of optical–microwave double resonance, for both microwave and optical detection, and compared the results with experimental data. Kasuga *et al.* (1976) have carried out microwave adiabatic rapid passage experiments on OCS, and determined T_1 for $J = 2 \leftarrow 1$ transition. Retallack and Lees (1976) carried out modulated microwave double resonance on the $v_2 = 1$ state of OCS, and observed $\Delta M = 2$ collision-induced transitions corresponding to nondipolar selection rules. Finally, Danyluk and King (1977), in an elegant two-photon sequential absorption experiment on I_2, have identified five new states of this molecule lying between 40,000 and 43,000 cm^{-1}.

Acknowledgments

This chapter reflects work supported in part by the National Science Foundation grant MPS-04733 (to J.I.S.) and by the donors of the Petroleum Research Fund of the American Chemical Society (to P.L.H.).

References

Abe, K., Myers, F., McCubbin, T. K., Jr., and Polo, S. R., 1974, *J. Mol. Spectry.* **50**:413.

Abouaf-Marguin, L., Dubost, H., and Legay, F., 1973, *Chem. Phys. Lett.* **22**:603.

Aldridge, J. P., Filip, H., Flicker, H., Holland, R. F., McDowell, R. S., Nereson, N. G., and Fox, K., 1975, *J. Mol. Spectry.* **58**:165.

Alimpiev, S. S., and Karlov, N. V., 1974, *Dokl. Akad. Nauk* **66**: 542.

Allamandola, L. J., and Nibler, J. W., 1974, *Chem. Phys. Lett.* **28**:335.

Allen, L., and Eberly, J. H., 1975, *Optical Resonance and Two-Level Atoms*, John Wiley & Sons, Inc., New York.

Ambartzumian, R. V., Letokhov, V. S., Makarov, G. N., and Puretzkiy, A. A., 1972, *Chem. Phys. Lett.* **16**:252.

Ambartzumian, R. V., Gorokhov, I. A., Letokhov, V. S., and Makarov, G. N., 1975a, *JETP Lett.* **22**:43.

Ambartzumian, R. V., Bekov, G. I., Letokhov, V. S., and Mishin, V. I., 1975b, *JETP Lett.* **21**:279.

Armstrong, J. J., and Gaddy, O. L., 1972, *IEEE J. Quantum Electron.* **QE8**:797.

Barger, R. L., Broida, H. P., Estin, A. J., and Radford, H. E., 1962, *Phys. Rev. Lett.* **9**:345.

Barnes, R. H., Moeller, C. E., Klucher, J. F., and Verber, C. M., 1974, *Appl. Phys. Lett.* **24**:610.

Bass, H. E., and Winter, T. G., 1972, *J. Chem. Phys.* **56**:3619.

Bass, M., Deutsch, T. F., and Weber, M. J., 1971, *Dye Lasers*, p. 269, Marcel Dekker, Inc., New York.

Bates, R. D., Knudtson, J. T., Flynn, G. W., and Ronn, A. M., 1970, *J. Chem. Phys.* **53**:3621.

Bates, R. D., Knudtson, J. T., Flynn, G. W., and Ronn, A. M., 1971, *Chem. Phys. Lett.* **8**:103.

Battaglia, A., Gozzini, A., and Polacco, E., 1959, *Nuovo Cimento* **14**:1076.

Beaulieu, A. J., 1970, *Appl. Phys. Lett.* **16**:504.

Bradley, D. J., Gale, G. M., and Smith, P. D., 1970, *J. Phys.* **B3**:L11.

Bradley, D. J., Ewart, P., Nicholas, J. V., Shaw, J. R. D., and Thompson, D. G., 1973a, *Phys. Rev. Lett.* **31**:263.

Bradley, D. J., Ewart, P., Nicholas, J. V., and Shaw, J. R. D., 1973b, *J. Phys.* **B6**:1594.

Brand, J. C. D., Hardwick, J. L., Pirkle, R. J., and Seliskav, C. J., 1973, *Can. J. Phys.* **51**:2184.

Brand, J. C. D., Chan, W. H., and Hardwick, J. L., 1975, *J. Mol. Spectry.* **56**:309.

Brechignac, P., Taieb, G., and Legay, F., 1975, *Chem. Phys. Lett.* **36**:242.

Breshears, W. D., 1973, *Chem. Phys. Lett.* **20**:429.

Breshears, W. D., and Blair, L. S., 1973, *J. Chem. Phys.* **49**:5824.

Bridges, W. B., 1964, *Appl. Phys. Lett.* **4**:128.

Brophy, J. H., Silver, J. A., and Kinsey, J. L., 1974, *Chem. Phys. Lett.* **28**:418.

Brossel, J., and Bitter, F., 1952, *Phys. Rev.* **86**:308.

Brown, S. R., 1974, *J. Chem. Phys.* **60**:1722.

Burak, I., Steinfeld, J. I., and Sutton, D. G., 1969a, *J. Quant. Spectry. Radiative Transfer* **9**:959.

Burak, I., Nowak, A. V., Steinfeld, J. I., and Sutton, D. G., 1969b, *J. Chem. Phys.* **51**:2275.

Burak, I., Houston, P. L., Sutton, D. G., and Steinfeld, J. I., 1971, *IEEE J. Quantum Electron.* **QE7**:73.

Burdett, J. K., and Poliakoff, M., 1974, *Chem. Soc. Rev.* **3**:293.

Burdick, B., 1967, *Advan. At. Mol. Phys.* **3**:73.

Butcher, R. J., Dennis, R. B., and Smith, S. D., 1975, *Proc. Roy. Soc. (London)* **A334**:541.

Butler, J. F., 1969, *Solid State Commun.* **7**:909.

Carroll, J. O., and Marcus, S., 1968, *Phys. Lett.* **A27**:590.

Chebotayev, V. P., and Letokhov, V. S., 1975, *Progr. Quantum Electron.* **4**:111.

Cheo, P. K., and Abrams, R. L., 1969, *Appl. Phys. Lett.* **14**:47.

Chu, F. Y., Freund, S. M., Johns, J. W. C., and Oka, T., 1973, *J. Mol. Spectry.* **48**:328.

Clyne, M. A. A., and Coxon, J. A., 1967, *Proc. Roy. Soc. (London)* **A298**:424.

Cohen, J. B., and Wilson, E. B., Jr., 1973a, *J. Chem. Phys.* **58**:456.

Cohen, J. B., and Wilson, E. B., Jr., 1973b, *J. Chem. Phys.* **58**:442.

Condon, E. U., 1928, *Phys. Rev.* **32**:858.

Condon, E. U., and Shortley, G. H., 1963, *The Theory of Atomic Spectra*, Cambridge University Press, New York.

Cook, J. M., Hills, G. W., and Curl, R. F., Jr., 1976, *Astrophys. J.* **207**:(in press).

Cox, A. P., Flynn, G. W., and Wilson, E. B., Jr., 1965, *J. Chem. Phys.* **42**:3094.

Curl, R. F., Jr., 1973, *J. Mol. Spectry.* **48**:165.

Curl, R. F., Jr., and Oka, T., 1973, *J. Chem. Phys.* **58**:4908.

Curl, R. F., Jr., Oka, T., and Smith, D. S., 1973, *J. Mol. Spectry.* **46**:518.

Dalton, L. R., and Dalton, L. A., 1973, *Magn. Resonance Rev.* **2**:361.

Daly, P. W., and Oka, T., 1970, *J. Chem. Phys.* **53**:3272.

Danyluk, M. D., and King, G. W., 1977, in press.

Davis, C. C., and King, T. A., 1975, *Advan. Quantum Electron.* **3**:168.

Dennis, R. B., Pidgeon, C. R., Smith, S. D., Wherrett, B. S., and Wood, R. A., 1972, *Proc. Roy. Soc. (London)* **A331**:203.

DiGiacomo, A., 1959, *Nuovo Cimento* **14**:1082.

Dobbs, G. M., Micheels, R. H., Steinfeld, J. I., Wang, J. H.-S., and Levy, J. M., 1975, *J. Chem. Phys.* **63**:1904.

Dobbs, G. M., Micheels, R., and Steinfeld, J. I., 1976, Proc. IX Int. Quantum Electronics Conf., *Opt. Commun.* **18**:72.

Douglas, A. E., and Huber, K. P., 1965, *Can. J. Phys.* **43**:74.

Drexhage, K. H., 1973, Structure and properties of laser dyes, in: *Dye Lasers* (F. P. Schäfer, ed.), p. 144, Springer-Verlag, Berlin.

Ducas, T. W., Littman, M. G., Freeman, R. R., and Kleppner, D., 1975, *Phys. Rev. Lett.* **35**:366.

Duong, H. T., Liberman, S., Pinaud, J., and Vialle, J.-L., 1974, *Phys. Rev. Lett.* **33**:339.

Elliot, R. J., Hayes, W., Jones, G. D., MacDonald, H. F., and Sennett, C. T., 1965. *Proc. Roy. Soc. (London)* **A289**:1.

Evenson, K. M., 1968, *Appl. Phys. Pett.* **12**:253.

Evenson, K. M., 1969, *Phys. Rev.* **178**:1.

Evenson, K. M., Dunn, J. L., and Broida, H. P., 1964, *Phys. Rev.* **A136**:1566.

Fabris, A. T., and Oka, T., 1972, *J. Chem. Phys.* **56**:3168.

Field, R. W., and Bergeman, T. H., 1971, *J. Chem. Phys.* **54**:2936.

Field, R. W., Bradford, R. S., Harris, D. O., and Broida, H. P., 1972a, *J. Chem. Phys.* **56**:4712.

Field, R. W., Bradford, R. S., Broida, H. P., and Harris, D. O., 1972b, *J. Chem. Phys.* **57**:2209.

Field, R. W., English, A. D., Tanaka, T., Harris, D. O., and Jennings, D. A., 1973, *J. Chem. Phys.* **59**:2191.

Field, R. W., Capelle, G. A., and Revelli, M. A., 1975, *J. Chem. Phys.* **63**:3328.

Ford, R. G., 1976, *J. Chem. Phys.* **65**: 354.

Fortin, R., 1971, *Can. J. Phys.* **49**:257.

Fourrier, M., and Redon, M., 1972, *Appl. Phys. Lett.* **21**:463.

Fourrier, M., and Redon, M., 1974, *J. Appl. Phys.* **45**: 1910.

Frankel, D. S., Jr., 1976a, *Opt. Commun.* **18**:31.

Frankel, D. S., Jr., 1976b, *J. Chem. Phys.* **65**:1696.

Frankel, D. S., Jr., and Steinfeld, J. I., 1975, *J. Chem. Phys.* **62**:3358.

Frankel, D. S., Jr., Steinfeld, J. I., Sharma, R. D., and Poulsen, L. L., 1974, *Chem. Phys. Lett.* **28**:485.

Frenkel, L., Marantz, H., and Sullivan, T., 1971, *Phys. Rev.* **A3**:1640.

Freund, S. M., and Oka, T., 1972, *Appl. Phys. Lett.* **21**:60.

Freund, S. M., Johns, J. W. C., McKellar, A. R. W., and Oka, T., 1973, *J. Chem. Phys.* **59**:3445.

Gallagher, T. F., Edelstein, S. A., and Hill, R. M., 1975a, *Phys. Rev.* **A11**: 1504.

Gallagher, T. F., Edelstein, S. A., and Hill, R. M., 1975b, *Phys. Rev. Lett.* **35**:644.

German, K. R., 1975, *J. Chem. Phys.* **62**:2584.

German, K. R., and Zare, R. N., 1969a, *Phys. Rev. Lett.* **23**:1207.

German, K. R., and Zare, R. N., 1969b, *Phys. Rev.* **186**:9.

German, K. R., and Zare, R. N., 1970, *Bull. Amer. Phys. Soc.* **15**:82.

Glorieux, P., and Macke, B., 1974, *Chem. Phys.* **4**:120.

Gordon, R. G., 1967, *J. Chem. Phys.* **46**:4399.

Gordon, R. G., and Steinfeld, J. I., 1976, Spectroscopic measurements of energy transfer by fluorescence and double resonance, in: *Molecular Energy Transfer* (R. D. Levine and J. Jortner, eds.), p. 67, Israel Universities Press, Jerusalem.

Gordon, R. G., Larson, P. E., Thomas, C. H., and Wilson, E. B., Jr., 1969, *J. Chem. Phys.* **50**:1388.

Gornik, W., Kaiser, D., Lange, W., Luther, J., Radloff, H.-H., and Schulz, H. H., 1973, *Appl. Phys.* **1**:285.

Green, B. D., and Steinfeld, J. I., 1975, unpublished results.

Häfele, H. G., 1974, *Appl. Phys.* **5**:97.

Hamilton, D. R., Knipp, J. K., and Kuper, J. B. H., 1947, *Klystrons and Microwave Triodes*, M.I.T. Radiation Laboratory Series No. 7, McGraw-Hill Book Company, New York.

Hänsch, T., 1972, *Appl. Opt.* **11**:895.

Haranath, P. B. V., and Rao, P. T., 1957, *Indian J. Phys.* **31**:368.

Haroche, S., Gross, M., and Silverman, M. P., 1974, *Phys. Rev. Lett.* **33**:1063.

Herbst, R. L., Fleming, R. N., and Byer, R. L., 1974, *Appl. Phys. Lett.* **25**:520.

Herzberg, G., 1944, *Atomic Spectra and Atomic Structure*, Dover Publications, Inc., New York.

Herzberg, G., 1945, *Molecular Spectra and Molecular Structure*, Vol. II: *Infrared and Raman Spectra of Polyatomic Molecules*, Van Nostrand Reinhold Company, New York.

Herzberg, G., 1950, *Molecular Spectra and Molecular Structure*, Vol. I: *Spectra of Diatomic Molecules*, Van Nostrand Reinhold Company, New York.

Herzberg, G., 1966, *Molecular Spectra and Molecular Structure*, Vol. III: *Electronic Spectra and Electronic Structure of Polyatomic Molecules*, Van Nostrand Reinhold Company, New York.

Hills, G. W., Cook, J. M., Curl, R. F., Jr., and Tittel, F. K., 1976, *J. Chem. Phys.* **65**:823.

Hinchen, J. J., 1975, *Appl. Phys. Letts.* **27**:672.

Hinchen, J. J., and Hobbs, R. H., 1976, *J. Chem. Phys.* **65**:2732.

Hoke, W. E., Bauer, D. R., Ekkers, J., and Flygare, W. H., 1976, *J. Chem. Phys.* **64**:5276.

Houston, P. L., and Steinfeld, J. I., 1975, *J. Mol. Spectry.* **54**:335.

Houston, P. L., Nowak, A. V., and Steinfeld, J. I., 1973, *J. Chem. Phys.* **58**: 3373.

Hudson, R. D., 1969, *Infrared System Engineering*, John Wiley & Sons, Inc., New York.

Hudson, R. D., and Hudson, J. W., 1975, *Infrared Detectors*, Halsted Press, New York.

Jacobs, R. R., Pettipiece, K. J., and Thomas, S. J., 1974a, *Appl. Phys. Lett.* **24**:375.

Jacobs, R. R., Thomas, S. J., and Pettipiece, K. J., 1974b, *IEEE J. Quantum Electron.* **QE10**:480.

Jacobs, R. R., Pettipiece, K. J., and Thomas, S. J., 1975, *Phys. Rev.* **A11**:54.

Javan, A., 1957, *Phys. Rev.* **107**:1579.

Jetter, H., Pearson, E. F., Norris, C. L., McGurk, J. C., and Flygare, W. H., 1973, *J. Chem. Phys.* **59**:1796.

Johnson, R. H., 1973, *IEEE J. Quantum Electron.* **QE8**:255.

Jones, D. G., Lambert, J. D., Saksena, M. P., and Stretton, J. L., 1969, *Trans. Faraday Soc.* **65**:965.

Jones, H., and Kohler, F., 1975, *J. Mol. Spectry.* **58**: 125.

Kano, S., Amano, T., and Shimizu, T., 1974, *Chem. Phys. Lett.* **25**:119.

Kano, S., Amano, T., and Shimizu, T., 1976, *J. Chem. Phys.* **64**:4711.

Kasper, J. V. V., and Pimentel, G. C., 1965, *Phys. Rev. Lett.* **14**:352.

Kasuga, T., Amano, T., and Shimizu, T., 1976, *Chem. Phys. Lett.* **42**:278.

Kibble, B. P., and Pancharatnam, S., 1965, *Proc. Phys. Soc. (London)* **86**:1351.

Kim, M. S., Smalley, R. E., and Levy, D. H., 1976, Paper RB10, Thirty-first Symposium on Molecular Structure and Molecular Spectroscopy, Columbus, Ohio.

King, G. W., 1964, *Spectroscopy and Molecular Structure*, Holt, Rinehart and Winston, Inc., New York.

Kompa, K. L., and Pimentel, G. C., 1967, *J. Chem. Phys.* **47**:857.

Kompa, K. L., Parker, J. H., and Pimentel, G. C., 1968, *J. Chem. Phys.* **49**:4257.

Kreiner, W. A., and Jones, H., 1974, *J. Mol. Spectry.* **49**:326.

Kugel, H. W., Leventhal, M., Murnick, D. E., Patel, C. K. N., and Wood, O. R., 1975, *Phys. Rev. Lett.* **35**:647.

Laubereau, A., Seilmeier, A., and Kaiser, W., 1975, *Chem. Phys. Lett.* **36**:232.

Lee, L. C., and Faust, W. L., 1971, *Phys. Rev. Lett.* **26**:648.

Legay, F., and Legay-Sommaire, N., 1964, *C. R. Acad. Sci. Paris* **259**:99.

Lemaire, J., Houriez, J., Bellet, J., and Thibault, J., 1969, *Compt. Rend.* **B268**:922.

Lemaire, J., Thibault, J., Herlemont, F., and Houriez, J., 1974, *Mol. Phys.* **27**:611.

Lengyel, B., 1966, *Introduction to Laser Physics*, John Wiley & Sons, Inc., New York.

Leone, S. R., and Moore, C. B., 1974, in *Chemical and Biological Applications of Lasers* (C. B. Moore ed.), p. 1, Academic Press, Inc., New York.

Letokhov, V. S., and Chebotayev, V. P., 1975, *Nonlinear Laser Spectroscopy*, Optical Sciences Series No. 4, Springer-Verlag, Berlin.

Levine, A. K., and DeMaria, A. J. (eds.), 1971, *Lasers*, Marcel Dekker, Inc., New York.

Levy, J. M., Wang, J. H.-S., Kukolich, S. G., and Steinfeld, J. I., 1972, *Phys. Rev. Lett.* **29**:395.

Levy, J. M., Wang, J. H.-S., Kukolich, S. G., and Steinfeld, J. I., 1973, *Chem. Phys. Lett.* **21**:598.

Magyov, G., 1974, *Appl. Opt.* **13**:25.

Marcus, R. A., and Liu, W. K., 1975, *J. Chem. Phys.* **63**:272, 290, 4564.

McClain, W. M., 1974, *Accts. Chem. Res.* **7**:129.

McIlrath, T. J., 1969, *Appl. Phys. Let.* **15**:41.

McIlrath, T. J., and Carlsten, J. L., 1973, *J. Phys.* **B6**:697.

Meyer, T. W., and Rhodes, C. K., 1974, *Phys. Rev. Lett.* **32**:637.

Meyer, T. W., Bischel, W. K., and Rhodes, C. K., 1974, *Phys. Rev.* **A10**:1433.

Mooradian, A., Brueck, S. R. J., and Blum, F. A., 1970, *Appl. Phys. Lett.* **17**:481.

Moore, C. E., 1958, *Atomic Energy Levels* (3 vols.), National Bureau of Standards Circular 467, Government Printing Office, Washington, D.C.

Nachshon, Y., and Coleman, P. D., 1974, *J. Chem. Phys.* **61**:2520.

Oka, T., 1966a, *J. Chem. Phys.* **45**:754.

Oka, T., 1966b, *J. Chem. Phys.* **45**:752.

Oka, T., 1967a, *J. Chem. Phys.* **47**:4852.

Oka, T., 1967b, *J. Chem. Phys.* **47**:13.

Oka, T., 1968a, *J. Chem. Phys.* **48**:4919.

Oka, T., 1968b, *J. Chem. Phys.* **49**:3135.

Oka, T., 1968c, *J. Chem. Phys.* **49**:4234.

Oka, T., 1973, *Advan. At. Mol. Phys.* **9**:127.

Olson, G. G., 1972, *Opt. Spectra* **6**(1):38.

Oppenheim, U. P., and Melman, P., 1971, *IEEE J. Quantum Electron.* **QE7**:426.

Orr, B. J., 1976, *Chem. Phys. Lett.* **43**:446.

Pantell, R. H., and Puthoff, H. E., 1969, *Fundamentals of Quantum Electronics*, John Wiley & Sons, Inc., New York.

Patel, C. K. N., 1964a, *Phys. Rev. Lett.* **12**:588.

Patel, C. K. N., 1964b, *Phys. Rev. Lett.* **13**:617.

Patel, C. K. N., and Shaw, E. D., 1970, *Phys. Rev. Lett.* **24**:451.

Patel, C. K. N., and Shaw, E. D., 1971, *Phys. Rev.* **B3**:1279..

Peterson, L. M., Lindquist, G. H., and Arnold, C. B., 1974, *J. Chem. Phys.* **61**:3480.

Peterson, O. G., Tuccio, S. A., and Snavely, B. B., 1970, *Appl. Phys. Lett.* **17**:245.

Poulsen, L. L., Houston, P. L., and Steinfeld, J. I., 1973, *J. Chem. Phys.* **58**: 3381.

Pratt, D. W., and Broida, H. P., 1969, *J. Chem. Phys.* **50**:2181.

Preses, J. M., and Flynn, G. W., 1977, *J. Chem. Phys.* **66**: 3112.

Pressley, R. J. (ed.), 1971, *Handbook of Lasers*, Chemical Rubber Company, Cleveland, Ohio.

Radio Corporation of America, 1970, *Photomultiplier Manual*, RCA, Inc., Harrison, N.J.

Redon, M., Gurel, H., and Fourrier, M., 1975, *Chem. Phys. Lett.* **30**:99.

Retallack, L. J., and Lees, R. M., 1976, *J. Chem. Phys.* **65**:3793.

Rhodes, C. K., Kelly, M. J., and Javan, A., 1968, *J. Chem. Phys.* **48**:3750.

Robinson, A. L., 1976, *Science* **192**:1323.

Ronn, A. M., and Lide, D. R., Jr., 1967, *J. Chem. Phys.* **47**:3669.

Ronn, A. M., and Wilson, E. B., Jr., 1967, *J. Chem. Phys.* **46**: 3262.

Röss, D. (ed.), 1969, *Lasers: Light Amplifiers and Oscillators*, Academic Press, Inc., New York.

Rousseau, D. L., 1975, *J. Mol. Spectry.* **58**:481.

Rousseau, D. L., and Williams, P. F., 1974, *Phys. Rev. Lett.* **33**:1368.

Runge, P. K., 1972, *Opt. Commun.* **5**:311.

Schäfer, F. P. (ed.), 1973, *Topics in Applied Physics*, Vol. I: *Dye Lasers*, Springer-Verlag, Berlin.

Scott, J. T., 1975, *Phys. Today* **28**(11):17.

Seibt, P. J., 1972, *J. Chem. Phys.* **57**:1343.

Series, G. W., 1959, *Rept. Progr. Phys.* **22**:280.

Sharma, R. D., and Brau, C. A., 1969, *J. Chem. Phys.* **50**:924.

Shimizu, T., and Oka, T., 1970a, *J. Chem. Phys.* **53**:2536.

Shimizu, T., and Oka, T., 1970b, *Phys. Rev.* **A2**:1177.

Shimoda, K., 1974, Infrared-microwave double resonance, in: *Laser Spectroscopy* (R. G. Brewer and A. Mooradian, eds.), Plenum Publishing Corporation, New York.

Shimoda, K., 1976, Double-Resonance spectroscopy of molecules by means of lasers, in: *Topics in Applied Physics* (H. Walther, ed.), Vol. 2, p. 197, Springer-Verlag, Berlin.

Shimoda, K., and Shimizu, T., 1972, *Progr. Quantum Electron.* **2**:45.

Shoemaker, R. L., Stenholm, S., and Brewer, R. G., 1974, *Phys. Rev.* **A10**:2037.

Siebert, D. R., Grabiner, F. R., and Flynn, G. W., 1974, *J. Chem. Phys.* **60**:1564.

Silvers, S. J., Bergeman, T. H., and Klemperer, W., 1970, *J. Chem. Phys.* **52**:4385.

Slater, J. C., 1950, *Microwave Electronics*, Van Nostrand Reinhold Company, New York.

Smalley, R. E., Wharton, L., and Levy, D. H., 1975, *J. Chem. Phys.* **63**:4977.
Smith, R. L., and Eck, T. G., 1970, *Phys. Rev.* **A2**:2179.
Solarz, R., and Levy, D. H., 1973, *J. Chem. Phys.* **58**:4026.
Solarz, R., and Levy, D. H., 1974, *J. Chem. Phys.* **60**:842.
Solarz, R., Levy, D. H., Abe, K., and Curl, R. F., Jr., 1974, *J. Chem. Phys.* **60**:1158.
Sorokin, P. P., Moruzzi, V. L., and Hammond, E. C., 1968, *J. Chem. Phys.* **48**:4726.
Spencer, D. J., Jacobs, T. A., Mirels, H., and Gross, R. W. F., 1969, *Int. J. Chem. Kinetics* **1**:493.
Steinfeld, J. I., 1974a, *Molecules and Radiation: An Introduction to Modern Molecular Spectroscopy*, Harper & Row, Publishers, New York.
Steinfeld, J. I., 1974b, Optical analogs of magnetic resonance spectroscopy, in: *Chemical and Biochemical Applications of Lasers* (C. B. Moore, ed.), Vol. 1, p. 103, Academic Press, Inc., New York.
Steinfeld, J. I and Jensen, C. C., 1976, in: *Tunable Lasers and Applications* (A. Mooradian, T. Jaeger, and P. Stokseth, eds.), p. 190, Springer-Verlag, Berlin.
Steinfeld, J. I., Burak, I., Sutton, D. G., and Nowak, A. V., 1970, *J. Chem. Phys.* **52**:5421.
Stephenson, J. C., 1973, *J. Chem. Phys.* **59**:1523.
Stevens, C. G., and Zare, R. N., 1975, *J. Mol. Spectry.* **56**:167.
Stevens, C. G., Swagel, M. W., Wallace, R., and Zare, R. N., 1973, *Chem. Phys. Lett.* **18**:465.
Stiefvater, O. L., 1975, *J. Chem. Phys.* **63**:2560.
Svanberg, S., and Belin, G., 1974, *J. Phys.* **B7**:L82.
Svanberg, S., Tsekeris, P., and Happer, W., 1973, *Phys. Rev. Lett.* **30**:819.
Takagi, K., Curl, R. F. Jr., and Su, R. T. M., 1975, *Appl. Phys.* **7**:181.
Takami, M., 1976, *Japan. J. Appl. Phys.* **15**:1063, 1889.
Takami, M., and Shimoda, K., 1972, *Japan. J. Appl. Phys.* **11**:1648.
Takami, M., and Shimoda, K., 1973a, *Japan. J. Appl. Phys.* **12**:394.
Takami, M., and Shimoda, K., 1973b, *Japan. J. Appl. Phys.* **12**:603.
Takami, M., and Shimoda, K., 1974, *Japan. J. Appl. Phys.* **13**:1699.
Takami, M., and Shimoda, K., 1976, *J. Mol. Spectry.* **59**:35.
Takami, M., Uehara, K., and Shimoda, K., 1973, *Japan. J. Appl. Phys.* **12**:924.
Tanaka, T., and Harris, D. O., 1976, *J. Mol. Spectry.* **56**:413.
Tanaka, T., English, A. D., Field, R. W., Jennings, D. A., and Harris, D. O., 1973, *J. Chem. Phys.* **59**:5217.
Tanaka, T., Abe, K., and Curl, R. F., Jr., 1974a, *J. Mol. Spectry.* **49**:310.
Tanaka, T., Field, R. W., and Harris, D. O., 1974b, *J. Chem. Phys.* **61**:3401.
Tanaka, T., Field, R. W., and Harris, D. O., 1975, *J. Mol. Spectry.* **56**:188.
Townes, C. H., and Schawlow, A., 1955, *Microwave Spectroscopy*, McGraw-Hill Book Company, New York.
Unland, M. L., and Flygare, W. H., 1966, *J. Chem. Phys.* **45**:2421.
Unland, M. L., Weiss, V., and Flygare, W. H., 1965, *J. Chem. Phys.* **42**:2138.
Webb, J. P., Webster, F. G., and Plourde, B. E., 1974, *Eastman Org. Chem. Bull.* **46**(3):1.
Wing, W. H., and Lamb, W. E., Jr., 1972, *Phys. Rev. Lett.* **28**:265.
Winnewisser, G., Shimizu, F. O., and Shimizu, T., 1972, Proc. VII Int. Quantum Electronics Conf., p. 87, Montreal, Quebec.
Wong, N.-B., Ikeda, T., and Harris, D. O., 1975, Paper TE-8, Thirtieth Symposium on Molecular Structure and Spectroscopy, Columbus, Ohio.
Wood, B. A., McNeish, A., Brignall, N. L., and Pidgeon, C. R., 1973, *Opt. Commun.* **8**:248.
Wood, O. R., and Schwarz, S. E., 1967, *Appl. Phys. Lett.* **11**:88.
Wood, O. R., Gordon, P. L., and Schwarz, S. E., 1969, *IEEE J. Quantum Electron.* **QE5**:502.
Wood, O. R., Burkhardt, E. G., Pollack, M. A., and Bridges, T. J., 1971, *Appl. Phys. Lett.* **18**:112.

Worlock, J. M., 1972, Two photon spectroscopy, in: *Laser Handbook* (F. T. Arecchi and E. O. Schulz-Dubois, eds.), North-Holland Publishing Co., Amsterdam.

Wu, F. Y., Grove, R. E., and Ezekiel, S., 1974, *Appl. Phys. Lettr.* **25**:73.

Yajima, T., 1961, *J. Phys. Soc. Japan* **16**:1709.

Yajima, T., and Shimoda, K., 1960, *J. Phys. Soc. Japan* **15**:1668.

Yardley, J. T., 1975, *Science* **190**:223.

Yardley, J. T., and Moore, C. B., 1963, *J. Chem. Phys.* **49**:1111.

Yariv, A., 1967, *Quantum Electronics*, John Wiley & Sons, Inc., New York.

Yogev, A., and Haas, Y., 1973, *Chem. Phys. Lett.* **21**:544.

Yuan, R. C. L., and Flynn, G. W., 1972, *J. Chem. Phys.* **57**:1316.

Yuan, R. C. L., and Flynn, G. W., 1973, *J. Chem. Phys.* **58**:649.

Yuan, R. C. L., Preses, J. M., Flynn, G. W., and Ronn, A. M., 1973, *J. Chem. Phys.* **59**: 6128.

Zito, R., Jr., and Schraeder, A. E., 1963, *Appl. Opt.* **2**:1323.

Coherent Transient Microwave Spectroscopy and Fourier Transform Methods

T. G. Schmalz and W. H. Flygare

2.1. Introduction

Transient experiments in gas-phase microwave spectroscopy of rotational transitions have developed rapidly in the past few years and show promise of remaining an area of ongoing theoretical and experimental research. The various phenomena involved are now sufficiently understood that it is possible to present a more-or-less unified treatment and comparison with the better-known transient experiments in magnetic resonance. Although isolated reports of the observation of microwave transient effects appear earlier in the literature (Dicke and Romer, 1955; Unland and Flygare, 1966; Hill *et al.*, 1967; Harrington, 1968), transient phenomena did not begin to attract widespread attention until about 1972 (Levy *et al.*, 1972; Macke and Glorieux, 1972, 1973, 1974, 1976; Wang *et al.*, 1973a; Brittain *et al.*, 1973; Amano and Shimizu, 1973; McGurk *et al.*, 1974b,c,d; Brown, 1974; Weatherly *et al.*, 1974; Dobbs *et al.*, 1975; Mäder *et al.*, 1975; Hoke *et al.*, 1975; Somers *et al.*, 1975; Coy, 1975).

Transient experiments normally involve observing the effects of bringing an ensemble of two-level quantum-mechanical systems into or out of resonance with high-power radiation in times short relative to the relaxation processes in the two-level system. Of course, transient experiments in

T. G. Schmalz and W. H. Flygare • Noyes Chemical Laboratory, University of Illinois, Urbana, Illinois

magnetic resonance have been familiar for two decades. Normally, the relaxation times in magnetic resonance experiments are 10^6–10^7 times longer than in gas-phase rotational systems. Thus, the radiation switching and speed of detection requirements in rotational-state spectroscopy are much more severe, which has led, in part, to the long delay in developing transient experiments in rotational microwave spectroscopy. The development of theory and experiment for rotational transient phenomena closely parallels that for vibrational transient phenomena (Hocker and Tang, 1969; Brewer and Shoemaker, 1971, 1972a,b; Loy, 1974), which are discussed in Chapter 3.

Transient experiments on rotation states are broadly classed as those involving coherent absorption and coherent spontaneous emission. These phenomena are most clearly defined in the limits where the radiation frequency is either within or outside the steady-state linewidth of a two-level transition. The steady-state linewidth includes the effects due to power saturation. For the most part, this discussion will be simplified to the normal case of negligible Doppler broadening in the steady-state rotational transition. *On-resonance* refers to the condition where the external radiation frequency is within the linewidth of the steady-state transition. *Off-resonance* refers to the condition where the radiation frequency is outside the linewidth of the steady-state transition. In these limiting regions *transient absorption* usually occurs immediately after the radiation has changed from the off-resonance to the on-resonance condition. *Transient emission* usually occurs immediately after the radiation has changed from the on-resonance to the off-resonance condition. In magnetic resonance, transient nutation is used to describe transient absorption, and free induction decay is used to describe spontaneous coherent transient emission. In the more complicated case where the radiation frequency is changed between two different points within the linewidth, the concepts of absorption and emission become less meaningful. However, in the limits described above, the descriptions are appropriate. This chapter will also describe *fast-passage* experiments in microwave spectroscopy, where a frequency is swept through a two-level resonance in a time short relative to the relaxation process.

In the following sections the theory necessary to understand the experiments which have been reported on microwave transients will be outlined. We also discuss the experimental progress in this field. The chapter begins with an examination of the measurement of transients and relates the measurement to the concepts of transient absorption and transient emission. A detailed description is next presented of the concepts of transient absorption and transient emission and the methods of extracting the relaxation times T_1 and T_2 from the measurements. T_1 is the relaxation time for the population difference in a two-level system, and T_2 is the relaxation of the radiation-induced polarization of the two-level system.

In the transient emission section the concept of $\pi/2$ and π pulses and the use of multiple-pulse experiments to measure T_1 and T_2 are also described. In all cases the analogies between the microwave experiments and the earlier measurements in nuclear magnetic resonance are emphasized. After that, fast-passage experiments, where a frequency is swept through a resonance in a short time relative to the relaxation times, are discussed. Two interesting limits in the fast-passage experiments are described, again with an attempt to compare with the more familiar experiments in nuclear magnetic resonance.

Next, an in-depth discussion is presented of the design and construction of a microwave Fourier transform spectrometer. The possible advantages of Fourier transform methods are discussed, including an analysis of the signal-to-noise gain which can be achieved. Examples are given of spectra obtained with the Fourier transform technique.

In the final section a microscopic description of T_1 and T_2 in terms of the rates of transfer of energy between rotational states is attempted. Using the concepts of state-changing and phase-changing collisions, it is shown what limits might be expected for the ratio of T_2 to T_1 for rotational states, and how these predictions correlate with the available experimental data.

2.2. Basic Theory and Experiment

The fundamental equations which govern all the transient phenomena we discuss in this chapter are the electric-dipole analogs of the Bloch equations of nmr (Bloch, 1946). The derivation of these equations has been given many times before (Feynman *et al.*, 1957; Wang *et al.*, 1973a; McGurk *et al.*, 1974a; Liu and Marcus, 1975a), including Chapter 1 of this text, so only the results will be summarized here (see also Appendix B for a more general result than given here). When a two-level quantum system interacts with an electric field of the form

$$E(t) = \varepsilon \cos(\omega t - kz) \tag{2-1}$$

with amplitude ε, frequency ω, and wave vector k, the motion of the Bloch vector, $\mathbf{r}$, is governed by the vector equation

$$\frac{d\mathbf{r}}{dt} = \boldsymbol{\omega} \times \mathbf{r} \tag{2-2}$$

in the absence of collisions. The Bloch vector $\mathbf{r}$ is given by the following combinations of density matrix elements:

$$\mathbf{r} = \begin{pmatrix} \rho_{ba} + \rho_{ab} \\ i(\rho_{ba} - \rho_{ab}) \\ \rho_{aa} - \rho_{bb} \end{pmatrix} \tag{2-3}$$

and ω is given in terms of the perturbation which couplies the lower state a and the upper state b, $V_{ab} = -\mu_{ab}\varepsilon \cos(\omega t - kz)$, plus the natural resonance frequency of the two-level system, $\omega_0 = (E_b - E_a)/\hbar$:

$$\boldsymbol{\omega} = \begin{vmatrix} (V_{ab} + V_{ba})/2\hbar \\ i(V_{ab} - V_{ba})/2\hbar \\ \omega_0 \end{vmatrix} \tag{2-4}$$

Upon transforming to an interaction representation (i.e., a reference frame rotating at frequency ω), in which the density matrix is defined by

$$\rho' = \exp\left[\frac{iS}{\hbar}\left(t - \frac{z}{c}\right)\right]\rho \exp\left[-\frac{iS}{\hbar}\left(t - \frac{z}{c}\right)\right] \tag{2-5}$$

where S is a diagonal matrix with eigenvalues E_a and $E_a + \hbar\omega$, and invoking the rotating wave approximation, which consists of dropping all high-frequency motions with respect to ω, Eq. (2-2) becomes

$$\frac{d\mathbf{r}'}{dt} = \boldsymbol{\omega}_{\text{eff}} \times \mathbf{r}' \tag{2-6}$$

with

$$\mathbf{r}' = \begin{vmatrix} \rho'_{ba} + \rho'_{ab} \\ i(\rho'_{ba} - \rho'_{ab}) \\ \rho'_{aa} - \rho'_{bb} \end{vmatrix} \tag{2-7}$$

and

$$\boldsymbol{\omega}_{\text{eff}} = \begin{vmatrix} -\omega_1 \\ 0 \\ \Delta\omega \end{vmatrix} \tag{2-8}$$

The effective strength of the interaction with the radiation is $\omega_1 = \mu_{ab}\varepsilon/\hbar$, where μ_{ab}, taken as real, is the dipole-moment matrix element coupling states a and b, and $\Delta\omega = \omega_0 - \omega$ is the difference between the resonant frequency of the system and the applied frequency.

Interactions between molecules are taken into account phenomenologically by adding linear damping terms to the right-hand side of Eq. (2-6), using some first-order relaxation theory for the density matrix, such as Redfield theory (Redfield, 1965; Liu and Marcus, 1975a). This gives, for the complete equation of motion of the system,

$$\frac{d\mathbf{r}'}{dt} = (\boldsymbol{\omega}_{\text{eff}} \times \mathbf{r}') - \frac{\hat{i}'r_1' + \hat{j}'r_2'}{T_2} - \frac{\hat{k}'(r_3' - \bar{r}_3')}{T_1} \tag{2-9}$$

where $\hat{i}', \hat{j}'$, and $\hat{k}'$ are unit vectors in the rotating frame and $\bar{r}_3'$ is the thermal equilibrium value of r'_3. It is Eq. (2-9) which must be solved under appropriate conditions to describe microwave transient experiments.

Although the Bloch vector–rotating frame formalism is convenient for deriving the basic equation of motion and for understanding the similarities among transient phenomena in various fields, we find it more convenient to work with physically real quantities when describing specific experiments. The physical quantities of interest in microwave transient experiments are the polarization, or macroscopic induced dipole moment, per unit volume and the population difference between levels a and b per unit volume (McGurk *et al.*, 1974a). The polarization is defined by

$$P = N \operatorname{Tr}(\mu\rho) \tag{2-10}$$

where N is the total number of molecules per unit volume, μ is the dipole moment operator, and Tr denotes the trace of a matrix representation. Expanding Eq. (2-10) using the transformation in Eq. (2-5) gives

$$P = N[\langle b|\mu|a\rangle\langle a|\rho'|b\rangle e^{i(\omega t - kz)} + \langle a|\mu|b\rangle\langle b|\rho'|a\rangle e^{-i(\omega t - kz)}] \tag{2-11}$$

Noting that the first term of Eq. (2-11) is the complex conjugate of the second term allows Eq. (2-11) to be rewritten as

$$P = (P_r + iP_i)e^{i(\omega t - kz)} + (P_r - iP_i)e^{-i(\omega t - kz)}$$
$$= 2P_r \cos(\omega t - kz) - 2P_i \sin(\omega t - kz) \tag{2-12}$$

Comparison of Eqs. (2-11) and (2-12) shows that

$$P_r = \frac{N\mu_{ab}}{2}(\rho'_{ab} - \rho'_{ba})$$
$$P_i = \frac{N\mu_{ab}}{2i}(\rho'_{ab} - \rho'_{ba}) \tag{2-13}$$

and comparison with Eq. (2-7) yields

$$P_r = \frac{N\mu_{ab}}{2}r'_1$$
$$P_i = \frac{N\mu_{ab}}{2}r'_2 \tag{2-14}$$

Thus, r'_1 and r'_2 are simply related to the real and imaginary (in-phase and out-of-phase) components of the macroscopic polarization.

The population difference is defined by

$$\Delta N = N(\rho_{aa} - \rho_{bb}) = N(\rho'_{aa} - \rho'_{bb}) \tag{2-15}$$

since the diagonal elements of the density matrix are not changed by the transformation in Eq. (2-5). Thus, it is readily apparent that

$$\Delta N = Nr_3' \tag{2-16}$$

Similarly,

$$\Delta N_0 = N\bar{r}_3' \tag{2-17}$$

is defined to be the equilibrium average of ΔN.

These definitions for P_r, P_i, and ΔN may now be substituted into Eq. (2-9) and the three vector components separated to give a set of three coupled first-order linear differential equations entirely in terms of physically observable quantities. Henceforward, the subscripts ab on μ are dropped. One has, then,

$$\frac{dP_r}{dt} + \Delta\omega P_i + \frac{P_r}{T_2} = 0$$

$$\frac{dP_i}{dt} - \Delta\omega P_r + \frac{1}{2}\mu\omega_1 \Delta N + \frac{P_i}{T_2} = 0 \tag{2-18}$$

$$\frac{d}{dt}\left(\frac{\mu}{2}\Delta N\right) + \omega_1 P_i + \frac{(\mu/2)(\Delta N - \Delta N_0)}{T_1} = 0$$

The roles of the phenomenological relaxation times T_1 and T_2 are now clear. The macroscopic polarization P, and hence the components P_r and P_i, relax to their equilibrium values of 0 with a relaxation time T_2. The population difference ΔN relaxes to its equilibrium value ΔN_0 with a relaxation time T_1.

In nmr T_1 is called the longitudinal or spin–lattice relaxation time (Abragam, 1961). It represents the rate at which energy is exchanged between the spin system and the surroundings. Since in microwave experiments energy is stored in the form of a population difference, the connection between these interpretations is clear. T_2 in nmr is called the transverse or spin–spin relaxation time (Abragam, 1961). It represents the rate at which the individual spins lose coherence. Very often in nmr the apparent T_2 contains a significant contribution from inhomogeneity in the magnetic field. If each spin feels a slightly different field at its local position, it will evolve in time slightly differently, soon leading to a loss of coherence. Many of the pulse methods developed in nmr are an attempt to overcome inhomogeneous broadening (Farrar and Becker, 1971). The analog of inhomogeneous broadening in microwave spectroscopy is Doppler broadening, where the interaction of each dipole with the field is slightly different, depending on its local velocity. In the experiments discussed here, Doppler broadening is usually negligible, so T_2 can be interpreted as a loss of coherent polarization due to molecular interactions.

Although Eqs. (2-18) are written entirely in terms of physically real quantities, none of them are, in fact, directly observed experimentally. No matter what the details of the experimental apparatus used to observe the transient phenomena, all eventually observe the current generated by some sort of nonlinear detector element. That current (in the square-law regime) is proportional to the square of the total electric field incident on the detector, averaged over all times faster than the response time of the detector. It is therefore necessary to calculate the electric field produced at the end of the sample cell due to the polarization of the sample.

To keep the analysis simple, we will at this point assume that the experiment is conducted in free space or in a waveguide far from cutoff. Then the group velocity will equal the phase velocity, and will be c, the speed of light, for any frequency wave. Consider a polarized microwave field of the form given in Eq. (1-1) traveling down the cell in the z direction and interacting with the gas. Without loss of generality, the field at the detector can be written as

$$E = 2\mathscr{E}_i \cos{(\omega t - kz)} + 2\mathscr{E}_r \sin{(\omega t - kz)} \tag{2-19}$$

where $k = 2\pi/\lambda$ is the wave vector and λ is the vacuum wavelength of the incident radiation. The coefficients $\mathscr{E}_i$ and $\mathscr{E}_r$ can be determined by solving the wave equation

$$\frac{\partial^2 E}{\partial z^2} = \frac{1}{c^2}\frac{\partial^2 E}{\partial t^2} + \frac{4\pi}{c^2}\frac{\partial^2 P}{\partial t^2} \tag{2-20}$$

subject to the boundary conditions at the end of the cell. Since it has already been assumed that the motions described by P_i and P_r are slow relative to ω, it is reasonable to adopt the slowly varying envelope approximation (Macomber, 1976), dropping terms in $\partial^2 P_i/\partial t^2$, $\partial^2 P_r/\partial t^2$, $\partial P_i/\partial t$, $\partial P_r/\partial t$, $\partial^2 \mathscr{E}_i/\partial z^2$, $\partial^2 \mathscr{E}_r/\partial z^2$, $\partial^2 \mathscr{E}_i/\partial t^2$, and $\partial^2 \mathscr{E}_r/\partial t^2$, which leads to the equations

$$\frac{\partial \mathscr{E}_i}{\partial z} + \frac{1}{c}\frac{\partial \mathscr{E}_i}{\partial t} = \frac{2\pi\omega}{c}P_i$$

$$\frac{\partial \mathscr{E}_r}{\partial z} + \frac{1}{c}\frac{\partial \mathscr{E}_r}{\partial t} = \frac{2\pi\omega}{c}P_r \tag{2-21}$$

Noting that dz/dt equals the velocity of the wave, here assumed to be c, allows one to use the chain rule to replace the left-hand sides of Eqs. (2-21) with total derivatives with respect to z. The equations can then be integrated along paths defined by $z = ct$.

To proceed exactly at this point, it would be necessary to replace t in Eqs. (2-21) with the retarded time $t' = t - z/c$, introducing a z dependence into the argument of P_r and P_i. However, if we restrict our attention to experiments conducted in sufficiently narrow bandwidths $\Delta\omega$ and over sufficiently short path lengths l that $\Delta\omega l/c \ll 2\pi$, which is almost always the case, we can approximate the z dependence of all electric fields by k, the wave vector of the incident field, and ignore the z dependence of the retarded time in P_r and P_i. Equations (2-21) may then be integrated directly subject to the boundary conditions $\mathscr{E}_i(0) = \varepsilon/2$ (the incident field) and $\mathscr{E}_r(0) = 0$. The result is (McGurk *et al.*, 1974a,c)

$$\mathscr{E}_i = \frac{\varepsilon}{2} + \frac{2\pi\omega l}{c} P_i$$

$$\mathscr{E}_r = \frac{2\pi\omega l}{c} P_r \tag{2-22}$$

where l is the length of the sample. The effect of approximating the z dependence of E by k is to ignore small phase differences in the emitted waves and thus to neglect phase cancellation effects which normally occur in Stark-switching experiments (the most common form of transient experiment). However, for typical experimental parameters, $\Delta\omega = 2\pi \times 5$ MHz and $l = 400$ cm, one has $\Delta\omega l/c = 0.42$, so that the approximation $\Delta\omega l/c \ll 2\pi$ is well satisfied and the phase shifts which are ignored are of no practical consequence. The treatment of Eqs. (2-21) when the approximations employed here are not valid will be explored in Section 2.6 in connection with pulse Fourier transform spectroscopy.

The signal at the detector is now given as

$$S = \frac{1}{2}\beta\left[\varepsilon^2 + 2\left(\frac{4\pi\omega l}{c}\right)\varepsilon P_i + \left(\frac{4\pi\omega l}{c}\right)^2 (P_i^2 + P_r^2)\right] \tag{2-23}$$

where β is the efficiency or conversion gain of the detector and all high frequencies have been averaged by the response characteristics of the diode. The first term is a dc or constant term representing the response of the detector to the incident microwave power. The last term is normally much smaller than the second term, so it is usual to ignore it and write the time-dependent response of the detector as

$$\Delta S(t) = \beta\left(\frac{4\pi\omega l}{c}\right)\varepsilon P_i \tag{2-24}$$

Thus, it is seen that the change in signal at the detector is directly proportional to the imaginary component of polarization, P_i. Equation (2-24) is valid no matter what the frequency offset between the resonance and the applied radiation field. The fact that the mathematical description of the signal is identical in all transient experiments should not, however, be allowed to obscure the different physics involved in different types of transient experiments.

If the microwave field is on-resonance, $\Delta S(t)$ is also proportional to the power absorbed by the molecules. For a nonsaturating field, P_i is negative, indicating that energy has been absorbed by the gas. However, in the course of transient absorption experiments, P_i may at times become positive, indicating that energy is being emitted by the gas. This phenomenon can be viewed as stimulated coherent emission or simply as negative absorption. It is an integral part of a transient nutation experiment. The steady-state value which P_i approaches is always negative, so that if the signal is averaged over the time of the transient nutation experiment, one always obtains the result that net absorption of energy by the gas has occurred. For this reason we refer to these experiments as transient absorption. Of course, absorption occurs even when the microwaves are not exactly on resonance. However, the magnitude of the absorption decreases as $\Delta \omega$ increases. When $\Delta \omega$ becomes sufficiently large and the molecules are initially polarized, the signal at the detector becomes dominated by spontaneous coherent emission.

Even for experiments conducted many half-widths off resonance, the signal at the detector continues to be given by Eq. (2-24) as long as the external microwave field is present. The signal at the detector will swing positive or negative as P_i becomes positive or negative. However, the time average of the signal may be either positive or negative, depending on the sign of the initial value of P_i. In these experiments it *does not* make sense to consider the periods of negative signal as absorption and those of positive signal as induced coherent emission. It is *not* correct to classify the experiment as net absorption or net emission on the basis of the average signal at the detector, because that average depends only on how one prepares the system initially.

Instead, these far off-resonance experiments following an initial polarization may be understood by considering the signal at the detector to result from the beat between the field emitted by the molecules and the microwave oscillator field. To understand this interpretation, consider what happens if the molecules are initially polarized and then the microwave field is turned off. The second term of Eq. (2-23), which normally dominates the signal, is zero. Instead, the signal is given by the third term of Eq. (2-23) as

$$\Delta S(t) = \frac{1}{2} \beta \left(\frac{4\pi\omega l}{c} \right)^2 (P_i^2 + P_r^2) \qquad (2\text{-}25)$$

This quantity is quite small; in fact, it has never been directly observed in microwave experiments. Nevertheless, it represents radiation by the molecules due to spontaneous coherent emission. Its value is always positive and it truly represents loss of energy by the molecules.

Consider now what happens if the microwaves are not turned off after the system is polarized but are instead conducted to the detector by a path external to the sample cell and then put directly onto the detector along with the spontaneous emission from the gas. According to Eqs. (2-19) and (2-22), the field at the end of the sample cell due to the emitting gas will be

$$E = 2\left(\frac{2\pi\omega l}{c}\right)P_i \cos(\omega t - kz) + 2\left(\frac{2\pi\omega l}{c}\right)P_r \sin(\omega t - kz) \quad (2\text{-}26)$$

while the microwave field will still be given by

$$E_m = \varepsilon \cos(\omega t - kz) \quad (2\text{-}27)$$

The signal at the detector will be proportional to the average of the square of the total field at the detector, i.e., to the square of the sum of E and E_m:

$$S = \frac{1}{2}\beta\left[\varepsilon^2 + 2\left(\frac{2\pi\omega l}{c}\right)\varepsilon P_i + \left(\frac{4\pi\omega l}{c}\right)^2(P_i^2 + P_r^2)\right] \quad (2\text{-}28)$$

Equation (2-28) is identical to Eq. (2-23). The time-dependent signal at the detector will be dominated by the second term and hence will still be given by Eq. (2-24). Yet quite clearly the value and sign of the detected signal have nothing to do with the absorption or emission of energy in the gas. Even when $\Delta S(t)$ is negative, it is quite clear that no absorption has taken place. The microwaves have not even passed through the gas! The signal at the detector is due entirely to the beat between the coherent spontaneously emitted field of the gas and the field of the microwave oscillator. Letting the microwaves pass through the gas does not change either the observed result or the interpretation (as long as the microwaves are sufficiently far off-resonance that negligible off-resonant absorption occurs). The observed results of these experiments, which are called free induction decays in nmr, we have termed transient emission since they result from a mixture of the emitted field of the molecules with the field of the external oscillator. Transient emission would continue to occur (though the observation would be different) even if the external field were absent. The conditions when off-resonance absorption is sufficiently small to make this interpretation valid have been investigated rather carefully and will be discussed later.

Before the different types of transient experiments are examined in more detail, the experimental methods which are common to most experiments will be discussed. The transient response (absorption or emission) can be observed either by bringing the microwave oscillator rapidly into or out of resonance with a two-level system or by using a fixed-frequency oscillator and using an electric field and the Stark effect to rapidly bring a two-level system into or out of resonance. A typical experimental apparatus for Stark switching is shown in Figure 2-1. The microwave source oscillator is phase-locked to a frequency synthesizer which allows any setting of the

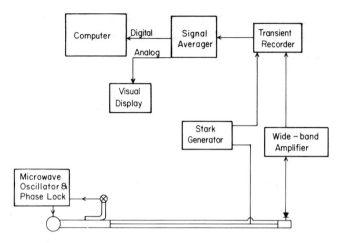

Figure 2-1. Block diagram of a typical apparatus for Stark-switching transient experiments on rotational states by microwave spectroscopy. The microwave field is held constant and the energy levels are switched in and out of resonance by the Stark field.

microwave frequency ν_M. The Stark generator applies high-voltage pulses of variable amplitude and duration to the Stark electrode in the waveguide. This Stark voltage gives rise to the Stark effect, where the zero-field resonance frequency is denoted as ν_0 and the resonance frequency of the two-level system with the Stark field on is denoted as ν_S. The Stark fields used normally have rise and fall times in the range 10–100 ns, considerably shorter than the relaxation times at normal pressures.

An alternative method of producing transients is to change rapidly the frequency of the oscillator into or out of resonance. While this method is useful for some experiments, it is difficult to control the accuracy of the starting and stopping frequencies. A more satisfactory zero- or static-field technique is to pass the microwave through a fast diode switch which is capable of 60- to 80-db attenuation with a switching time on the order of 1 ns and a pulse duration lower limit of 10–20 ns. The switches will handle relatively high powers. However, currently available switches will not work above 18 GHz.

After producing the transient effect with Stark switching or diode switching, the transient response is detected in a diode detector and the resultant signal is amplified. Care must be exercised that the detector–amplifier combination is broad–band enough to pass, with high fidelity, the transient signals. There is no problem in using a sufficiently wide banded amplifier. However, the silicon crystal diode detectors, which are normally used, have limitations that are troublesome in passing the transient signals. Usually, one must use a square-wave-modulated microwave to test Si diode detectors and selectively choose one which has a bandwidth adequate for the transient signals.

Passing on now to the recorder part of Figure 2-1, we note that commercially available digital transient recorders offer time bases as short as 10 ns with 8- to 10-bit recording accuracy over each of 512 to 1028 points on the time scale (see Sections 1.3 and 2.4). These recorders, which record a single transient signal, then require about a 20-ms period to transfer the data to a storage-averaging device. Thus, the sensitivity of the apparatus is limited by the relatively long transfer time between the transient recorder and the storage averager. Another method of recording data is to replace the transient recorder with a commercial boxcar integrator, which also can give time resolution of 10 ns with a large amplitude of dynamic range. Boxcar integrators record only a single time window of minimum period (10 ns to 1 s) on each transient experiment. The time window then is stepped forward in each successive experiment to obtain the complete transient signal. A boxcar integrator using a 10-ns window with 1000-point capability would require at least $1000 \times (10 \text{ ns}) \times 1000 = 10 \text{ ms}$ to make a single recording. This is somewhat more efficient than the transient recorder, but both instruments lack real time-recording capability.

Both the transient recorder (digital) and boxcar integrator (analog) are inherently slow because they have such a large dynamic range capability in the recorded signal. This works well with strong signals, where the signal-to-noise ratio of a single transient is large and real time averaging is not needed. However, if the signal is buried in the noise, real time averaging may become essential. Recent experiments have made use of analog-to-digital converter–averager devices which use only 1-bit accuracy on a single transient. This technique sacrifices no information if the signal is buried in the noise. The 1-bit transient recorder can be integrated in a manner which allows near-real-time averaging and significant improvements in the signal-to-noise ratio. The data can be transferred to a computer for Fourier transforming, further averaging, or other data analysis.

2.3. Transient Absorption

Transient absorption experiments are initially the easiest to understand because of their close relationship to conventional steady-state microwave spectroscopy. Steady-state absorption experiments differ from transient experiments only in that the radiation is brought slowly into resonance. Steady-state absorption is usually described in terms of the absorption coefficient, defined by

$$I = I_0 e^{-\gamma l} \cong I_0 (1 - \gamma l) \tag{2-29}$$

where γ is the absorption coefficient, I is the outgoing radiation intensity, I_0 is the incident intensity, and l is the path length. The right-hand relation in Eq. (2-29) assumes that the medium is weakly absorbing. Using Eqs. (2-19) and (2-22) for the electric field at the detector and Eq. (2-1) for the incident electric field, and neglecting the small terms in P_i^2 and P_r^2 allows one to solve Eq. (2-29) for γ in terms of P_i. The result is

$$\gamma = -\frac{8\pi\omega}{c}\frac{P_i}{\varepsilon} \tag{2-30}$$

so that γ, P_i, and ΔS [given in Eq. (2-24)] give equivalent descriptions of the absorption experiment.

In normal microwave spectroscopy, the frequency of the incident oscillator is swept through the resonance in a period of time long relative to the relaxation processes. The gas is then always in equilibrium with the radiation. Under these conditions, Eqs. (2-18) can be solved by setting $dP_r/dt = dP_i/dt = d \, \Delta N/dt = 0$, giving

$$P_r = \frac{\frac{1}{2}\mu\omega_1(\Delta\omega)\,\Delta N_0}{(1/T_2)^2 + (T_1/T_2)\omega_1^2 + (\Delta\omega)^2}$$

$$P_i = \frac{-\frac{1}{2}\mu\omega_1(1/T_2)\,\Delta N_0}{(1/T_2)^2 + (T_1/T_2)\omega_1^2 + (\Delta\omega)^2} \tag{2-31}$$

$$\Delta N = \frac{\Delta N_0[(1/T_2)^2 + (\Delta\omega)^2]}{(1/T_2)^2 + (T_1/T_2)\omega_1^2 + (\Delta\omega)^2}$$

P_i in Eq. (2-31) gives the standard Lorentzian line shape for the absorption coefficient of Eq. (2-30) with the half-width at half-height given by

$$\Delta\nu_{1/2} = \frac{\Omega_p}{2\pi} = \frac{1}{2\pi}[(1/T_2)^2 + (T_1/T_2)\omega_1^2]^{1/2} \tag{2-32}$$

where $\Omega_p/2\pi$ indicates the power-saturated linewidth. If low power radiation is used, $(1/T_2)^2 \gg (T_1/T_2)\omega_1^2$, and the normal half-width at half-height of the Lorentzian (Townes and Schawlow, 1955) is obtained, $\Delta\nu_{1/2} = (1/2\pi)(1/T_2)$.

The most common form of the transient absorption experiments is the on-resonant case, where a two-level system is brought into resonance, $\Delta\omega = 0$, in a time short relative to the relaxation processes. When $\Delta\omega = 0$, Eqs. (2-18) can be solved exactly (McGurk *et al.*, 1974a) giving [see Appendix A for general methods of solving Eqs. (2-18)]

$$P_r(t) = P_r(t_i)e^{-(t-t_i)/T_2}$$

$$P_i(t) = \frac{\mu\omega_1\,\Delta N_0}{2}(1/\Omega_p^2)\{-(1/T_2)+e^{-(t-t_i)/T}$$

$$\times\{(1/T_2)\cos\Omega_0(t-t_i)-[(T_1/T_2)(\omega_1^2/\Omega_0)$$

$$+(1/T_2)(\phi/\Omega_0)]\sin\Omega_0(t-t_i)\}\}$$ (2-33)

$$+e^{-(t-t_i)/T}\left\{P_i(t_i)\left[\cos\Omega_0(t-t_i)-\frac{\phi}{\Omega_0}\sin\Omega_0(t-t_i)\right]\right.$$

$$\left.-\frac{\mu\omega_1}{2\Omega_0}[\Delta N(t_i)-\Delta N_0]\sin\Omega_0(t-t_i)\right\}$$

$$\Delta N(t) =$$

$$\Delta N_0\frac{(T_1/T_2)\omega_1^2 e^{-(t-t_i)/T}[\cos\Omega_0(t-t_i)+(1/\Omega_0 T)\sin\Omega_0(t-t_i)+(1/T_2)^2]}{\Omega_p^2}$$

$$+e^{-(t-t_i)/T}\left\{P_i(t_i)(2\omega_1/\mu\Omega_0)\sin\Omega_0(t-t_i)\right.$$

$$\left.+[\Delta N(t_i)-\Delta N_0]\left[\cos\Omega_0(t-t_i)+\frac{\phi}{\Omega_0}\sin\Omega_0(t-t_i)\right]\right\}$$

where $P_r(t_i)$, $P_i(t_i)$, and $\Delta N(t_i)$ contain the appropriate initial conditions at the initial time, t_i, where $t_i \leq t$ and

$$\frac{1}{T} = \tfrac{1}{2}(1/T_1+1/T_2)$$

$$\phi = \tfrac{1}{2}(1/T_2-1/T_1)$$ (2-34)

$$\Omega_0 = [\omega_1^2-\tfrac{1}{4}(1/T_2-1/T_1)^2]^{1/2}$$

Ω_p is defined in Eq. (2-32). When $\omega_1^2 < \tfrac{1}{4}(1/T_2-1/T_1)^2$, Eqs. (2-33) must be modified by replacing the sine and cosine functions with hyperbolic functions and setting $\Omega_0 = [\tfrac{1}{4}(1/T_2-1/T_1)^2-\omega_1^2]^{1/2}$ (McGurk *et al.*, 1974a). Equations (2-33) are general and only require the initial conditions to completely specify the behavior of the system. Normally, one finds that $1/T_1 \cong 1/T_2$, and therefore it is expected that

$$\Omega_0 \cong \omega_1$$ (2-35)

Usually, the initial conditions (before the on-resonant absorption) are obtained from the far off-resonance limit of the steady-state results in Eqs. (2-31), where $(\Delta\omega)^2 \gg (1/T_2)^2$ and $(\Delta\omega)^2 \gg (T_1/T_2)\omega_1^2$. Under these initial conditions one has $P_r(t_i=0) = P_i(t_i=0) = 0$ and $\Delta N(t_i=0) = \Delta N_0$ from Eqs. (2-31). Substituting these results and the condition in Eq. (2-35) into Eqs.

(2-33) gives

$$0 \le t;\ \Delta\omega = 0,\ P_r(0) = P_i(0) = 0,\ \Delta N(0) = \Delta N_0:$$

$$P_r(t) = 0$$

$$P_i(t) = \frac{\mu\omega_1 \Delta N_0}{2}$$

$$\times \frac{e^{-t/T}[(1/T_2)\cos\omega_1 t - [(T_1/T_2)\omega_1 + (1/T_2)(\phi/\omega_1)]\sin\omega_1 t] - 1/T_2}{\Omega_p{}^2}$$

$$\Delta N(t) = \Delta N_0 \frac{(T_1/T_2)\omega_1{}^2 e^{-t/T}[\cos\omega_1 t + (1/\omega_1 T)\sin\omega_1 t] + (1/T_2)^2}{\Omega_p{}^2}$$

$$(2\text{-}36)$$

Figure 2-2 illustrates schematically the relationship between the absorption coefficient given by Eq. (2-30) with the second of Eqs. (2-36) for P_i and the molecular-energy-level system. The low-power, long-time limit of Eqs. (2-36) gives back Eqs. (2-31), as, of course, it should. At low power the approach to equilibrium is smooth. But at higher powers, the $\sin\omega_1 t$ and $\cos\omega_1 t$ terms in Eq. (2-36) become important, and oscillations occur in the approach to equilibrium. At long times the absorption coefficient settles down to some constant value, but its magnitude is significantly reduced at higher powers (power saturation).

Of equal interest is the behavior of the absorption coefficient when the interaction with the radiation is stopped before this constant long-time value is reached. Under the conditions of high-power radiation, where

$$\omega_1 T_1 \gg 1 \qquad \omega_1 T_2 \gg 1 \qquad\qquad (2\text{-}37)$$

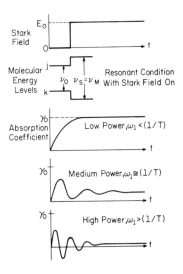

Figure 2-2. Schematic showing the transient effects in the absorption of radiation when a Stark field is used to bring two levels which are initially far off-resonance suddenly into resonance with the external oscillator with frequency ν_M. ν_0 indicates the zero-field resonance frequency and ν_S indicates the Stark-field resonance frequency which in this experiment is equal to the external oscillator frequency ν_M. The oscillations in the absorption coefficient are proportional to the power factor, ω_1, as shown in Eq. (2-36). γ_0 is the long time limit of Eq. (2-36). Note the lower steady-state absorption ($t \to \infty$) with increasing radiation power.

Eqs. (2-36) reduce to

$$P_r(t) = 0$$

$$P_i(t) = -\frac{\mu}{2} \Delta N_0 e^{-t/T} \sin \omega_1 t \qquad (2\text{-}38)$$

$$\Delta N(t) = \Delta N_0 e^{-t/T} \cos \omega_1 t$$

And, if the interaction is terminated in a time short relative to the relaxation time,

$$t \ll T \qquad (2\text{-}39)$$

Eqs. (2-38) reduce to the normal equations that are used to describe π and $\pi/2$ pulse experiments:

$$P_r(t) = 0$$

$$P_i(t) = -\frac{\mu}{2} \Delta N_0 \sin \omega_1 t \qquad (2\text{-}40)$$

$$\Delta N(t) = \Delta N_0 \cos \omega_1 t$$

The $\pi/2$ pulse, where

$$\omega_1 t = \pi/2 \qquad (2\text{-}41)$$

leads to maximum imaginary polarization, P_i, and zero population difference between the two levels.

$$P_r(t_{\pi/2}) = 0$$

$$P_i(t_{\pi/2}) = -\frac{\mu}{2} \Delta N_0 \qquad (2\text{-}42)$$

$$\Delta N(t_{\pi/2}) = 0$$

At a later time we have the π pulse condition, where

$$\omega_1 t = \pi \qquad (2\text{-}43)$$

which leads to zero polarization and a population inversion,

$$P_r(t_\pi) = 0$$

$$P_i(t_\pi) = 0 \qquad (2\text{-}44)$$

$$\Delta N(t_\pi) = -\Delta N_0$$

The $\pi/2$ and π pulse conditions will be discussed shortly; however, first the solutions to Eqs. (2-18) for off-resonance conditions will be examined. Whenever the conditions

$$\frac{1}{\omega_1}(1/T_2 - 1/T_1) \ll 1 + \frac{(\Delta\omega)^2}{\omega_1^2}$$

$$\frac{1}{\omega_1^2}(1/T_2 - 1/T_1)^2 \ll 1 + \frac{(\Delta\omega)^2}{\omega_1^2}$$

(2-45)

are satisfied, Eqs. (2-18) can be solved approximately, as shown in Appendix A, by dropping all terms of order $\omega_1(1/T_2 - 1/T_1)/[\omega_1^2 + (\Delta\omega)^2]$ and higher order with respect to 1. As mentioned before, it will be shown that normally $T_1 \cong T_2$ for rotational systems and the conditions in Eq. (2-45) are expected to be satisfied even for small $\Delta\omega$. The resultant solution to Eqs. (2-18) for $P_i(t)$ takes the form

$$
\begin{aligned}
P_i(t) = &-\frac{\mu\omega_1 \Delta N_0}{2}\frac{1/T_2}{\Omega_p^2 + (\Delta\omega)^2}\\
&+\frac{\mu\omega_1 \Delta N_0}{2}\frac{(\Delta\omega)^2(1/T_2 - 1/T_1)}{\Omega_1^2[(T_1/T_2)\omega_1^2 + (\Delta\omega)^2]}\\
&\times \exp\left[-\frac{(1/T_2)\omega_1^2 + (1/T_1)\Delta\omega^2}{\Omega_1^2}(t - t_1)\right]\\
&+\left\{\left[P_i(t_1) + \frac{\mu\omega_1 \Delta N_0}{2}\frac{1/T_2}{\Omega_p^2 + (\Delta\omega)^2} - \frac{\mu\omega_1 \Delta N_0}{2}\right.\right.\\
&\times\left.\frac{(\Delta\omega)^2(1/T_2 - 1/T_1)}{\Omega_1^2[(T_1/T_2)\omega_1^2 + (\Delta\omega)^2]}\right]\cos\Omega_1(t - t_1)\\
&+\left[\Delta\omega P_r(t_1) - \frac{\mu\omega_1(\Delta N(t_1) - \Delta N_0)}{2} - \frac{\mu\omega_1 \Delta N_0}{2}\frac{(T_1/T_2)\omega_1^2 + (\Delta\omega)^2}{\Omega_p^2 + (\Delta\omega)^2}\right.\\
&\left.\left.-\frac{\mu\omega_1 \Delta N_0}{2}\frac{(1/T_2 - 1/T_2)^2(\Delta\omega)^4}{\Omega_1^4[(T_1/T_2)\omega_1^2 + (\Delta\omega)^2]}\right]\Omega_1^{-1}\sin\Omega_1(t - t_1)\right\}\\
&\times\exp\left[-\frac{(\Delta\omega)^2(1/T_2)}{\Omega_1^2}(t - t_1)\right]\exp\left[-\frac{(\omega_1)^2\frac{1}{2}(1/T_2 + 1/T_1)}{\Omega_1^2}(t - t_1)\right]
\end{aligned}
$$

with

$$\Omega_1 = [\omega_1^2 + (\Delta\omega)^2]^{1/2}$$

(2-46)

Just as Eqs. (2-33) become simpler when the initial conditions and details of the observation are considered, Eq. (2-46) may also be simplified in certain important cases. If one considers the off-resonance transient absorption case described by Eq. (2-46) under the far off-resonance initial conditions and assuming $\Delta\omega \gg \Omega_p$, the largest term is

$$\Delta\omega \gg \Omega_p, \quad P_r(t_i = 0) = P_i(t_1 = 0) = 0, \quad \Delta N(t_1 = 0) = \Delta N_0:$$

$$P_i(t) = -\frac{\mu\omega_1 \Delta N_0}{2\,\Delta\omega} e^{-t/T_2} \sin \Delta\omega t \qquad (2\text{-}47)$$

Equation (2-47) predicts an oscillatory approach to equilibrium. The period of the oscillation increases and the amplitude decreases with increasing frequency separation from resonance. This behavior should be contrasted with the behavior of the equations describing emission which are derived in the next section.

The off-resonant solutions for $\Delta N(t)$ and $P_r(t)$ have not been included because they are long and complicated and normally not needed. Any off-resonant observation will only involve $P_i(t)$. The need for transient off-resonant initial conditions is probably small. Most interesting experiments involve off-resonant steady-state initial conditions or on-resonant transient initial conditions. However, if off-resonant transient initial conditions are needed for $P_r(t)$ and $\Delta N(t)$, the equations corresponding to $P_i(t)$ above can be easily derived using Eq. (2-45) and the methods outlined in Appendix A. Equation (2-46) can be reduced to the second of Eqs. (2-33) when $\Delta\omega$ is equal to zero if one recognizes that the condition in Eq. (2-45) allows one to neglect ϕ in Eq. (2-34) with respect to ω_1 and to set Ω_0 in Eq. (2-34) equal to ω_1. Equation (2-46) also reduces correctly to the exact solution of Eqs. (2-18), which can be found when $T_1 = T_2$ (McGurk *et al.*, 1974a). The $T_1 = T_2$ result for $P_i(t)$, valid for any $\Delta\omega$, under the initial conditions of $P_r(0) = P_i(0) = 0$ and $\Delta N(0) = \Delta N_0$ is given below:

$$0 \leq t, \quad \Delta\omega \neq 0, \quad T_1 = T_2, P_r(0) = P_i(0) = 0, \quad \Delta N(0) = \Delta N_0:$$

$$P_i(t) = \frac{\mu\omega_1 \Delta N_0}{2} \frac{e^{-t/T_2}[(1/T_2)\cos\Omega_1 t - \Omega_1 \sin\Omega_1 t] - 1/T_2}{(1/T_2)^2 + (\Omega_1)^2} \qquad (2\text{-}48)$$

Equation (2-48) provides the simplest description of the system when T_1 is known to be equal to T_2, but in practice it is not so useful, because one is often interested in measuring T_1 and T_2 to find out if they are, in fact, equal. Nevertheless, a detailed analysis of the absorption coefficient given by Eq. (2-30) with Eq. (2-48) for P_i will now be presented because Eq. (2-48) most clearly illustrates the important features of transient absorption. Figure 2-3 shows plots of the absorption coefficient for various values of $\Omega_1 = [\omega_1{}^2 + (\Delta\omega)^2]^{1/2}$. The general features of the absorption coefficient when $\Omega_1 \ll 1/T_2$ show a gradual increase which levels off at the steady-state result, as shown in Figures 2-3a and b. Then, as $\Omega_1 \cong 1/T_2$, the sinusoidal character

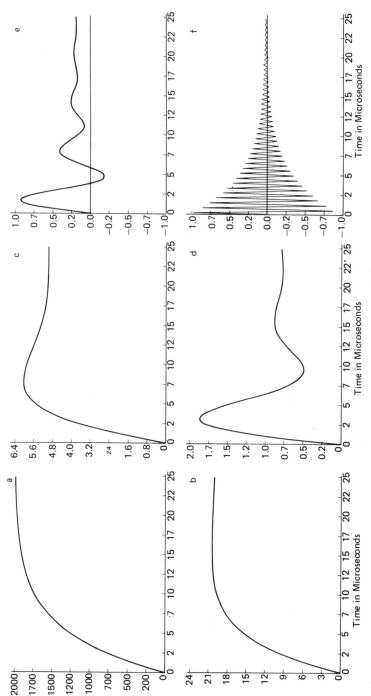

Figure 2-3. Plots of the transient absorption coefficient in Eq. (2-30) as a function of time using $P_i(t)$ from Eq. (2-48). The initial conditions are far off-resonance and the time $t = 0$ indicates the time when the two-level system is brought into resonance with the microwave field. The figures from a through f show increasing oscillations due to increasing $\Delta\omega$ or ω_1 in $\Omega_1 = [\omega_1^2 + (\Delta\omega)^2]^{1/2}$

begins to show up in Figures 2-3c and d. Finally, as $\Omega_1 > 1/T_2$, stronger oscillations take place and the absorption coefficient swings negative as shown in Figures 2-3e and f. These periods of negative absorption, or emission, return to positive values of absorption again and the final result levels off to a steady-state absorption at a considerably lower level than in Figure 2-3a or b. Figures 2-3e and f show clearly the onset of the conditions for $\pi/2$ and π pulses. The $\pi/2$ pulse described in Eqs. (2-42) is achieved at the first maximum in $P_i(t)$, and the π pulse described in Eqs. (2-44) is achieved the first time the $P_i(t)$ function achieves the $P_i(t) = 0$ condition. Of course, rigorously, Eqs. (2-37) and (2-39) must be satisfied for $\pi/2$ and π pulses, which requires a response similar to Figure 2-3f to obtain true $\pi/2$ and π excitation.

Notice that when $T_1 = T_2$, nutation may be induced either by staying on-resonance and increasing the power (ω_1) or by keeping the power constant and moving off-resonance. When T_1 is not equal to T_2, P_i must be given by either Eq. (2-33) or (2-46). The behavior of the absorption coefficient will in either case be qualitatively similar to that shown in Figure 2-3. As already shown in Figure 2-2, with Eq. (2-33) describing on-resonance absorption, nutation will result from increasing the power. With Eq. (2-46) describing off-resonance absorption, nutation may result either from increasing the power or from increasing the frequency offset beyond the steady-state low-power linewidth, $1/T_2$.

Experimental on-resonant transient absorption signals with off-resonance steady-state initial conditions are shown in Figure 2-4. The results in Figure 2-4 were fit with Eqs. (2-24) and (2-33) to obtain values of T_1 and T_2 for the $J = 0 \rightarrow J = 1$ rotational transition in the OCS molecule (McGurk et al., 1974b). Typical off-resonant transient absorption signals at frequencies of 1 and 2 MHz are shown in Figure 2-5. The decreased amplitude of the higher-frequency off-resonant signal is evident (2 MHz relative to 1 MHz). Of course, P_i in Eq. (2-47) goes to zero as $\Delta\omega \rightarrow \infty$. We also note the important result that the condition in Eq. (2-47), $\Delta\omega \gg \Omega_p$, is sufficient to ensure that the decay of the signal depends only on T_2. Transient nutation in nmr was first studied by Torrey (1949), who demonstrated the "wiggles" in the approach to equilibrium and the variation of the signal as one moves off-resonance. The solutions obtained in this section are formally identical to those of Torrey, and the experimental effects are in all cases exactly analogous.

One note of caution should be added at this point. For experiments conducted in waveguides, the amplitude of the electric field is not constant across the sample. To use any of the formulas in this section, it is necessary to replace ω_1 (which depends on ε) with the appropriate functional form and integrate over the field distribution in the cell (Mäder et al., 1975).

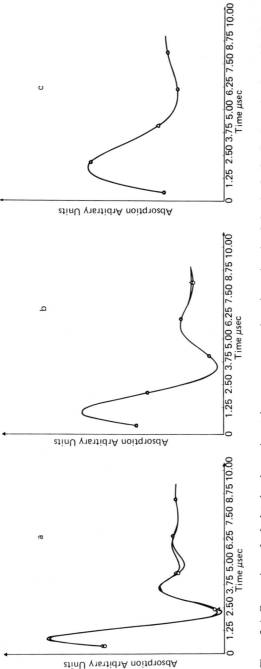

Figure 2-4. Comparison of calculated and experimental on-resonant transient absorption signals for the $J = 0 \to J = 1$ rotational transition in pure OCS. The observed and calculated absorption coefficients [from Eq. (2-30)] are plotted here with $P_i(t)$ from Eqs. (2-33). According to $P_i(t)$ in Eqs. (2-33), the observed response contains a combination of T_1 and T_2 processes and both relaxation times were extracted from the data (see McGurk *et al.*, 1974b).

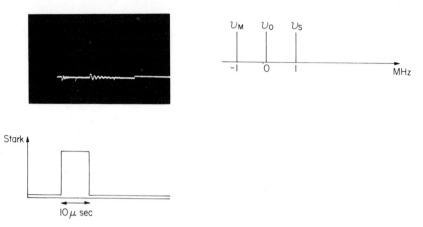

Figure 2-5. Off-resonant transient absorption with the far off-resonant initial conditions $P_r(0) = P_i(0) = 0$, $\Delta N(0) = \Delta N_0$ in the $J = 0 \rightarrow J = 1$ rotational transition in OCS. In this experiment the microwave oscillator is set at ν_M and the energy levels are Stark-switched between ν_0 (Stark field off) and ν_S (Stark field on). When the Stark field is turned on, a 2-MHz off-resonant transient absorption signal is observed. When the Stark field is switched off, a 1-MHz off-resonant transient absorption signal is observed. These off-resonant absorption signals have amplitudes which fall off with $\Delta\omega$ and the signals decay with the T_2 relaxation time according to Eq. (2-47) (from McGurk *et al.*, 1974c).

2.4. Transient Emission

In our previous discussion, the signal at the detector was examined when the initial conditions were far off-resonance where no initial polarization or change in equilibrium population is obtained [$P_r(0) = P_i(0) = 0$ and $\Delta N(0) = \Delta N_0$). In all cases, a net decrease of signal at the detector or a net absorption of radiation energy was found and we referred to the phenomena as transient absorption. In this section initial conditions in which the system is polarized and the population difference is driven to a nonequilibrium condition will be examined. Under these conditions, there is spontaneous coherent emission of radiation by the system. We call this phenomenon transient emission. There is also the possibility of creating the initial conditions with a different oscillator with different power and frequency than is used for the subsequent transient experiments. However, for the Stark switching methods discussed here, the same oscillator with constant frequency and power is used to produce the initial conditions and is also present during the observation of the subsequent transient effects. The change in the resonant or off-resonant condition is obtained by Stark-switching the molecular-energy levels.

There are a large number of initial conditions and subsequent observing situations. If the initial conditions are created off-resonance in steady state with a high-power oscillator and the subsequent observations are made on-resonance, the observed signal at the detector is obtained by substituting $P_i(t)$ in Eq. (2-33) into Eq. (2-24) using Eqs. (2-31) for the initial conditions. If the subsequent on-resonance observation is made with a lower-power oscillator, the terms involving $P_r(t_i)$, $P_i(t_i)$, and $\Delta N(t_i)$ in $P_i(t)$ in Eq. (2-33) may be more important than the first terms leading to a signal dominated by emission.

A more interesting and much more common situation is to create the polarization on-resonance and observe the subsequent transient effects with the oscillator off-resonance. Figure 2-6 presents schematically the energy-level scheme and types of signals encountered in these experiments. Equation (2-46), which was derived for off-resonant absorption, can be used in Eq. (2-24) to predict the signal at the detector in this experiment as well as long as the condition in Eq. (2-45) is satisfied. This is because the Bloch equations show no formal or mathematical difference when emission or absorption is being described. We emphasize again that this does not mean that the physics is identical in the two experiments. Different terms in

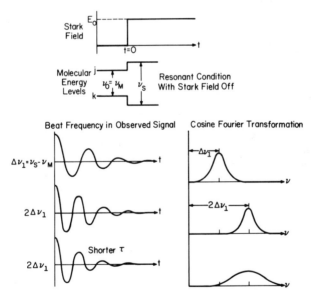

Figure 2-6. Schematic showing the transient effects observed during spontaneous coherent emission of radiation after a Stark field is used to turn off the resonance interaction. ν_0 is the zero-field resonance frequency equal in this experiment to the external oscillator frequency ν_M. The molecules undergo coherent emission at ν_S, the resonant frequency with the Stark field on. The response curves are beats or difference frequencies between the emitting frequency ν_S and the reference oscillator at $\nu_0 = \nu_M$. The response curves are described by Eq. (2-50) and the on-resonant steady-state initial conditions on $P_i(0)$ from Eqs. (2-31). The corresponding cosine Fourier transforms are also shown. τ is the relaxation time.

Eq. (2-46) are dominant in the two regimes, and their interpretation is quite different. Figure 2-6 also shows the Fourier transforms of the emission signals, which will be discussed in more detail in Section 2.6.

Equation (2-46) is rather complicated as it stands. However, it may be simplified by noting that Eq. (2-45) may be satisfied in three ways. It will be satisfied when $\Delta\omega$ is made very large, as is often the case in emission experiments, and it will be satisfied if T_1 is nearly equal to T_2, as is often the case in studying rotational states. It may also be satisfied if ω_1 is much larger than $1/T_2$ or $1/T_1$. The latter case is not a very useful one. If we ignore this case, all the terms in $(1/T_2 - 1/T_1)$ in Eq. (2-46) will be negligible. Equation (2-46) may then be considerably simplified to

$$P_i(t) = \frac{\mu\omega_1 \Delta N_0}{2}[\Omega_p^2 + (\Delta\omega)^2]^{-1}\{-(1/T_2) + e^{-(t-t_i)/T_2}$$

$$\times\{(1/T_2)\cos\Omega_1(t-t_i) - [(T_1/T_2)(\omega_1^2/\Omega_1) + (\Delta\omega)^2/\Omega_1]$$

$$\times\sin\Omega_1(t-t_i)\}\}$$

$$+ e^{-(t-t_i)/T_2}\left\{P_i(t_i)\cos\Omega_1(t-t_i) + \left[\frac{\Delta\omega}{\Omega_1}P_r(t_i)\right.\right.$$

$$\left.\left. - \frac{\mu\omega_1}{2}(\Delta N(t_i) - \Delta N_0)\frac{1}{\Omega_1}\right]\sin\Omega_1(t-t_i)\right\} \tag{2-49}$$

Equation (2-49) is equivalent to Eq. (12) of McGurk *et al.* (1974d). It should be remembered that Eq. (2-49) requires either that $(\Delta\omega)^2 \gg \omega_1^2$ and $(\Delta\omega)^2 \gg (1/T_2)^2$, or that $T_1 \cong T_2$. Equation (2-49) may then be used along with Eq. (2-24) and the on-resonance steady-state initial conditions given in Eq. (2-31), the on-resonance transient initial conditions given in Eq. (2-36), or the $\pi/2$, π pulse initial conditions given in Eqs. (2-42) and (2-44) to predict the signal at the detector.

Substituting first the on-resonance steady-state initial conditions obtained from eqs (2-31) into Eq. (2-49) and setting $t_i = 0$ gives

$$P_i(t) = \frac{\mu\omega_1 \Delta N_0}{2}\left\{\frac{e^{-t/T_2}[(1/T_2)\cos\Omega_1 t - (\bar{\Omega}_1^2/\Omega_1)\sin\Omega_1 t] - 1/T_2}{\Omega_p^2 + (\Delta\omega)^2}\right.$$

$$\left. + \frac{e^{-t/T_2}[(T_1/T_2)(\omega_1^2/\Omega_1)\sin\Omega_1 t - (1/T_2)\cos\Omega_1 t]}{\Omega_p^2}\right\} \tag{2-50}$$

$$\bar{\Omega}_1 = [(T_1/T_2)\omega_1^2 + (\Delta\omega)^2]^{1/2}$$

Ω_1 is defined in Eq. (2-46) and Ω_p is defined in Eq. (2-32).

The first term in brackets corresponds to the off-resonant transient absorption with $P_i(0) = P_r(0) = 0$ and $\Delta N(0) = \Delta N_0$ initial conditions considered previously beginning with Eq. (2-46). The second term in brackets represents the onset of the transient emission terms. As we move farther

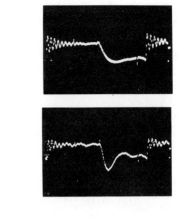

Figure 2-7. On-resonant transient absorption (Stark field off) and transient emission (Stark field on) in the $J = 0 \rightarrow J = 1$ rotational transition of OCS. The lower of the two curves is taken at a higher microwave power. On-resonant transient absorption is also shown in Figure 2-4 and the signals shown here are described by Eqs. (2-24) and (2-36). The off-resonant coherent transient emission signal taken with the Stark field on is described by Eq. (2-24) and $P_i(t)$ for transient emission in Eq. (2-51) (from McGurk *et al.*, 1974c).

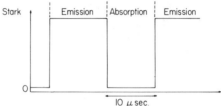

off-resonance with our external oscillator (during the period of observation), the second term in Eq. (2-50), representing emission, will start to exceed the first term, representing absorption.

In the limit of far off-resonance, only the second part of the second term remains, giving

$$P_i(t) = -\frac{\mu\omega_1 \Delta N_0}{2T_2\Omega_p^2}e^{-t/T_2}\cos\Omega_1 t \qquad (2\text{-}51)$$

This is the result for far off-resonance transient emission following on-resonance steady state initial conditions. Note that the amplitude of this term is independent of $\Delta\omega$. An example of this case is given in Figure 2-7, which shows transient absorption when the Stark field is off and transient emission when the Stark field is on. When the Stark field is on, the two-level system is far off-resonance. After the Stark field is switched off, the two-level system is brought into resonance, with the microwave radiation field leading to a decrease in signal at the detector. This is the on-resonant transient absorption described by the signal in Eq. (2-24), with $P_i(t)$ in Eq. (2-36) as also shown in Figure 2-4. After several relaxation times, the absorption levels off to the steady-state value. These $t > T_2$ on-resonant steady-state absorption conditions provide the initial conditions for $P_i(t)$ at the point when the Stark field is switched on and the two-level system is taken far off-resonance. The subsequent off-resonance signal is described by substituting $P_i(t)$ in Eq. (2-51) into Eq. (2-24). The cos $\Delta \omega t$ transient emission signal is clearly evident. If the observation is made instead off-resonance by only 10 half-widths at half-height of the power-saturated steady-state line, $\Delta \omega = 10 \Omega_p$, the other terms in Eq. (2-50) will also contribute to the observed signal at a level of about 10% of the dominant transient emission signal. An

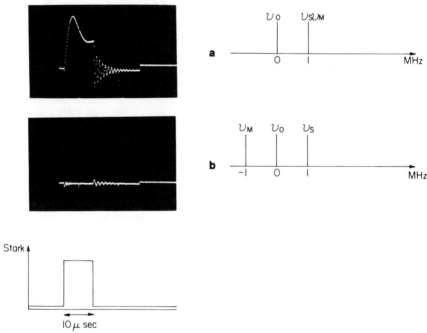

Figure 2-8. Experimental results showing the relative contributions of off-resonant absorption and coherent emission. In these experiments, ν_0 is the zero-Stark-field resonant frequency, ν_S is the Stark field on-resonant frequency, and ν_M is the fixed microwave oscillator frequency. The upper curve shows on-resonant transient absorption (Stark field on) followed by the off-resonant transient signal (Stark field off). This upper curve is similar to Figure 2-7 with the Stark field on and off periods reversed. The field-off signal in the upper curve is a combination of a dominant coherent transient emission signal [Eq. (2-51)] along with a weaker off-resonance transient absorption signal [see discussion following Eq. (2-50)]. The second signal shown in this figure shows the magnitude of the off-resonant transient absorption contribution [Eqs. (2-50) and (2-48)] to the total off-resonant signal shown in the upper curve. The off-resonant frequency $\Delta \omega$ is approximately 10 steady-state half-widths (at half-height), and the relative amplitudes of the total field-off signal in the upper curve to the field off signal in the lower curve are in the ratio 1:0.12 (from McGurk *et al.*, 1974c).

example showing the total off-resonance signal compared to the off-resonance transient absorption signal alone is shown in Figure 2-8. The upper curve shows on-resonance transient absorption (Stark field on) followed by off-resonant transient emission (Stark field off) as described in Figure 2-7. The lower curve (also shown in Figure 2-3) shows the contribution due to the off-resonant transient absorption processes. Comparing the field-off (off-resonant) results shows that the transient absorption signal at 1 MHz contributes about 10% to the off-resonant signal in the upper curve: the conclusion is that the off-resonant signal in the upper curve is dominated by the off-resonant emission part of Eq. (2-50).

Of course, much larger spontaneous coherent emission signals can be obtained by using initial conditions other than the on-resonance steady-state conditions. Whenever $\Delta\omega \gg \Omega_p$, Eq. (2-49) simplifies to

$$P_i(t) = e^{-t/T_2}[P_i(0) \cos \Delta\omega t + P_r(0) \sin \Delta\omega t] \qquad (2\text{-}52)$$

Equation (2-52) is identical to Eq. (132) of McGurk *et al.* (1974a). For on-resonant initial conditions, $P_r(0) = 0$. Equations (2-42) for an on-resonant $\pi/2$ pulse provide the initial conditions leading to the largest possible emission signal. Substituting Eqs. (2-42) into Eq. (2-52) leads to

$$P_i(t) = -\frac{\mu \, \Delta N_0}{2} \cos \Delta\omega t \; e^{-t/T_2} \qquad (2\text{-}53)$$

Equation (2-53), along with Eq. (2-24), will give the largest possible emission signal at the detector. On the other hand, Eqs. (2-44) show that when $\Delta\omega \gg \Omega_p$, there will be no emission following a π pulse. Figure 2-8 shows experimentally the emission signal observed following a $\pi/2$ or π pulse. The nmr analogs of these experiments were first performed by Hahn (1950a).

The pure emission signal given by Eq. (2-53) may be used to measure T_2 directly. However, the entire steady-state spectrum may also be recovered from this emission signal by Fourier-transforming from the time domain to the frequency domain. Fourier transform techniques are likely to become increasingly important in microwave transient experiments, as they have in nmr. They will be explored more fully in Section 2.6.

In the preceding sections, a great many approximate solutions to the basic equations describing microwave transient phenomena have been derived. In only a few cases do Eqs. (2-18) give an exact solution; even then, the exact solutions are often complicated, and approximation is needed to put the solutions into a simple form which clearly demonstrates the dominant phenomena taking place. In treating emission and fast passage, exact solutions are generally not even possible. Naturally, one must be concerned that these approximate solutions do not oversimplify the physics and ignore important effects. Fortunately, there is a general numerical technique capable of providing an accurate solution to Eqs. (2-18) under any experimental conditions. These exact solutions may then be compared with the solution predicted by the approximate equations used here. If agreement is obtained, it seems safe to assume that the approximations involved are sufficiently good.

Equations (2-18) are a set of three coupled first-order differential equations. They may be solved numerically to essentially any accuracy desired by an Adams predictor-corrector technique. A program to obtain such solutions is in operation in this laboratory and has been used to investigate many of the approximations which have been employed. As an example, we cite the following. The relative contributions of off-resonance absorption and emission to the imaginary component of the polarization have been analyzed as a function of the frequency offset from resonance ($\Delta\omega$), the effective external field strength (ω_1), and the unsaturated linewidth ($1/T_2$). The condition $(\Delta\omega) \geq 10[\omega_1{}^2 + (1/T_2)^2]^{1/2}$ is found to be sufficient to reduce the deviation from a pure emission result to less than 10%. This is in agreement with the experimental results shown in Figure 2-8. Other examples of the use of numerical techniques will be presented at appropriate places in the following sections.

Although a transient emission experiment itself provides a measure only of T_2, a variant of the experiment can be used to extract T_1. Consider a system to which a π pulse is applied at $t = 0$. Equation (2-44) shows that no emission will follow the pulse but that the population will be completely inverted. After the pulse, if the molecules are shifted far out of resonance, P_r and P_i will remain zero. ΔN will decay according to the equation

$t \geq t_\pi$:

$$\Delta N(t) = \Delta N_0(1 - 2e^{-(t-t_\pi)/T_1}) \qquad (2\text{-}54)$$

Suppose now that after an interval τ, a $\pi/2$ pulse is applied to the system. The result is to produce a polarization according to Eq. (2-33), but with the initial conditions $P_i = P_r = 0$, and ΔN [From Eq. (54)] $= \Delta N_0(1 - 2e^{-\tau/T_1})$. The imaginary component of the polarization at the end of the $\pi/2$ pulse is then

$t_{\pi/2} = t_\pi + \tau + \frac{1}{2}t_\pi$:

$$P_i(t_{\pi/2}) = -\frac{\mu\,\Delta N_0}{2}(1 - 2e^{-\tau/T_1}) \qquad (2\text{-}55)$$

After the end of the $\pi/2$ pulse, if the molecules are again shifted far out of resonance, they will undergo coherent spontaneous emission. The imaginary component of the polarization will obey Eq. (2-49) with the initial condition given by Eq. (2-55). Substituting Eq. (2-55) into Eq. (2-49) and assuming that $\Delta\omega \gg \Omega_p$ gives

$t \geq t_{\pi/2}$:

$$P_i(t) = -\frac{\mu \, \Delta N_0}{2}(1 - 2e^{-\tau/T_1})e^{-(t-t_{\pi/2})/T_2} \cos \Delta\omega (t - t_{\pi/2}) \quad (2\text{-}56)$$

Equation (2-56) looks just like the expression in Eq. (2-53) for coherent spontaneous emission except that the intensity of the emission is multiplied by the factor $(1 - 2e^{-\tau/T_1})$. A measure of the amplitude of the transient emission signal as a function of the delay time τ between the π pulse and the $\pi/2$ pulse will thus permit the extraction of T_1 (Mäder *et al.*, 1975). This experiment is the exact analog of the π, $\pi/2$ pulse train experiment in nmr (Farrar and Becker, 1971; Macomber, 1976).

In the microwave experiments described here, it is not always possible to satisfy the conditions in Eqs. (2-37) and (2-39), which lead to true $\pi/2$ and π pulses with the final result in Eq. (2-56). Therefore, one must return to Eqs. (2-36) and examine the nature of $P_i(t)$ and $\Delta N(t)$ under less than ideal conditions. For the experiments described here, it is necessary that P_i oscillate at least once back to its initial zero value. It can be shown (Mäder *et al.*, 1975) from Eqs. (2-36) for P_i that this will be true as long as $\omega_1 T_1 > 3.7$, again assuming that $\phi/\omega_1 = (1/2\omega_1)(1/T_2 - 1/T_1) \ll 1$.

According to this discussion, we require at a time t_1 that the polarization, $P_r(t_1)$ and $P_i(t_1)$, go to zero. Under these conditions, where Eqs. (2-37) and (2-39) are not satisfied, one could not expect $\Delta N(t_1)$ to achieve a full population inversion as in the case of a true π pulse. Instead, one expects that

$$-\Delta N_0 \leq \Delta N(t_1) \leq \Delta N_0 \quad (2\text{-}57)$$

The more nearly the conditions in Eqs. (2-37) and (2-39) can be achieved at $t = t_1$, the more nearly will $\Delta N(t_1)$ approach $-\Delta N_0$, the true population inversion.

At times $t \geq t_1$, the microwaves are taken far out of resonance, where $\Delta\omega \gg \omega_1$ and $\Delta\omega \gg 1/T$, and the transient emission solutions to Eqs. (2-18) are valid, giving the following result for the population difference:

$t \geq t_1$:

$$\Delta N(t) = \Delta N_0 + [\Delta N(t_1) - \Delta N_0]e^{-(t-t_1)/T_1} \quad (2\text{-}58)$$

The initial conditions $P_r(t_1) = P_i(t_1) = 0$ assure that P_r and P_i will remain zero and $\Delta N(t_1)$ in Eq. (2-25) is obtained from Eq. (2-36) and is limited by Eq. (2-57).

We are now interested in the change in signal at the detector after the two-level system is again brought into resonance with the radiation field at a time $t_2 > t_1$. The value of $P_i(t)$ for times after t_2 is given by Eq. (2-33), where the initial conditions are obtained from Eq. (2-58) for $t = t_2$. Making the appropriate substitution gives $P_i(t)$ for times greater than t_2.

$t \geq t_2$:

$$P_i(t) = \frac{\mu \omega_1 \Delta N_0}{2T_2}$$

$$\times \frac{e^{-(t-t_2)/T}\{\cos \omega_1(t-t_2) - [\omega_1 T_1 + (\phi/\omega_1)] \sin \omega_1(t-t_2)\} - 1}{(1/T_2)^2 + (T_1/T_2)\omega_1{}^2}$$

$$+ \frac{\mu[\Delta N_0 - \Delta N(t_1)]}{2} e^{-(t-t_2)/T} e^{-(t_2-t_1)/T_1} \sin \omega_1(t-t_2) \qquad (2\text{-}59)$$

It is now necessary to consider the relative change in signal ΔS at the detector for the times t' ($t_1 > t' > 0$) and t'' ($t'' > t_2$). When $t'' = t' + t_2$, one has, with Eqs. (2-24), (2-36), and (2-59):

$$\frac{\Delta S(t') - \Delta S(t'')}{\Delta S(t')} = g(t') \left[\frac{\Delta N_0 - \Delta N(t_1)}{\Delta N_0} \right] e^{-\tau/T_1} \qquad (2\text{-}60)$$

where

$$g(t') = \frac{[1/(\omega_1 T_1) + \omega_1 T_1] \sin \omega_1 t'}{[\omega_1 T_1 + (\phi/\omega_1)] \sin \omega_1 t' - \cos \omega_1 t' + e^{t'/T}} \qquad (2\text{-}61)$$

$$\tau = t_2 - t_1$$

In Eq. (2-60) we have the important result that the relative change in signal at t' and t'' (when $t'' = t' + t_2$) decays only with the relaxation time T_1 as a function of the delay τ. Note that if Eqs. (2-37) and (2-39) are satisfied and if $t' = t_{\pi/2}$ and $t_1 = t_\pi$, $g(t') = 1.0$, and if $t' \ll T$, Eq. (2-60) reduces to

$$\frac{\Delta S(t') - \Delta S(t'')}{\Delta S(t')} = 2e^{-\tau/T_1} \qquad (2\text{-}62)$$

In summary, it has been shown that it is not necessary to observe the emission after the second pulse to extract T_1. Equation (2-60) shows that the height of the second pulse relative to the first pulse depends only on the factor $e^{-\tau/T_1}$. Therefore, as long as the height of the second pulse at some definite position along the curve can be found as a function of the delay time between pulses, T_1 can be measured. Figure 2-9 shows the experimentally observed height of the second pulse as a function of delay time for the $J = 0 \rightarrow J = 1$ transition in OCS. Of course, the first pulse in these experiments is not a true π pulse, owing to such complicating factors as the field

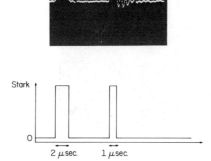

Figure 2-9. Experimental demonstration of responses following resonant polarization of the OCS $J = 0 \rightarrow J = 1$ transition by π and $\pi/2$ pulses after 2 and 1 μs, respectively. The 2-μs π pulse excitation corresponds to terminating the on-resonance stimulation, shown in Figure 2f, for instance, at the first $P_i(t) = 0$ point. No further signal is seen after the 2-μs π pulse (from McGurk *et al.*, 1974b).

distribution in the waveguide, finite rise and fall times for the Stark pulse, and relaxation processes that are not completely negligible during the pulse. All these factors have been investigated numerically and have been found not to affect the measurement of T_1 as long as the first pulse ends when P_i is zero. In the experiments shown in Figure 2-10, the length of the first pulse was determined empirically simply by adjusting until no emission was observed.

It should also be noticed that Eq. (2-55) predicts that for sufficiently short delay times τ, the $\pi/2$ pulse will actually have the opposite sign from

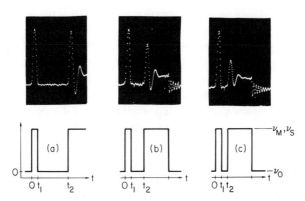

Figure 2-10. Diagram and oscilloscope photographs showing the pulse sequence signals as obtained from Stark switching the $J = 0 \rightarrow J = 1$ transition in OCS at 5.5 mTorr pressure. The delay times $\tau = t_2 - t_1$ between the pulses are (a) 13.0 μs, (b) 5.5 μs, (c) 2.5 μs. The length of the first pulse is $t_1 = 2.4$ μs and the length of the second pulse is 10 μs. ν_M denotes the microwave frequency, ν_S the Stark shifted frequency, and ν_0 the zero-field frequency of the $M = 0$, $J = 0 \rightarrow 1$ transition of OCS. In (b) and (c), the transient emission, which occurs after the second pulse is switched off, is also observed (from Mäder *et al.*, 1975).

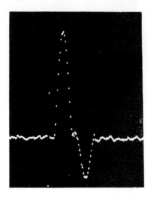

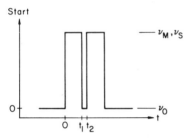

Figure 2-11. Diagram and oscilloscope photograph for the pulse sequence signal in the $J = 0 \rightarrow J = 1$ transition in OCS at 1.5 mTorr pressure. The population inversion as achieved by the first pulse is clearly demonstrated by the second pulse signal to show an induced transient emission. The length of both pulses is 2 μs, the delay between the pulses is 0.5 μs, and ν_S, ν_0, and ν_M are defined in Figure 2-10 (from Mäder *et al.*, 1975).

the initial π pulse. This corresponds to the application of the $\pi/2$ pulse while the population is still partially inverted. This effect is illustrated in Figure 2-11, where the population inversion is clearly in evidence.

The pulse method described above appears to be a quicker and more accurate method of obtaining T_1 than transient nutation. T_1 is still much harder to measure than T_2 because many pulse delays must be used at each pressure to map out the pressure-dependent part of T_1. Nevertheless, the combination of free induction decays and delayed pulses appears to be capable of providing reliable measures of T_1 and T_2 for many molecular systems.

Another pulse sequence experiment should also be discussed at this point. When some source of inhomogeneous broadening, such as Doppler broadening, is present, it will not be possible to measure the true T_2 from free induction decay. If the inhomogeneous broadening is sufficient to dominate the relaxation of the polarization (or if a source of severe inhomogeneous broadening such as an inhomogeneous Stark field is present), the true T_2 may be recovered from a $\pi/2$, π pulse echo experiment. First, let us write out the expressions for P_r and P_i during the pulse sequence, ignoring the inhomogeneous broadening. At the end of the $\pi/2$ pulse, one has, from Eqs. (2-42),

$$P_r(t_{\pi/2}) = 0$$

$$P_i(t_{\pi/2}) = -\frac{\mu \Delta N_0}{2}$$

(2-63)

Between the pulses Eq. (2-52) and the corresponding equation for P_r show that the polarization will evolve as

$$P_r(t) = -\left(-\frac{\mu \Delta N_0}{2}\right) \sin\left[\Delta\omega(t - t_{\pi/2})\right] e^{-(t-t_{\pi/2})/T_2}$$

$$P_i(t) = \left(-\frac{\mu \Delta N_0}{2}\right) \cos\left[\Delta\omega(t - t_{\pi/2})\right] e^{-(t-t_{\pi/2})/T_2}$$

(2-64)

After an interval τ, a π pulse is applied to the system. This results in leaving P_r unchanged, but changing the sign of P_i. Following the π pulse the polarization will again evolve as in Eq. (2-52).

$$P_r(t) = e^{-(t-t_\pi)/T_2}\left\{\frac{\mu \Delta N_0}{2} \sin(\Delta\omega t) e^{-\tau/T_2} \cos\left[\Delta\omega(t - t_\pi)\right]\right.$$

$$\left. -\frac{\mu \Delta N_0}{2} \cos(\Delta\omega t) e^{-\tau/T_2} \sin\left[\Delta\omega(t - t_\pi)\right]\right\}$$

$$P_i(t) = e^{-(t-t_\pi)/T_2}\left\{\frac{\mu \Delta N_0}{2} \cos(\Delta\omega\tau) e^{-\tau/T_2} \cos\left[\Delta\omega(t - t_\pi)\right]\right. \quad (2\text{-}65)$$

$$\left. +\frac{\mu \Delta N_0}{2} \sin(\Delta\omega t) e^{-\tau/T_2} \sin\left[\Delta\omega(t - t_\pi)\right]\right\}$$

$$t_\pi = t_{\pi/2} + \tau + 2t_{\pi/2}$$

It will be assumed here that Doppler broadening is the only source of inhomogeneous broadening present. The extension to other types of broadening is straightforward. McGurk *et al.* (1974a) have shown that for molecules moving with velocity v, Doppler broadening may be treated simply by replacing the frequency difference $\Delta\omega$ in Eqs. (2-64) and (2-65) with the Doppler-shifted frequency difference

$$\Delta\omega' = \omega_0 - \omega(1 - v/c) = \Delta\omega + \omega' \quad (2\text{-}66)$$

and integrating over the Maxwell distribution of velocities,

$$P(v) = \left(\frac{M}{2\pi k_B T}\right)^{1/2} e^{-Mv^2/2k_B T} = \frac{q}{\sqrt{\pi}} e^{-q^2\omega'^2} \quad (2\text{-}67)$$

After some algebra, this yields the following expressions for the imaginary components of the polarization before and after the π pulse:

$t_{\pi/2} < t < t_{\pi/2} + \tau$:

$$P_i(t) = -\frac{\mu \, \Delta N_0}{2} \cos \left[\Delta \omega (t - t_\pi) \right] e^{-(t-t_\pi)/T_2} e^{-(t-t_\pi)^2/4q^2} \tag{2-68}$$

$t > t_\pi$:

$$P_i(t) = \frac{\mu \, \Delta N_0}{2} \cos \left[\Delta \omega (t - 2\tau - 3t_{\pi/2}) \right] e^{-(t-3t_{\pi/2})/T_2} e^{-(t-2\tau-3t_{\pi/2})^2/4q^2} \tag{2-69}$$

Since it has been assumed that the system is in the Doppler-broadened regime ($\frac{1}{4}q^2 \gg 1/T_2$), it is clear from Eq. (2-68) that the polarization between the pulses will quickly damp to zero, producing no transient signal. However, Eq. (2-69) shows that at a time $t_{\text{echo}} = 2\tau + 3t_{\pi/2}$ *after* the π pulse, the exponential in q^2 will have a value of unity and a transient signal will occur. The strength of the signal will be proportional to

$$P(t_{\text{echo}}) = \frac{\mu \Delta N_0}{2} e^{-2\tau/T_2} \tag{2-70}$$

Thus, a series of echo experiments with different delay times τ will map out an exponential in T_2 even in the presence of Doppler broadening.

Echo techniques have not yet found much application in microwave spectroscopy because, at the usual pressures and frequencies, linewidths are not dominated by Doppler broadening. In the millimeter wave region, on the other hand, Doppler broadening is often very important and echo techniques may play a greater role. Macke and Glorieux (1974) have recently reported the use of an inhomogeneous Stark field to create the needed polarization dephasing artificially, and clearly demonstrate the existence of the echo. Again, of course, echo techniques are widely known and used in magnetic resonance (Hahn, 1950b; Farrar and Becker, 1971; Macomber, 1976) to overcome inhomogeneous (though usually not of Doppler origin) broadening.

2.5. Fast Passage

Fast passage is a phenomenon which is quite distinct from the absorption and emission we have discussed so far. It arises when a microwave field is swept through a resonance in a time short compared to the relaxation time. One might expect very little change in the polarization or population difference since the molecules are in resonance for only a very short time. Nevertheless, a surprisingly large polarization can be induced in the molecules, which leads to an appreciable free induction decay at the end of the sweep. Fast passage can thus be thought of as polarization during the sweep followed by emission after the sweep. In some versions of the experiment, significant population excitation is possible as well.

Equations (2-18) are capable of providing a description of fast passage (McGurk *et al.*, 1974c). However, the types of solution used for transient nutation and free induction decay are not applicable, since they all assume a constant value of $\Delta\omega$. If we assume a constant value for the sweep speed given by $d(\Delta\omega)/dt = \alpha$, and assume that the sweep is completed in a time short compared to the relaxation time so that collisional damping terms can be neglected ($\alpha T_2^2 \gg 1$), Eqs. (2-18) can be rewritten as

$$\alpha \frac{dP_r}{d\,\Delta\omega} + \Delta\omega P_i = 0$$

$$\alpha \frac{dP_i}{d\,\Delta\omega} - \Delta\omega P_r + \frac{\mu\omega_1}{2}\,\Delta N = 0 \qquad (2\text{-}71)$$

$$\alpha \frac{d}{d\,\Delta\omega}\left(\frac{\mu\,\Delta N}{2}\right) - \omega_1 P_i = 0$$

Equations (2-71) can be solved for sufficiently low powers ($\alpha/\omega_1^2 \gg 1$) by assuming that the population difference remains constant and at equilibrium throughout the sweep through resonance. ΔN may then be replaced by ΔN_0 in the second of Eqs. (2-71) and the first two equations solved to give

$$P_r(t_1) = \frac{\mu\omega_1}{2}\Delta N_0 \left(\frac{\pi}{\alpha}\right)^{1/2}\left\{\sin\left(\frac{(\Delta\omega_f)^2}{2\alpha}\right)\left[\mp C\left(\left|\frac{\Delta\omega_i}{(\pi\alpha)^{1/2}}\right|\right) \pm C\left(\left|\frac{\Delta\omega_f}{(\pi\alpha)^{1/2}}\right|\right)\right] \right.$$
$$\left. - \cos\left(\frac{(\Delta\omega_f)^2}{2\alpha}\right)\left[\mp S\left(\left|\frac{\Delta\omega_i}{(\pi\alpha)^{1/2}}\right|\right) \pm S\left(\left|\frac{\Delta\omega_f}{(\pi\alpha)^{1/2}}\right|\right)\right]\right\}$$

$$P_i(t_1) = -\frac{\mu\omega_1}{2}\Delta N_0 \left(\frac{\pi}{\alpha}\right)^{1/2}\left\{\cos\left(\frac{(\Delta\omega_f)^2}{2\alpha}\right)\left[\mp C\left(\left|\frac{\Delta\omega_i}{(\pi\alpha)^{1/2}}\right|\right) \pm C\left(\left|\frac{\Delta\omega_f}{(\pi\alpha)^{1/2}}\right|\right)\right] \right.$$
$$\left. + \sin\left(\frac{(\Delta\omega_f)^2}{2\alpha}\right)\left[\mp S\left(\left|\frac{\Delta\omega_i}{(\pi\alpha)^{1/2}}\right|\right) \pm S\left(\left|\frac{\Delta\omega_f}{(\pi\alpha)^{1/2}}\right|\right)\right]\right\}$$
$$(2\text{-}72)$$

Here t_1 is the time at the end of the sweep, $\Delta\omega_f$ is the value of $\Delta\omega$ at the end of the sweep, and $\Delta\omega_i$ is the value of $\Delta\omega$ before the sweep. The upper signs in Eqs. (2-72) are for $\Delta\omega_i$ or $\Delta\omega_f$ positive, and the lower signs are for negative values. The functions $C(a)$ and $S(a)$ are Fresnel sine and cosine integrals, defined by

$$C(a) = \int_0^a \cos{(\pi x^2/2)} \, dx$$

$$(2\text{-}73)$$

$$S(a) = \int_0^a \sin{(\pi x^2/2)} \, dx$$

Equations (2-72) may be simplified somewhat by noting that for sufficiently large values of $\Delta\omega_i$ and $\Delta\omega_f$ the Fresnel integers C and S may be replaced by their asymptotic values of $\frac{1}{2}$. Equations (2-72) then become

$$P_r(t_1) = \frac{\mu\omega_1}{2} \Delta N_0 \left(\frac{\pi}{\alpha}\right)^{1/2} \left[\sin\left(\frac{(\Delta\omega_f)^2}{2\alpha}\right) - \cos\left(\frac{(\Delta\omega_f)^2}{2\alpha}\right)\right]$$

$$(2\text{-}74)$$

$$P_i(t_1) = -\frac{\mu\omega_1}{2} \Delta N_0 \left(\frac{\pi}{\alpha}\right)^{1/2} \left[\sin\left(\frac{(\Delta\omega_f)^2}{2\alpha}\right) + \cos\left(\frac{(\Delta\omega_f)^2}{2\alpha}\right)\right]$$

Equations (2-74), along with $\Delta N(t_1) \approx \Delta N_0$, can be used as initial conditions along with Eqs. (2-49) and (2-24) to predict the transient emission signal at the detector after the end of the sweep. The beat frequency will be $\Delta\omega_f$ because the microwaves are at ω_f while the molecules emit at ω_0. The strength of the free induction decay is seen to depend on $\omega_1/\sqrt{\alpha}$ while the phase depends on $(\Delta\omega_f)^2/2\alpha$.

The basic approximation used in deriving the solution for linear fast passage—that the population difference can be decoupled from the polarization—has also been checked numerically. The deviation of the exact solution from that given by Eq. (2-72) becomes greater as the ratio $\varkappa\mathscr{E}/\sqrt{\alpha}$ becomes greater. For the largest value of $\varkappa\mathscr{E}$ investigated, 1×10^6 rad/s, and the smallest value of the sweep speed α investigated, 4×10^{13} rad/s^2, the deviation was never more than 5%. For almost any ratio $\varkappa\mathscr{E}/\sqrt{\alpha}$ smaller than 0.1, the agreement is essentially exact.

The experimental arrangement used to observe fast passage is essentially identical to that described earlier. The linear sweep rate needed is obtained by purposely detuning the Stark pulse generator to provide long, smooth, leading and trailing edges. The molecular resonance is therefore swept in frequency during the rise and fall time of the Stark pulse.

If the microwaves are first tuned to a frequency slightly above the transition frequency, ω_0, and the Stark field is turned on with a sufficient strength, the line will first come into resonance during the rise time (assuming a positive Stark shift) and then be taken on through resonance to its Stark-shifted frequency. It is polarized by the sweep and then emits at the Stark-shifted frequency. When the Stark field is turned off, the transition is once more swept through resonance and emission is observed at the zero-field frequency. Figure 2-12 provides a clear picture of the differences

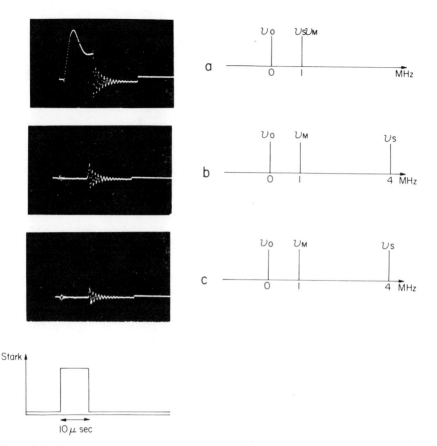

Figure 2-12. Demonstration of transient emission, and emission following fast passage in the $J = 0 \to J = 1$ transition in OCS. ν_0 is the zero-Stark-field resonant frequency, ν_S is the Stark field on-resonant frequency, and ν_M is the fixed microwave oscillator frequency. (a) The Stark pulse brings the $J = 0 \to J = 1$, $\Delta M = 0$ transition into resonance leading to transient absorption for the period of 10 μs. After the Stark pulse is switched off, the response is a combination of weak off-resonant absorption and the beat between the microwave oscillator and the coherent emission from the polarized molecules. (b) The rise and fall of the Stark pulse sweeps the molecular resonance through the microwave frequency (fast passage), which polarizes the gas leading to 3- and 1-MHz coherent transient emission signals, respectively. The sweep speed for the fall of the pulse is 4.8 MHz/μs. (c) The same experiment as in (b) is repeated with a sweep speed of 12.5 MHz/μs. The data are from McGurk *et al.* 1974c.

in the various transient phenomena discussed so far. Figure 2-12a shows transient absorption followed by transient emission as the molecules are first brought into resonance and then shifted out of resonance by the Stark field. Figures 2-12b and c show the emission signals following fast passage for two different values of the sweep speed.

As seen from Figure 2-12, the magnitude of the emission following fast passage can be quite large. Equations (2-74) and (2-53) can be used to compare the magnitude of the free induction decay (transient emission) following a $\pi/2$ pulse (the maximum obtainable by pulse methods) and that following fast passage.

$$\frac{|\Delta S(\text{fast passage})|}{|\Delta S(\pi/2)|} = \omega_1 \left(\frac{2\pi}{\alpha}\right)^{1/2} \tag{2-75}$$

For typical experimental values of $\omega_1 = 10^6$ rad/s and $\alpha = 10^{14}$ rad/s^2, this ratio is about 0.25. Thus, fast passage can be expected to be a significant effect whenever a transition is swept through resonance. One must be careful in designing a transient experiment that other Stark components than the one being studied are not inadvertently polarized in this fashion.

The large polarization obtainable with fast passage also has ramifications for Fourier transform spectroscopy. Any line in the band $\omega_i \leq \omega_0 \leq \omega_f$ will be polarized by the fast-passage sweep. Fast passage thus provides a means of saturating all transitions in a large band for Fourier transform work without the need of an unreasonably high power source.

Another version of the fast-passage experiment is called in nmr adiabatic fast passage (as opposed to the above-described linear fast passage). This occurs when the sweep through resonance is fast compared to the relaxation time ($\alpha T_2^2 \gg 1$), but the power is sufficiently large that α/ω_1^2 is much smaller than 1. Equations (2-71) are still valid, but it is not possible to proceed by setting $\Delta N \approx \Delta N_0$. Instead, one finds that $dP_i/dt \approx 0$. Under these conditions, the second of Eqs. (2-71) can be differentiated with respect to $\Delta\omega$, yielding the following relation between P_r and ΔN:

$$\frac{\mu}{2}\frac{d\,\Delta N}{d\,\Delta\omega} = \frac{1}{\omega_1}\left(P_r + \Delta\omega\frac{dP_r}{d\,\Delta\omega}\right) \tag{2-76}$$

Equation (2-76) can be substituted into the last of Eqs. (2-71) and that result, in turn, substituted into the first of Eqs. (2-71) to give the following differential equation for P_r:

$$\left(\frac{\omega_1}{\Delta\omega} + \frac{\Delta\omega}{\omega_1}\right)\frac{dP_r}{d\,\Delta\omega} + \frac{1}{\omega_1}P_r = 0 \tag{2-77}$$

Equation (2-77) can be solved by elementary methods and the result put back into Eq. (2-76) to give the complete solution for adiabatic fast passage:

$$P_r(t_1) = \frac{-(\mu/2)\omega_1\,\Delta N_0}{\sqrt{\omega_1^2 + (\Delta\omega_f)^2}}$$

$$P_i(t_1) = 0 \tag{2-78}$$

$$\Delta N(t_1) = \frac{-(\Delta\omega_f)\,\Delta N_0}{\sqrt{\omega_1^2 + (\Delta\omega_f)^2}}$$

Note that in the limit $(\Delta\omega_f)^2 \gg \omega_1^2$ these equations reduce to

$$P_r(t_1) = P_i(t_1) = 0$$
$$\Delta N(t_1) = -\Delta N_0 \qquad (2\text{-}79)$$

Thus, adiabatic fast passage (in the limit) amounts to inverting the population without polarizing the molecules. It is thus the fast-passage analog of a π pulse and may be used instead of high-power pulses in pulse train experiments, just as linear fast passage may be used to produce polarization.

Fast-passage effects are well known in magnetic resonance (Bloembergen *et al.*, 1948; Jacobsohn and Wangsness, 1948; Ernst, 1966). Once again, both the equations obtained and the experimental results in microwave spectroscopy are exact analogs of the same phenomena in nmr. Fast passage in infrared transient spectroscopy is discussed in Section 3.9.1.

2.6. Fourier Transform Microwave Spectroscopy

Fourier transform methods have long been used in magnetic resonance, and more recently in infrared spectroscopy, to obtain higher signal-to-noise ratios when dealing with weak spectra. Many of the same types of experimental situations are encountered in microwave spectroscopy. In this section the development of Fourier transform microwave spectroscopy is outlined. The technique uses a train of intense microwave pulses with carrier frequency $\nu_p = \omega_p/2\pi$, pulse length t_p, and period between pulses t_0. The radiation pulses travel along the z direction in a waveguide in the dominant TE_{10} mode. The short pulses are of sufficient duration and radiation power to produce a polarization in the sample gas which will begin to emit after the radiation power has passed.

To make the analysis tractable, certain conditions must be imposed on the exciting pulses. Consider the observation of all lines in a band $\pm\Delta\omega$ around the carrier frequency ω_p. It is assumed that the pulse length is short relative to the relaxation times for the system, $t_p \ll T_2 \cong T_1$, and that the time between pulses is long relative to the polarization relaxation time, $t_0 \gg T_2$. It is also assumed that the pulse is intense enough to satisfy

$$\varkappa_j \mathscr{E} = \frac{\mu_j \varepsilon}{\hbar} \gg |\Delta\omega_j| = |\omega_j - \omega_p| \qquad (2\text{-}80)$$

For all transitions j within the bandwidth $\Delta\omega$. ω_j denotes the natural resonance frequency of the jth transition and μ_j is the transition dipole for that transition. For convenience, we have defined $\mathscr{E} = \varepsilon/2$.

Unlike Stark-switching experiments, in which all molecules in the absorption cell are brought simultaneously into and out of resonance, in pulse experiments polarization is created at different times in different parts of the waveguide as the pulse moves through the sample. The coherent field at the detector due to the emission from all these molecules must be calculated, taking proper account of the phase of the emitted field from different parts of the cell. The Fourier transform method shows its full power only for large bandwidths in long cells. Therefore, the approximation used in Section 2.2 to avoid an exact treatment of the z dependence of the emitted field, valid when $\Delta \omega l/c \ll 2\pi$, will have to be abandoned. In addition, the effect of the waveguide on the form of the emitted field will have to be considered explicitly.

Before investigating the Fourier transform experiment it seems useful to gather together several results concerning wave transmission in waveguides (Feynman *et al.*, 1964; Townes and Schawlow, 1955). Consider first a microwave field propagating in an empty waveguide. The electric field must satisfy the wave equation

$$\nabla^2 E_m - \frac{1}{c^2} \frac{\partial^2 E_m}{\partial t^2} = 0 \tag{2-81}$$

where ∇^2 in place of $\partial^2/\partial z^2$ accounts for the variation of the electric field across the long dimension of the waveguide. The requirement that E_m be zero at the walls allows one to write down the x dependence of the field. For transmission in the TE_{10} mode, E_m may be written as

$$E_m = 2\mathscr{E}_0 \sin (\pi x/d) \cos (\omega t - kz) \tag{2-82}$$

Here, k represents the wave vector in the waveguide, not in free space, and d is the broad dimension of the waveguide. Substituting Eq. (2-82) into Eq. (2-81) yields the result

$$k = \left(\frac{\omega^2}{c^2} - \frac{\pi^2}{d^2} \right)^{1/2} \tag{2-83}$$

Now consider a cell containing a polarizable gas. The wave equation is modified to

$$\nabla^2 E - \frac{1}{c^2} \frac{\partial^2 E}{\partial t^2} = \frac{4\pi}{c^2} \frac{\partial^2 P}{\partial t^2} \tag{2-84}$$

It will be assumed that the emitted field, E_e, also propagates the in TE_{10} mode, so the total electric field can be written as

$$E = E_m + E_e = \sin (\pi x/d)[2\mathscr{E}_0 \cos (\omega t - kz)$$
$$+ 2\mathscr{E}_i \cos (\omega_0 t - k_0 z) + 2\mathscr{E}_r \sin (\omega_0 t - k_0 z)] \tag{2-85}$$

Substituting Eq. (2-85) into Eq. (2-84) gives

$$\nabla^2 E_m - \frac{1}{c^2}\frac{\partial^2 E_m}{\partial t^2} - \frac{\pi^2}{d^2}E_e + \frac{\partial^2 E_e}{\partial z^2} - \frac{1}{c^2}\frac{\partial^2 E_e}{\partial t^2} = \frac{4\pi}{c^2}\frac{\partial^2 P}{\partial t^2} \qquad (2\text{-}86)$$

Since E_m satisfies the homogeneous wave equation [Eq. (2-81)], the first two terms cancel, and one has left an equation for the emitted field only.

We are interested only in a single mode, so Eq. (2-86) must be projected onto the function describing the x dependence of that mode. Since E_e is assumed to be in the TE$_{10}$ mode, it, of course, projects fully. However, the right-hand side of Eq. (2-86) must be multiplied by $2/d \sin(\pi x/d)$ and integrated from zero to d.

Suppose that the polarization can be written in the form

$$P = 2\tilde{P}_r \cos(\omega_0 t - k_0 z) - 2\tilde{P}_i \sin(\omega_0 t - k_0 z) \qquad (2\text{-}87)$$

where ω_0 and k_0 are the frequency and wave vector characteristic of the emitted field. If the coefficients $\tilde{P}_r$ and $\tilde{P}_i$ are slowly varying in time and space relative to ω_0 and k_0, a standard approximation is the slowly varying envelope approximation (Macomber, 1976) which neglects time derivatives of $\tilde{P}_r$ and $\tilde{P}_i$ with respect to ω_0. This allows one to set

$$\frac{4\pi}{c^2}\frac{\partial^2 P}{\partial t^2} = -\frac{4\pi\omega_0^2}{c^2}P \qquad (2\text{-}88)$$

It is possible to show (Macomber, 1976) that a consistent approximation is to neglect terms involving second derivatives of $\mathscr{E}_i$ and $\mathscr{E}_r$ on the left-hand side of Eq. (2-86). Writing $\Omega_0 = (\omega_0 t - k_0 z)$ and $\bar{P}_r$ and $\bar{P}_i$ for the projected integrals over x of $\tilde{P}_r$ and $\tilde{P}_i$, this leaves

$$-\frac{\pi^2}{d^2}(2\mathscr{E}_i)\cos\Omega_0 - \frac{\pi^2}{d^2}(2\mathscr{E}_r)\sin\Omega_0 - k_0^2(2\mathscr{E}_i)\cos\Omega_0$$

$$-k_0^2(2\mathscr{E}_r)\sin\Omega_0 + 4k_0\frac{\partial\mathscr{E}_i}{\partial z}\sin\Omega_0 - 4k_0\frac{\partial\mathscr{E}_r}{\partial z}\cos\Omega_0$$

$$+\frac{\omega_0^2}{c^2}(2\mathscr{E}_i)\cos\Omega_0 + \frac{\omega_0^2}{c^2}(2\mathscr{E}_r)\sin\Omega_0 + \frac{4\omega_0}{c^2}\frac{\partial\mathscr{E}_i}{\partial t}\sin\Omega_0$$

$$-\frac{4\omega_0}{c^2}\frac{\partial\mathscr{E}_r}{\partial t}\cos\Omega_0 = -\frac{4\pi\omega_0^2}{c^2}(2\bar{P}_r\cos\Omega_0 - 2\bar{P}_i\sin\Omega_0) \qquad (2\text{-}89)$$

Equation (2-83) shows that the terms linear in $\mathscr{E}_i$ and $\mathscr{E}_r$ cancel exactly. The coefficients of $\cos \Omega_0$ and $\sin \Omega_0$ are thus separable into two first-order linear differential equations.

$$-\frac{4\pi\omega_0^2}{c^2}(2\bar{P}_r) = -4k_0\frac{\partial \mathscr{E}_r}{\partial z} - \frac{4\omega_0}{c^2}\frac{\partial \mathscr{E}_r}{\partial t}$$

$$-\frac{4\pi\omega_0^2}{c^2}(-2\bar{P}_i) = 4k_0\frac{\partial \mathscr{E}_i}{\partial z} + \frac{2\omega_0}{c^2}\frac{\partial^2 \mathscr{E}_i}{\partial t^2}$$
(2-90)

We have now only to introduce some definitions. The phase velocity for a wave of frequency ω_0 is, by definition,

$$v_{0p} = \frac{\omega_0}{k_0}$$
(2-91)

It represents the rate at which a point of constant phase moves down the guide. The group velocity, given by

$$v_{0g} = \frac{c^2}{v_{0p}}$$
(2-92)

is the rate that energy or information is transmitted by the wave. If the cutoff wavelength of the waveguide is designated as ω_c, it is easy to show (Feynman et al., 1964) that

$$v_{0p} = \frac{c}{\sqrt{1-(\omega_c/\omega_0)^2}}$$
(2-93)

a result that will be needed later. Using Eqs. (2-91) and (2-92), it is straightforward to reduce Eqs. (2-90) to

$$\frac{2\pi\omega_0}{v_{0g}}\bar{P}_r = \frac{\partial \mathscr{E}_r}{\partial z} + \frac{1}{v_{0g}}\frac{\partial \mathscr{E}_r}{\partial t}$$

$$\frac{2\pi\omega_0}{v_{0g}}\bar{P}_i = \frac{\partial \mathscr{E}_i}{\partial z} + \frac{1}{v_{0g}}\frac{\partial \mathscr{E}_i}{\partial t}$$
(2-94)

Equations (2-94) are identical to the free-space result except that the speed of light is everywhere replaced by v_{0g}, the group velocity of the emitted field in the guide.

Let us now return to our analysis of the pulse Fourier transform experiment. Consider the polarization created at a point z' along the waveguide. If the pulse enters the waveguide at $t = 0$, its leading edge will reach the point z' at a time $t_1 = z'/v_g$, where v_g is the group velocity of the pulse. At that moment, the molecules at z' will begin to undergo pulse absorption, as described in Eqs. (2-40). At a time $t_2 = z'/v_g + t_p$, the end of the pulse will reach z', the molecules will cease absorbing, and from Eqs. (2-40) they will have the following polarization:

$$[P_r(t_2, z')]_j = 0$$

$$[P_i(t_2, z')]_j = -\frac{\hbar \varkappa_j}{4}(\Delta N_0)_j \sin(\varkappa_j \mathscr{E} t_p)$$

(2-95)

The degree of polarization of each transition j may be different, but the polarization does not depend on $\Delta \omega_j$ since, by hypothesis, $\varkappa_j \mathscr{E} \gg \Delta \omega_j$ for all j in $\Delta \omega$.

After the pulse, the polarized molecules will begin to emit. The emission will, of course, be at the resonance frequency ω_j, so the polarization at $t > t_2$ will be proportional to the initial polarization, will oscillate at $\Delta \omega_j$, and will decay as e^{-t/T^2} as in Eq. (2-22), but it will begin in time and space at t_2 and z'. One has then

$$[P_r(t, z')]_j = \frac{\hbar \varkappa_j}{4}(\Delta N_0)_j \sin(\varkappa_j \mathscr{E} t_p) \sin[\Delta \omega_j(t - t_2)]e^{-(t-t_2)/T_{2j}}$$

$$[P_i(t, z')]_j = -\frac{\hbar \varkappa_j}{4}(\Delta N_0)_j \sin(\varkappa_j \mathscr{E} t_p) \cos[\Delta \omega_j(t - t_2)]e^{-(t-t_2)/T_{2j}}$$

(2-96)

Equations (2-96) are functions of z' through the quantity t_2. In the actual experiment, one must remember that the amplitude of the electric field is not constant across the waveguide but varies as in Eq. (2-82). Thus, P_r and P_i are also implicitly functions of x.

Equations (2-94) must now be used to calculate the emitted field due to the polarization. P_r and P_i give the coefficients of $\cos(\omega t - kz)$ and $\sin(\omega t - kz)$ in the total polarization. But to solve Eqs. (2-94) the polarization must be given in terms of ω_j and k_j, the frequency and wave vector for the emitted field, as in Eq. (2-87). The coefficients $\tilde{P}_r$ and $\tilde{P}_i$ are easily found in terms of P_r and P_i.

$$\tilde{P}_r = P_r \cos(\Delta \omega_j t - \Delta k_j z) + P_i \sin(\Delta \omega_j t - \Delta k_j z)$$

$$\tilde{P}_i = P_i \cos(\Delta \omega_j t - \Delta k_j z) - P_r \sin(\Delta \omega_j t - \Delta k_j z)$$

(2-97)

where $\Delta k_j = k_j - k$. Equations (2-96) and (2-97) must now be combined, integrated over $\sin(\pi x/d)$, and inserted into Eq. (2-94). Since the only x dependence in $\tilde{P}_r$ and $\tilde{P}_i$ is that of the electric field in $\sin(\varkappa_j \mathscr{E} t_p)$ in Eq. (2-96), the x integral can be done directly. It is easy to show (Mäder *et al.*, 1975) that the result is merely to replace the sine function in Eqs. (2-96) with twice the first-order Bessel function, $J_1(\varkappa_j \mathscr{E}_0 t_p)$.

We now turn our attention to the solution of Eqs. (2-94). Using the chain rule, one can write the total derivative of $\mathscr{E}_r$ (or $\mathscr{E}_i$) with respect to z as

$$\frac{d\mathscr{E}_r}{dz} = \frac{\partial \mathscr{E}_r}{\partial z} + \frac{dt}{dz}\frac{\partial \mathscr{E}_r}{\partial t}$$

(2-98)

The right-hand sides of Eqs. (2-94) look just like the right-hand side of Eq. (2-98) except that, for the jth transition, $1/v_{jg}$ appears in place of dt/dz. Therefore, they can be replaced by the total derivative with respect to z and integrated along any path for which $dz/dt = v_{jg}$. This is accomplished by replacing the time in the left-hand sides of Eqs. (2-94) by the retarded time, $t' = t - [(z - z')/v_{jg}]$ and integrating over z' (Feynman *et al.*, 1964; Lorrain and Corson, 1970). Physically, this operation takes account of the fact that polarization at a point z' cannot affect the field at another point z until a time $(z - z')/v_{jg}$ later, because information cannot be carried through space any faster than the group velocity, v_{jg}.

After some rather tedious algebra, Eqs. (2-96), (2-97), and (2-98), along with the integration over x and the expression for the retarded times, can all be substituted into Eq. (2-94) and collected together to yield

$$(\mathscr{E}_r)_j = \frac{4\pi\omega}{v_g}\left(-\frac{\hbar\varkappa_j}{4}\right)(\Delta N_0)_j J_1(\varkappa_j \mathscr{E}_0 t_p) e^{-(t-\bar{\imath})/T_{2j}} I_1$$

$$(\mathscr{E}_i)_j = \frac{4\pi\omega}{v_g}\left(-\frac{\hbar\varkappa_j}{4}\right)(\Delta N_0)_j J_1(\varkappa_j \mathscr{E}_0 t_p) e^{-(t-\bar{\imath})/T_{2j}} I_2$$

$$(2\text{-}99)$$

Equations (2-99) are only valid for $t > \bar{\imath}$ where $\bar{\imath} = t_p + z/v_g$ is the time at which the trailing edge of the pulse reaches z. The requirement from Eq. (2-95) that $\varkappa_j \mathscr{E} t_p$ be of order unity to create a significant polarization has been used along with Eq. (2-80) to drop terms in $\Delta\omega_j t_p$. In addition, the assumption $\Delta\omega/\omega_j \ll 1$ has been used to replace ω_j/v_{jg} with ω/v_g in the amplitude factor only. The integrals I_1 and I_2 are given by

$$I_1 = \int_0^l \exp\left\{\frac{1}{T_{2j}}\left[(z - z')\left(\frac{1}{v_{jg}} - \frac{1}{v_g}\right)\right]\right\} \sin\left[\left(\frac{\Delta\omega_j}{v_g} - \Delta k_j\right)z'\right] dz'$$

$$I_2 = \int_0^l \exp\left\{\frac{1}{T_{2j}}\left[(z - z')\left(\frac{1}{v_{jg}} - \frac{1}{v_g}\right)\right]\right\} \cos\left[\left(\frac{\Delta\omega_j}{v^g} - \Delta k_j\right)z'\right] dz'$$

$$(2\text{-}100)$$

Equations (2-91)–(2-93) can next be used to evaluate the arguments of the integrals. After more tedious algebra, and again assuming $\Delta\omega_j/\omega_j \ll 1$, one gets to the not-at-all-obvious result:

$$\frac{\Delta\omega_j}{v_g} - \Delta k_j = 0 \qquad (2\text{-}101)$$

We thus have the important result that I_1, and hence $\mathscr{E}_r$, are zero. The integral in I_2 can be carried out, giving

$$I_2 = \exp\left[\frac{z}{T_{2j}}\left(\frac{1}{v_{jg}} - \frac{1}{v_g}\right)\right]\left(-\frac{T_{2j}}{1/v_{jg} - 1/v_g}\right)\left\{\exp\left[-\frac{l}{T_{2j}}\left(\frac{1}{v_{jg}} - \frac{1}{v_g}\right)\right] - 1\right\}$$

$$(2\text{-}102)$$

In order that there not be loss of signal intensity, it is necessary that

$$\left| \frac{z}{T_{2j}} \left(\frac{1}{v_{jg}} - \frac{1}{v_g} \right) \right|$$

be less than unity. With this restriction, we can expand the exponentials in Eq. (2-102) to give

$$I_2 = \left(-\frac{T_{2j}}{1/v_{jg} - 1/v_g} \right) \left\{ \exp\left[\frac{1}{T_{2j}} (z-l) \left(\frac{1}{v_{jg}} - \frac{1}{v_g} \right) \right] - \exp\left[\frac{z}{T_{2j}} \left(\frac{1}{v_{jg}} - \frac{1}{v_g} \right) \right] \right\}$$

$$\cong \left(-\frac{T_{2j}}{1/v_{jg} - 1/v_g} \right) \left[1 + \frac{z-l}{T_{2j}} \left(\frac{1}{v_{jg}} - \frac{1}{v_g} \right) - 1 - \frac{z}{T_{2j}} \left(\frac{1}{v_{jg}} - \frac{1}{v_g} \right) \right]$$

$$= l \tag{2-103}$$

Putting these results into Eqs. (2-99) gives as the final result for the amplitude of the emitted field,

$$\mathscr{E}_r = 0$$

$$\tag{2-104}$$

$$\mathscr{E}_i = \frac{4\pi\omega l}{v_g} \left(-\frac{\hbar \varkappa_j}{4} \right) (\Delta N_0)_j J_1(\varkappa_j \mathscr{E}_0 t_p) e^{-(t-\bar{\imath})/T_{2j}}$$

There is no loss of field intensity due to phase cancellation in this type of experiment, even in long cells with large bandwidths, as long as the difference in transit time down the waveguide of the pulse and emission is shorter than the relaxation time of the system being studied.

The electric fields from all j transitions can be detected simultaneously by beating them with a local oscillator at frequency ω_{LO} having a fixed phase relationship to ω_p, and integrating over the Doppler distribution (when necessary). From Eqs. (2-85) and (2-104), the resultant signal is

$$\Delta S(t) = \frac{8\sqrt{\pi}\omega l q \beta}{v_g} \sum_j \left(-\frac{\hbar \varkappa_j}{4} \right) (\Delta N_0)_j J_1(\varkappa_j \mathscr{E}_0 t_p) e^{-(t-\bar{\imath})/T_{2j}}$$

$$\times \int_{-\infty}^{\infty} e^{-q^2\omega'^2} \cos\left[(\omega_j - \omega_{LO} + \omega')t - \phi_j(z) \right] d\omega' \tag{2-105}$$

where the amplitude of the local oscillator has here been absorbed into the conversion efficiency of the mixer, β, and all phase factors have been collected together in $\phi_j(z)$.

The cosine Fourier transform in Eq. (2-105) beginning at $t = \bar{t}$ then gives the following signal as a function of frequency:

$$\Delta S(\omega) = \frac{4\sqrt{\pi}\omega l q \beta}{v_g} \sum_j \left(-\frac{\hbar \varkappa_j}{4}\right) (\Delta N_0)_j J_1(\varkappa_j \mathscr{E}_0 t_p)$$

$$\times \left[\cos \phi_j(z) \int_{-\infty}^{\infty} \frac{e^{-q^2\omega'^2(1/T_{2j})}\,d\omega'}{(1/T_{2j})^2 + (\omega_j - \omega_{LO} - \omega + \omega')^2}\right.$$

$$\left. + \sin \phi_j(z) \int_{-\infty}^{\infty} \frac{e^{-q^2\omega'^2(\omega_j - \omega_{LO} - \omega + \omega')}\,d\omega'}{(1/T_{2j})^2 + (\omega_j - \omega_{LO} - \omega + \omega')^2}\right] \quad (2\text{-}106)$$

This is a frequency-dependent [through $\phi_j(z)$, which depends on k_j and hence ω_j] linear combination of the absorption and dispersion spectrum. The pure absorption spectrum can be regained from Eq. (2-106) by numerical methods if the sine Fourier transform is also available. It can be seen that the spectrum in the frequency domain can be obtained without power saturation in either the pressure or Doppler-broadened limit.

Equation (2-106) shows that the intensities of each of the transitions depend on $[P_i]_j$ as given in Eq. (2-95) (modified with the Bessel function in place of the sine function) and are accordingly a function of the pulse length t_p. We have therefore to distinguish between two situations:

1. The first case is where we are interested in recovering one or more relatively weak lines with similar $\varkappa_j$ in a noisy spectrum. In this case we do not need exact amplitude relations. t_p is then roughly adjusted to the first maximum of the Bessel function, and the spectrum can be obtained with optimum signal-to-noise ratio but slightly distorted amplitudes of the lines according to their $\varkappa_j$.
2. In the second case, we are interested in the relation of the amplitudes of lines with different $\varkappa_j$. The pulse length is then decreased sufficiently below the first maximum of the Bessel function for the strongest line. $[P_i]_j$ can then be approximated by

$$[P_i]_j = -\frac{\hbar \varkappa_j^2 (\Delta N_0)\mathscr{E}_0 t_p}{4} \quad (2\text{-}107)$$

and the intensities of the lines will be proportional to $\varkappa_j^2$, which is the same as in conventional steady-state spectroscopy for unsaturated transitions.

We will now describe in some detail the microwave Fourier transform spectrometer constructed by Ekkers and Flygare (1976). The following requirements for their instrument working in the frequency range from 4 to 8 GHz with a maximum bandwidth of 50 MHz were carefully considered.

1. In order to saturate a transition with a dipole transition moment $\mu_{ab} \cong 1$ D somewhere in a band of 50 MHz, the pulse power must be sufficiently large to fulfill Eq. (2-80) where $\Delta\omega/2\pi = 25$ MHz. The waveguide formula

$$P = E_0^2 \frac{a \cdot b}{480 \cdot \pi} \frac{\lambda}{\lambda_g}$$

   allows the calculation of the necessary power P in watts. E_0 is the maximum field amplitude in volts/cm, a and b are the dimensions of the waveguide in centimeters, and λ and λ_g are the free-space wavelength and the waveguide wavelength, respectively. For a C-band guide, we need pulse powers greater than 10 W. To maximize Eq. (2-95) the pulse length has to be less than 10 ns.

2. After the pulse is switched off, the power emitted from the excited molecules is (for a moderately strong transition) 80–100 dB below the pulse power. This means that the on/off ratio of the pulses has to be considerably higher than 100 db.

3. After the signal is detected as a beat against the local oscillator frequency, the signal has to be brought into a suitable frequency range to make analog-to-digital conversion possible. The maximum frequency of 50 MHz implies a sample rate of 10 ns per point. The responses after each pulse have to be in phase to make averaging possible. The averaging process has to be fast compared to the time during which the signal is sampled in order to make the deadtime between the pulses as short as possible.

The Ekkers–Flygare spectrometer is shown in Figure 2-13. The pulse sequence for the four PIN switches is shown in Figure 2-14. The delay of switch 2 is variable compared to switch 1. Power from the master oscillator (MO) reaches the 20-W continuous-wave (cw) TWT amplifier only when the first two switches are open. At the input of the amplifier the pulse has an on/off ratio of 80 db, a 10-ns switching time, and a length which is continuously variable from 1 μs down to zero. At the output of the TWT amplifier we obtain a pulse of nearly 20 W peak power with a signal-to-noise ratio of about 50 db. This noise, which is also present when the pulse is off, is cut down by another 80 db by switch 3. Any noise from the high power source is now negligible during the detection of the weak signals. Switch 4, with an isolation of 80 db, protects the mixer crystal from the high-power pulse. Switches 3 and 4 also have transition times of 10 ns.

The sample cell consists of an empty 4-m C-band waveguide. The local oscillator (LO) is kept 60 MHz above the MO by a phase-lock loop. The detector is a balanced mixer with an IF amplifier for 60 MHz and a bandwidth of 50 MHz. The signal band from 35 to 85 MHz is subsequently amplified and mixed down to 0–50 MHz with a variable-frequency oscillator

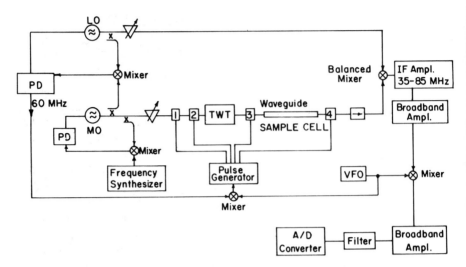

Figure 2-13. Pulsed microwave Fourier transform spectrometer from 4 to 8 GHz. The numbers 1, 2, 3, and 4 designate the PIN switches as described in the text. PD stands for phase detector and includes a phase-locking system.

(VFO) set at 85 MHz. After mixing, the signal is amplified in a second stage and all frequencies above 50 MHz are cut off by a low-pass filter. As noted above, the emission signal has a fixed phase relation to the MO. To allow averaging, the pulses have to be triggered from the phase-stable 60-MHz difference frequency between the MO and the LO. This 60 MHz must be

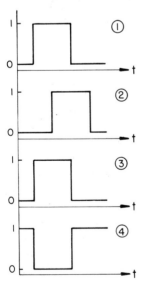

Figure 2-14. Pulse sequence for PIN switches 1 to 4, which produce a 10-ns high-power microwave pulse and at the same time protects the detector during the pulse-on time. Each of the four pulses is approximately 1 μs long. The repetition rate for all four pulses is the period T. The triggering occurs always at the same phase of the incoming high-frequency signal, but the number of cycles between the pulses is variable and determines T.

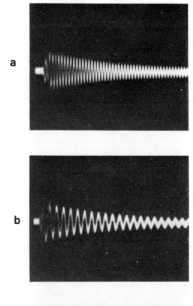

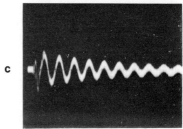

Figure 2-15. Oscilloscope traces of the emission signal at 5 mTorr of the $1_{11} \rightarrow 1_{10}$ transition in CH_2O after a single pulse. In Figure 2-15a the carrier frequency from the MO is exactly on resonance (4823.1 MHz) with the absorption line. The beat frequency of 25 MHz is determined by the VFO (see Figure 2-13). In Figures 2-15b and c, the carrier was moved 10 to 20 MHz away from the resonance. It is evident that the amplitude of the emission remains constant. The beat frequency changes from 25 to 5 MHz from Figures 2-15a to c (from Ekkers and Flygare, 1976).

mixed with the same signal from the VFO (85 MHz) which was used for the actual signal conversion. From this difference signal (25 MHz) the pulse generator is triggered. The repetition rate is adjustable to obtain a variable period between the pulses. Figure 2-15 shows the signal of 5 mTorr of formaldehyde (CH_2O) at 4823.1 MHz after a single pulse. In Figure 2-15a the MO is exactly on-resonance, which leads to a beat at the midband frequency of 25 MHz. In Figures 2-15b and c the MO frequency (the carrier frequency ω_p of the pulses) was moved 10 and 20 MHz, respectively, off-resonance. The fact that the amplitude remains constant over the complete frequency band confirms that the assumption made in Eq. (2-80) is fulfilled. Figure 2-15 shows that a band of at least 40 MHz (20 MHz on each side of the carrier frequency) can be covered by the pulse train.

After the low-pass filter the signals which lie in a frequency range of 0–50 MHz are fed into the analog-to-digital converter. Whereas AD converters with 10-ns resolution are ｝ ₂sently available, commercial instruments have only very slow repetition rates, owing to the fact that each recording has to be stored in a magnetic core memory before the gathering of new data. For any reasonable number of data points this leads to repetition rates well below 500 Hz or deadtimes of 20 ms between each recording. This leads to an unacceptable decrease in the signal-to-noise ratio.

As mentioned previously, these signals are normally buried in the balanced mixer noise and, therefore, 1-bit-accuracy analog-to-digital conversion can be used. One of several versions of 1-bit analog-to-digital converters that we have developed will be described here. This is a typical method of near-real-time averaging. The AD converter consists of a 1-bit comparator and a 512-bit bipolar random-access memory for the storage of 512 points. The sampling rate is 51.2 MHz, which gives a resolution of slightly less than 20 ns in the time domain. The resulting data collection time is 10 μs. After the collection cycle, the instrument enters the averaging mode, in which the 512 one-bit words of the buffer memory are added to the previously collected data in eight parallel channels. The data are then stored in 16 RAM memories of the same type. This allows the storage of the 512 points with a resolution of 16 bits. One averaging cycles takes 20 μs, which leads, together with the data-collection time, to a maximum repetition rate of 30 kHz. The actual repetition rate at which the instrument is run is determined externally by the variable repetition rate of the microwave pulses. Each data-collection cycle of the analog-to-digital converter is therefore started by trigger pulses from the pulse generator. It is important to mention here that the 51.2-MHz clock for the converter has to be in phase with the incoming signals. Unless this is the case, the signal-to-noise ratio will rapidly decrease for signals with high frequencies. As a clock we therefore use a simple RC oscillator which is triggered at the beginning of each data-collection cycle. The shortest possible time for the averager memory to fill up to 16 bits is approximately 2 s, after which the data are read into a PDP8/e laboratory computer, where further averaging can be done.

After sufficient averaging, the signals in the time domain are transformed to the frequency domain in the computer by the fast Fourier transform method of Cooley and Tuckey (1965). The resolution Δf of the obtained spectrum in the frequency domain is given by

$$\Delta f = (n \cdot \Delta t)^{-1} \tag{2-108}$$

where n is the number of points in the time domain and Δt their resolution. With a sample rate of 51.2 MHz and 512 points, we obtain a resolution of 100 kHz per point in the frequency domain. The width F of the calculated spectrum in the frequency domain is given by

$$F = (2 \cdot \Delta t)^{-1} \tag{2-109}$$

which means that with this sampling rate, the local oscillator frequency must be kept near the center of the 50-MHz band, since the width of the calculated spectrum is only 25 MHz. In a single experiment, one is left with the uncertainty of whether the detected line is above or below the reference frequency. This ambiguity can be removed by repeating the experiment with a slightly different local oscillator frequency and noting the direction of shift. In addition, incoming signals in the range 25–50 MHz are folded back into the range 0–25 MHz, owing to the slow sampling rate. This problem can be relieved by increasing the resolution of the AD converter to 10 ns.

According to Eq. (2-106), the phase of the Fourier-transformed signal in the frequency domain changes over the bandwidth of the instrument. This is, however, not the only phase problem which occurs. In addition, a frequency-dependent phase shift is introduced, owing to the fact that the data collection can normally only begin a certain time ($\sim$100 ns) after the pulse is switched off. This and other phase shifts due to the use of band-limiting filters are well known in nmr and can easily be corrected with a digital computer. The true absorption spectrum, $S_a(\omega)$, is obtained as a linear combination of the cosine and the sine Fourier transforms of the signal in the time domain: $S_{\cos}(\omega)$ and $S_{\sin}(\omega)$, respectively:

$$S_a(\omega) = S_{\cos}(\omega) \cos \phi - S_{\sin}(\omega) \sin \phi \qquad (2\text{-}110)$$

where the phase factor ϕ is a linear function of the frequency, which is normally fitted after both transforms have been calculated.

Figure 2-16 shows the frequency-domain spectra obtained with our pulsed Fourier transform spectrometer for $C^{(13)}H_2O$ in natural abundance displayed on an oscilloscope. The formaldehyde pressure was approximately 1 mTorr. The spectra cover 25 MHz and each frequency point corresponds to 100 kHz. The displayed line corresponds to the $1_{11} \rightarrow 1_{10}$ rotational transition of $C^{(13)}H_2O$ at 4593.3 MHz. The carrier frequency, f_{MO}, was kept at 4-MHz off-resonance in the upper spectrum and 21-MHz off-resonance in the lower spectrum. Both spectra were obtained after an averaging time of 15 s and an optimum exponential filter was used in the digital conversion. The line has an absorption coefficient of 6×10^{-8} cm^{-1}. The signal-to-noise ratio (peak signal amplitude to rms noise amplitude) obtained is approximately 50:1.

For high-resolution studies, Δf has to be decreased according to Eq. (2-108). This can be done by increasing Δt and/or adding zeros at the end of the recorded spectra in the time domain.

Figure 2-17 shows the averaged signal in the time domain for the $1_{11} \rightarrow 1_{10}$ rotational transition in CD_2O at 6083 MHz for very low pressures and $T = -77°C$. The signal was digitized with a sampling rate of 1 μs per point and zeros up to 1024 points were added. Figure 2-18 shows the Fourier-transformed and phase-corrected spectrum in the frequency

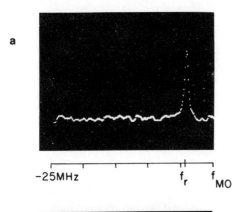

a

−25MHz f_r f_{MO}

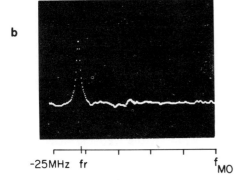

b

−25MHz fr f_{MO}

Figure 2-16. Fourier transform spectra of $C^{(13)}H_2O$ in natural abundance. In both cases, 25 MHz are covered. f_r corresponds to the resonance frequency of the $1_{11} \rightarrow 1_{10}$ rotational transition at 4593.3 MHz. f_{MO} is the carrier frequency of the microwave pulses and is set to 4597 MHz for Figure 2-16a and 4614 MHz for Figure 2-16b, respectively. The averaging time is 15 s (from Ekkers and Flygare, 1976).

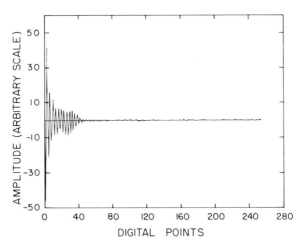

Figure 2-17. The transient emission signal of 0.2 mTorr of CD_2O averaged over 10^4 pulses converted at a range of 1 μs per point. 256 points are shown here (from Ekkers and Flygare, 1976).

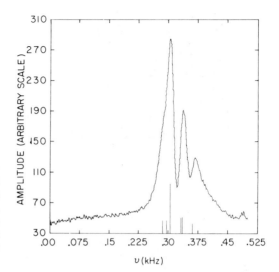

Figure 2-18. The Fourier transformed spectrum obtained from the emission in Figure 2-17 after a phase correction according to Eq. (2-110). The markers indicate the frequencies measured with a molecular beam spectrometer (from Ekkers and Flygare, 1976).

domain. In this case the fast Fourier transform method yields 512 points with a resolution of 1 kHz. The splitting of the lines occurs due to the nuclear quadrupole interaction. The frequencies obtained from these measurements agree very well with those obtained from molecular beam measurements (Tucker and Tomasevich, 1973).

At this point it seems useful to consider the bandwidth, resolution, and sensitivity limits of Fourier transform spectroscopy in some depth. The commonly quoted expressions for bandwidth and resolution are, respectively,

$$B = 1/t_p, \qquad \Delta T = t_0 \qquad (2\text{-}111)$$

Here it is assumed that the radiation pulse train is the limiting factor in the experiment. The first of Eqs (2-111) merely represents the spectral width which can be polarized by the radiation since Eq. (2-95) requires $x_j \mathscr{E} t_p \approx 1$, which gives $B \approx x_j \mathscr{E}$. The second of Eqs (2-111) assumes real-time signal averaging since the data collection time after one pulse, which determines resolution, will end with the onset of the next pulse.

In practice, however, bandwidth and resolution are usually not given by Eqs. (2-111). One is almost always limited to a finite number of channels, n, in which data can be stored and averaged. The usable bandwidth F is limited by the sampling theorem, which requires that a periodic signal be sampled at least twice per cycle, leading to Eq. (2-109), where Δt is the time resolution between points. The total time for which data may be accumulated is then $n \cdot \Delta t$, leading directly to Eq. (2-108) for the attainable frequency resolution. There is thus a trade-off between resolution and bandwidth depending on the time interval, Δt.

There is also a trade-off between resolution and sensitivity. Since the total obtainable signal depends on how often the experiment is repeated, the longer the data collection time per experiment, the lower the sensitivity. This is analogous to the situation encountered in frequency-domain spectroscopy with Stark modulation. The higher the Stark modulation frequency, the higher the sensitivity, but the greater the artificial broadening. Of course, increasing the modulation frequency beyond the width of the line does not lead to further improvement in sensitivity just as repetition of the time-domain experiment in a time shorter than the relaxation time does not increase the sensitivity, because the molecules have not had a chance to relax from the previous polarization.

Finally, let us consider the optimum sensitivity of the Fourier transform spectrometer by calculating the signal-to-noise ratio of a pulse Fourier transform spectrometer relative to a conventional absorption spectrometer. We will follow closely the treatment given by Ernst and Anderson (1966). Suppose that a spectral range F has to be investigated in a total time T_t. Both experiments will use a superheterodyne detection system with a balanced mixer. The noise is assumed to be white with a power density P_0 per spectral unit. The signal-to-noise (S/N) ratio is defined as the ratio of the peak signal amplitude to the rms noise amplitude.

It is intuitively easy to understand that the pulse Fourier transform spectrometer leads to an improvement in the S/N ratio as compared to the conventional steady-state experiment. In the steady-state experiment, the frequency is swept over a frequency range and the spectrum is recorded as a function of the frequency. At a certain time we look at one single frequency and recover only one point of the spectrum under investigation. Since, in many cases, a microwave spectrum is not very dense, most of the time is spent by recording the base line between the spectral lines.

In contrast, the time domain collects data from all frequencies at all times. Of course, it also collects noise from all frequencies at all times as well, but since the signals are coherent while the noise is random, the signal will increase faster than the noise. This intuitive argument is quantified as follows.

The expression for $(S/N)_s$ in the steady-state experiment is available in the literature (Ernst and Anderson, 1966). A single-scan, true slow-passage experiment is assumed. It is also assumed that the balanced bridge is perfectly balanced and that $T_1 = T_2$. To reduce the noise bandwidth, a low-pass filter is used. The maximum S/N ratio is obtained with a linear matched filter (Goldman, 1953), and under conditions of partial saturation,

$$\varkappa^2 \mathscr{E}^2 = 2(1/T_2)^2 \tag{2-112}$$

Stark modulation, which must usually be used for base-line stabilization, leads to an additional reduction in the S/N ratio by a factor of $2/\pi$. The optimum S/N ratio for the single-scan experiment is then given by

$$(S/N)_s{}^2 = \frac{2\omega^2 l^2 \beta^2 \hbar^2 x^2 \, \Delta N_0{}^2 \, T_t}{3\sqrt{3} R P_0 c^2} \frac{1}{F} \frac{T_t}{T_2} \qquad (2\text{-}113)$$

where R is the resistance across which the signal is measured.

In the Fourier transform experiment, the total signal-to-noise ratio is given as

$$(S/N)_P{}^2 = \left(\frac{T_t}{T} \cdot s_{max}\right)^2 \Big/ \left(RP\frac{T_t}{T}\right) \qquad (2\text{-}114)$$

$P(T_t/T)$ is the total noise accumulated during the time T_t and it is assumed that data are collected continuously in periods of time T. Thus, T_t/T is the number of pulse experiments in T_t. s_{max} is the maximum Fourier component of the signal after proper filtering. To obtain the maximum possible S/N ratio, the use of a matched filter is again necessary. The filtering is done in the time domain by multiplying the pulse response by $\exp(-t/T_2)$. This corresponds to a matched filter in the frequency domain.

To obtain the maximum signal amplitude in the frequency domain, the signal in the time domain, Eq. (2-105), must be multiplied by $\exp(-t/T_2)$ and Fourier-transformed. Ignoring the Doppler distribution, maximizing the $J_1(x\mathscr{E}_0 t_p)$ function in Eq. (2-105), and selecting $\phi_j = 0$, all of which correspond to maximum signal, and selecting the resonance frequency component gives

$$s_{max} = \frac{\sqrt{2}}{\sqrt{T}} \frac{T_2}{4} \frac{4\pi\omega l\beta}{c} \left(-\frac{\hbar x \, \Delta N_0}{4}\right) \qquad (2\text{-}115)$$

The noise is also filtered, leading to

$$P = P_0 \frac{T_2}{2T} \qquad (2\text{-}116)$$

In both Eqs. (2-115) and (2-116), it has been assumed that $T \gg T_2$ so that $\exp(-T/T_2)$ can be ignored with respect to 1. Substituting Eqs. (2-115) and (2-116) into Eq. (2-114) gives the maximum attainable S/N in the pulse experiment:

$$(S/N)_P{}^2 = \frac{T_t T_2}{4 R P_0 T} \left(\frac{4\pi\omega l\beta}{c}\right)^2 \left(-\frac{\hbar x \, \Delta N_0}{4}\right)^2 \qquad (2\text{-}117)$$

The signal-to-noise gain in the pulse experiment relative to the steady-state experiment can now be calculated.

$$\frac{(S/N)_P}{(S/N)_s} = \left[\frac{F}{2(\frac{1}{2}\pi T_2)} \frac{T_2}{T} \frac{3\pi(3)^{1/2}}{8}\right]^{1/2} \qquad (2\text{-}118)$$

Equation (2-118) is only valid at large T. At smaller T the equation will be modified by exponential factors which cause the last two factors to approach 1. A typical experimental value is $T = 3T_2$, which yields

$$\frac{(S/N)_P}{(S/N)_2} \approx \left[\frac{F}{2(\frac{1}{2}\pi T_2)}\right]^{1/2} \qquad (2\text{-}119)$$

Equation (2-119) indicates that the gain in S/N ratio for the pulse method is essentially given by the square root of the ratio of the total sweep width to a characteristic linewidth in the spectrum. This result is analogous to the result obtained by Ernst and Anderson (1966) for the nmr case. The actual gain in sensitivity in microwave spectroscopy depends on $|\langle a|\mu|b\rangle_i|$, the value of the dipole transition moment because the degree of polarization obtained from Eq. (2-95) or Eq. (2-105) depends on $\varkappa_j$. Equation (2-119) assumes that the maximum of this function can be reached, but this is not always the case. With this limitation in mind, the following points may serve to summarize the advantages of the pulse method.

1. For an average dipole transition moment of 1 D, a spectral range of 50 MHz (± 25 MHz on each side of the carrier) can be polarized. If we assume a typical linewidth of 500 kHz full width, then according to Eq. (2-119), a factor of 10 can be gained in S/N. This may be decisive in the detection of transitions with very small population differences (e.g., rare isotopic species, vibrational satellites).

2. For transitions with very small dipole moments, the spectral range which can be polarized is also smaller. The pulse method, however, may still allow the detection of a weak line which requires a prohibitively high Stark voltage in the steady-state for modulation.

3. For high-resolution spectra, the pulse method has the advantage that no power broadening occurs. The optimum S/N ratio for the steady-state experiment, on the other hand, was obtained under conditions which lead to a line broadening of $\sqrt{3}$.

2.7. Molecular Interpretation of T_1 and T_2

In this section we attempt to relate T_1 and T_2 to fundamental, presumably microscopic, laws. Since both T_1 and T_2 are introduced into the theory used here phenomenologically, there is no assurance that either will have a simple interpretation on the molecular level. Yet it is the possibility of such an interpretation as a means of obtaining molecular information that is the primary motivating factor for the measurement of T_1 and T_2. A large body of low-power linewidth data is in the literature, and there have been many attempts to treat microwave linewidths theoretically (see Rabitz, 1974, for a recent review). Much of this theory can be used to gain insight into the nature of T_2. However, most theoretical treatments of line widths either explicitly or implicitly assume that T_1 is equal to T_2. Only recently has attention been drawn to the calculation of T_1 (Hoke *et al.*, 1976). Theoretical investigations of the microscopic meaning of T_1 are almost nonexistent, owing in large measure to the lack of reliable measurements of T_1 before the onset of transient experiments.

There is now available extensive data from transient experiments on the behavior of both T_1 and T_2 for two molecular systems: broadening of the low-J rotational transitions of OCS by OCS and several other gases, and self-broadening of the inversion doublets of ammonia. The relevant experimental data are summarized in Table 2-1. The behavior of these two systems is quite different. It thus seems that an examination of the differences between T_1 and T_2 is now warranted, with a view toward constructing a theory capable of explaining the information which has recently been gained experimentally about the relaxation behavior of OCS and NH_3.

Many authors have assumed that T_1 is approximately equal to T_2 for gas-phase rotational and rotational–vibrational transitions in microwave and infrared spectroscopy. The rationalization of this viewpoint runs something like the following. T_1 measures the relaxation of the population difference between two rotational levels characterized by different J quantum numbers. T_2 measures the decay of the induced polarization in the two-level system. Certainly, a J changing collision would be expected to contribute to T_2 as well as T_1. On the other hand, weak collisions which simply interrupt the phase of the oscillating dipole without changing the energy state of the molecule would be expected to contribute to T_2 but not to T_1. Therefore, $1/T_2 \geq 1/T_1$. But rotational energy levels are so close together that virtually any collision strong enough to interrupt the phase of oscillation would also be strong enough to change J. Therefore, $T_2 \cong T_1$. A molecular interpretation of T_1 and T_2 will now be presented which will show that the above argument is invalid. It will prove necessary to consider the final states of molecules suffering collisions, not just the strength of the collision, to properly·understand the differences between T_2 and T_1. The attempt which will be made here will be nonformal, making no reference to general theories of scattering, classical or otherwise. Instead, it will use simple kinetic ideas to try to gain some insight into the nature of molecular relaxation. At the end, contact will be made with more rigorous theoretical treatments of these problems.

From a classical kinetic viewpoint, T_1 turns out to be easier to understand than T_2. The rate of change of the population difference between two levels can be expressed exactly in terms of the populations of the states accessible to the system and the transition rates between states. Let the two-level system be designated as lower state a and upper state b. Let all the

Table 2-1   Relaxation Data for OCS and NH$_3$

Line	Collision partner	$1/2\pi pT_2$ [a]	$1/2\pi pT_1$ [a]	$1/2\pi pT$ [a,b]	Method	Reference
OCS						
$J = 0 \to 1$	OCS	6.45 ± 0.15	—	—	Emission	Hill et al. (1967)
		4.82 ± 1.00	6.12 ± 0.94	5.98	Absorption	Brittain et al. (1973)
		5.13 ± 0.50	—	—	Absorption	McGurk et al. (1974b)
		5.83 ± 0.05	—	—	Emission	McGurk et al. (1974b)
	He	3.14 ± 0.08	5.89 ± 0.64	—	Emission, pulse	Mäder et al. (1975)
	CH$_3$F	12.25 ± 0.32	3.14 ± 0.31	—	Emission, pulse	Mäder et al. (1975)
	OCS	5.99 ± 0.02	12.57 ± 1.43	—	Emission, pulse	Mäder et al. (1975)
$J = 0 \to 1, M = 0$	OCS		6.06 ± 0.03		Emission, pulse	Coy (1975)
$J = 1 \to 2$	OCS			6.03	Absorption	Brittain et al. (1973)
	He	5.25 ± 0.50	—	—	Absorption[c]	Wang et al. (1973a)
	Ar	1.96 ± 0.30	—	—	Absorption[c]	Wang et al. (1973a)
		1.67 ± 0.20	—	—	Absorption[c]	Wang et al. (1973a)
$J = 1 \to 2, M = 0$	OCS	5.87 ± 0.43	—	—	Emission	McGurk et al. (1974d)
$J = 1 \to 2, M = \pm 1$	OCS	5.83 ± 0.43	—	—	Emission	McGurk et al. (1974d)
^{15}NH$_3$						
$(J, K) = (3,3), (^{14}$NH$_3)$	NH$_3$	—	25.4 ± 2.0	28.6	Absorption	Brittain et al. (1973)
$(8,7), (^{14}$NH$_3)$	NH$_3$	24.5 ± 2.0	—	—	Double resonance	Dobbs et al. (1975)
$(1,1)$	NH$_3$	19.6 ± 1.0	—	—	Emission	McGurk et al. (1974d)
$(2,2)$		23.7 ± 0.7	—	—	Emission	McGurk et al. (1974d)
$(3,3)$		24.8 ± 0.8	—	—	Emission	McGurk et al. (1974d)
$(3,2)$		19.2 ± 1.4	—	—	Emission	McGurk et al. (1974d)
$(4,4)$		23.4 ± 2.0	—	—	Emission	McGurk et al. (1974d)
$(1,1)$		22.4 ± 0.5	31.8 ± 1.6	—	Emission, pulse	Hoke et al. (1975)
$(2,1)$		15.4 ± 0.5	17.5 ± 1.6	—	Emission, pulse	Hoke et al. (1975)
$(2,2)$		24.7 ± 0.5	35.0 ± 1.6	—	Emission, pulse	Hoke et al. (1975)
$(3,2)$		19.1 ± 0.5	30.2 ± 1.6	—	Emission, pulse	Hoke et al. (1975)
$(3,3)$		25.0 ± 0.5	44.6 ± 3.2	—	Emission, pulse	Hoke et al. (1975)
$(4,3)$		21.3 ± 0.5	36.6 ± 1.6	—	Emission, pulse	Hoke et al. (1975)
$(4,4)$		25.0 ± 0.5	46.2 ± 3.2	—	Emission, pulse	Hoke et al. (1975)

[a]Units are MHz/Torr.
[b]$1/T = \frac{1}{2}(1/T_1 + 1/T_2)$.
[c]These experiments are contaminated by fast-passage effects, but that does not affect the observed relaxation time.

other states be indicated by i. Then N_a, N_b, and N_i will represent the populations (per unit volume) of the states and R_{ab}, R_{ai}, etc., will represent the transition rates. T_1 is defined by the equation

$$\frac{d \, \Delta N}{dt} = -\frac{\Delta N - \Delta N_0}{T_1}$$

$$\Delta N = N_a - N_b \tag{2-120}$$

Explicitly summing all possible transitions gives

$$\frac{dN_a}{dt} = \dot{N}_a = -N_a R_{ab} - \sum_i N_a R_{ai} + N_b R_{ba} + \sum_i N_i R_{ia}$$

$$\frac{dN_b}{dt} = \dot{N}_b = -N_b R_{ba} - \sum_i N_b R_{bi} + N_a R_{ab} + \sum_i N_i R_{ib} \tag{2-121}$$

At equilibrium $\dot{N}_a = \dot{N}_b = 0$, so

$$0 = -N_a{}^0 R_{ab} - \sum_i N_a{}^0 R_{ai} + N_b{}^0 R_{ba} + \sum_i N_i{}^0 R_{ia}$$

$$0 = -N_b{}^0 R_{ba} - \sum_i N_b{}^0 R_{bi} + N_a{}^0 R_{ab} + \sum_i N_i{}^0 R_{ib} \tag{2-122}$$

where superscript 0 indicates the equilibrium population. Combining Eqs. (2-121) and (2-122) gives

$$\frac{d \, \Delta N}{dt} = -2(N_a - N_a{}^0)R_{ab} + 2(N_b - N_b{}^0)R_{ba} - \sum_i (N_a - N_a{}^0)R_{ai}$$

$$+ \sum_i (N_b - N_b{}^0)R_{bi} + \sum_i (N_i - N_i{}^0)R_{ia} - \sum_i (N_i - N_i{}^0)R_{ib} \tag{2-123}$$

To write Eq. (2-123) in a form which can be compared with Eq. (2-120) requires some approximation. It is assumed that the relaxation behavior of states a and b is not much different, and therefore that there is little net transfer of population out of the states a and b during the course of the experiment. (See Appendix B for a discussion of how to treat the problem when the relaxations from states a and b are different.) When the relaxations from states a and b are the same, we can write

$$N_a + N_b \sim N_a{}^0 + N_b{}^0$$

$$N_i \cong N_i{}^0 \qquad (i \neq a, b) \tag{2-124}$$

Under these conditions Eq. (2-123) becomes

$$\frac{d \, \Delta N}{dt} = -\left(R_{ab} + R_{ba} + \tfrac{1}{2}\sum_i R_{ai} + \tfrac{1}{2}\sum_i R_{bi}\right)[(N_a - N_b) - (N_a{}^0 - N_b{}^0)] \tag{2-125}$$

and comparison with Eq. (2-120) yields

$$\frac{1}{T_1} = R_{ab} + R_{ba} + \tfrac{1}{2} \sum_i R_{ai} + \tfrac{1}{2} \sum_i R_{bi} \qquad (2\text{-}126)$$

Notice that transitions connecting states a and b contribute twice as heavily to T_1 as other transitions. This is because such transitions increase the population of one state and decrease that of the other, changing ΔN by 2. This should be contrasted with the expression for T_2.

It is more difficult to decide which collisions should be counted as contributing to T_2. It seems that any collision changing the state of a polarized molecule will contribute to T_2. In addition, it is possible that a collision could destroy the phase coherence of a molecule but not change its energy (though there is no clear experimental evidence that this ever occurs). On the other hand, only the first collision of this type should count, since once the molecule has lost its phase memory, it cannot recover it.

The quantity to be examined is the magnitude of the polarization given by

$$P = (P_r^2 + P_i^2)^{1/2} \qquad (2\text{-}127)$$

Equation (2-18) gives

$$\frac{dP_r}{dt} = -\frac{P_r}{T_2} - \Delta\omega P_i$$

$$\frac{dP_i}{dt} = -\frac{P_i}{T_2} + \Delta\omega P_r \qquad (2\text{-}128)$$

Differentiating Eq. (2-127) with respect to t and combining with Eqs. (2-128) gives

$$\frac{dP}{dt} = -\frac{P}{T_2} \qquad (2\text{-}129)$$

As indicated above, the total polarization will be proportional to the total number of molecules initially polarized which have not yet suffered a collision. One has then

$$P \propto \tilde{N}$$

$$\tilde{N} = \tilde{N}_a + \tilde{N}_b \qquad (2\text{-}130)$$

and from Eq. (2-129),

$$\frac{d\tilde{N}}{dt} = -\frac{\tilde{N}}{T_2} \qquad (2\text{-}131)$$

where $\tilde{N}$, etc., represent numbers of polarized molecules. It is assumed that the polarized molecules, which are actually in a superposition state, can be thought of as being in either state a or b at any given time, with a distribution depending on the nature of the superposition state. The time dependence of $\tilde{N}_a$ and $\tilde{N}_b$ is given by

$$\tilde{N}_a(t) = \tilde{N}_a(0) - \int_0^t \gamma_a \tilde{N}_a(t')\, dt'$$

$$\tilde{N}_b(t) = \tilde{N}_b(0) - \int_0^t \gamma_b \tilde{N}_b(t')\, dt' \qquad (2\text{-}132)$$

where

$$\gamma_a = R_{aa} + R_{ab} + \sum_i R_{ai}$$

$$\gamma_b = R_{ba} + R_{bb} + \sum_i R_{bi}$$

R_{aa} and R_{bb} represent the rates of "phase-changing collisions" and the quantities γ_a and γ_b are thus the total collision rates for molecules in states a and b, respectively.

Taking derivatives of Eqs. (2-132) and combining gives

$$\frac{d\tilde{N}}{dt} = -\left(R_{aa} + R_{ab} + \sum_i R_{ai}\right)\tilde{N}_a - \left(R_{bb} + R_{ba} + \sum_i R_{bi}\right)\tilde{N}_b \qquad (2\text{-}133)$$

Again, some approximation is needed to put Eq. (2-133) into a form like Eq. (2-131). For rotational energy levels which are much closer together than kT, it is reasonable to assume that the relaxation behavior of the two levels is not much different. Therefore, it should be a reasonable approximation to rewrite Eq. (2-133) as

$$\frac{d\tilde{N}}{dt} = -\left(\tfrac{1}{2}R_{aa} + \tfrac{1}{2}R_{ab} + \tfrac{1}{2}R_{bb} + \tfrac{1}{2}R_{ba} + \tfrac{1}{2}\sum_i R_{ai} + \tfrac{1}{2}\sum_i R_{bi}\right)(\tilde{N}_a + \tilde{N}_b) \qquad (2\text{-}134)$$

Comparing with Eq. (2-131) yields

$$\frac{1}{T_2} = \tfrac{1}{2}R_{aa} + \tfrac{1}{2}R_{bb} + \tfrac{1}{2}R_{ab} + \tfrac{1}{2}R_{ba} + \tfrac{1}{2}\sum_i R_{ai} + \tfrac{1}{2}\sum_i R_{bi} \qquad (2\text{-}135)$$

Substracting Eq. (2-126) for $1/T_1$ from Eq. (2-135) for $1/T_2$ gives

$$1/T_2 - 1/T_1 = \tfrac{1}{2}(R_{aa} + R_{bb} - R_{ab} - R_{ba}) \qquad (2\text{-}136)$$

Thus, the difference between T_1 and T_2 is given entirely in terms of transition probabilities involving the states in the two-level system being considered. From Eq. (2-136) it is not obvious that there is any necessary relation between T_1 and T_2. If there are very weak collisions, it is possible that R_{aa} and R_{bb} could be much larger than R_{ab} or R_{ba}. On the other hand, it is possible that weak collisions could be dominated by the longest-range force. For molecules with permanent electric dipoles, this would be the dipole–dipole interaction, which is governed by $\Delta J = \pm 1$ selection rules. If this were the case, R_{ab} and R_{ba} could be much larger than R_{aa} and R_{bb} and $1/T_1$ would be larger than $1/T_2$. Inversion transitions in NH_3 are also strongly dipole-allowed, and this is the apparent reason for the $1 \leq T_2/T_1 \leq 2$ observation in the inversion levels in $^{15}NH_3$. However, if one introduces the assumption of "strong collisions," one finds that $T_1 = T_2$. This can perhaps be seen most clearly by rearranging Eq. (2-136) to give

$$1/T_2 - 1/T_1 = \tfrac{1}{2}\left[\left(\gamma_a - \sum_i R_{ai}\right) + \left(\gamma_b - \sum_i R_{bi}\right)\right] - R_{ab} - R_{ba}$$

$$(2\text{-}137)$$

Strong collisions are interpreted to mean those sufficiently strong to leave the molecule in any of a large number of final states regardless of its initial state. Under these conditions, R_{aa}, R_{ab}, R_{ba}, and R_{bb} will all be much smaller than the total collisional rates, γ_a and γ_b, defined in Eq. (2-132) and $\gamma_a \cong \sum_i R_{ai}$, $\gamma_b \cong \sum_i R_{bi}$. Then the difference between T_1 and T_2 will be much smaller than either T_1 or T_2, or $T_1 \cong T_2$. This assumption would appear to be especially reasonable in the microwave region where $kT \gg \hbar\omega_0$. It seems to offer the best explanation for the observed equalities of T_1 and T_2 in OCS.

We have now justified the assertion made several times in the derivations in earlier sections that T_1 is on the order of T_2. The preceding arguments would lead us to conclude that in most rotational systems where many final states are accessible to kinetic collisions, T_1 will be equal to T_2. Even in such systems as ammonia, where direct transitions are strongly favored, T_2/T_1 has a maximum value of 2 because direct collisions contribute to T_2 (singly) as well as T_1 (doubly). Only in a system which is dominated by pure phase-changing collisions, in which $1/T_2$ could be much larger than $1/T_1$, is it possible for the two relaxation times to be of different orders of magnitude. No examples of systems of this type have yet been found.

These ideas can be employed to analyze other types of rotational relaxation experiments. Kukolich and coworkers (Kukolich *et al.*, 1973; Ben-Reuven and Kukolich, 1973; Wang *et al.*, 1973b) have used a beam maser technique to examine a number of systems where T_1-type cross sections are apparently quite different from T_2-type cross sections for some polar gases. One of the largest differentials between the T_1- and T_2-type scattering processes was found in the $J = 1 \rightarrow 2$ OCS transition for the OCS–CH_3F collision pair where the T_1-type cross section was considerably larger than the T_2-type cross section. These results on the $J = 1 \rightarrow 2$ OCS transition are quite different from the result obtained by Mäder *et al.* (1975) for the OCS–CH$_3$F pair where $T_1 \cong T_2$ for the $J = 0 \to J = 1$ OCS transition. Assuming that the relative T_1 and T_2 processes are the same in the OCS–CH$_3$F system for both the $J = 0 \to J = 1$ and $J = 1 \to J = 2$ OCS transitions, we must conclude that the difference between the microwave result and the beam maser results lies in the difference between the two experiments. In the beam maser experiments, only forward-scattered (at small angles) molecules were analyzed for the T_1- and T_2-type scattering cross sections. These low-angle products would result from grazing collisions, which would in turn be expected to be dominated by long-range forces (Mäder *et al.*, 1975; Liu and Marcus, 1975b). These long-range forces would be dipole–dipole forces in highly polar molecules, and it is precisely this kind of force which could lead to $1/T_1 > 1/T_2$. Comparing the microwave results indicates that the forward-scattering processes have larger T_1-type cross sections than T_2-type cross sections, but the average isotropic T_1- and T_2-type scattering cross sections are nearly identical. Thus, larger-angle scattering cross sections would probably lead to larger T_2-type scattering cross sections. In summary, we would expect forward-scattered particles to experience stronger T_1 processes and back-scattered particles to favor stronger T_2 processes in polar gases. These conclusions could be checked by doing an angular-dependent study in the beam maser experiment.

The interested reader is referred to recent work by Liu and Marcus (1975a,b) where the validity of a Bloch-equation-type treatment of molecular relaxation is examined. They derive a general description of molecular relaxation under the impact approximation. With the additional assumption that the relaxation behavior of the two levels in the observed transition is not much different, they are able to obtain expressions for T_1 and T_2 in terms of molecular relaxation matrix elements which can be calculated (semiclassically) from first principles. The theoretical understanding of T_1 and T_2 which now seems to be developing offers the potential in the next few years for a fruitful collaboration between theory and experiment in the investigation of gas-phase molecular relaxation processes.

2.8. Conclusion

In this chapter we have summarized the main concepts in microwave rotational transient experiments which lead to a measurement of the phenomenological relaxation times T_1 and T_2 which enter the electric dipole analogs of the Bloch equations. When electromagnetic radiation interacts with a molecule through a two-level electric dipole interaction, two things happen to an ensemble of these molecules. First, a macroscopic polarization is produced which has a relaxation time T_2. Second, a nonthermal equilibrium population distribution is produced which decays back to equilibrium with a relaxation time T_1. All observations of transient phenomena involve the interplay of the polarization and population differences through their coupling in the electric dipole Bloch equations.

We center our discussion of microwave–molecule resonant transient phenomena around two main processes: *transient absorption,* which is normally called transient nutation in magnetic resonance experiments, and *transient emission,* which is normally called the free induction decay in magnetic resonance experiments.

Transient absorption occurs when a two-level system is driven from a condition of equilibrium population difference and negligible polarization to a new state (in a short time relative to the relaxation processes) which contains a macroscopic polarization and a nonequilibrium population difference. During this process, energy is being taken from the radiation to produce a higher-energy system. Thus, we refer to this process, which has involved a net absorption of energy as transient absorption. The rates and behavior of the system as it moves toward its new state of macroscopic polarization and nonthermal equilibrium are determined by solving the coupled differential equations leading to the extraction (by comparison with experiment) of T_1 and T_2. For transient absorption as described here, the signal at the detector is given in Eq. (2-24) and absorption coefficient is given in Eq. (2-30) in terms of the imaginary component of macroscopic polarization. In all cases a net absorption is observed.

Transient emission occurs when the system is taken from a condition of interaction with the radiation where the system is polarized and in nonthermal equilibrium, to a condition where the external radiation is either removed or at least taken far off-resonance and out of interaction with the molecular two-level system. Again we consider the situation where the switching time is short relative to the relaxation processes. Under these conditions, the system relaxes from a condition of higher to lower energy after the radiation–molecule interaction is terminated. Some of the energy stored in the molecules is released by spontaneous coherent emission. In the process described above, the signal at the detector after the radiation–molecule interaction is terminated is still given by Eq. (2-24); however, the signal at the detector in Eq. (2-24) now arises from a beat between the radiation field coherently emitted from the system with the radiation field of the reference microwave oscillator. The phase and nature of the beat-frequency signal depend on the details of the initial preparation of the system. Of course, in the absence of the reference oscillator, the system still coherently emits the radiation field, and this signal could also be observed at the detector. Because this process would continue in the absence of the reference oscillator, we have called this process transient emission.

We have also discussed fast-passage experiments, where the microwave frequency sweeps completely through a two-level resonance in a time short relative to the relaxation time. Under these conditions, a high degree of polarization may be created without appreciably disturbing the equilibrium population difference. It is also possible to appreciably alter the population distribution with fast-passage techniques. After fast passage, the process of spontaneous coherent emission is observed.

The signal obtained as a beat between the coherently emitted molecular field and the reference oscillator can in all cases be Fourier-transformed to give the spectrum of the original transitions which were polarized. This demonstration of microwave Fourier transform spectroscopy can lead to the same advantages as experienced in nuclear magnetic resonance. Details of the design of a Fourier transform microwave spectrometer are presented in Section 2.6.

In Section 2.7 we provided a microscopic interpretation of T_1 and T_2 in terms of the transition probabilities between states. In the limit of strong collisions, where relaxation from one state to a large number of states is possible, we find that $T_1 = T_2$, which is the most reasonable explanation for the observation that $T_1 = T_2$ in a number of molecular systems. On the other hand, if specific selection rules favor transfer from one to the other state involved in the two-level interaction, T_1 may be shorter than T_2. In order for T_1 to be longer than T_2, the molecules would have to experience collisions which relax the polarization without relaxing the populations. We are hopeful that continued theoretical and experimental study will lead to an improved understanding of molecular relaxation processes.

Appendix A. Solution of the Bloch Equations

We illustrate here a general technique for solving the Bloch equations [Eqs. (2-18)], with specific application to the approximate off-resonance solution presented in Eq. (2-46). Before doing so, it is convenient to rewite Eqs. (2-18) in terms of the following reduced variables:

$$\tau = \omega_1 t, \qquad \alpha = \frac{1}{\omega_1 T_1}$$

$$\beta = \frac{1}{\omega_1 T_2}, \qquad \delta = \frac{\Delta\omega}{\omega_1} \qquad \text{(2-A1)}$$

$$M = \frac{\mu \, \Delta N}{2}, \qquad M_0 = \frac{\mu \, \Delta N_0}{2}$$

With these substitutions Eqs. (2-18) become

$$\frac{dP_r}{d\tau} + \beta P_r + \delta P_i = 0$$

$$\frac{dP_i}{d\tau} + \beta P_i - \delta P_r + M = 0 \qquad (2\text{-}A2)$$

$$\frac{dM}{d\tau} + \alpha M - P_i = \alpha M_0$$

Equations (2-A2) may be solved by Laplace transform techniques. Taking Laplace transforms of Eqs. (2-A2) gives

$$(u + \beta)\bar{P}_r + \delta\bar{P}_i = r_0 M_0$$

$$-\delta\bar{P}_r + (u + \beta)\bar{P}_i + \bar{M} = i_0 M_0 \qquad (2\text{-}A3)$$

$$-\bar{P}_i + (u + \alpha)\bar{M} = \frac{\alpha M_0}{u} + m_0 M_0$$

where $\bar{P}_r$, $\bar{P}_i$, and $\bar{M}$ are the Laplace transforms and $r_0 M_0$, $i_0 M_0$, and $m_0 M_0$ are the initial values of P_r, P_i, and M, respectively.

Equations (2-A3) may be solved formally and the result inverse-Laplace-transformed to give the following general solution to Eqs. (2-A2):

$$f = Ae^{-a\tau} + Be^{-b\tau}\cos s\tau + \frac{c}{s}e^{-b\tau}\sin s\tau + D \qquad (2\text{-}A4)$$

where f may stand for any of the components P_r/M_0, P_i/M_0, or M/M_0 and for $f = P_r/M_0$:

$$A = \frac{-g(-a)}{a[(b-a)^2 + s^2]}$$

$$B = -(A + D) + r_0$$

$$C = aA + bB - \beta r_0 - \delta i_0$$

$$D = \frac{\alpha\delta}{\alpha(\beta^2 + \delta^2) + \beta}$$

$$g(u) = r_0 u[1 + (u + \alpha)(u + \beta)] + \delta\alpha + \delta u m_0 - i_0 \delta u(u + \alpha) \qquad (2\text{-}A5)$$

for $f = P_i/M_0$:

$$A = \frac{-g(-a)}{a[(b-a)^2+s^2]}$$

$$B = -(A+D)+i_0$$

$$C = aA+bB-m_0+\delta r_0-\beta i_0$$

$$D = \frac{-\alpha\beta}{(\alpha(\beta^2+\delta^2)+\beta}$$

$$g(u) = r_0\delta u(u+\alpha)+i_0 u(u+\alpha)(u+\beta)-(\alpha+m_0 u)u+\beta)$$

for $f = M/M_0$:

$$A = \frac{-g(-a)}{a[(b-a)^2+s^2]}$$

$$B = -(A+D)+m_0$$

$$C = aA+bB+\alpha(1-m_0)+i_0$$

$$D = \frac{\alpha(\beta^2+\delta^2)}{\alpha(\beta^2+\delta^2)+\beta}$$

$$g(u) = r_0\delta u+i_0 u(u+\beta)+(\alpha+m_0 u)[(u+\beta)^2+\delta^2]$$

In addition, consideration of the boundary conditions gives the relations

$$2b+a = 2\beta+\alpha$$
$$b^2+s^2+2ab = 2\alpha\beta+\beta^2+\delta^2+1 \tag{2-A6}$$

Thus, the solution of Eqs. (2-A2) is reduced to finding the constant a where $-a$ is a real negative root of the equation

$$(u+\alpha)(u+\beta)^2+u+\beta+\delta^2(u+\alpha) = 0 \tag{2-A7}$$

Equation (2-A7) may be put into slightly different form by letting $z = u+\beta$. Then Eq. (2-A7) can be rearranged to give

$$z = (\beta-\alpha)\left(1-\frac{1}{1+\delta^2+z^2}\right) \tag{2-A8}$$

Clearly, $z \leq \beta-\alpha$. The second condition in Eq. (2-45) can be rewritten with Eqs. (2-A1) as

$$(\beta-\alpha)^2 \ll 1+\delta^2 \tag{2-A9}$$

Thus, an approximate solution to Eq. (2-A8) under the condition of Eq. (2-A9) is easily seen to be

$$z \cong (\beta - \alpha)\left(1 - \frac{1}{1+\delta^2}\right) \qquad (2\text{-A}10)$$

This gives

$$a = \alpha + \frac{\beta - \alpha}{1+\delta^2} \qquad (2\text{-A}11)$$

The recovery of Eq. (2-46) is now simply algebra. From Eqs. (2-A6),

$$b = \beta - \frac{1}{2}\frac{\beta - \alpha}{1+\delta^2}$$
$$s^2 = 1+\delta^2 \qquad (2\text{-A}12)$$

Calculating the coefficients from the second of Eqs. (2-A5), noting that Eq. (2-A9) allows one to drop terms or order $(\beta - \alpha)/(1+\delta^2)$ with respect to 1, substituting into Eq. (2-A4), and changing back to the original notation with Eq. (2-A1) yields Eq. (2-46). General off-resonant expressions for P_r and ΔN can be derived by the same technique simply by using the appropriate coefficients from Eqs. (2-A5). Notice also that both the exact on-resonance solution ($\delta = 0$) and the $T_1 = T_2$ ($\alpha = \beta$) solution of the Bloch equations are easily seen from Eq. (2-A8) since in either case $z = 0$ is clearly a solution of Eq. (2-A8).

Appendix B. Two-State Relaxation Processes

The Bloch equations [Eqs. (2-18)] are derived by assuming that a single relaxation time, T_1, is adequate to describe the return of the population difference to equilibrium. This assumption is valid if the relaxation behavior of the two levels in the two-level system is not much different. In that case the total population of the two-level system does not change much even when the molecules are redistributed from one state to the other. The approximations used in Section 2.7 to derive expressions for T_1 and T_2 are then also valid. However, a more general treatment of relaxation is possible when the above assumption is invalid. Initially, four independent quantities are needed to describe the density matrix for a two-level system: the magnitude of the off-diagonal element, its phase, and each of the two diagonal elements. Three of these quantities can be represented by P_r, P_i, and ΔN. The fourth quantity needed is then $N_T = N_a + N_b$, the total population of the two-level system. For convenience, the Bloch vector notation of

Section 2.2 is used. The quantities r_1, r_2, and r_3 have already been defined. The new component is given by

$$r_4 = \rho_{aa} + \rho_{bb} = \rho'_{aa} + \rho'_{bb}$$
$$= N_T/N$$

(2-B1)

The quantities appearing above are all defined in Section 2.2. The thermal equilibrium value of r_4 is denoted by $\bar{r}_4$.

If one now assumes no more than that the relaxation is first order, a set of coupled differential equations for the four components of $\vec{r}'$ can be written down. For these equations to be solvable, it is necessary that the populations of all levels other than a and b remain near equilibrium. If this is not the case, one does not have a true two-level system and all levels whose populations do change must be included explicitly in the analysis (e.g., in double resonance experiments; see Chapter 1). But for a true two-level system it can be shown (Liu and Marcus, 1975a) that r'_1, r'_2, r'_3, and r'_4 obey the following equations:

$$\frac{dr'_1}{dt} + \Delta\omega r'_2 + \frac{r'_1}{T_2} = 0$$

$$\frac{dr'_2}{dt} - \Delta\omega r'_1 + \omega_1 r'_3 + \frac{r'_2}{T_2} = 0$$

$$\frac{dr'_3}{dt} - \omega_1 r'_1 + \alpha_1(r'_3 - \bar{r}_3) + \alpha_2(r'_4 - \bar{r}_4) = 0$$

$$\frac{dr'_4}{dt} + \gamma_1(r'_4 - \bar{r}_4) + \gamma_2(r'_3 - \bar{r}_3) = 0$$

(2-B2)

α_1, α_2, γ_1, and γ_2 are relaxation parameters whose nature will be examined shortly.

General transient solutions to Eqs. (2-B2) are not known. However, solutions can be found in two interesting cases. The steady-state solutions can be found with any standard technique for solving simultaneous algebraic equations, after setting all the derivatives equal to zero. The interesting quantity is r'_2, which gives P_i and hence the steady-state absorption coefficient [Eq. (2-30)]. The result is

$$(r'_2)_{\text{steady state}} = \frac{-\omega_1(1/T_2)\bar{r}_3}{(1/T_2)^2 + (\Delta\omega)^2 + \omega_1^2 x}$$

$$x = \frac{(1/T_2)\gamma_1}{\alpha_1\gamma_1 - \alpha_2\gamma_2}$$

(2-B3)

Equation (2-B3) is identical to the result in Eq. (2-31) except for the appearance of x in place of T_1/T_2. Thus, the low-power linewidth is given by

$\Delta\omega = \frac{1}{2}\pi T_2$, even in the case of a two-level system with dissimilar relaxation behavior for the two levels, but the broadening of the line upon power saturation is altered. It is easily seen that putting $\alpha_1 = 1/T_1$ and $\alpha_2 = 0$ in Eqs. (2-B2) results in the usual Bloch equations [Eqs. (2-18)], and the same substitution in Eq. (2-B3) gives $x = T_1/T_2$, as it should.

Before examining the meaning of this result, let us derive the solution to Eqs (2-B2), which govern free decay of the population difference. In this case, the coupling through ω_1 is zero and the last two of Eqs. (2-B2) are separable from the first two. One has

$$\frac{dr'_3}{dt} + \alpha_2(r'_3 - \bar{r}'_3) + \alpha_2(r'_4 - \bar{r}'_4) = 0$$

$$\frac{dr'_4}{dt} + \gamma_1(r'_4 - \bar{r}'_4) + \gamma_2(r'_3 - \bar{r}'_3) = 0$$

(2-B4)

Solving the first equation for r'_4, differentiating with respect to t, and substituting the results into the second equation gives

$$\frac{d^2 r'_3}{dt^2} + (\alpha_1 + \gamma_1)\frac{dr'_3}{dt} + (\alpha_1\gamma_1 - \alpha_2\gamma_2)(r'_3 - \bar{r}'_3) = 0 \qquad (2\text{-B5})$$

The roots of the characteristic equation corresponding to Eq. (2-B5) are

$$\lambda_\pm = \frac{1}{2}\{-(\alpha_1 + \gamma_1) \pm [(\alpha_1 - \gamma_1)^2 + 4\alpha_2\gamma_2]^{1/2}\} \qquad (2\text{-B6})$$

so Eq. (2-B5) has the solution

$$r'_3 - \bar{r}'_3 = Ae^{\lambda_+ t} + Be^{\lambda_- t} \qquad (2\text{-B7})$$

Consideration of the initial conditions gives

$$r'_3 - \bar{r}'_3 = \frac{(\alpha_1 + \lambda_-)[r'_3(0) - \bar{r}'_3] + \alpha_2[r'_4(0) - \bar{r}'_4]}{\lambda_- - \lambda_+} e^{\lambda_+ t}$$

$$+ \frac{(\alpha_1 + \lambda_+)[r'_3(0) - \bar{r}'_3] + \alpha_2[r'_4(0) - \bar{r}'_4]}{\lambda_+ - \lambda_-} e^{\lambda_- t} \qquad (2\text{-B8})$$

$r'_3(0)$ and $r'_4(0)$ are the values of r'_3 and r'_4 at the beginning of the free decay. For pulse stimulation one would ordinarily have $r'_4(0) = \bar{r}'_4$. Thus, we have the result that the free decay of the population difference is the *sum of two exponentials*. If both of these decays can be measured in a π, τ, $\pi/2$ pulse experiment, relaxation information about both states a and b can be extracted simultaneously.

Finally, let us consider the interpretation of α_1, α_2, γ_1, and γ_2. Liu and Marcus (1975a) give expressions for these quantities in terms of relaxation supermatrix elements. All the types of supermatrix elements appearing in those expressions have straightforward classical interpretations. The notation of Section 2.7 may therefore be adopted to write

$$\alpha_1 = \tfrac{1}{2}\left(2R_{ab} + 2R_{ba} + \sum_i R_{ai} + \sum_i R_{bi}\right) = \frac{1}{T_1}$$

$$\alpha_2 = \tfrac{1}{2}\left(2R_{ab} - 2R_{ba} + \sum_i R_{ai} - \sum_i R_{bi}\right)$$

$$\gamma_1 = \tfrac{1}{2}\left(\sum_i R_{ai} + \sum_i R_{bi}\right)$$

$$\gamma_2 = \tfrac{1}{2}\left(\sum_i R_{ai} - \sum_i R_{bi}\right)$$

(2-B9)

The wealth of detailed relaxation information which can be gathered from combinations of experiments on systems with complicated relaxation behavior is thus potentially very large. For example, the sum of the two exponential relaxation rates in Eq. (2-B8) when compared to Eq. (2-135) for $1/T_2$ allows a direct measure of the rate of phase-changing collisions.

ACKNOWLEDGMENT

We gratefully acknowledge the support of the National Science Foundation. We are also grateful to R. T. Hofmann for general discussions concerning this work.

References

Abragam, A., 1961, *The Principles of Nuclear Magnetism*, Oxford University Press, Inc., New York.
Amano, T., and Shimizu, T., 1973, *J. Phys. Soc. Japan* **35**:237.
Ben-Reuven, A., and Kukolich, S. G., 1973, *Chem. Phys. Lett.* **23**:376.
Bloch, F., 1946, *Phys. Rev.* **70**:460.
Bloembergen, N., Purcell, E. M., and Pound, R. V., 1948, *Phys. Rev.* **73**:679.
Brewer, R. G., and Shoemaker, R. L., 1971, *Rev. Phys. Lett.* **27**:631.
Brewer, R. G., and Shoemaker, R. L., 1972a, *Phys. Rev. Lett.* **28**:1430.
Brewer, R. G., and Shoemaker, R. L., 1972b, *Phys. Rev.* **A6**:2001.
Brittain, A. H., Manor, P. J., and Schwendeman, R. H., 1973, *J. Chem. Phys.* **58**:5735.
Brown, S. R., 1974, *J. Chem. Phys.* **60**:1722.
Cooley, J., and Tuckey, J., 1965, *Math. Comput.* **19**:297.
Coy, S. L., 1975, *J. Chem. Phys.* **63**:5145.
Dicke, R. H., and Romer, R. H., 1955, *Rev. Sci. Instr.* **26**:915.
Dobbs, G. M., Micheels, R. H., Steinfeld, J. I., Wang, J. H.-S., and Levy, J. M., 1975, *J. Chem. Phys.* **63**:1904.

Ekkers, J., and Flygare, W. H., 1976, *Rev. Sci. Instr.* **47**:448.

Ernst, R. R., 1966, *Advan. Magn. Resonance* **2**:1.

Ernst, R. R., and Anderson, W. A., 1966, *Rev. Sci. Instr.* **37**:93.

Farrar, T. E., and Becker, E. D., 1971, *Pulse and Fourier Transform NMR*, Academic Press, Inc., New York.

Feynman, R. P., Hellwarth, R. W., and Vernon, F. L., Jr., 1957, *J. Appl. Phys.* **28**:49.

Feynman, R. P., Leighton, R. B., and Sands, M., 1964, *Lectures on Physics*, Vol. 3, Addison-Wesley Publishing Company Inc., Reading, Mass.

Goldman, S., 1953, *Information Theory*, Prentice-Hall, Inc., Englewood Cliffs, N.J.

Hahn, E. L., 1950a, *Phys. Rev.* **77**:297.

Hahn, E. L., 1950b, *Phys. Rev.* **80**:580.

Harrington, H., 1968, Symposium on Molecular Structure and Spectroscopy, Columbus, Ohio.

Hill, R. M., Kaplan, D. E., Herman, G. F., and Ichiki, S. K., 1967, *Phys. Rev. Lett.* **18**:105.

Hocker, B., and Tang, E. L., 1969, *Phys. Rev.* **184**:356.

Hoke, W. E., Ekkers, J., and Flygare, W. H., 1975, *J. Phys. Chem.* **63**:4075.

Hoke, W. E., Bauer, D. R., Ekkers, J., and Flygare, W. H., 1976, *J. Chem. Phys.* **64**:5276.

Jacobsohn, B. A., and Wangsness, R. K., 1948, *Phys. Rev.* **73**:942.

Kukolich, S. G., Wang, J. H.-S., and Oates, D. E., 1973, *Chem. Phys. Lett.* **20**:519.

Levy, J. M., Wang, J. H.-S., Kukolich, S. G., and Steinfeld, J. I., 1972, *Phys. Rev. Lett.* **29**:395.

Liu, W. K., and Marcus, R. A., 1975a, *J. Chem. Phys.* **63**:272.

Liu, W. K., and Marcus, R. A., 1975b, *J. Chem. Phys.* **63**:290.

Lorrain, P., and Corson, D. R., 1970, *Electromagnetic Fields and Waves*, W. H. Freeman and Company, Publishers, San Francisco.

Loy, M. T., 1974, *Phys. Rev. Lett.* **32**:814.

Macke, B., and Glorieux, P., 1972, *Chem. Phys. Lett.* **14**:85.

Macke, B., and Glorieux, P., 1973, *Chem. Phys. Lett.* **18**:91.

Macke, B., and Glorieux, P., 1974, *Chem. Phys.* **4**:120.

Macke, B., and Glorieux, P., 1976, *Chem. Phys. Lett.* **40**:287.

Macomber, J. D., 1976, *The Dynamics of Spectroscopic Transitions*, John Wiley & Sons, Inc., New York.

Mäder, H., Ekkers, J., Hoke, W. E., and Flygare, W. H., 1975, *J. Chem. Phys.* **62**:4380.

McGurk, J. C., Schmalz, T. G., and Flygare, W. H., 1974a, *Advan. Chem. Phys.* **25**:1.

McGurk, J. C., Hofmann, R. T., and Flygare, W. H., 1974b, *J. Chem. Phys.* **60**:2922.

McGurk, J. C., Schmalz, T. G., and Flygare, W. H., 1974c, *J. Chem. Phys.* **60**:4181.

McGurk, J. C., Mäder, H., Hofmann, R. T., Schmalz, T. G., and Flygare, W. H., 1974d, *J. Chem. Phys.* **61**:3759.

Rabitz, H., 1974, *Ann. Rev. Phys. Chem.* **25**:155.

Redfield, A. G., 1965, *Advan. Magn. Resonance* **1**:1.

Somers, R. M., Poehler, T. O., and Wagner, P. E., 1975, *Rev. Sci. Instr.* **46**:719.

Torrey, H. C., 1949, *Phys. Rev.* **76**:1059.

Townes, C., and Schawlow, A., 1955, *Microwave Spectroscopy*, McGraw-Hill Book Company, New York.

Tucker, K. D., and Tomasevich, G. R., 1973, *J. Mol. Spectry.* **48**:475.

Unland, M. L., and Flygare, W. H., 1966, *J. Chem. Phys.* **45**:2421.

Wang, J. H.-S., Levy, J. M., Kukolich, S. G., and Steinfeld, J. J., 1973a, *Chem. Phys.* **1**:141.

Wang, J. H.-S., Oates, D. E., Ben-Reuven, A., and Kukolich, S. G., 1973b, *J. Chem. Phys.* **59**:5268.

Weatherly, T. L., Williams, Q., and Tsai, F., 1974, *J. Chem. Phys.* **61**:2925.

Coherent Transient Infrared Spectroscopy

R. L. Shoemaker

3.1. Introduction

Ever since the advent of the laser in 1960, there has been considerable interest in optical analogs of the many transient phenomena seen in pulsed nuclear magnetic resonance experiments on spin systems. We refer to these phenomena as coherent transient effects and the interest in them is twofold. First, they exhibit very clearly the dynamics involved in the interaction of radiation and matter, and thus can be used to obtain a deeper understanding of this interaction. Second, it is found that the coherent transient effects can be used as a probe to study collisional decay processes in considerable detail as well as to obtain other spectroscopic information.

An important experimental breakthrough occurred in 1964 when Kurnit *et al.* observed photon echoes (the optical analog of spin echoes) in ruby. Between 1964 and 1971 a number of other laser experiments were done, but progress was hindered by difficulties associated with the pulsed laser sources used. These problems were alleviated in 1971 with the introduction of Stark switching, which allows the use of stable continuous-wave (cw) laser sources. By now, virtually every effect seen in magnetic resonance has been observed in the optical region along with several new phenomena which have no nmr analogs. Active research in this area is presently being carried on in a number of laboratories, and the literature of the field has grown to significant proportions.

R. L. Shoemaker • Optical Sciences Center, The University of Arizona, Tucson, Arizona

In this chapter we examine the basic theoretical and experimental aspects of coherent transient effects in the optical region. To keep the discussion manageable, we restrict our attention to low-pressure atomic and molecular gases and exclude any detailed discussion of transient effects in solids. We also exclude from discussion all effects which arise from the propagation of laser beams through optically thick media (samples for which the linear absorption of the laser beam is large). Specifically, this means that self-induced transparency, self-phase modulation, pulse compression, and all other pulse–reshaping effects are not treated.

Particular emphasis is placed on new features which occur in the optical region as compared to coherent transient effects in the microwave or radio-frequency regions. Many of these new features are due to the fact that optical transitions in low-pressure gases are dominated by Doppler broadening. Thus, we examine in considerable detail the effects of thermal motion in transient experiments. Among other things it will be seen that echo techniques are required to extract much of the information available in these experiments. Furthermore, the Doppler broadening is generally so large that only a portion of the Doppler linewidth is excited in an experiment. This complicates the mathematics, but it also enables us to examine new phenomena, such as velocity-changing collisions, within the Doppler profile.

The general plan of the chapter is as follows. We begin with a thorough discussion of the basic theoretical concepts and approximations needed to treat coherent transient phenomena. Next, a general discussion of the various experimental techniques which may be used is presented, followed by specific discussions of the various phenomena which have been observed. An attempt is made to first provide a simplified treatment of each transient effect so that its basic nature may be understood. More realistic theoretical descriptions are then discussed, along with the actual experiments. In addition, a short discussion of superradiance and its relation, if any, to coherent transient experiments is included in Section 3.7, since the term "superradiance" appears in many treatments of coherent transient effects.

In the theoretical treatment of coherent transient problems it is a practical necessity to use the density matrix approach rather than dealing directly with the Schrödinger equation. This situation arises because one needs to consider experiments which involve the interaction of radiation with gases. In such experiments we never know the exact quantum-mechanical state of each molecule in the gas. Rather we know only the *distribution* of molecules in various states. We do not know how long any particular molecule has been in a given state, or indeed, even which state it is in. Thus, we have less than complete information about the system, and no single wavefunction will suffice to describe it since the specification of a wavefunction presumes complete information about the system. Instead, the gas must be described by a statistical ensemble whose members have differing wavefunctions. While it is very awkward to handle these ensembles explicitly, the density matrix formalism can handle them easily since it is basically a statistical mechanical tool. In fact, the name "density matrix" arose from the correspondence between it and the classical distribution function for the density of states in phase space.

The density matrix also turns out to be especially convenient for the description of collisions and other decay processes in gases. This feature is important since collisions are a major topic in this chapter. A proper treatment of collisional phenomena begins with the many-body Hamiltonian for the collisionally interacting molecules in the gas. However, if the impact approximation (discussed below) is valid, the results one obtains from this approach are essentially the same as those obtained from a far simpler phenomenological treatment (Liu and Marcus, 1975). In this chapter we use basically this simpler approach. A proper treatment does have the advantage of providing a microscopic interpretation of the decay constants, but the detailed mathematics are rather complicated and we do not attempt such a treatment here.

In the phenomenological approach one assumes that the interaction between the molecules and the radiation field may be treated as if the molecules were isolated and noninteracting. Collisions are then introduced as decay processes which simply inject or remove molecules from a given state and which occur on a time scale that is instantaneous compared to the time scales of interest. Such a treatment will be valid if we restrict ourselves to dilute gases and consider only radiation fields which are not too far from resonance. Stating things more precisely, we want the duration of a collision τ_c and the time between collisions τ to satisfy the inequality $\tau_c/\tau \ll 1$. We also want the detuning of the laser frequency Ω from the molecular resonance of interest to satisfy $|\Omega - \omega_0| \ll 1/\tau_c$. The reason for the latter inequality is that one has terms in the density matrix which oscillate at $\Omega - \omega_0$ and we need to have the collision duration be instantaneous compared to the time scale of this oscillation. One can then regard the collisions as a simple interruption of the interaction between the field and the molecules. The neglect of τ_c compared to τ and $1/(\Omega - \omega_0)$ is often referred to as the impact approximation. Since collision durations are typically a few picoseconds, the approximation is usually valid for gas pressures of less than 1 atm and detunings of approximately 100 GHz or less from the transition(s) of interest. Thus, it will certainly hold for the experimental situations discussed in this chapter where we are dealing with resonance phenomena in low-pressure gases.

A second approximation which will always be made is to neglect absorber–absorber collisions. What is meant by this is the following. We are concerned with resonance phenomena involving transitions between only a few (generally just two) of the many energy levels populated in a molecular gas. Absorber molecules are defined as the small fraction of molecules which are in these active levels and capable of absorbing radiation. Since the probability of one absorber molecule colliding with another absorber rather than a molecule in some other level is thus very small, it is neglected. The approximation is usually valid in a low-pressure atomic gas also. Although nearly all the atoms are in one level, the ground state, only a small fraction of these have the proper velocity to be Doppler-shifted into resonance with the optical field and thus be active absorbers. The reason one wants to neglect absorber–absorber collisions is to simplify the mathematics. When an absorbing molecule collides with an inactive molecule, the population of the active levels is reduced by one. However, if two absorber molecules collide, the active-level population could be reduced by two, and inclusion of this possibility would complicate the theory.

A final approximation which we make in our treatment is to assume that the molecular translational motion may be treated classically. This approximation may sound rather innocuous, but there may exist experimental situations where it is not valid. The problem arises when one considers collisions in which the state of a colliding molecule remains unchanged but its velocity is altered. A careful consideration of this situation shows that a consistent theoretical treatment will require quantization of the molecular translation in some cases. This point is discussed in Section 3.10, where we consider velocity-changing collisions in some detail. Throughout the rest of the chapter we assume that the molecular translational motion may be described classically.

3.2. Density and Population Matrices

3.2.1. Basic Theory

Our formal treatment begins with the familiar Schrödinger equation for an isolated molecule,

$$\mathscr{H}\Psi = i\hbar\frac{\partial}{\partial t}\Psi \tag{3-1}$$

$$\mathscr{H} = \sum_j \frac{1}{2m_j}\left[-i\hbar\,\boldsymbol{\nabla}_j - \frac{e_j}{c}\mathbf{A}(\mathbf{R}+\mathbf{r}_j, t)\right]^2 + V(\mathbf{r}_j)$$

where the vector potential **A** takes into account any electromagnetic fields present, such as a laser beam. We have chosen the Coulomb gauge so that the scalar potential is zero and $\mathbf{E} = -(1/c)(\partial\mathbf{A}/\partial t)$. **R** is the center-of-mass coordinate of the molecule, and r_j denotes the position of the jth nucleus or electron with respect to the center of mass. Note that the Hamiltonian we have written down is a *semiclassical* Hamiltonian because we are treating the electromagnetic field as a classical field. We will consistently adopt this semiclassical viewpoint throughout, treating all optical fields, both incident and emitted, as classical fields. Since we will be dealing with absorption or emission from phased arrays of molecules, this approach will be completely satisfactory. One should beware of using this approach in cases where spontaneous emission is being observed, however, as incorrect results may be obtained. An example is resonance fluorescence in a strong field, where semiclassical theory gives quantitative predictions which are at variance with both experiment and quantum-electrodynamical theories in which the emitted radiation field is quantized (Schuda *et al.*, 1974; Mollow, 1969).

The first step in simplifying the Hamiltonian in Eq. (3-1) is to note that all radiation fields having wavelengths longer than the far uv satisfy the inequality, $\lambda \gg$ molecular dimensions. Thus, we may neglect any variation in $\mathbf{A}$ over molecular dimensions and set $\mathbf{A}(\mathbf{R}+\mathbf{r}_j, t) = \mathbf{A}(\mathbf{R}, t)$. This is known as the *dipole approximation*, since it leads to a dipolar interaction between the molecule and the field. In many texts the next step is to assume that the electromagnetic field is sufficiently weak to allow neglect of the $\mathbf{A} \cdot \mathbf{A}$ term in the Hamiltonian. However, considering the high intensities available with lasers, it is not immediately obvious that such an approximation is valid. Fortunately, the approximation is unnecessary. We may obtain the desired result instead by a simple unitary transformation (Fiutak, 1963; Sargent *et al.*, 1974). We set

$$\theta \equiv \frac{1}{c\hbar} \sum_j \mathbf{A}(\mathbf{R}, t) \cdot \mathbf{r}_j \tag{3-2}$$

and multiply the Schrödinger equation by $e^{i\theta}$, to obtain

$$e^{-i\theta} \mathcal{H} e^{i\theta} e^{-i\theta} \Psi = i\hbar e^{-i\theta} \frac{\partial}{\partial t} e^{i\theta} e^{-i\theta} \Psi \tag{3-3}$$

where we have also used the fact that $e^{i\theta} e^{-i\theta} = 1$. Note that Eq. (3-3) is simply another Schrödinger equation,

$$\mathcal{H}' \Psi' = i\hbar \left(\frac{\partial}{\partial t} \right)' \Psi' \tag{3-4}$$

with transformed operators and wavefunctions,

$$\mathcal{H}' = e^{-i\theta} \mathcal{H} e^{i\theta}, \qquad \left(\frac{\partial}{\partial t} \right)' = e^{-i\theta} \frac{\partial}{\partial t} e^{i\theta}, \qquad \Psi' = e^{-i\theta} \Psi \tag{3-5}$$

Now the transformed Schrödinger equation (3-4) is just as valid as and entirely equivalent to the original Schrödinger equation, provided that when we calculate matrix elements, all operators are transformed according to $\mathcal{O}' = \exp(-i\theta)\mathcal{O}\exp(i\theta)$. The reason for the equivalence is that all observables remain unchanged, since all matrix elements are unchanged under the transformation, i.e., $\langle\Psi_1'|\mathcal{O}'|\Psi_2'\rangle = \langle\Psi_1|\mathcal{O}|\Psi_2\rangle$. Note that in practice we may generally ignore the fact that we have made a transformation, since only in the very rare case where we need matrix elements of an operator explicitly containing $\partial/\partial t$ or $\partial/\partial x_j$ do we have $\mathcal{O}' \neq \mathcal{O}$.

Let us now see what is gained by the transformation above. By straightforward algebra we readily find that

$$\mathcal{H}' = e^{-i\theta}\mathcal{H}e^{i\theta} = \sum_j \left[\frac{-\hbar^2}{2m_j}\nabla_j^2 + V(\mathbf{r}_j)\right] \equiv \mathcal{H}_0$$

and

$$\left(\frac{\partial}{\partial t}\right)' = \frac{i}{c\hbar}\frac{\partial\mathbf{A}}{\partial t}\cdot\sum_j e_j\mathbf{r}_j + \frac{\partial}{\partial t}$$

(3-6)

Substituting this into Eq. (3-4) and using the relation $\mathbf{E} = -(1/c)(\partial\mathbf{A}/\partial t)$, we obtain

$$\left[\mathcal{H}_0 - \mathbf{E}(\mathbf{R}, t)\cdot\sum_j e_j\mathbf{r}_j\right]\Psi' = i\hbar\frac{\partial}{\partial t}\Psi'$$

(3-7)

which is the desired Schrödinger equation containing a dipolar molecule–field interaction. Note that the Hamiltonian above is valid for arbitrarily large field strengths, since we have made only the dipole approximation to obtain this result.

We now introduce the density matrix by expanding Ψ' in terms of the stationary-state eigenfunctions of $\mathcal{H}_0$, i.e.,

$$\Psi' = \sum_i c_i(t)\Psi_i(\mathbf{r}_1, \ldots, \mathbf{r}_N)$$

(3-8)

where the Ψ_i are solutions of $\mathcal{H}_0\Psi_i = E_i\Psi_i$. The density matrix ρ' is defined to be the matrix with elements*

$$\rho_{ij}' = c_i c_j^*$$

(3-9)

and its equation of motion is readily derived from the equation of motion for the probability amplitudes c_i to be

$$\dot{\rho}_{ij}' = -\frac{i}{\hbar}\sum_n (H_{in}\rho_{nj}' - \rho_{in}'H_{nj})$$

(3-10)

where $H_{in} = \langle\Psi_i|\mathcal{H}_0 - \mathbf{E}(\mathbf{R}, t)\cdot\sum_j e_j\mathbf{r}_j|\Psi_n\rangle$ and the summation on n is over all eigenstates of the molecule. If we similarly define the matrix of any operator

---

*The symbol ρ' is used here for the density matrix in order to distinguish it from the population matrix ρ, which will be introduced shortly.

$\mathcal{O}$ to be the matrix with elements $\mathcal{O}_{ij} = \langle \Psi_i | \mathcal{O}' | \Psi_j \rangle$, a very useful relation can be found for the average value of an operator,

$$\langle \mathcal{O} \rangle = \langle \Psi' | \mathcal{O}' | \Psi' \rangle$$

$$= \sum_{i,j} \rho'_{ij} \mathcal{O}_{ji} \tag{3-11}$$

The elements of the density matrix have simple physical interpretations. ρ'_{ii} is clearly just the probability for being in level i, since $\rho'_{ii} = |c_i|^2$. ρ_{ij}, on the other hand, is proportional to the average of the transition dipole moment between states i and j. To see this, consider a situation where a molecule has a wavefunction which is a mixture of only two states, Ψ_a and Ψ_b. Equation (3-11) then tells us that the average value of the dipole moment operator (i.e., the average value of the oscillating dipole moment which exists in the molecule) is just $\mu_{ab}\rho'_{ba} + \mu_{ba}\rho'_{ab} = 2\mu_{ab} \, \text{Re} \, (\rho'_{ab})$. To obtain the last expression we have assumed that $\mu_{ab} = \mu_{ba}$ and used the fact that $(\rho'_{ba})^* = \rho'_{ab}$. Here μ_{ab} is the transition dipole matrix element for the transition $a \to b$.

The density matrix defined in Eq. (3-9) is for a single molecule or group of identical molecules whose wavefunction is known and is often called a pure-case density matrix. Thus, as it stands, it is not adequate to describe real gases since a gas consists of an ensemble of molecules having differing wavefunctions. Furthermore, we have not yet accounted for the effects of collisions. Within the impact approximation discussed earlier, collisions may be regarded as instantaneously injecting or removing molecules from a given level, and thus can be included in our formalism by requiring that when a molecule enters some state α at time $t = t_0$ via a collision, we set

$$\rho'_{ij}(t = t_0) = \delta_{i\alpha}\delta_{j\alpha} \tag{3-12}$$

Collisional decay out of the state is then taken into account by simply adding a term $-\gamma_{ij}\rho'_{ij}$ to the density matrix equation of motion so that we have

$$\dot{\rho}'_{ij} = -\frac{i}{\hbar} \sum_n (H_{in}\rho'_{nj} - \rho'_{in}H_{nj}) - \gamma_{ij}\rho'_{ij} \tag{3-13}$$

Implicit in Eq. (3-12) is the entirely reasonable assumption that any collision hard enough to change the molecular state will also completely destroy any average dipole moment existing in the system prior to the collision [$\rho'_{ij}(t = t_0) = 0$ if $i \neq j$]. This assumption will not necessarily be true for collisions which change only the molecular velocity, and such collisions are not included in the present formalism. They are discussed separately in Section 3.10.

Since Eqs. (3-12) and (3-13) describe only a single molecule or group of identical molecules which entered some state α at time t_0, we must perform an ensemble average to obtain a description of the behavior of the entire gas.

This could be done by first solving the equation of motion (3-13) and then doing an ensemble average, but it is much more convenient to do the average first. We accomplish this by summing the pure-case density matrices to obtain a new ensemble averaged density matrix. To better understand exactly how and why this is done, let us first consider which quantities will vary from one molecule to another and hence must be averaged or summed over. First, if we examine the gas at some instant of time t, we find that different molecules will have had their last collision at varying times t_0 and will have entered various states α as a result of the collision. Also, that last collision will have taken place at different positions in space $\mathbf{R}_0$, and the molecules will have varying velocities $\mathbf{v}$. The pure-case density matrices are thus a function of all these variables, $\rho'_{ij}(\alpha, \mathbf{R}_0, \mathbf{v}, t - t_0)$. We now want to sum over these variables to obtain a new density matrix and its equation of motion. If we have an optical field present, however, we cannot do this for one of the quantities, the velocity. The reason is that molecules with different velocities see different Doppler shifts of the optical field. As a result they have differing Hamiltonians and matrix elements H_{ij}, leading to different density matrix equations of motion. Thus, the velocity summation must be performed at the end of the problem and the new density matrix will only contain summations over $\alpha, \mathbf{R}_0$, and t_0.

Throughout this chapter we consider the optical field to be a uniform plane wave. In this case the density matrix is no longer a function of x_0, y_0, v_x, or v_y (the optical field is assumed to propagate in the z direction), which simplifies the mathematics considerably. If we associate one pure-case density matrix $\rho'_{ij}(\alpha, z_0, v_z, t - t_0)$ with each molecule, our new density matrix can be defined schematically as

$$\rho_{ij}(z, v_z, t) = \sum_{\substack{\text{all molecules} \\ \text{with velocity } v_z}} \rho'_{ij}(\alpha, z_0, v_z, t - t_0) \qquad (3\text{-}14)$$

where v_z is the z component of the molecular velocity. As we have written it, $\rho_{ij}(z, v_z, t)$ is not really an ensemble averaged density matrix, because we have summed rather than averaged to obtain ρ_{ij}. Instead, it might better be called a *population matrix*, since $\rho_{ii}(z, v_z, t)$ is just the population (in molecules/cm^3) of level i at the point z for molecules having velocity v_z (Sargent *et al.*, 1974; Lamb, 1976). We will use this nomenclature throughout the chapter.

In order to explicitly write out the summation in Eq. (3-14), we need to introduce the quantity $\lambda_\alpha(v_z)$, which is the rate (molecules/s/cm^3) at which molecules with velocity v_z enter the state α due to collisions. Thus, in a time interval dt_0, $\lambda_\alpha(v_z) \, dt_0$ molecules with velocity v_z enter state α. $\lambda_\alpha(v_z)$ is not a function of time or space because we assume the collisional rates to be unaffected by any optical fields. Summing over all states α, entry times t_0, and initial positions z_0, we have as the explicit form of Eq. (3-14),

$$\rho_{ij}(z, v_z, t) = \sum_\alpha \lambda_\alpha(v_z) \int_{-\infty}^{t} dt_0 \int_{0}^{L} dz_0 \rho'_{ij}(\alpha, z_0, v_z, t-t_0) \delta[z - z_0 - v_z(t-t_0)]$$

(3-15)

This equation is our working definition of the population matrix. The delta function in the integral ensures that we only sum over those molecules whose space–time entry points (z_0, t_0) are such that they reach the point z at time t. Hence, the population matrix element $\rho_{ij}(z, v_z, t)$ represents those molecules having velocity v_z which are at z at time t regardless of their initial α, z_0, or t_0. The integral over z_0 in Eq. (3-15) is trivial and yields

$$\rho_{ij}(z, v_z, t) = \sum_\alpha \lambda_a(v_z) \int_{-\infty}^{t} dt_0 \rho'_{ij}(\alpha, z_0 = z - v_z(t-t_0), v_z, t-t_0)$$

(3-16)

An equation of motion for the population matrix may now be obtained quite simply. We just differentiate Eq. (3-16) with respect to t, recalling that

$$\frac{d}{dt} \int_{c}^{t} f(t, t_0)\, dt_0 = f(t, t) + \int_{c}^{t} \frac{\partial}{\partial t} f(t, t_0)\, dt_0$$

(3-17)

Thus,

$$\dot{\rho}_{ij}(z, v_z, t) = \sum_\alpha \lambda_\alpha \rho'_{ij}(\alpha, z_0 = z, v_z, 0)$$

$$+ \sum_\alpha \lambda_\alpha \int_{-\infty}^{t} \frac{\partial}{\partial t} \rho'_{ij}(\alpha, z_0 = z - v_z(t-t_0), v_z, t-t_0)\, dt_0$$

(3-18)

Using Eq. (3-12) we see that the first term on the right-hand side is just $\lambda_i \delta_{ij}$. The second term can also be simplified, but care must be taken in differentiating ρ'_{ij}. In addition to the $t - t_0$ time dependence, z_0 is an implicit function of time, and hence by the chain rule we obtain two terms for the time derivative of ρ_{ij}, a $\dot{\rho}'_{ij}$ term and a term $v_z(\partial \rho'_{ij}/\partial z_0)$. By $\dot{\rho}'_{ij}$ we mean the time derivative of ρ'_{ij} with respect to its fourth variable, $t - t_0$. The second term $v_z(\partial \rho'_{ij}/\partial z_0)$ arises from the time derivative of ρ'_{ij} with respect to its second variable, z_0. We now have

$$\dot{\rho}_{ij}(z, v_z, t) = \lambda_i \delta_{ij} + \sum_\alpha \lambda_\alpha \int_{-\infty}^{t} \left(\dot{\rho}'_{ij} + v_z \frac{\partial \rho'_{ij}}{\partial z_0} \right) dt_0$$

(3-19)

Further simplification is obtained by noting that the spatial derivative of ρ'_{ij} is negligible compared to $\dot{\rho}'_{ij}$. One can readily see this by considering, for example, the response of a sample to an optical field pulse. Let the pulse duration be Δt so that its length in space is $l = c \, \Delta t$. If passage of the pulse through the sample causes ρ'_{ij} to change by $\Delta\rho'_{ij}$, then $\dot{\rho}'_{ij} \sim \Delta\rho'_{ij}/\Delta t$. The spatial derivative, on the other hand, is $\partial\rho'_{ij}/\partial z_0 \sim \Delta\rho'_{ij}/l$. Using $l = c\,\Delta t$ we then have $v_z(\partial\rho'_{ij}/\partial z_0) \sim (v_z/c)(\Delta\rho'_{ii}/\Delta t)$, which is negligible compared to ρ'_{ij} because $v_z/c \sim 10^{-5}$. Hence, we may drop the spatial derivative in Eq. (3-19). This approximation is valid for nearly any problem in which only traveling-wave optical fields are present. For a standing wave, however, the optical field has nodes every half-wavelength, so $\partial\rho'_{ij}/\partial z_0$ is large regardless of the value of $\dot{\rho}'_{ij}$.

We now substitute the right-hand side of Eq. (3-13) for $\dot{\rho}'_{ij}$ in (3-19). Since H_{ij} and γ_{ij} are not functions of α or t_0, the population matrix definition, Eq. (3-16), can then be used to obtain

$$\dot{\rho}_{ij}(z, v_z, t) = \lambda_i \delta_{ij} - \frac{i}{\hbar}\sum_n [H_{in}\rho_{nj}(z, v_z, t) - \rho_{in}(z, v_z t)H_{nj}] - \gamma_{ij}\rho_{ij}(z, v_z, t) \tag{3-20}$$

This is the desired equation of motion for the population matrix.

The pumping rates λ_i can be expressed in terms of the equilibrium populations of the energy levels. To see this, consider a diagonal population matrix element ρ_{ii} in the case where no optical field is present. The Hamiltonian matrix H_{ij} is then diagonal, so the second term in Eq. (3-20) is zero and we have

$$\dot{\rho}_{ii} = \lambda_i - \gamma_{ii}\rho_{ii} \tag{3-21}$$

Since no optical fields are present, the system will eventually reach thermal equilibrium. The net population of the energy levels will then be constant, so $\dot{\rho}_{ii} = 0$ and

$$\lambda_i = \gamma_{ii}\rho_{ii} = n_i(v_z)\gamma_{ii} \tag{3-22}$$

Here $n_i(v_z)$ is the equilibrium population of level i for molecules with velocity v_z.

The optical field must now be specified explicitly. In this chapter we take it to be the linearly polarized plane wave

$$\mathbf{E}(z, t) = \hat{x}[E_0(z, t)\cos(\Omega t - kz)] \tag{3-23}$$

We need to consider only the simplest possible time dependence for $E_0(z, t)$, the case of rectangular pulses where E_0 is suddenly turned on to a constant value at some time t_i, and then is instantaneously turned off back to zero at the end of the pulse. For example, the pulses used to produce three of the basic transient effects are shown in Figure 3-1. The behavior of the gas sample for any given pulse sequence is calculated in a piecewise fashion, by solving the population matrix equations for each time interval $t_i - t_{i-1}$, using the solution at the end of the previous interval as an initial condition. Thus,

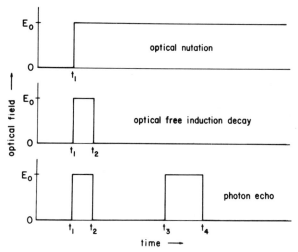

Figure 3-1. Typical pulse sequences used in transient-effect experiments.

we need only solve the population matrix equations for the case of a fixed-frequency constant-amplitude field with arbitrary initial conditions, and any transient effect can be calculated. The one exception is adiabatic inversion, where one has a constant-amplitude field combined with a frequency sweep. This is a special case and will be treated separately in Section 3.9. Unless otherwise noted, we assume throughout the rest of the chapter that Ω is a constant.

According to Eq. (3-7), the Hamiltonian in the presence of an optical field is

$$\mathcal{H} = \mathcal{H}_0 - \mathbf{E}' \cdot \sum_j e_j \mathbf{r}_j \qquad (3\text{-}24)$$

Note that $\mathbf{E}'$ in Eq. (3-24) is not given by Eq. (3-23), since $\mathbf{E}$ is the optical field in a laboratory-fixed coordinate frame. $\mathbf{E}'$ is the optical field seen by the molecule, i.e., the field in a moving coordinate frame in which the molecule is at rest. Since v_z/c is very small for molecular thermal velocities, we will make a nonrelativistic transformation to obtain $\mathbf{E}'$. The proper transformation relations for this case are

$$z' = z - v_z t$$
$$\Omega' = \Omega - k v_z \qquad (3\text{-}25)$$
$$k' = k, \qquad t' = t$$

These relations follow immediately from the Galilean transformation relations and the fact that the total phase of a wave is rigorously invariant for any transformation between inertial frames, since it can be associated with a mere counting of wavefronts, i.e., $(\Omega't' - k'z') = (\Omega t - kz)$ always (Jackson, 1962). The relation $k' = k$ amounts to saying that the wavelength of light is unchanged on going to the moving frame. Although this is not rigorously correct since $k' = \Omega'/c$ in a proper relativistic theory, taking $k' = k$ leads to negligible errors and is much more convenient.* From Eqs. (3-23)–(3-25) the Hamiltonian is now

$$\mathcal{H} = \mathcal{H}_0 - E_0(z', t) \cos (\Omega't - kz') \sum_j e_j X_j \qquad (3\text{-}26)$$

Here X_j is the projection of $\mathbf{r}_j$ along the space-fixed x axis.

Next, let us suppose that the optical field is resonant, or nearly resonant, with only one isolated nondegenerate transition. We denote the upper level of the transition by a and the lower level by b. The relevant matrix elements H_{ij} are then

$$H_{aa} = \hbar W_a, \qquad H_{bb} = \hbar W_b$$

$$H_{ab} = H_{ba} = -\mu_{ab} E_0 \cos (\Omega't - kz') \qquad (3\text{-}27)$$

where

$$\mu_{ab} = \mu_{ba} = \int \Psi_a^*(\Sigma_j e_j X_j)\Psi_b \, d\tau$$

Here $\hbar W_a$ and $\hbar W_b$ are the energies of the upper and lower states, respectively, and μ_{ab} is the transition dipole matrix element for the transition $a \to b$. In general, $\mu_{ab} = \mu_{ba}^*$, but for linearly polarized radiation, μ_{ab} can always be made real by a proper choice of phase for the wavefunctions ψ_a and ψ_b. Using Eqs. (3-20), (3-22), and (3-27), the relevant population matrix equations become

$$\dot{\rho}_{aa} = n_a\gamma_a + i\kappa E_0[\cos (\Omega't - kz')](\rho_{ba} - \rho_{ab}) - \gamma_a\rho_{aa}$$

$$\dot{\rho}_{bb} = n_b\gamma_b + i\kappa E_0[\cos (\Omega't - kz')](\rho_{ab} - \rho_{ba}) - \gamma_b\rho_{bb}$$

$$\dot{\rho}_{ab} = (-i\omega_0 - \gamma_{ab})\rho_{ab} + i\kappa E_0[\cos (\Omega't - kz')](\rho_{bb} - \rho_{aa}) \qquad (3\text{-}28)$$

$$\rho_{ba} = \rho_{ab}^*$$

Here $\kappa = \mu_{ab}/\hbar$, and γ_a and γ_b are shorthand notation for γ_{aa} and γ_{bb}, the decay rates of the level populations; $\omega_0 = W_a - W_b$ is the resonant frequency of the $a \to b$ transition. We have assumed that no other transitions are close to resonance with the optical field, so we need not consider any other matrix elements ρ_{ij}, as they will remain essentially at their thermal equilibrium values $\rho_{ij} = 0$, $i \neq j$; $\rho_{ii} = n_i$).

The physical situation described by Eqs. (3-28) is illustrated in Figure 3-2. We have two active levels coupled by the radiation field. These levels are part of a manifold of other levels and communicate with them via

---

*If we let $k' = \Omega'/c$, then we must also allow for time dilation. One cannot use both $k' = \Omega'/c$ and $t' = t$, because then $(\Omega't' - k'z') \neq (\Omega t + kz)$.

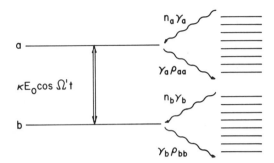

Figure 3-2. Idealized energy-level configuration for transient-effect experiments.

collisional pumping and decay, which occur at the rates $n_a\gamma_a$, $n_b\gamma_b$ and $\gamma_a\rho_{aa}$, $\gamma_b\rho_{bb}$, respectively.

Equations (3-28) cannot be solved exactly, owing to the presence of the cosine terms. However, by making what is known as the "rotating wave approximation," they can be cast into a form that does allow exact solutions for many cases of interest. To make the rotating wave approximation (RWA), we rewrite $\cos(\Omega't - kz')$ as a sum of exponentials and note that one of the rotating phasors $\exp[\pm i(\Omega't - kz')]$ has a negligible effect compared to the other. A simple way to show this is to consider a perturbation theory solution of Eqs. (3-28). We write the population matrix elements as a power-series expansion in the interaction strength κE_0. Thus,

$$\rho_{ij} = \rho_{ij}^{(0)} + \rho_{ij}^{(1)} + \rho_{ij}^{(2)} + \cdots \tag{3-29}$$

where $\rho_{ij}^{(n)}$ denotes the contribution to ρ_{ij} which is proportional to $(\kappa E_0)^n$. Substituting Eq. (3-29) into Eqs. (3-28) and equating terms of equal power in κE_0, we find that the first-order perturbation theory contribution to ρ_{ab} is given by

$$\dot{\rho}_{ab}^{(1)} + i\omega_0\rho_{ab}^{(1)} = i\frac{\kappa E_0}{2}[e^{i(\Omega't-kz')} + e^{-i(\Omega't-kz')}](\rho_{bb}^{(0)} - \rho_{aa}^{(0)}) \tag{3-30}$$

Here we have set $\gamma_{ab} = 0$ to simplify the algebra and have written $\cos(\Omega't - kz')$ as a sum of exponentials. If we consider the system to be initially in thermal equilibrium with the optical field turned on at $t = 0$, it is trivial to show that $\rho_{ab}^{(0)} = 0$ and $\rho_{bb}^{(0)} - \rho_{aa}^{(0)} = n_b - n_a$. Multiplying Eq. (3-30) by the integrating factor $\exp(i\omega_0 t)$, we then find that

$$\frac{d}{dt}[\rho_{ab}^{(1)} e^{i\omega_0 t}] = i\frac{\kappa E_0}{2}(n_b - n_a)[e^{-ikz'} e^{i(\Omega'+\omega_0)t} + e^{ikz'} e^{-i(\Omega'-\omega_0)t}] \tag{3-31}$$

Notice that the exponentials arising from $\cos(\Omega't - kz')$ have produced two driving terms in Eq. (3-31), one oscillating at twice an optical frequency $(\Omega' + \omega_0)$, the other oscillating at the difference frequency $(\Omega' - \omega_0)$. However, the sum frequency term produces a negligible effect on $\rho_{ab}^{(1)}$, as one can see by simply integrating Eq. (3-31). Using the fact that $\rho_{ab}^{(1)}(0) = 0$, we obtain

$$\rho_{ab}^{(1)}(t) = \frac{\kappa E_0}{2}(n_b - n_a)e^{-i\omega_0 t}\left[e^{-ikz'}\frac{e^{i(\Omega' + \omega_0)t} - 1}{\Omega' + \omega_0} - e^{ikz'}\frac{e^{-i(\Omega' - \omega_0)t} - 1}{\Omega' - \omega_0}\right] \quad (3\text{-}32)$$

Clearly the first term in the brackets is negligible, since the very large number $\Omega' + \omega_0$ appears in the denominator. The rotating-wave approximation consists of neglecting the sum frequency term in (3-31) so that we obtain

$$\rho_{ab}^{(1)} = \frac{\kappa E_0(n_b - n_a)}{2(\Omega' - \omega_0)}[e^{i(\Omega' - \omega_0)t} - 1]e^{-i(\Omega't - kz')} \quad (3\text{-}33)$$

If the expression for $\rho_{ab}^{(1)}$ is substituted into the equations for $\rho_{aa}^{(2)}$ and $\rho_{bb}^{(2)}$, ($\rho_{aa}^{(1)}$ and $\rho_{bb}^{(1)}$ are zero), we find that the exponentials arising from $\cos(\Omega't - kz')$ again produce driving terms at twice an optical frequency and at a difference frequency. As before, the sum frequency term gives a negligible contribution. Furthermore, this argument may be repeated for all higher orders of perturbation theory. This suggests that we may make the RWA even when Eqs. (3-28) are solved nonperturbatively, and this is, in fact, the case. An alternative approach to justifying the RWA can be made using the vector model introduced in Section 3.4.2. This approach is discussed in Allen and Eberly (1975) and can be used both to make the RWA plausible and also to get an idea of the effects produced when it is not made. The principal effect of not making the RWA is to introduce a small shift of the resonance frequency known as the Bloch–Siegert shift. This shift is given by $\Delta\omega_0 = \kappa^2 E_0^2/4\omega_0$ and is typically <1 Hz for experiments in the optical region, a completely negligible amount.

Formally, it is convenient to introduce the RWA into Eqs. (3-28) as follows. We note from Eq. (3-33) that ρ_{ab} oscillates at the optical field frequency Ω' just as one would expect from classical considerations. Thus, let us remove the expected Ω' dependence by setting

$$\rho_{ab} \equiv \tilde{\rho}_{ab}e^{-i(\Omega't - kz')} \quad (3\text{-}34)$$

so that $\tilde{\rho}_{ab}$ represents a slowly varying amplitude for ρ_{ab}. On introducing this definition into Eqs. (3-28), and then making the RWA, i.e., ignoring terms which oscillate at twice the optical frequency, one finds that

$$\dot{\rho}_{aa} = n_a\gamma_a + i\frac{\kappa E_0}{2}(\tilde{\rho}_{ba} - \tilde{\rho}_{ab}) - \gamma_a\rho_{aa}$$

$$\dot{\rho}_{bb} = n_b\gamma_b + i\frac{\kappa E_0}{2}(\tilde{\rho}_{ab} - \tilde{\rho}_{ba}) - \gamma_b\rho_{bb} \quad (3\text{-}35)$$

$$\dot{\tilde{\rho}}_{ab} = [i(\Omega' - \omega_0) - \gamma_{ab}]\tilde{\rho}_{ab} + i\frac{\kappa E_0}{2}(\rho_{bb} - \rho_{aa})$$

$$\tilde{\rho}_{ba} = \tilde{\rho}_{ab}^*$$

This is a very important set of equations, which will form the basis for most of our further work. Note that Eqs. (3-35) are a set of coupled first-order linear differential equations with constant coefficients; all oscillating terms have been removed by Eq. (3-34) together with the RWA. This makes possible fairly simple analytic solutions, although not for all cases, as we will see.

3.2.2. Physical Interpretation and Applicability

Since Eqs. (3-35) are basic to the rest of the chapter, it may be useful to briefly review their physical meaning as well as the scope of their applicability. Equations (3-35) describe the behavior of the population matrix elements ρ_{aa}, ρ_{bb}, and $\tilde{\rho}_{ab}$ when an optical field is present. As was discussed earlier, ρ_{aa} and ρ_{bb} are simply the number of molecules/cm^3 at z in levels a and b, respectively, which have velocity v_z. $\tilde{\rho}_{ab}$ is related to $\langle \mu_{ab} \rangle$, the average value of the oscillating dipole moment/cm^3 induced in molecules having velocity v_z, as can be seen by using Eqs. (3-11) and (3-34), which yield

$$\langle \mu_{ab} \rangle = \mu_{ab}\rho_{ba} + \mu_{ba}\rho_{ab} = \mu_{ab} 2 \operatorname{Re}\left[\tilde{\rho}_{ab}\, e^{-i(\Omega' t - kz')}\right] \qquad (3\text{-}36)$$

Thus, the population matrix elements have a rather direct physical significance. The oscillating dipole moment and hence $\tilde{\rho}_{ab}$ is particularly important, since this quantity is the one used to calculate the observable absorption and emission of radiation. The connection is discussed in detail in Section 3.3.1.

Equations (3-35) provide a valid description for more situations than one might guess at first glance. As derived, for example, the equations apply only to an isolated, nondegenerate transition. The molecular transitions that one usually encounters, on the other hand, are degenerate in M_J, the component of rotational angular momentum along a space-fixed axis. However, with only slight extension, Eqs. (3-35) will describe the behavior of such transitions also, since, in the absence of any other static fields, the only field present which can define a space-fixed axis is the optical field itself. Hence, the optical field and the axis of quantization are collinear and we have $\Delta M_J = 0$ selection rules. This means that each lower-state M_J sublevel is optically connected only to the excited-state sublevel having the same M_J. Each of the $M_J \rightarrow M_J$ transitions interacts with the optical field independently, so each may be treated as a two-level system using Eqs. (3-35), and the absorption or emission due to all the sublevels summed at the end of the problem. This analysis will also apply if a static external field is applied along a direction parallel to the electric field vector of the light. If a static electric or magnetic field is applied in a direction perpendicular to **E**, however,

$\Delta M_J = \pm 1$ selection rules will obtain and Eqs. (3-35) will not in general be applicable.

It is also interesting to briefly consider which processes are described by the phenomenological decay rates γ_a, γ_b, and γ_{ab}. As originally introduced in Eq. (3-13), they described the effects of collisions in which the molecules change their rotational state and are thus removed from level a or level b. We will refer to such collisions as state-changing collisions. The upper-state decay rate γ_a need not be the same as the lower-state decay rate γ_b. γ_{ab}, however, is given by $\frac{1}{2}(\gamma_a + \gamma_b)$. To see this, we note that the pure-case density matrix elements ρ'_{aa} and ρ'_{bb} are given by $|c_a|^2$ and $|c_b|^2$, the squares of the probability amplitudes for each level. Since these decay with rates γ_a and γ_b, the probability amplitudes themselves decay as $\exp(-\frac{1}{2}\gamma_a t)$ and $\exp(-\frac{1}{2}\gamma_b t)$. Thus, the pure-case density matrix element $\rho'_{ab} = c_a c_b^*$ decays as $\exp[-\frac{1}{2}(\gamma_a + \gamma_b)t]$. The population matrix is just a sum of pure-case density matrices, so it behaves in the same way. It should be noted, however, that if the dominant collisional decay channel is one in which a molecule in state a collides and ends up in state b, or vice versa, γ_{ab} will not be given by the average of the level decay rates. In fact, Eqs. (3-35) will then not be valid, since the pumping terms $n_a \gamma_a$ and $n_b \gamma_b$ do not properly describe this kind of direct coupling. When the correct pumping terms are inserted, one finds the effective level decay rate is $\gamma_{\text{eff}} = (\gamma_a + \gamma_b)$, with γ_{ab} given by $\gamma_{ab} = \frac{1}{2}\gamma_{\text{eff}}$. This situation was discussed in detail in Chapter 2. It is not of great interest here, since in the infrared or visible region direct collisional coupling between the levels involves vibrational or electronic relaxation processes, and these are generally slow compared to rotational relaxation.

The discussion above should serve to emphasize the fact that although we consider only two levels as far as the optical field is concerned, we are by no means discussing a simple two-level system as far as collisional relaxation is concerned. It is essential that there be a manifold of other levels present to which levels a and b are collisionally coupled if Eqs. (3-35) are to be valid.

In general, one expects the collisional decay rates to be a function of the velocity v_z, since, on the average, molecules with large v_z have greater speed $[(v_x^2 + v_y^2 + v_z^2)^{1/2}]$ than molecules with small v_z, and hence one expects them to make more collisions per unit time, leading to a larger decay rate. Actually, a detailed analysis of the collisional process shows that the speed dependence of collisional decay rates is a function of the intermolecular potential. In particular, no speed dependence is predicted for a dipole–dipole interaction, while an increasing amount of speed dependence should occur for higher-order multipole interactions. The speed dependence of collisional decay rates has been studied in NH_3 by Mattick et al. (1973) and in CH_3F by Grossman et al. (1977).

Another situation where Eqs. (3-35) may not be valid is when the $a \rightarrow b$ transition involves $K \neq 0$ states in symmetric top molecules. Here weak collisions which change only M_J have been shown to have large cross sections. If this process becomes a significant fraction of the total collisional relaxation it can destroy our assumption that each $M_J \to M_J$ transition can be treated as an independent two-level system.

Collisions in which the rotational state changes are not the only type of collision which can occur. One can also have phase-interrupting collisions, in which the level populations ρ_{aa} and ρ_{bb} remain unchanged while ρ_{ab} is destroyed. To see how this can happen, consider a group of N identical molecules which have been prepared in a superposition of states (by an optical pulse, for example), and let each molecule be described by a pure-case density matrix ρ'. At the end of the pulse we will have a nonzero ρ'_{ab} which is the same for all N molecules. Furthermore, if we ignore state-changing collisions, ρ_{ab} after the pulse will be given by $\rho'_{ab} = C \exp(-i\omega_0 t)$, where C is a constant.* Thus, all the molecules are initially oscillating in phase with each other and the population matrix element ρ_{ab} is just $\rho_{ab} = N\rho'_{ab}$, since the ρ'_{ab} all add up in phase. Now suppose that collisions occur in which molecules have their phase of oscillation interrupted or changed. The molecules which collided will have $\rho'_{ab} = C \exp[-i(\omega_0 t + \phi)]$, where ϕ is a random phase which varies from one molecule to another. This will reduce the size of ρ_{ab} since there will be partial cancellations when the ρ'_{ab} are added up. Hence, the phase-interrupting collisions will cause ρ_{ab} to decay even though the level populations remain unchanged.

Phase-interrupting collisions manifest themselves as an increase in γ_{ab} over the value $\frac{1}{2}(\gamma_a + \gamma_b)$ that it would have if only state-changing collisions occurred. They are common in electronic transitions, where one often finds $\gamma_{ab} \gg \gamma_a, \gamma_b$, but a definitive observation of phase-interrupting collisions in vibration–rotation transitions has yet to be made. A detailed discussion of this point may be found in Section 3.10.1.

Although we introduced γ_a, γ_b, and γ_{ab}, as collisional decay rates, the γ's can also be used to describe two other processes, spontaneous emission and transit-time effects. Spontaneous emission results in a change of state, and thus its behavior as far as the population matrix equations are concerned is identical to that for state-changing collisions: γ_a and γ_b need not be equal, but $\gamma_{ab} = \frac{1}{2}(\gamma_a + \gamma_b)$. Also, the same *caveat* discussed in connection with state-changing collisions applies here. If spontaneous emission from level a to level b is the dominant relaxation process, Eqs. (3-35) will not be applicable unless the pumping terms are modified. Dominant $a \to b$ spontaneous emission is most likely to arise when one looks at atomic transitions in which the ground state is the lower level. Transient effects in this situation have been discussed by Schenzle and Brewer (1976).

---

*This may be easily shown by solving Eq. (3-10) for a two-level molecule with no optical field and an initial condition of $\rho'_{ab} = 0$.

Transit-time effects can also be described by the γ's. These effects are a result of the fact that in an actual experiment the optical field which excites the molecules is not a uniform plane wave but rather is a laser *beam* having a definite size and cross-sectional area. Since molecules have velocity components perpendicular to the direction of propagation of the beam, they can move into or out of the beam during the course of an experiment, and the time required for them to do this is called the transit time. Molecules which move out of the beam are effectively lost as far as the experiment is concerned, since one generally only observes the absorption and emission which take place in the small solid angle defined by the laser beam. Note also that as some molecules move out of the beam, new molecules move into the beam. These are accounted for in Eqs. (3-35) by the $n_i\gamma_i$ pump terms, which ensure a thermal-equilibrium Boltzmann distribution for the incoming molecules.

It is by no means obvious that the movement of molecules out of the beam will cause an exponential decay of the population matrix elements. However, the fact is that over the range of times presently accessible to experiment, the ρ_{ii}'s do appear to decay exponentially. This has been verified directly in the time domain as well as in the frequency domain where Lamb dip line shapes appear to remain Lorentzian (thus implying exponential decay) even when the linewidth is transit-time-limited (Hall, 1971, 1973, 1975). A proper theoretical treatment of the transit-time problem is very complicated since laser beams generally have a Gaussian intensity profile, so that each molecule samples different parts of this profile as it moves across the beam. Progress has recently been made on this problem in the frequency domain, though (Bordé *et al.*, 1976), and non-Lorentzian lineshapes are predicted under certain conditions.

In this chapter we will take our cue from the experimental results to date, and treat transit-time effects as an exponential decay. Since the decay is caused by molecules moving out of the beam, one expects that $\gamma_a = \gamma_b = \gamma_{ab}$ when transit-time broadening is dominant. Furthermore, in low-pressure gases, the transit-time decay will be independent of pressure, while collisional decay processes are directly proportional to pressure. This feature allows one to distinguish between transit-time and collisional effects by plotting the total decay rate against pressure. The transit-time decay will then appear as the intercept at zero pressure.

3.3. Absorption and Emission of Radiation

3.3.1. Polarization and Reduced Wave Equations

In the previous section we developed the equations which govern the microscopic behavior of molecules interacting with an optical field. It still remains to relate the microscopic variables (the population matrix elements ρ_{ij}) to the observables (the absorbed or emitted radiation). This topic will be taken up in the present section.

The link between microscopic and macroscopic quantities is provided by Eq. (3-36). This equation relates $\langle \mu_{ab} \rangle$, the quantum-mechanical average value of the oscillating dipole moment (per cm^3), to the off-diagonal population matrix element ρ_{ab}. Now a quantum-mechanical average value may be regarded as either the average value of many measurements on a single molecule or as the average value of a single measurement on an ensemble of molecules. We choose the latter interpretation and identify the average oscillating dipole moment (per cm^3) with $\langle \mu_{ab} \rangle$:

$$\bar{\mu}_{ab}(z, v_z, t) \equiv \langle \mu_{ab} \rangle = 2\mu_{ab} \, \mathrm{Re} \left[\tilde{\rho}_{ab} e^{-i(\Omega't - kz')} \right] \tag{3-37}$$

Since this is the dipole-moment density due to molecules with velocity component v_z, the polarization of the medium, i.e., the macroscopic dipole moment per unit volume, is a summation over all molecular velocities:

$$P(z, t) = \int_{-\infty}^{\infty} \bar{\mu}_{ab}(z, v_z, t) \, dv_z \tag{3-38}$$

Note that to calculate $P(z, t)$ using this equation we must transform $\bar{\mu}_{ab}$ from a coordinate system moving with the molecular velocity v_z (since $\tilde{\rho}_{ab}$ and hence $\bar{\mu}_{ab}$ is calculated in this coordinate system) back to the laboratory frame. This transformation is trivial, however, because we usually need only replace Ω' by $\Omega - kv_z$ and $\Omega't - kz'$ by $\Omega t - kz$.

Here we are assuming that a single nondegenerate transition is being excited. If we have a degeneracy in M_J present as discussed earlier, Eq. (3-38) must be modified by including an additional summation over the various $M_J \rightarrow M_J$ components of the transition.

We now must relate the polarization of the medium to the absorbed or emitted radiation. This connection is provided by Maxwell's equations. In a dilute gas where no free charges are present, we have (in mks units)

$$\left. \begin{array}{ll} \boldsymbol{\nabla} \cdot \mathbf{D} = 0, & \boldsymbol{\nabla} \times \boldsymbol{\mathscr{E}} = -\dfrac{\partial \mathbf{B}}{\partial t} \\[2ex] \boldsymbol{\nabla} \cdot \mathbf{B} = 0, & \boldsymbol{\nabla} \times \mathbf{H} = \dfrac{\partial D}{\partial t} \end{array} \right\} \tag{3-39}$$

with

$$\begin{aligned} \mathbf{D} &= \varepsilon_0 \boldsymbol{\mathscr{E}} + \mathbf{P} \\ \mathbf{B} &= \mu_0 \mathbf{H} \end{aligned} \tag{3-40}$$

These equations may be cast in a more useful form by taking the curl of the

$\nabla \times \mathscr{E}$ equation and substituting the time derivative of the $\nabla \times \mathbf{H}$ equation for the right-hand side, using Eqs. (3-40) as needed. We obtain

$$\nabla \times \nabla \times \mathscr{E} + \mu_0 \varepsilon_0 \frac{\partial^2 \mathscr{E}}{\partial t^2} = -\mu_0 \frac{\partial^2 \mathbf{P}}{\partial t^2} \tag{3-41}$$

This is a wave equation which describes the electric field $\mathscr{E}$ propagating through a medium whose polarization is given by $\mathbf{P}$. In general, Eq. (3-41) together with the population matrix equations (3-35) form a coupled set of equations which must be solved self-consistently; i.e., the electric field $\mathscr{E}$ which is driven by $\mathbf{P}$ in Eq. (3-41) is the same field which must be inserted in the population matrix equations used to calculate $\mathbf{P}$ in the first place. In pulse propagation problems or any other situation involving optically thick media, this must be done somehow. There are some special cases where analytic solutions may be obtained, but in general numerical solutions are required (cf. Sargent *et al.*, 1974; Allen and Eberley, 1975). We are interested in optically thin samples of dilute gases, however. This allows us to simplify the problem enormously, because by definition an incident field $\mathbf{E}$ passes through an optically thin sample essentially unchanged. Thus, any field reradiated by the molecules as a result of their polarization is so small that it can be neglected compared to $\mathbf{E}$ when the polarization itself is calculated.

In practice, then, we will calculate $\mathbf{P}$ using only the incident field $\mathbf{E}$. That $\mathbf{P}$ is then regarded as a known, fixed source term in the wave equation, and the field $\mathscr{E}$ is regarded as being only the field *reradiated* by the molecules, rather than the total field. A calculation of $\mathscr{E}$ then yields both the absorption or emission and the index of refraction for the sample. This procedure will be justified if $|\mathscr{E}| \ll |\mathbf{E}|$. Notice that for a sample to be optically thin, it is not sufficient to simply have the absorption small; the index of refraction must also be near unity. A thin piece of glass illuminated by visible light, for example, is by no means an optically thin sample! Although the light is not attenuated appreciably by the glass, the reradiated fields are so large that the apparent velocity of light through the glass is only $0.667c$ (the index of refraction is 1.5). For dilute gases, however, we need not worry about this, since their index of refraction is always near unity if the absorption is small.

Even though we take $\mathbf{P}$ to be a known source term in Eq. (3-41), this wave equation still is not soluble as it stands and must be simplified. The first step in the simplification is to note that since we have taken the incident optical field which produces $\mathbf{P}$ to be a linearly polarized plane wave, $\mathbf{P}$, and hence $\mathscr{E}$, will be plane waves also:

$$\mathbf{P} = \hat{x}[P_c(z, t) \cos (\Omega t - kz) + P_s(z, t) \sin (\Omega t - kz)]$$

$$\mathscr{E} = \hat{x}[\mathscr{E}_c(z, t) \cos (\Omega t - kz) + \mathscr{E}_s(z, t) \sin (\Omega t - kz)] \tag{3-42}$$

This assumes that the gas is not birefringent or optically active. Note that we allow $\mathbf{P}$ and $\mathscr{E}$ to have both in-phase and out-of-phase components, since we do not know in advance what the phase of **P** relative to the incident field $\mathbf{E} = \hat{x}[E_0 \cos{(\Omega t - kz)}]$ will be.

The plane-wave amplitudes P_c, P_s, $\mathscr{E}_c$, and $\mathscr{E}_s$ are functions of z and t only. Thus, $\mathscr{E}$ is not a function of x or y and one finds that $\nabla \times \nabla \times \mathscr{E} = -\partial^2 \mathscr{E}/\partial z^2$. The wave equation then becomes

$$\frac{\partial^2 \mathscr{E}}{\partial z^2} - \frac{1}{c^2}\frac{\partial^2 \mathscr{E}}{\partial t^2} = \mu_0 \frac{\partial^2 \mathbf{P}}{\partial t^2} \tag{3-43}$$

since $c = (\mu_0 \varepsilon_0)^{-1/2}$. If Eqs. (3-42) are now substituted into Eq. (3-43) and the derivatives worked out explicitly, one obtains two equations for the amplitudes only,

$$\frac{\partial^2 \mathscr{E}_c}{\partial z^2} - \frac{1}{c^2}\frac{\partial^2 \mathscr{E}_c}{\partial t^2} - 2k\left(\frac{\partial \mathscr{E}_s}{\partial z} + \frac{1}{c}\frac{\partial \mathscr{E}_s}{\partial t}\right) = \mu_0\frac{\partial^2 P_c}{\partial t^2} + 2\Omega\mu_0\frac{\partial P_s}{\partial t}\Omega^2\mu_0 P_c$$

$$\frac{\partial^2 \mathscr{E}_s}{\partial z^2} - \frac{1}{c^2}\frac{\partial^2 \mathscr{E}_s}{\partial t^2} + 2k\left(\frac{\partial \mathscr{E}_c}{\partial z} + \frac{1}{c}\frac{\partial \mathscr{E}_c}{\partial t}\right) = \mu_0\frac{\partial^2 P_s}{\partial t^2} - 2\Omega\mu_0\frac{\partial P_c}{\partial t} - \Omega^2\mu_0 P_s \tag{3-44}$$

This result follows from the fact that Eq. (3-43) holds for all z and t, so that the sin and cos terms may be equated separately. We have also used the fact that $k^2 = \Omega^2/c^2$.

The wave equation as given by (3-44) is now in a form which can be simplified even further by making what is often called the slowly varying envelope approximation. It consists of neglecting the second derivatives of the amplitudes $\mathscr{E}_c$, $\mathscr{E}_s$, and all the derivatives of P_c and P_s. This procedure is justified on the grounds that these amplitudes are "slowly varying," where the term "slowly varying" here means that the electric field and polarization amplitudes vary slowly in time compared to the optical frequency Ω, and slowly in space compared to an optical wavelength. This condition is easily satisfied for all situations discussed in this chapter. On making the slowly varying envelope approximation, Eqs. (3-44) become

$$\frac{\partial \mathscr{E}_c}{\partial z} + \frac{1}{c}\frac{\partial \mathscr{E}_c}{\partial t} = -\frac{\Omega}{2\varepsilon_0 c}P_s$$

$$\frac{\partial \mathscr{E}_s}{\partial z} + \frac{1}{c}\frac{\partial \mathscr{E}_s}{\partial t} = \frac{\Omega}{2\varepsilon_0 c}P_c \tag{3-45}$$

A somewhat simple-minded but intuitive justification of this approximation is given in Appendix A. A more rigorous justification can be made using the formal Green's function solutions of Eqs. (3-44) and (3-45). When these solutions are compared, we find that they are identical provided that $\partial P_c/\partial t \ll \Omega P_s$ (and $\partial P_s/\partial t \ll \Omega P_c$), and that a negligible amount of $\mathscr{E}_c$ and $\mathscr{E}_s$ are scattered in the backward direction (Crisp, 1969). The polarization inequalities follow from the slowly varying assumption (see Appendix A), while backward scattering only occurs in the rare situation where $\mathscr{E}_c$ and $\mathscr{E}_s$ vary rapidly and have phase and group velocities in the medium which are greatly different (Crisp, 1969).

The reduced wave equations given in Eqs. (3-45) are now easily soluble for the case of an optically thin medium since P_s and P_c are known functions which we calculate from the microscopic equations of motion. Furthermore, P_s and P_c are then only functions of $(t-t_1)$, where t_1 is the time at which an experiment is begun (e.g., in an optical nutation experiment, t_1 is the time at which the step function pulse is applied; in a photon echo experiment it is the time at which the first pulse is applied). The only z dependence of the polarization is in the boundary condition t_1. To solve Eq. (3-45) we make a change of variables. Instead of writing $\mathscr{E}_c$ and P_s in terms of the variables (z, t), suppose that we express them in terms of (z, t_R), where

$$t_R = t - z/c \tag{3-46}$$

is called the retarded time. This combination of variables greatly simplifies Eqs. (3-45) because

$$\frac{\partial \mathscr{E}_c(z, t_R)}{\partial z} = \frac{\partial \mathscr{E}_c(z, t)}{\partial z} \frac{\partial z}{\partial z} + \frac{\partial \mathscr{E}_c(z, t)}{\partial t} \frac{\partial (t_R + z/c)}{\partial z}$$

$$= \left(\frac{\partial}{\partial z} + \frac{1}{c} \frac{\partial}{\partial t} \right) \mathscr{E}_c(z, t) \tag{3-47}$$

Equation (3-47) is just an expression of the chain rule for partial derivatives. The reduced wave equations are now

$$\frac{\partial}{\partial z} \mathscr{E}_c(z, t_R) = -\frac{\Omega}{2\varepsilon_0 c} P_s(t_R + z/c - t_1)$$

$$\frac{\partial}{\partial z} \mathscr{E}_s(z, t_R) = \frac{\Omega}{2\varepsilon_0 c} P_c(t_R + z/c - t_1) \tag{3-48}$$

Notice that by using the retarded time we have effectively transformed into a rest frame for pulses of light; i.e., the transformation removes the time dependence of $\mathscr{E}_c$ and $\mathscr{E}_s$, owing to the fact that the field is propagating through the sample. Note, for example, that a pulse whose envelope is $\mathscr{E}_c(t_R, z)$ has $\mathscr{E}_c = constant$ if it propagates through a region where $P_s = 0$. This device of transforming into the retarded time frame is widely used in electrodynamics.

Notice that the transformation, while simplifying the differential equations, has apparently also introduced a z dependence into P_s and P_c since $t - t_1 = t_R + z/c - t_1$. However, we have not yet specified the boundary condition t_1. We said that t_1 is the time at which the first pulse is applied. But it takes time for light to propagate through the sample, so t_1 must be different for different positions z. If we let the sample be located between $z = 0$ and $z = L$, and let τ_1 be the time at which the first pulse reaches $z = L$, then the boundary condition for the polarization at position z is $t_1 = \tau_1 - (L - z)/c$. Putting this into Eq. (3-48) we see that the z dependence of P_c and P_s disappears, since

$$t_R + z/c - t_1 = t_R - \tau_1 + L/c$$

Thus, Eqs. (3-48) may be integrated trivially to give

$$\mathscr{E}_c(L, t_R) - \mathscr{E}_c(0, t_R) = -\frac{\Omega L}{2\varepsilon_0 c} P_s(t_R - \tau_1 + L/c)$$

$$\mathscr{E}_s(L, t_R) - \mathscr{E}_s(0, t_R) = \frac{\Omega L}{2\varepsilon_0 c} P_c(t_R - \tau_1 + L/c)$$

(3-49)

Since the reradiated field propagates in the same direction as the driving field, there is no reradiated field in front of the sample. Hence, $\mathscr{E}_c(0, t_R) = \mathscr{E}_s(0, t_R) = 0$. If we now transform back out of the reduced time frame, we have

$$\mathscr{E}_c(L, t) = -\frac{\Omega L}{2\varepsilon_0 c} P_s(t - \tau_1)$$

$$\mathscr{E}_s(L, t) = \frac{\Omega L}{2\varepsilon_0 c} P_c(t - \tau_1)$$

(3-50)

Notice how simple this result is. Given a polarization $P_s(t - t_1)$, the output field $\mathscr{E}_c$ leaving the end of the sample is just proportional to $P_s(t - t_1)$ evaluated at $z = L$ ($t_1 = \tau_1$ at $z = L$).

Having obtained the reradiated field $\mathscr{E}$, we see that the total field leaving the sample is

$$E_T = (E_0 + \mathscr{E}_c) \cos(\Omega t - kL) + \mathscr{E}_s \sin(\Omega t - kL)$$

(3-51)

and the corresponding intensity (in mks units) is

$$I = c\varepsilon_0 \langle E_T \rangle_{\text{av}}^2$$

$$= \tfrac{1}{2}c\varepsilon_0(E_0{}^2 + 2E_0\mathscr{E}_c + \mathscr{E}_c{}^2 + \mathscr{E}_s{}^2)$$

(3-52)

In Eq. (3-52), the symbol $\langle E_T \rangle_{\text{av}}$ denotes the time average of E_T over several cycles of the optical field. Since $\mathscr{E}_c$ and $\mathscr{E}_s$ are much smaller than E_0, the terms $\mathscr{E}_c{}^2$ and $\mathscr{E}_s{}^2$ may be neglected and we have

$$I = I_0 + c\varepsilon_0 E_0 \mathscr{E}_c$$

(3-53)

This equation gives the absorption or emission of radiation by the sample. It, together with Eqs. (3–37), (3–38), (3–42), and (3-50), provides the desired link between the microscopic variables and the observable quantity, which is

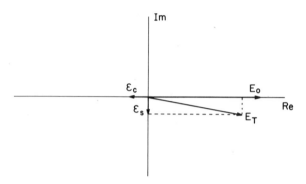

Figure 3-3. Phasor diagram for an optically thin sample showing relation between input field E_0, reradiated fields $\mathscr{E}_c$ and $\mathscr{E}_s$, and total field E_T leaving sample.

the transmitted intensity. Notice that only the in-phase component $\mathscr{E}_c$ contributes to absorption when we have an optically thin sample.

The effect of $\mathscr{E}_s$, the reradiated field component 90° out of phase with the driving field, is to produce an index of refraction which differs from unity. This can be seen quite simply by referring to Figure 3-3. We see on the phasor diagram that $\mathscr{E}_c$ reduces the size of E_T if the medium is absorbing, while the major effect of $\mathscr{E}_s$ is to retard the phase of E_T. A phase retardation is nothing more than a reduction of the apparent velocity of light through the sample, however, and this is what the refractive index n describes. More precisely,

$$n = 1 + \frac{c}{\Omega L} \frac{\mathscr{E}_s}{E_0} \tag{3-54}$$

for an optically thin medium.* Since index of refraction effects are not usually measured in transient-effect experiments, we will not consider them in further detail.

3.3.2. Steady-State Absorption: An Example

In order to illustrate calculational procedures outlined thus far, we consider next the well-known example of steady-state absorption by an isolated nondegenerate transition. Our starting point is the population matrix equations (3-35). An examination of these differential equations shows that they form a set of damped equations (due to the $\gamma_{ij}\rho_{ij}$ terms) which contain *no* oscillatory driving terms. Thus, if a constant-amplitude

---

*An excellent discussion of reradiated fields and the origin of the index of refraction can be found in Feynman *et al.* (1963).

optical field E_0 is on for a sufficiently long time, each population matrix element will eventually settle down to a constant value. But this means that $\dot{\rho}_{aa}$, $\dot{\rho}_{bb}$, and $\dot{\rho}_{ab}$ are all zero, so that Eqs. (3-35) become the set of algebraic equations

$$0 = n_a\gamma_a + i\frac{\kappa E_0}{2}(\tilde{\rho}_{ba} - \tilde{\rho}_{ab}) - \gamma_a\rho_{aa}$$

$$0 = n_b\gamma_b + i\frac{\kappa E_0}{2}(\tilde{\rho}_{ab} - \tilde{\rho}_{ba}) - \gamma_b\rho_{bb} \qquad (3\text{-}55)$$

$$0 = [i(\Omega' - \omega_0) - \gamma_{ab}]\tilde{\rho}_{ab} + i\frac{\kappa E_0}{2}(\rho_{bb} - \rho_{aa})$$

These are readily solved by dividing the first two equations by γ_a and γ_b, respectively, and subtracting them to obtain

$$\rho_{aa} - \rho_{bb} = (n_a - n_b) + i\frac{\kappa E_0}{2}(\tilde{\rho}_{ba} - \tilde{\rho}_{ab})\left(\frac{1}{\gamma_a} + \frac{1}{\gamma_b}\right)$$

$$\tilde{\rho}_{ab} = \frac{(\kappa E_0/2)(\rho_{aa} - \rho_{bb})}{(\Omega' - \omega_0) + i\gamma_{ab}} \qquad (3\text{-}56)$$

which yields

$$\rho_{aa} - \rho_{bb} = (n_a - n_b)\frac{(\Omega' - \omega_0)^2 + \gamma_{ab}^2}{(\Omega' - \omega_0)^2 + (\gamma_{ab}/\gamma_{av})\kappa^2 E_0^2 + \gamma_{ab}^2}$$

$$\tilde{\rho}_{ab} = \frac{\frac{1}{2}\kappa E_0(n_a - n_b)[(\Omega' - \omega_0) - i\gamma_{ab}]}{(\Omega' - \omega_0)^2 + (\gamma_{ab}/\gamma_{av})\kappa^2 E_0^2 + \gamma_{ab}^2} \qquad (3\text{-}57)$$

with

$$\frac{1}{\gamma_{av}} = \frac{1}{2}\left(\frac{1}{\gamma_a} + \frac{1}{\gamma_b}\right)$$

We see here that $\tilde{\rho}_{ab}$ exhibits a power-broadened Lorentzian denominator as expected. Note also that the population difference $\rho_{aa} - \rho_{bb}$ remains near its thermal equilibrium value of $(n_a - n_b)$ when the optical field is small, but approaches zero for large fields (κE_0 large). Substituting the expression for $\tilde{\rho}_{ab}$ into Eq. (3-37), we obtain

$$\bar{\mu}_{ab}(z, v_z, t) = \frac{\mu_{ab}\kappa E_0(n_a - n_b)}{\delta^2 + \Gamma^2}[\delta \cos(\Omega't - kz') - \gamma_{ab}\sin(\Omega't - kz')] \quad (3\text{-}58)$$

where $\delta = \Omega' - \omega_0$ and $\Gamma^2 = (\gamma_{ab}/\gamma_{av})\kappa^2 E_0^2 + \gamma_{ab}^2$. In this equation $\bar{\mu}_{ab}$ is the oscillating dipole moment (per cm^3) due to molecules with velocity component v_z. To obtain the polarization, we rewrite $\bar{\mu}_{ab}$ in terms of laboratory-frame quantities to explicitly display the velocity dependence and integrate over v_z as indicated in Eq. (3-38). Velocity dependences here exist only in $\delta = \Omega - kv_z - \omega_0$ and in n_a and n_b. n_a, for example, is the thermal equilibrium number of molecules (per cm^3) in level a having velocity v_z, i.e.,

$$n_a = \frac{N_a e^{-v_z^2/\bar{u}^2}}{\bar{u}\sqrt{\pi}} \tag{3-59}$$

where

$$\bar{u} = \sqrt{\frac{2k_B T}{m}}$$

Here we have assumed a Maxwell–Boltzmann distribution of velocities with $\bar{u}$ being the most probable molecular speed and N_a being the total number of molecules/cm^3 in level a. n_b is given by an equation identical to Eq. (3-59).

We now have

$$P(z, t) = \frac{\mu_{ab}\kappa E_0(N_a - N_b)}{\bar{u}\sqrt{\pi}} \int_{-\infty}^{\infty} e^{-v_z^2/\bar{u}^2} \left[\frac{\delta}{\delta^2 + \Gamma^2} \cos(\Omega t - kz) \right.$$

$$\left. - \frac{\gamma_{ab}}{\delta^2 + \Gamma^2} \sin(\Omega t - kz) \right] dv_z \tag{3-60}$$

At the power levels usually encountered in cw lasers, the Lorentzian response $1/\delta^2 + \Gamma^2)$ due to a single-velocity group of molecules is a much narrower function of velocity than the Gaussian velocity distribution $\exp(-v_z^2/\bar{u}^2)$. Thus, to a good approximation we may pull the Gaussian outside the integral and evaluate it at the peak of the Lorentzian [i.e., at $v_z = (\Omega - \omega_0)/k$], since it is a slowly varying function compared to the Lorentzian factors. We then see that the integral of the first term vanishes since $\delta/(\delta^2 + \Gamma^2)$ is an odd function of v_z, while the second term gives $\gamma_{ab}\pi/k\Gamma$. Thus,

$$P(z, t) = \left[\frac{\mu_{ab}\kappa E_0\sqrt{\pi}}{k\bar{u}} \frac{\gamma_{ab}}{\Gamma}(N_b - N_a)e^{-(\Omega-\omega_0)^2/k^2\bar{u}^2} \right] \sin(\Omega t - kz) \tag{3-61}$$

To find the absorption we now use the reduced wave equations (3-45). On comparison with Eqs. (3-42) we see that the amplitude P_s in those equations is just the term in brackets in (3-61) while $P_c = 0$. Hence, from Eq. (3-50),

$$\mathscr{E}_c(z = L) = -\frac{\mu_{ab}\kappa E_0\Omega\sqrt{\pi}L}{2c\varepsilon_0 k\bar{u}} \frac{\gamma_{ab}}{\Gamma}(N_b - N_a)e^{-(\Omega-\omega_0)^2/k^2\bar{u}^2} \tag{3-62}$$

The wave equation for $\mathscr{E}_s$ gives $\mathscr{E}_s = 0$, since the driving term P_c is zero.

The total field leaving the end of the sample is then $E = (E_0 + \mathscr{E}_c)\cos(\Omega t - kz)$, and the corresponding intensity is

$$I = \tfrac{1}{2} c\varepsilon_0 (E_0 + \mathscr{E}_c)^2$$

$$\approx I_0 - \frac{\sqrt{\pi}\hbar\Omega\kappa^2 E_0^2 L\gamma_{ab}(N_b - N_a)}{2k\bar{u}\Gamma} e^{-(\Omega-\omega_0)^2/k^2\bar{u}^2} \tag{3-63}$$

This is the well-known result for the absorption due to a Doppler-broadened transition in an optically thin medium, with the second term of Eq. (3-63) being just $\alpha L I_0$, where α is the absorption coefficient. Notice that the absorption profile is a broad Gaussian typically many MHz wide, whereas the response due to a single-velocity group of molecules is a narrow Lorentzian of width Γ, which is typically 1 MHz or less [see Eq. (3-58)]. At any given incident frequency Ω, only a small fraction of the molecules in levels a and b are interacting resonantly—the velocity group v_z such that $\Omega - kv_z - \omega_0 = 0$. However, as the frequency is scanned, other velocity groups move into resonance, so we cannot see the narrow resonant response of any one velocity group. The Gaussian can thus be thought of as a superposition of the Lorentzian responses due to each velocity group.

Our treatment here gives an interesting picture of the absorption process. The optical field gives rise to a polarization which reradiates a field 180° out of phase with the incident field. The resulting destructive interference produces a field coming out of the sample which contains less energy than the incident field. Although this may seem to be a peculiar way of describing absorption, it is a useful picture, since among other things it suggests the existence of an effect called "optical free induction decay." Suppose that we have a sample which is absorbing light in steady state. If we suddenly cut off the driving field so that $E_0 = 0$, the derivation we have just done says that the sample will then *emit* radiation, the field $\mathscr{E}_c$, which was 180° out of phase with E_0. This phenomenon will be discussed in detail in Section 3.7.

Finally, notice that there are apparently no dispersive (refractive index) effects associated with the absorption; since the phase of the total field was not retarded by passage through the sample, it remained $\cos(\Omega t - kz)$. This result is not quite correct. Dispersive effects are taken into account in our theory by $\mathscr{E}_s$, and we found $\mathscr{E}_s = 0$. However, this was the result of an approximation. In Eq. (3-60), we pulled the Gaussian factor $\exp(-v_z^2/\bar{u}^2)$ out of the integral for P, and this caused P_c and hence $\mathscr{E}_s$ to vanish. This is only a good approximation for the factor $\delta/(\delta^2 + \Gamma^2)$ near Doppler line center, where index-of-refraction effects are, in fact, very small. Had we not removed the Gaussian, we would have obtained a correct description of the index.

3.4. Solutions of the Population Matrix Equations

3.4.1. Introduction

In Section 3.2.1 we developed a set of population matrix equations (3-35) which describe the behavior of an isolated, nondegenerate transition when an optical field is present. Unfortunately, explicit closed-form solutions to these equations can only be obtained for the steady-state case discussed in the previous section. To obtain solutions to the transient problem, we are forced to make further simplifications.

To see where the problem lies, we begin by rewriting Eqs. (3-35) in a somewhat more convenient form. We notice that the $\tilde{\rho}_{ab}$ equation is coupled only to the population difference $(\rho_{aa} - \rho_{bb})$. When we subtract the ρ_{aa} and ρ_{bb} equations to obtain an equation for this difference, however, we find that it is, in turn, coupled to the population sum $(\rho_{aa} + \rho_{bb})$. Explicitly, the equations become

$$\dot{\rho}_{aa} + \dot{\rho}_{bb} = n_a\gamma_a + n_b\gamma_b - \tfrac{1}{2}(\gamma_a + \gamma_b)(\rho_{aa} + \rho_{bb}) - \tfrac{1}{2}(\gamma_a - \gamma_b)(\rho_{aa} - \rho_{bb})$$

$$\dot{\rho}_{aa} - \dot{\rho}_{bb} = n_a\gamma_a - n_b\gamma_b - i\kappa E_0(\tilde{\rho}_{ab} - \tilde{\rho}_{ba}) - \tfrac{1}{2}(\gamma_a + \gamma_b)(\rho_{aa} - \rho_{bb})$$

$$\qquad\qquad - \tfrac{1}{2}(\gamma_a - \gamma_b)(\rho_{aa} + \rho_{bb}) \qquad\qquad\qquad (3\text{-}64)$$

$$-i\dot{\tilde{\rho}}_{ab} = [(\Omega' - \omega_0) + i\gamma_{ab}]\tilde{\rho}_{ab} - \frac{\kappa E_0}{2}(\rho_{aa} - \rho_{bb})$$

To obtain the relaxation terms in the form shown, we have added and subtracted $\tfrac{1}{2}\gamma_a\rho_{bb} + \tfrac{1}{2}\gamma_b\rho_{aa}$ in the first two equations. Now this set of three differential equations is actually four equations, because $\tilde{\rho}_{ab}$ is a complex quantity and is coupled to the $\rho_{aa} - \rho_{bb}$ equation only by its imaginary part, $2i\,\mathrm{Im}\,\tilde{\rho}_{ab} = \tilde{\rho}_{ab} - \tilde{\rho}_{ba}$. Thus, we may convert Eqs. (3-64) into a set of differential equations involving real quantities only by writing $\tilde{\rho}_{ab}$ in terms of its real and imaginary parts. Let

$$2\tilde{\rho}_{ab} \equiv u - iv \qquad\qquad (3\text{-}65)$$

and also for convenience set

$$w \equiv \rho_{aa} - \rho_{bb}$$
$$\qquad\qquad\qquad\qquad (3\text{-}66)$$
$$s \equiv \rho_{aa} + \rho_{bb}$$

On substituting these definitions into Eqs. (3-64) and separately equating the real and imaginary parts of the $\tilde{\rho}_{ab}$ equation, we find that

$$\dot{u} = (\Omega' - \omega_0)v - \gamma_{ab}u$$
$$\dot{v} = -(\Omega' - \nu_0)u + \kappa E_0 w - \gamma_{ab}v$$
$$\dot{w} = n_a\gamma_a - n_b\gamma_b - \kappa E_0 v - \tfrac{1}{2}(\gamma_a + \gamma_b)w - \tfrac{1}{2}(\gamma_a - \gamma_b)s \qquad (3\text{-}67)$$
$$\dot{s} = n_a\gamma_a + n_b\gamma_b - \tfrac{1}{2}(\gamma_a + \gamma_b)s - \tfrac{1}{2}(\gamma_a - \gamma_b)w$$

This set of equations is completely equivalent to our original set of population matrix equations (3-35), but is more convenient to use for transient problems. Being a set of linear differential equations with constant (real) coefficients, they can in principle be solved by any one of several standard techniques. In addition, the u and v which appear in Eqs. (3-67) are more directly related to the polarization amplitudes P_c and P_s than $\tilde{\rho}_{ab}$ is. To see this, we simply substitute the definition of u and v, Eq. (3-65), into Eqs. (3-37) and (3-38), obtaining

$$P(z, t) = \mu_{ab} \int_{-\infty}^{\infty} u \, dv_z \cos{(\Omega t - kz)} - \mu_{ab} \int_{-\infty}^{\infty} v \, dv_z \sin{(\Omega t - kz)}$$

$$(3\text{-}68)$$

for the polarization. On comparison with Eq. (3-42), we find

$$P_c = \mu_{ab} \int_{-\infty}^{\infty} u \, dv_z, \qquad P_s = -\mu_{ab} \int_{-\infty}^{\infty} v \, dv_z \qquad (3\text{-}69)$$

The quantities u and v (when multiplied by μ_{ab}) are thus exactly the in- and out-of-phase components of the polarization for molecules of a given velocity v_z. This is a very convenient property.

Despite the fact that standard techniques are available to solve Eqs. (3-67), when one begins to actually do this, one finds that the solutions can only be expressed in terms of the roots of a characteristic quartic equation which is associated with the set of four differential equations. While general expressions do exist for the roots of a quartic, they are so complicated when written out explicitly that the result is worthless.

In order to obtain usable solutions of Eqs. (3-67), we must look for special situations in which the equations simplify. In particular, let us examine under what conditions one or more of the equations can be uncoupled from the rest of the set, since this will then reduce the characteristic quartic equation to a cubic or quadratic one. We see that there are three cases in which the equations simplify: (A) $\Omega' - \omega_0 = 0$, (B) $\kappa E_0 = 0$, and (C) $\gamma_a - \gamma_b = 0$.

Let us consider these three cases in turn. In case A we suppose that the molecules are excited exactly on resonance so that $\Omega' - \omega_0 = 0$. This will uncouple the $\dot{u}$ equation, giving a decaying exponential solution for u and a set of three coupled equations for v, w, and s. This case is not of great interest here, since we are interested in Doppler-broadened transitions where we cannot satisfy the condition $\Omega' - \omega_0 = \Omega - kv_z - \omega_0 = 0$ simultaneously for all molecular velocities v_z.

Case B is more interesting. Suppose that we suddenly turn the optical field off, so that $\kappa E_0 = 0$. This uncouples the $\dot{v}$ and $\dot{w}$ equations, and the set of four equations falls apart into two sets of two coupled equations which are trivial to solve. This case describes the behavior of systems following an optical pulse, and thus can be used in the treatment of optical free induction decay, photon echoes, and delayed optical nutation. Since the solutions are useful, let us work them out explicitly. Suppose that the optical field is suddenly turned off at time $t = t_k$. The $\dot{u}$ and $\dot{v}$ equations are then

$$\dot{u} = \delta v - \gamma_{ab} u$$
$$\dot{v} = -\delta u - \gamma_{ab} v \tag{3-70}$$

where we have set $\delta = \Omega' - \omega_0$. These equations can almost be solved by inspection. The decay terms can be removed by setting $u = \tilde{u} \exp(-\gamma_{ab}t)$ and $v = \tilde{v} \exp(-\gamma_{ab}t)$ to obtain $\dot{\tilde{u}} = \delta \tilde{v}$ and $\dot{\tilde{v}} = -\delta \tilde{u}$. But these equations have simple $\sin \delta t$ and $\cos \delta t$ solutions, so that

$$u = (A \sin \delta t + B \cos \delta t) e^{-\gamma_{ab}t}$$
$$v = (A \cos \delta t - B \sin \delta t) e^{-\gamma_{ab}t} \tag{3-71}$$

where A and B are constants determined by the initial conditions. If we let $u = u(t_k)$, $v = v(t_k)$ at $t = t_k$, we obtain

$$u(t) = e^{-\gamma_{ab}(t-t_k)} [u(t_k) \cos \delta(t-t_k) + v(t_k) \sin \delta(t-t_k)]$$
$$v(t) = e^{-\gamma_{ab}(t-t_k)} [v(t_k) \cos \delta(t-t_k) - u(t_k) \sin \delta(t-t_k)] \tag{3-72}$$

The equations for w and s are most easily solved by adding and subtracting the last two of Eqs. (3-67) to regain our original equations for ρ_{aa} and ρ_{bb}. For $\kappa E_0 = 0$, these are

$$\dot{\rho}_{aa} = n_a \gamma_a - \gamma_a \rho_{aa}$$
$$\dot{\rho}_{bb} = n_b \gamma_b - \gamma_b \rho_{bb} \tag{3-73}$$

The equations are uncoupled and clearly have exponential solutions. Solving them and inserting the boundary conditions as before, we find

$$\rho_{aa}(t) = n_a + e^{-\gamma_a(t-t_k)} [\rho_{aa}(t_k) - n_a]$$
$$\rho_{bb}(t) = n_b + e^{-\gamma_b(t-t_k)} [\rho_{bb}(t_k) - n_b] \tag{3-74}$$

The sum and difference of these solutions gives w and s.

3.4.2. Optical Bloch Equations

Case C is by far the most useful and well-known special case of the population matrix equations, and the remainder of this section will be devoted to it. We assume that the upper- and lower-state decay rates are equal; i.e., $\gamma_a = \gamma_b$. Then $\gamma_a - \gamma_b = 0$, and the $\dot{w}$ and $\dot{s}$ equations become uncoupled, with the $\dot{s}$ equation giving just $s = constant$ (assuming the usual case in which $\rho_{aa} + \rho_{bb} = n_a + n_b$ at the start of an experiment). The remaining three equations become

$$\dot{u} = (\Omega' - \omega_0)v - \gamma_{ab}u$$

$$\dot{v} = -(\Omega' - \omega_0)u + \kappa E_0 w - \gamma_{ab}v \qquad (3\text{-}75)$$

$$\dot{w} = -\kappa E_0 v - \gamma[w - (n_a - n_b)]$$

where $\gamma = \gamma_a = \gamma_b$. These are the famous "optical Bloch equations," so called because they are exactly analogous to the Bloch equations of nmr. In fact, Eqs. (3-75) will become the Bloch equations if we simply interpret ω_0 as being $-gH_0$ with g the gyromagnetic ratio and H_0 a static magnetic field, and interpret κE_0 as being gH_1, where H_1 is a small oscillating magnetic field of frequency Ω' applied at right angles to H_0. The quantities u, v, and w are then the components of the magnetization due to the nuclear spins (in a coordinate frame rotating at frequency Ω'), and γ and γ_{ab} are usually written as $1/T_1$ and $1/T_2$, respectively, where T_1 and T_2 are the longitudinal and transverse spin relaxation times.*

The correspondence between the Eqs. (3-75) and the Bloch equations should not come as a complete surprise, since our equations describe transitions between two levels in a molecule and the Bloch equations describe transitions between the two levels of spin $\frac{1}{2}$ particles. Although in our case the levels are connected via an electric-dipole transition, while in nmr the levels are connected via a magnetic-dipole transition, this does not alter the fundamental form of the equations.

The correspondence with nmr is valuable in two respects. First, since Eqs. (3-75) and the Bloch equations have the same functional form, a great deal is known about the solutions of (3-75) in many different situations. Pulsed nmr is a fairly mature subject, and many years have been devoted to examining various solutions of the Bloch equations. This work may be used to considerable advantage both in analyzing optical coherent transient-effect experiments and in looking for new optical effects. Second, the nmr correspondence allows us to develop a geometrical picture for transitions between two levels in a molecule. This picture, which is often called the vector model, is useful pedagogically since it allows one to visualize the behavior of the system during transient-effect experiments.

Before discussing either the vector model or solutions of the optical Bloch equations, however, we should consider the conditions under which the equations will be valid. As we saw, the optical Bloch equations are a special case of the population matrix equations (3-67), the case where the upper- and lower-state populations relax with equal rates, $\gamma_a = \gamma_b$. While many systems do have these rates equal, there are several others which

---

*For a discussion of the Bloch equations for nuclear spins, see any elementary textbook on nmr (e.g., Slichter, 1963).

definitely do not, and one must beware of applying the optical Bloch equations indiscriminately to any two-level system. In the discussion at the end of Section 3.2.2 we saw that γ_a and γ_b contain contributions from several relaxation processes, the principal ones being transit time, collisional relaxation, and spontaneous emission. The transit time gives no problem, since the molecule physically passes out of the beam giving $\gamma_a = \gamma_b$ always. Collisional relaxation is another story, however. In the infrared region, the relevant question is whether collisions which change the rotational state occur with the same probability in excited vibrational levels as in the ground vibrational levels. In most cases one expects the answer to be yes. The vibrational motions of the nuclei are rapid compared to the time required for a collision, and hence the vibrational motion will be largely averaged away during a collision. This results in the collisional interaction between two molecules being independent of vibrational state. There are exceptions, however. The most dramatic example is ammonia where the ν_2 vibrational motion causes the excited-state dipole moment to be averaged away with the result that $\gamma_a \simeq 0.2\gamma_b$ for $\nu_2 = 0 \to 1$ transitions (Loy, 1974). Owing to the very limited number of studies made to date, it is difficult to say exactly where else a substantial vibrational dependence of the collisional interaction will occur. Possible candidates would be symmetric molecules in which the excited-state vibrational motion breaks the symmetry to produce a dipole moment, or very light simple molecules where resonant exchange of rotational energy could play a role. In any event, one should be wary of assuming a priori that the collisional contributions to γ_a and γ_b are equal. In the visible and uv, spontaneous emission tends to become the dominant decay process, so we need to consider that also. Here one must examine each transition individually. If the lower state of the transition is the ground electronic state, for example, we will have no decay of that state due to spontaneous emission, whereas the upper level may have a large decay rate. For transitions between two excited electronic states, on the other hand, one could have $\gamma_a \simeq \gamma_b$ for some pairs of states and the optical Bloch equations would be valid. In general, though, the two decay rates will not be equal for atomic transitions and the full set of population matrix equations (3-67) should be used.

Assuming that the optical Bloch equations are applicable, we now consider how one can formulate a geometrical picture for optical transitions known as the vector model (Feynman et al., 1957). Our clue is the fact that, if we neglect relaxation, the nmr Bloch equations and the classical equations of motion for a spinning particle in a magnetic field are identical (Slichter, 1963). The classical equation is

$$\frac{d\mathcal{M}}{dt} = \mathcal{M} \times g\mathbf{H} \tag{3-76}$$

where $\mathcal{M}$ is the magnetic moment of the particle. This is a familiar gyroscope equation from elementary physics, which says that a magnetic field **H** exerts a torque on $\mathcal{M}$ given by $\mathcal{M} \times g\mathbf{H}$. The torque is clearly always perpendicular to both $\mathcal{M}$ and **H**, causing $\mathcal{M}$ to precess in a cone about **H**. We can obtain this same simple picture for the optical Bloch equations if we define the vectors

$$\mathbf{M} \equiv (u, v, w), \qquad \mathbf{\Omega} = (\kappa E_0, 0, \Omega' - \omega_0) \qquad (3\text{-}77)$$

Neglecting relaxation, Eqs. (3-75) become

$$\frac{d\mathbf{M}}{dt} = \mathbf{M} \times \mathbf{\Omega} \qquad (3\text{-}78)$$

This is the vector model for optical transitions. It says that the vector **M**, which has components u, v, and w, precesses in a cone about an effective field $\mathbf{\Omega}$. This situation is illustrated in Figure 3-4. We have labeled the axis system as I, II, and III here to emphasize the fact that the vectors **M** and $\mathbf{\Omega}$ move in an abstract vector space, not the physical space in which we exist. This does not detract from the usefulness of the vector model, however, since our purpose in introducing it is only to enable a visualization of the behavior of optical Bloch equation solutions.

Notice that the components of both **M** and $\mathbf{\Omega}$ have direct physical meaning. u and v are the in- and out-of-phase components of the polarization for molecules of the velocity group v_z being considered (we define a velocity group v_z as the set of molecules in the sample which have velocities between v_z and $v_z + dv_z$), while w is the population difference $\rho_{aa} - \rho_{bb}$ for this group. The driving field $\mathbf{\Omega}$ always lies in the I–III plane and has nonvanishing components κE_0 and $\Omega' - \omega_0$. κE_0 is the strength of the

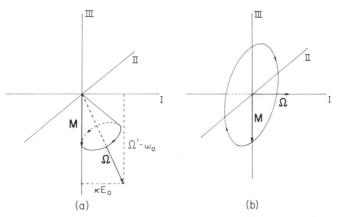

Figure 3-4. The vector model of a transition. (a) The driving vector $\mathbf{\Omega}$ causes **M** to precess in a cone about $\mathbf{\Omega}$. $\mathbf{\Omega}$ has components κE_0 and $\Omega' - \omega_0$ along the I and III axes, respectively. (b) At exact resonance ($\Omega' - \omega_0 = 0$) the cone of precession flattens into a circle in the II–III plane.

interaction between the molecules and the optical field, while $\Omega' - \omega_0$ is the detuning of the field from resonance. It is important to keep in mind that each velocity group has its own vector $\mathbf{M}(v_z)$ and effective driving field $\mathbf{\Omega}(v_z)$ with $\mathbf{M}(v_z)$ for one velocity group moving differently than $\mathbf{M}(v_z)$ for other groups, since $\mathbf{\Omega}$ depends on v_z through $\Omega' - \omega_0 = \Omega - kv_z - \omega_0$.

The behavior of $\mathbf{M}$ in various situations will be discussed in detail when specific coherent transient effects are treated. Here we only note that in simple situations $\mathbf{M}$ behaves in a manner consistent with our intuition. For example, in thermal equilibrium where no optical field is present, $\kappa E_0 = 0$, so $\mathbf{\Omega}$ points along the III axis or is zero. But u and v are also zero in thermal equilibrium, so $\mathbf{M}$ points straight down along the $-$III axis ($\rho_{bb} > \rho_{aa}$ in equilibrium). Then $\mathbf{M} \times \mathbf{\Omega} = 0$ and the system does nothing, as we expect. Now turn on a field E_0. This introduces a component of $\mathbf{\Omega}$ along the I axis and $\mathbf{M}$ will begin to move in a cone about $\mathbf{\Omega}$ as shown in Figure 3-4a. As it does this it develops polarization components u and v and the population difference w changes. This is exactly what we expect for a system which is absorbing or emitting radiation.

The vector model gives a very useful picture of the transition process. Consider a single-velocity group of molecules which are all initially in the lower of the two levels. If we turn on an exactly resonant optical field as shown in Figure 3-4b, $\mathbf{\Omega}$ will lie on the I axis, and $\mathbf{M}$ will start to precess in a circle in the II-III plane. As $\mathbf{M}$ turns, its v component (the imaginary part of ρ_{ab}) initially grows while w shrinks to zero, and then v shrinks while w grows until $v = 0$. This means that the molecules are evolving in unison through various superpositions of ground and excited states. At the point in time where $v = 0$, we have $\mathbf{M}$ pointing straight up along the III axis, and hence every molecule is now in the excited state. This is an interesting method of producing a complete population inversion. Note that it takes a definite amount of time to make a complete transition, and that the transition process could be halted any time before it is complete by simply cutting off the optical field. By doing so one can obtain any desired superposition of states. This picture of a transition is in sharp contrast to the impression one sometimes gets from elementary perturbation-theory treatments—that transitions occur as instantaneous quantum jumps functional one level to another. Such a picture simply cannot describe coherent transient effects. This point is discussed at length by Macomber (1976).

So far, we have ignored relaxation processes, but it is easy to see qualitatively what their effect is. u and v tend exponentially toward zero with a rate constant γ_{ab}, while w tends toward $n_a - n_b$ with a rate constant γ. This damps the precessional motion of $\mathbf{M}$ about $\mathbf{\Omega}$ so that eventually $\mathbf{M}$ becomes a constant if $\mathbf{\Omega}$ remains constant.

The reader should note that the definitions we have used to obtain the vector model may differ somewhat from definitions given elsewhere. In particular, one often finds that the equivalents of our u and v are initially defined as the real and imaginary parts of $2\rho_{ab}$ rather than $2\tilde{\rho}_{ab}$, as we have done. In addition, $\boldsymbol{\Omega}$ is then written as $\boldsymbol{\Omega}' = (2\kappa E_0 \cos{(\Omega't - kz')}, 0, -\omega_0)$. The resulting equation has the same form: $\dot{\mathbf{M}}' = \mathbf{M}' \times \boldsymbol{\Omega}'$. Our vector model equations may be recovered from this by transforming into a coordinate system rotating about the III axis at $\Omega't - kz'$ and making the rotating-wave approximation. The nice feature about doing the problem this way is that it does give one a good intuitive understanding of the rotating-wave approximation (cf. Allen and Eberly, 1975).

3.4.3. Matrix Solution of the Optical Bloch Equations

Although the vector model is very useful in obtaining a physical picture of two-level system behavior, for quantitative calculations we still need analytic solutions of the optical Bloch equations (3-75). Unfortunately, tractable closed-form solutions for these equations cannot be found. However, there are several useful special cases where they can be solved. One method of obtaining these solutions is the Laplace transform technique which was used by Torrey (1949) in his original presentation of the full range of Bloch equation solutions. There are other techniques which can equally well be used. In this chapter we choose to solve the equations with a matrix method first introduced into nmr by E. T. Jaynes (1955). Although this initially involves some mathematics which may be unfamiliar, the technique is extremely powerful and produces results in a form which is very useful for actual calculations. We begin by writing Eqs. (3-75) in matrix notation. Let

$$\mathbb{M} \equiv \begin{pmatrix} u \\ v \\ w \end{pmatrix}, \qquad \mathbb{B} \equiv \begin{pmatrix} 0 & -\delta & 0 \\ \delta & 0 & -\kappa E_0 \\ 0 & \kappa E_0 & \Delta\gamma \end{pmatrix}, \qquad \mathbb{A} \equiv \begin{pmatrix} 0 \\ 0 \\ \gamma(n_a - n_b) \end{pmatrix} \qquad (3\text{-}79)$$

where $\delta = \Omega' - \omega_0$ and $\Delta\gamma = \gamma - \gamma_{ab}$. With these definitions the optical Bloch equations become

$$\dot{\mathbb{M}} = -(\gamma_{ab} + \mathbb{B})\mathbb{M} + \mathbb{A} \qquad (3\text{-}80)$$

Here $(\gamma_{ab} + \mathbb{B})$ means the matrix $\mathbb{B} = \gamma_{ab}\mathbb{1} + \mathbb{B}$, where $\mathbb{1}$ is the unit matrix. This equation is very similar in spirit to the vector model equations except that here we have written the vector product $\boldsymbol{\Omega} \times \mathbf{M}$ as an operator operating on the vector $\mathbf{M}$ (which we write as a column matrix, $\mathbb{M}$), and the relaxation is included.

We now solve the matrix equation (3-80) exactly as if it were a scalar equation; i.e., we multiply it by the integrating factor $\exp{(\gamma_{ab} + \mathbb{B})t}$ to get

$$e^{(\gamma_{ab} + \mathbb{B})t}\dot{\mathbb{M}} + (\gamma_{ab} + \mathbb{B})e^{(\gamma_{ab} + \mathbb{B})t}\mathbb{M} = e^{(\gamma_{ab} + \mathbb{B})t}\mathbb{A}$$

or   (3-81)

$$\frac{d}{dt}[e^{(\gamma_{ab} + \mathbb{B})t}\mathbb{M}] = e^{(\gamma_{ab} + \mathbb{B})t}\mathbb{A}$$

Integrating this and multiplying through by $\exp{[-(\gamma_{ab} + \mathbb{B})t]}$, we obtain

$$\mathbb{M}(t) = (\gamma_{ab} + \mathbb{B})^{-1}\mathbb{A} + e^{-(\gamma_{ab} + \mathbb{B})t}\mathbb{C}\mathbb{A} \qquad (3\text{-}82)$$

where $\mathbb{C}$ is a constant of integration which must be determined from the initial conditions. In order to be completely general, we let $\mathbb{M} = \mathbb{M}(t_k)$ at time $t = t_k$. This gives

$$\mathbb{C}\mathbb{A} = e^{(\gamma_{ab} + \mathbb{B})t_k}[\mathbb{M}(t_k) - (\gamma_{ab} + \mathbb{B})^{-1}\mathbb{A}] \qquad (3\text{-}83)$$

so that the complete formal solution for $\mathbb{M}(t)$ becomes

$$\mathbb{M}(t) = e^{-(\gamma_{ab} + \mathbb{B})(t - t_k)}\mathbb{M}(t_k) + [1 - e^{-(\gamma_{ab} + \mathbb{B})(t - t_k)}](\gamma_{ab} + \mathbb{B})^{-1}\mathbb{A} \quad (3\text{-}84)$$

It is not obvious that one can manipulate matrix equations in the way that we have just done. However, in the first part of Appendix B we show that one can do this and that Eq. (3-84) is correct. The meaning of the matrix $\exp{(-\mathbb{B}t)}$ is also discussed there. One finds it is simply a square matrix having the same dimensions as $\mathbb{B}$ itself.

Before we explicitly evaluate the elements of $\exp{(-\mathbb{B}t)}$, let us briefly consider the physical meaning of the solutions for $\mathbb{M}(t)$ given in Eq. (3-84). The expression for $\mathbb{M}(t)$ consists of two terms. The first term is the more important one, since it represents the coherent response of the molecules to a field which is turned on or off at $t = t_k$. It dies away at long times, leaving only the second term, which is the contribution of molecules which enter state a or b after $t = t_k$. To see this we simply note that the second term is unaffected if we set $\mathbb{M}(t_k) = 0$; i.e., assume there are no molecules in states a or b at $t = t_k$. For many transient-effect experiments we can ignore the second term, as it will just show the growth of a steady-state background. In this situation, then, we can use the simpler equation

$$\mathbb{M}(t) = e^{-(\gamma_{ab} + \mathbb{B})t'}\mathbb{M}(t_k) \qquad (3\text{-}85)$$

where $t' = t - t_k$.

In those cases where the steady-state background *is* of interest, the factor $(\gamma_{ab} + \mathbb{B})^{-1}\mathbb{A}$ must be evaluated. This may be accomplished by directly inverting the matrix $(\gamma_{ab} + \mathbb{B})$ using determinants and then multiplying by $\mathbb{A}$, but there is really no need to do this, as we already know what $(\gamma_{ab} + \mathbb{B})^{-1}\mathbb{A}$ is. If we set $t - t_k = \infty$ in (3-84), the time-dependent exponentials go to zero and we have $\mathbb{M}(t = \infty) = (\gamma_{ab} + \mathbb{B})^{-1}\mathbb{A} = constant$. Thus, $(\gamma_{ab} + \mathbb{B})^{-1}\mathbb{A}$ is just the steady-state value of $\mathbb{M}$, and we have calculated this already in Section 3.3.2, where we found $\tilde{\rho}_{ab}$ and $\rho_{aa} - \rho_{bb}$ in steady state [cf. Eqs. (3-57)]. Using those equations and the definitions of u, v, and w, one obtains

$$\mathbb{M}(\infty) = (\gamma_{ab} + \mathbb{B})^{-1}\mathbb{A} = \frac{n_a - n_b}{\delta^2 + (\gamma_{ab}/\gamma)\kappa^2 E_0^2 + \gamma_{ab}^2} \begin{pmatrix} \delta\kappa E_0 \\ \gamma_{ab}\kappa E_0 \\ \delta^2 + \gamma_{ab}^2 \end{pmatrix} \quad (3\text{-}86)$$

With the steady-state term denoted by $\mathbb{M}(\infty)$, Eq. (3-84) may be rewritten more compactly as

$$\mathbb{M}(t) = e^{-(\gamma_{ab} + \mathbb{B})t'}[\mathbb{M}(t_k) - \mathbb{M}(\infty)] + \mathbb{M}(\infty) \quad (3\text{-}87)$$

Again, $t' = t - t_k$.

The formal solutions (3-84) or (3-87) are still of no use, though, unless we can evaluate the exponential factor. Now

$$e^{-(\gamma_{ab} + \mathbb{B})t'} = e^{-\gamma_{ab}t'}e^{-\mathbb{B}t'} \quad (3\text{-}88)$$

as shown in Appendix B. Hence our problem reduces to one of evaluating $\exp(-\mathbb{B}t')$. While it is not obvious how this can be done, we show in the second part of Appendix B that, in fact,

$$e^{-\mathbb{B}t'} = a_0\mathbb{1} + a_1\mathbb{B} + a_2\mathbb{B}^2 \quad (3\text{-}89)$$

where a_0, a_1, and a_2 are scalar functions of t', and the coefficients are determined from the equations

$$\dot{a}_0 = -\delta^2 \Delta\gamma a_2$$
$$\dot{a}_1 = g^2 a_2 - a_0 \quad (3\text{-}90)$$
$$\dot{a}_2 = -\Delta\gamma a_2 - a_1$$

where $g^2 = \delta^2 + \kappa^2 E_0^2$.

At this point it may appear that we have made no progress, since our "solution" of the optical Bloch equations has produced another set of differential equations. However, Eqs. (3-90) are simpler in form, homogeneous, and obey very simple initial conditions. Thus, we have indeed made progress. Now a tractable closed-form solution for Eqs. (3-90) cannot be obtained in general because it must be expressed in terms of the roots of a cubic equation. This is not a failing of our method, but rather a reflection of the fact that the optical Bloch equations themselves have no tractable closed-form solutions. Nonetheless, there are several useful special cases for which the equation can be solved. The most useful and widely used by far is the case $\gamma_{ab} = \gamma$ (analogous to the $T_1 = T_2$ case in nmr), which is discussed below. Another useful solution can be obtained when the optical field is large so that $|\kappa E_0| \gg |\Delta\gamma|$. This case is treated in the third part of Appendix B.

When $\gamma_{ab} = \gamma$, we have $\Delta\gamma = 0$, and Eqs. (3-90) become

$$\dot{a}_0 = 0, \qquad \dot{a}_1 = g^2 a_2 - a_0, \qquad \dot{a}_2 = -a_1 \qquad (3\text{-}91)$$

The initial conditions on these equations are $a_0(0) = 1$, $a_1(0) = 0$, $a_2(0) = 0$ at $t' = t - t_k = 0$. This follows from Eq. (3-89) and the fact that $\exp(-\mathbb{B}t') = \mathbb{1}$ at $t' = 0$. Equations (3-91) are now trivial to solve. Clearly, $a_0 = 1$ and $\ddot{a}_1 = g^2 \dot{a}_2 - \dot{a}_0$, which gives $\ddot{a}_1 = -g^2 a_1$. On solving this harmonic oscillator equation and inserting the result into $\dot{a}_2 = -a_1$, one finds

$$a_0 = 1, \qquad a_1 = -\frac{1}{g}\sin gt, \qquad a_2 = \frac{1}{g^2}(1 - \cos gt) \qquad (3\text{-}92)$$

when the initial conditions are used. Equation (3-89) then becomes

$$e^{-\mathbb{B}t'} = \begin{pmatrix} 1 - \dfrac{\delta^2}{g^2}(1-\cos gt') & \dfrac{\delta}{g}\sin gt' & \dfrac{\delta\kappa E_0}{g^2}(1-\cos gt') \\[3mm] -\dfrac{\delta}{g}\sin gt' & \cos gt' & \dfrac{\kappa E_0}{g}\sin gt' \\[3mm] \dfrac{\delta\kappa E_0}{g^2}(1-\cos gt') & -\dfrac{\kappa E_0}{g}\sin gt' & 1 - \dfrac{\kappa^2 E_0^{\,2}}{g^2}(1-\cos gt') \end{pmatrix} \qquad (3\text{-}93)$$

This is the explicit expression for $\exp(-\mathbb{B}t')$ that we desired. Equations (3-86), (3-87), and (3-93) provide a complete analytic solution of the optical Bloch equations for those cases where we may take $\gamma_{ab} = \gamma$.

The condition $\gamma_{ab} = \gamma$ implies that all relaxation times are equal. Although this is a rather restrictive requirement, there are a number of systems that do satisfy it. We have already considered the conditions under which one may use a single γ for the level lifetimes so that the optical Bloch equations apply. We now require in addition that $\gamma_{ab} = \gamma$. Looking back to the discussion at the end of Section 3.2.2, it is easy to see what this implies. All relaxation rates are equal for transit-time effects, so they give no problem. For spontaneous emission and collisions which change the state of a molecule, we found that $\gamma_{ab} = \frac{1}{2}(\gamma_a + \gamma_b)$. Since $\gamma_a = \gamma_b = \gamma$ has already been assuming in writing down the optical Bloch equations, we may take $\gamma_{ab} = \gamma$ for these processes also. This leaves only phase-interrupting collisions to consider. We saw in Section 3.2.2 that these processes caused ρ_{ab} to decay with no change in ρ_{aa} or ρ_{bb}. Thus, when phase-interrupting collisions are important, we will have $\gamma_{ab} > \frac{1}{2}(\gamma_a + \gamma_b)$, and hence $\gamma_{ab} \neq \gamma$. For molecular transitions in the infrared, phase-interrupting collisions have not yet been found to be important, so it will normally (but not necessarily!) be safe to assume that all relaxation rates are equal and the solution shown in Eq. (3-93) can be used. For atomic transitions, however, one normally finds that phase-interrupting collisions are important and (3-93) should not be used.

The reason these collisions can be important in atoms is that atoms do not usually have the manifold of nearby electric-dipole-connected levels that molecules do. Hence, rather hard collisions can occur where the atomic state does not change but the phase is interrupted. In molecules such hard collisions will nearly always change the state.

It may seem that the matrix approach is an unnecessarily complicated way to solve the optical Bloch equations for the simple $\gamma_{ab} = \gamma$ case. The effort is amply repaid, however, as soon as one begins to consider multiple pulse sequences. Consider the photon echo pulse sequence shown in Figure 3-1, for example. The behavior of the system during each time interval is described by an equation identical to (3-87), i.e.,

$$\mathbb{M}(t_2) = e^{-\mathbb{B}_2}[\mathbb{M}(t_1) - \mathbb{M}_2(\infty)] + \mathbb{M}_2(\infty)$$

$$\mathbb{M}(t_3) = e^{-\mathbb{B}_3}[\mathbb{M}(t_2) - \mathbb{M}_3(\infty)] + \mathbb{M}_3(\infty) \tag{3-94}$$

$$\mathbb{M}(t_4) = e^{-\mathbb{B}_4}[\mathbb{M}(t_3) - \mathbb{M}_4(\infty)] + \mathbb{M}_4(\infty)$$

where $\mathbb{B}_i = (\gamma_{ab} + \mathbb{B}_i)(t_i - t_{i-1})$. $\mathbb{B}_i$ and $\mathbb{M}_i(\infty)$ are $\mathbb{B}$ and $\mathbb{M}(\infty)$, with the optical field strength appropriate for the interval $t_i - t_{i-1}$ inserted. Equations (3-94) allow us to calculate $\mathbb{M}$ at the end of the second pulse given the state of the system at the beginning of the pulse train, i.e., $\mathbb{M}(t_1)$. Substituting $\mathbb{M}(t_2)$ from the first equation into the second and then inserting $\mathbb{M}(t_3)$ into the third equation, we find that

$$\mathbb{M}(t_4) = e^{-\mathbb{B}_4} e^{-\mathbb{B}_3} e^{-\mathbb{B}_2}[\mathbb{M}(t_1) - \mathbb{M}_2(\infty)] + e^{-\mathbb{B}_4} e^{-\mathbb{B}_3}[\mathbb{M}_2(\infty) - \mathbb{M}_3(\infty)]$$

$$+ e^{-\mathbb{B}_4}[\mathbb{M}_3(\infty) - \mathbb{M}_4(\infty)] + \mathbb{M}_4(\infty) \tag{3-95}$$

Ordinarily, the $\mathbb{M}_i(\infty)$ terms in Eq. (3-95), which give the response of molecules entering state a or b *after* the start of the first pulse, do not produce any effect of interest and we can use simply

$$\mathbb{M}(t_4) = e^{-\mathbb{B}_4} e^{-\mathbb{B}_3} e^{-\mathbb{B}_2} \mathbb{M}(t_1) \tag{3-96}$$

Thus, we can find the net response of the system to any arbitrary sequence of pulses by simply multiplying together 3×3 matrices, one for each interval of time during which a pulse is on or off. This is a very convenient way to handle the problem, especially when computer calculations are required. Note that Eqs. (3-95) and (3-96) apply to any solution of the Bloch equations, not just to the $\gamma_{ab} = \gamma$ case we have been discussing.

When one does have $\gamma_{ab} = \gamma$, however, the problem may be simplified even further. Then $\exp(-\mathbb{B}t')$ is given by Eq. (3-93), and this matrix is nothing other than a three-dimensional rotation matrix. This may be seen by noting that if we set $\gamma_{ab} = \gamma = 0$, $\mathbb{M}(\infty) = (\gamma_{ab} + \mathbb{B})^{-1} \mathbb{A} = 0$ because $\mathbb{A} = 0$ while $\exp(-\mathbb{B}t')$ in (3-93) is unchanged. The matrix solution (3-87) then becomes

$$\mathbb{M}(t) = e^{-\mathbb{B}t'} \mathbb{M}(t_k) \tag{3-97}$$

But this solution must be identical to the solution of the vector model equation $\dot{\mathbf{M}} = \mathbf{M} \times \mathbf{\Omega}$ because that equation is simply the optical Bloch equations for the case $\gamma_{ab} = \gamma = 0$ written in vector form. Since the solution of the vector model equation is one in which $\mathbf{M}$ precesses in a cone about $\mathbf{\Omega}$, the operator $\exp(-\mathbb{B}t')$ must also describe a rotation of M about an axis given by $\mathbf{\Omega} = (\kappa E_0, 0, \Omega' - \omega_0)$. Furthermore, the angle of the rotation is just $gt' = \sqrt{\kappa^2 E_0^2 + (\Omega' - \omega_0)^2} \, t'$, as can be seen by noting that in (3-93) $\exp(-\mathbb{B}t') = \mathbb{1}$ for $gt' = 2n\pi$, n an integer.

We have just shown that $\exp(-\mathbb{B}t')$ is the matrix operator for rotation through an angle gt' about $\mathbf{\Omega}$. This is very convenient because there exists a two-dimensional representation of the rotation matrix which allows us to reduce the problem from one of multiplying 3×3 matrices to one of multiplying 2×2 matrices (Jaynes, 1955). It may be easier to understand how this is possible if we first consider a two-dimensional example. The rotation of a two-dimensional vector through an angle θ is given by

$$\begin{pmatrix} u' \\ v' \end{pmatrix} = \begin{pmatrix} \cos\theta & -\sin\theta \\ \sin\theta & \cos\theta \end{pmatrix} \begin{pmatrix} u \\ v \end{pmatrix} \tag{3-98}$$

But this formula also has a one-dimensional representation: one simply uses complex numbers. Let $z = u + iv$. A rotation by θ is then represented by

$$z' = e^{i\theta} z \tag{3-99}$$

Given a three-dimensional vector, one can do the same thing except that a generalization of complex numbers is required. The appropriate object is called a quaternion and it has its own unique law of multiplication (cf. Corben and Stehle, 1960). Rather than deal with these unfamiliar objects, however, we note that quaternions can be represented by 2×2 matrices whose elements are ordinary complex numbers. This allows one to establish a correspondence between three-dimensional rotation operators and 2×2 matrices. We state without proof the end result, which is as follows: To every vector M having components u, v, and w, we may associate a 2×2 matrix

$$\mathsf{m} = \begin{pmatrix} w & u - iv \\ u + iv & -w \end{pmatrix} \tag{3-100}$$

and to every rotation of M generated by a matrix $\mathbb{R}$ [$\mathbb{R} = \exp(-\mathbb{B}t')$ in our case] i.e.,

$$\mathsf{M}' = \mathbb{R}\mathsf{M} \tag{3-101}$$

there corresponds a transformation of m,

$$\mathsf{m}' = \mathsf{r}\mathsf{m}\mathsf{r}^{-1} \tag{3-102}$$

If $\mathbb{R}$ represents a rotation through an angle θ about an axis given by a unit vector $\hat{n}$, then

$$\mathbf{r} = \begin{pmatrix} \alpha & \beta \\ -\beta^* & \alpha^* \end{pmatrix} \tag{3-103}$$

where

$$\alpha = \cos\frac{\theta}{2} - in_z \sin\frac{\theta}{2}$$

$$\beta = n_y \sin\frac{\theta}{2} - in_x \sin\frac{\theta}{2}$$

α and β are called the Cayley–Klein parameters for the rotation. Finally, the matrix $\mathbb{R}$ can be written in terms of the elements of $\mathbf{r}$ as

$$\mathbb{R} = \begin{vmatrix} \mathrm{Re}\,(\alpha^2 - \beta^2) & -\mathrm{Im}\,(\alpha^2 + \beta^2) & \mathrm{Re}\,(-2\alpha\beta) \\ \mathrm{Im}\,(\alpha^2 - \beta^2) & \mathrm{Re}\,(\alpha^2 + \beta^2) & \mathrm{Im}\,(-2\alpha\beta) \\ \mathrm{Re}\,(2\alpha^*\beta) & \mathrm{Im}(2\alpha^*\beta) & \alpha\alpha^* - \beta\beta^* \end{vmatrix} \tag{3-104}$$

where Re and Im denote real and imaginary parts of the indicated quantities. It is not our intent to prove the above statements here. Rigorous discussions may be found in Corben and Stehle (1960), and in Goldstein (1950). We merely want to use the results as a time-saving calculational device.

The first step in utilizing the 2×2 representation is to note that we are interested in rotations about the vector $\mathbf{\Omega}$ by an angle gt'. Thus, α and β in (3-103) are given by

$$\alpha = \cos\,(gt'/2) - i\frac{\delta}{g}\sin\,(gt'/2)$$

$$\beta = -i\frac{\kappa E_0}{g}\sin\,(gt'/2) \tag{3-105}$$

Now suppose that we want to calculate the response of a system characterized initially by $\mathbf{M}(t_1)$ to a sequence of pulses. For concreteness, consider again two pulses as we did in Eqs. (3-94)–(3-96). As we saw there, the response is determined by the matrix product $\exp\,(-\mathbb{B}_4)\exp\,(-\mathbb{B}_3)\exp\,(-\mathbb{B}_2)$. We may calculate this by computing instead the product $\mathbf{r}_T = \mathbf{r}_4\mathbf{r}_3\mathbf{r}_2$, which, written out explicitly, is

$$\mathbf{r}_T = \begin{pmatrix} \alpha & \beta \\ -\beta^* & \alpha^* \end{pmatrix} = \begin{pmatrix} \alpha_4 & \beta_4 \\ -\beta_4^* & \alpha_4^* \end{pmatrix}\begin{pmatrix} \alpha_3 & \beta_3 \\ -\beta_3^* & \alpha_3^* \end{pmatrix}\begin{pmatrix} \alpha_2 & \beta_2 \\ -\beta_2^* & \alpha_2^* \end{pmatrix} \tag{3-106}$$

The elements α_i and β_i appearing in each matrix $\mathbf{r}_i$ are given by (3-105) using the same optical field strengths and time intervals t' that were used in each matrix $\mathbb{B}_i$. We then put the resultant α and β into (3-104), which gives the matrix $\mathbb{R} = \exp(-\mathbb{B}_4) \exp(-\mathbb{B}_3) \exp(-\mathbb{B}_2)$. Usually we do not need to know the entire matrix $\mathbb{R}$. For example, if we only want to know the out-of-phase polarization v which results from the pulse sequence, and initially the system is in thermal equilibrium so that $\mathbb{M}(t_1) = (0, 0, n_a - n_b)$, then only $\text{Im}(-2\alpha\beta)$ need be calculated.

The usefulness and relative simplicity of this matrix approach will become clearer when photon echoes are discussed in Section 3.8.

3.5. Experimental Techniques

3.5.1. Pulsed Laser Experiments

Before discussing the various coherent transient effects in detail, we briefly turn our attention to the methods used to observe them. Transient effects are the response of a sample to pulses of light, and hence a pulsed laser is the obvious device to use as a light source. With such a source, transient-effect experiments are in principle extremely simple. The laser pulse just passes through the sample and hits a detector, which sees both the transmitted laser pulse and any emission from the sample.

The first optical coherent transient-effect experiment in gases was performed in 1968 by Patel and Slusher, who observed photon echoes in SF_6 gas excited by CO_2 laser pulses. Their experiment was inspired by the earlier observations of photon echoes in ruby by Kurnit *et al.* in 1964. The four-year time gap is somewhat surprising but was partly due to the belief that photon echoes could not be observed in gases because the molecular motion would prevent any rephasing of the dipoles. Scully *et al.* demonstrated theoretically that this belief was false early in 1968. The success of Patel and Slusher stimulated experiments by other workers, and there followed in quick succession papers reporting optical nutation (Hocker and Tang, 1968), photon echoes in an atomic vapor (Bölger and Diels, 1968), adiabatic inversion (Treacy and De Maria, 1969), and optical free induction decay (Cheo and Wang, 1970).

While these experiments were all successful in observing coherent transient effects, it proved to be very difficult to make accurate quantitative measurements. One of the difficulties can be appreciated by considering the laser pulses required for three basic experiments. For an optical nutation experiment one would like to have a constant-amplitude optical field which is turned on suddenly in step-function fashion. For an optical free induction decay experiment one would like to have a laser pulse which is suddenly turned off. Finally, for a photon echo experiment, one needs two laser pulses whose amplitudes are adjustable and whose time separation can be precisely controlled and varied. All three of these pulse sequences were illustrated in Figure 3-1. We notice that none of the pulses just described look much like the pulses usually obtained from pulsed lasers, as these tend to be short pulses with a fast rise and an exponential tail. This is one major difficulty with pulsed laser experiments: it is hard to generate the required pulses. They can be obtained, but it requires specially modified lasers and considerable effort.

A second problem associated with pulsed lasers is not quite so obvious, but nonetheless very important. This is the difficulty of observing the transient effects in the presence of the intense laser pulses. If the interpretation of coherent transient experiments is to be kept simple and unambiguous, optically thin samples must be used. Furthermore, note that an optically thin sample was defined as one in which the optical field E_0 passes through the sample essentially unchanged. The requirement is on the field E_0 rather than the intensity because E_0 is the quantity which appears in the population matrix equations of motion. If we say that a change in E_0 of 5–10% is sufficiently small that E_0 can still be regarded as a constant, then the field $\mathscr{E}_c$ emitted by the molecules will be on the order of $0.07E_0$. However, detectors measure intensities, not fields, and $\mathscr{E}_c^2 \sim 0.005E_0^2$. Thus, in an optically thin sample, the emitted transient effect signals will be extremely small compared to the laser pulses and can easily be entirely masked by them. For this reason, most pulsed laser experimenters have, in fact, used optical densities on the order of unity so that the transient effects could be seen. This greatly complicates the interpretation of their experiments. The observation problem can be alleviated somewhat by a dual-beam arrangement. One splits the laser beam into two and sends only one beam through the sample, as Treacy and De Maria (1969) have done. We then measure only the difference between the two beams, thus eliminating the laser pulses. This technique can greatly improve matters, although it does place severe requirements on the detectors as they must be very stable, have identical transient responses, and have a large dynamic range.

3.5.2. Stark Switching

In 1971 a new method of observing coherent transient effects was introduced by Brewer and Shoemaker. The technique, called Stark switching, eliminates the pulsed laser problems just discussed and allows one to make quantitative measurements on samples that are truly optically thin. The basis of the technique is the fact that coherent transient effects will be produced whenever the resonant *interaction* between the molecules and the optical field is pulsed. While turning the radiation on and off will certainly do this, there are other methods which will accomplish the same thing. In the

Stark-switching technique one makes use of the fact that most dipolar molecules have a Stark effect; i.e., their resonance frequencies ω_0 can be varied by applying an electric field across the sample. If the electric field is pulsed, the molecules can be switched in and out of resonance with a fixed-frequency laser beam, and this has the same effect as turning on and off the optical field. Thus, transient-effect experiments using Stark switching are still conceptually very simple. The experimental setup is shown schematically in Figure 3-5. One passes a stable cw laser beam through the sample and applies a pulsed electric field (often called the Stark field) across it by means of a pair of metal plates. The coherent transient effects are observed by monitoring both the transmitted light and any molecular emission with a photodetector. In a typical experiment, one might use a 1-W cw laser and a 30-cm-long sample cell containing a few milliTorr of gas. At these pressures collisional relaxation times are on the order of a few microseconds, so one would like to use Stark field pulses with rise and fall times of 50–100 ns and have a detector with a ~10-MHz bandwidth. The output signal can either be displayed directly on an oscilloscope, or the results of many experiments can be averaged by a boxcar integrator and displayed on an x-y recorder.

We stated above that switching the resonance frequency has the same effect as turning the optical field on and off. One can see that this is so by considering the vector model introduced in Section 3.4.2. The vector **M** having components u, v, and w describes the state of the molecules and this vector precesses in a cone around the vector $\boldsymbol{\Omega} = (\kappa E_0, 0, \Omega' - \omega_0)$, The motion was illustrated in Figure 3-4. The first point to notice is that if we switch off the optical field so $E_0 = 0$, then $\boldsymbol{\Omega} = (0, 0, \Omega' - \omega_0)$ and **M** precesses around the III axis. This motion does not indicate that the state of the molecule is changing, since both the magnitude of the polarization (which is proportional to $\sqrt{u^2 + v^2}$) and the population difference w are unchanged by this motion. It rather indicates simply that the oscillating dipole moment is oscillating at a frequency other than Ω'. If we now turn on a nearly resonant field such that $\kappa E_0 \gg \Omega' - \omega_0$, the molecules will begin to change their state, since **M** will now precess about the I axis. While we can interrupt the interaction by turning off the field, it should be clear that a nearly equivalent act is to shift the molecular resonance frequency by an amount $\Delta\omega_0$ so that

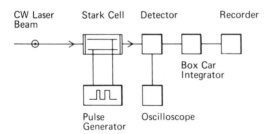

CW Laser Beam   Stark Cell   Detector   Recorder

Box Car Integrator

Pulse Generator   Oscilloscope

Figure 3-5. Block diagram of the experimental setup for observing transient effects using Stark switching.

we have $\kappa E_0 \ll \Omega' - (\omega_0 + \Delta\omega_0)$. $\boldsymbol{\Omega}$ then lies nearly along the III axis, and **M** simply precesses about the III axis with no significant change in molecular state occurring. This is exactly what would happen if E_0 had been turned off instead. The only difference between turning off the field and switching out of resonance is that in the latter case we do not quite interrupt the molecule–field interaction completely because $\boldsymbol{\Omega}$ still has a small component along the I axis. However, we can make $\boldsymbol{\Omega}$ lie as close to the III axis as we wish by increasing the frequency shift $\Delta\omega_0$ sufficiently.

Let us now consider the advantages of observing coherent transient effects with Stark switching rather than pulsed lasers. First, the shape and duration of the Stark pulses can be controlled easily and precisely, as they are generated electronically. It is extremely difficult to obtain this kind of control with a pulsed laser. Furthermore, it is easy to build pulse generators which produce perfectly repeatable Stark pulses at very high repetition rates. This allows one to average many experiments together using a signal averager or boxcar integrator and thereby obtain a greatly improved signal-to-noise ratio. Repeatable, high-repetition-rate laser pulses, on the other hand, cannot be achieved without considerable difficulty. Another advantage is that there is no problem in distinguishing the molecular response from the excitation pulses as there is when pulsed lasers are used. With Stark switching the laser beam is cw, so it produces just a dc signal in the detector. The only ac signals present are just those due to the transient effects one wants to observe! The final advantage is slightly more subtle. When molecules emit radiation after being shifted out of resonance, the emission occurs at the natural resonance frequency ω_0 of the molecules, which is, of course, now shifted from the laser frequency. This point will be verified in detail when we study optical free induction decay in Section 3.7. Since the laser is always on, the frequency-shifted emission and the laser radiation will both hit the detector and produce a heterodyne beat signal there. One can see how this works by noting that a detector is a square-law device, i.e., it detects the square of the total optical field E_T hitting it. If the molecules which are switched out of resonance emit a field $\mathscr{E}_c \cos\omega_0 t$, then $E_T = E_0 \cos\Omega t + \mathscr{E}_c \cos\omega_0 t$. On squaring this and averaging over several optical periods (the detector does not respond directly at optical frequencies), we obtain a signal of $\frac{1}{2}E_0^2 + E_0\mathscr{E}_c \cos(\Omega - \omega_0)t + \frac{1}{2}\mathscr{E}_c^2$. The heterodyne beat detected in a Stark switching experiment is the $E_0\mathscr{E}_c$ term, which is much larger (typically by a factor of $\sim 10^2$ in an optically thin sample) that the term $\frac{1}{2}\mathscr{E}_c^2$, which is what one would have to detect using a pulsed laser. Thus, we have a heterodyne detection scheme for molecular emission, and this produces an enormous gain in sensitivity.

Taken together, these experimental advantages make the observation of coherent transient effects far easier than had previously been possible. Effects that were nearly unobservable before can be studied in detail, and one can begin to think about using the transient effects as a powerful tool to study collisional relaxation processes. Of course, the Stark switching technique does have the drawback of requiring transitions with a large Stark effect, but this still leaves a reasonably large class of molecules accessible to study.

One subtle point has been glossed over in the treatment of Stark switching thus far. We assume in our theoretical treatments that the Stark field is turned on and off instantaneously, and that the only effect of this is to instantaneously shift the resonant frequency ω_0. Actually, this cannot be, because an electric field which is turned on instantaneously would have frequency components at microwave or optical frequencies and cause transitions between rotational energy levels. Hence, in practice, we must turn the Stark field on or off slowly enough so that the energy levels simply change adiabatically. At the same time we want the field to be turned on or off quickly enough so that the state of the molecules remain unchanged; i.e., we do not want any appreciable excitation or decay to occur during the rise or fall time T_R of the electric field. Fortunately, it is easy to satisfy both of these conditions. We need γ, $\kappa E_0 \ll 1/T_R \ll \omega_{\rm rot}$ where $\omega_{\rm rot}$ is the lowest frequency allowed rotational transition for the molecules. $\omega_{\rm rot}$ is typically 10^{10} Hz or more, while γ and κE_0 are typically 10^6–10^7 Hz or less. Thus, we want T_R to be $\sim 10^{-8}$ s, an easily achievable figure.

We also note that when a Stark field is suddenly turned on, the voltage does not appear instantaneously everywhere on the plates. It takes time for a voltage signal to propagate along them. If we do the sensible thing and connect the plates to the pulse generator at the front of the cell (where the laser beam enters), the plates will act like a stripline conductor and the Stark voltage will propagate down them at a speed which is nearly c. Thus, the boundary conditions discussed in Section 3.3.1 hold for Stark switching also.

In practice, there is one other potential difficulty with Stark switching that we have not yet considered. The problem is that we have assumed an entire Doppler-broadened transition can be shifted in and out of resonance, and this ordinarily requires extremely large electric fields. Fortunately, there is no need to use such large fields. When an optical field falls within the linewidth of a Doppler-broadened transition, not all the molecules in the coupled levels interact resonantly. Instead, only those molecules whose velocity v_z is such that they are Doppler-shifted into resonance with the optical field interact. Roughly speaking, we may regard molecules which have $|\Omega - kv_z - \omega_0| \equiv |\Omega' - \omega_0| \leq \kappa E_0$ as being resonant, since, according to the vector model, only these molecules will have a driving vector Ω which lies near the I axis and thus be induced to make transitions. We are assuming here that the optical field is strong so that $\kappa E_0 \gg \gamma$. Molecules with other velocities have $|\Omega' - \omega_0| \gg \kappa E_0$ and do not interact resonantly. Since κE_0 is

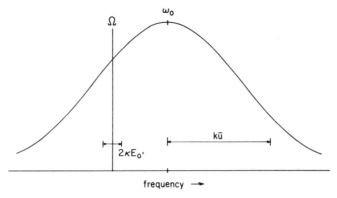

Figure 3-6. A strong optical field excites molecules whose laboratory-frame resonant frequencies lie within $\Omega \pm \kappa E_0$. The width of this group of resonant molecules is generally small compared to the Doppler width $k\bar{u}$.

typically 20 MHz, whereas the Doppler width $k\bar{u}$ is 250 MHz or more,[*] only a small fraction of the molecules interact resonantly. This is illustrated in Figure 3-6. For convenience, the x-axis is labeled in terms of laboratory-frame frequencies $\omega = \omega_0 + kv_z$ rather than velocity. If we now apply a pulsed electric field just large enough to shift the molecular resonance frequency by, say, $\Delta\omega_0 = 5$–$10\kappa E_0$, the group of resonant molecules will be shifted completely out of resonance, since for them $\Omega' - \omega_0 + \Delta\omega_0 \gg \kappa E_0$. Thus, we need shift only a fraction of the Doppler width in a Stark-switching experiment to observe transient effects. However, there is a complication. When we use small electric fields and stay within the Doppler linewidth, each time we shift some velocity group of molecules v' into resonance by turning on the electric field, we will also shift another velocity group of molecules v *out* of resonance. This situation is illustrated in Figure 3-7 for a step-function pulse. Clearly, two transient effects will always occur simultaneously each time we turn the field on or off; one from some group of molecules which is shifted into resonance, another from a group of molecules which is shifted out of resonance. Fortunately, this is not nearly as severe a problem as one might think, because the two effects can be distinguished. Molecules which are moved out of resonance radiate at their natural resonance frequency, which is now shifted from the laser frequency and gives rise to a heterodyne beat signal. Molecules which are moved into resonance, on the other hand, produce absorption or stimulated emission at the laser frequency. There are other features which differ as well. Separation of the two effects will be considered in more detail in later sections as each transient effect is discussed.

[*]These are angular frequencies. To obtain ordinary frequency, divide by 2π.

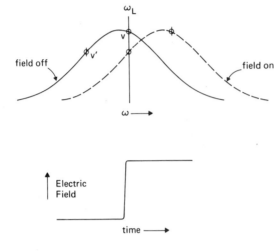

Figure 3-7. A Stark field which is suddenly turned on shifts ω_0 and produces two transient effects simultaneously. One velocity group of molecules v' is shifted into resonance and a second velocity group v is shifted out of resonance.

Stark switching has been used to study every observable coherent transient effect by now, and most of the experimental results discussed in this chapter have been obtained using the technique. Although the number of molecules which can be studied is limited, it is an extremely powerful tool in those cases where it is applicable. We expect that due to its simplicity and great sensitivity it will continue to be used for many experiments in the future.

3.5.3. Frequency Switching

Clearly, one can only study molecules which have a permanent dipole moment with Stark switching. If we want to study transient effects in atoms or nonpolar molecules, some other method must be used. An excellent technique exists for this case also. It is called laser frequency switching and was first demonstrated by J. L. Hall in 1973. The idea here is very simple. Instead of switching the resonance frequency ω_0, we switch the laser frequency Ω. If we look back at the population matrix equations (3-67), we see that Ω and ω_0 enter only as the difference $\Omega' - \omega_0 = \Omega - kv_z - \omega_0$. Thus, switching the laser frequency Ω is exactly equivalent to switching ω_0 and identical theoretical treatments apply.

To change Ω we need a means of switching the frequency of a cw laser. This can be accomplished by placing an electro-optic crystal inside the laser cavity. The crystal is oriented so that when a voltage is applied across it, its index of refraction changes. A change in the index changes the round-trip optical path length for light circulating in the laser cavity, and this changes the laser frequency because the frequency is constrained to be an integral multiple of $c/2nL$, where n is the index of refraction and L is the length of the laser cavity. This method of laser frequency modulation is a well-known technique in optical communications. Very rapid frequency changes are possible with intracavity FM, because unlike internal amplitude modulation, intracavity FM is not limited by the narrow laser cavity bandwidth (Yariv, 1964; Kiefer *et al.*, 1972). Hence, by applying voltage pulses to the electro-optic crystal, the laser can be switched between two frequencies. If the output beam is passed through a sample cell, coherent transient effects will be observed and all the advantages discussed in connection with Stark switching will apply here also. Since the laser frequency changes are small for reasonable applied voltages (a few MHz/100 V applied field is typical), one generally works entirely within the Doppler linewidth. This gives two transient effects simultaneously, as discussed above for the Stark-switching case.

It is rather difficult to set up a frequency-switched laser, since it involves building a stable laser with an intracavity modulator. However, once built it does offer several advantages over a Stark-switching setup. An obvious one is that it is more versatile than Stark switching, since any molecular or atomic transition which can be brought into resonance with a laser can be studied. Another advantage is that in dealing with transitions having several M_J components, all the M_J components are shifted by the same amount when frequency switching is used. This simplifies the experimental interpretation compared to a similar Stark-switching experiment where the M_J components shift by differing amounts, because the Stark effect is different for each M_J sublevel. Finally, since one is not constrained by having to get the laser beam undistorted through a pair of Stark plates, one can expand the laser beam in frequency switching to any desired diameter. This is a desirable feature because it minimizes transit-time contributions to the decay of the coherent transient effects.

It appears that frequency switching will be used rather extensively in the future. To date, the technique has been used in three laboratories. J. L. Hall (1973) did the first experiments using a He–Ne laser at 3.39 μm. More recently, R. G. Brewer and Genack (1976) have succeeded in frequency-switching a dye laser. This development will allow detailed studies in a number of atomic vapors. In addition, some preliminary experiments have been done in our laboratory using a frequency-switched CO_2 laser.

3.6. Optical Nutation

In this and subsequent sections we turn our attention to a detailed discussion of specific coherent transient effects. The first effect is called optical nutation, and it is conceptually very simple. Suppose that we have a sample which is initially in thermal equilibrium, and suddenly we turn on, stepwise, a resonant optical field $E_0 \cos(\Omega t - kz)$. Molecules starting out in the lower level will absorb radiation and be driven to the upper state, but once there they will immediately be stimulated to emit radiation. This drives them back down to the lower state, where they can again absorb radiation and repeat the process. Thus, the molecules are driven in unison back and forth between the ground and excited states, giving rise to an alternating absorption and emission of radiation which is called optical nutation. This phenomenon may be illustrated very graphically using the vector model. Initially u and v are zero, so **M** points straight down along the $-$III axis. When an exactly resonant optical field is turned on, Ω points along the I axis and causes **M** to precess in a disk lying in the II–III plane as shown in Figure 3-8a. As **M** precesses, the population difference w changes, reflecting the fact that we are driving the system up to the excited state and back down again. Also, the v component of **M**, which is proportional to the out-of-phase component of the oscillating dipole moment induced in the molecules, changes sign depending on whether **M** is moving upward or downward. This sign determines whether the molecules are absorbing or emitting radiation.

Of course, only one velocity group in the Doppler-broadened line will be exactly resonant with the optical field. All other velocities will see an Ω which lies somewhere between the I and III axes, and **M** for these molecules will precess in a cone about Ω as shown in Figure 3-8b. Molecules which are very far off-resonance will hardly be affected at all since Ω is then nearly parallel to the III axis, which is the initial direction of **M**. Also, relaxation damps the precessional motion of **M** so that it eventually becomes constant

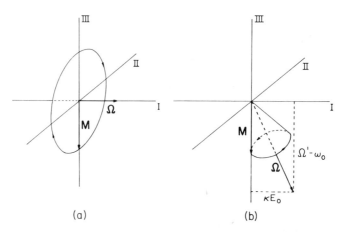

(a)   (b)

Figure 3-8. Optical nutation. (a) A resonant optical field causes **M** to precess in a disk at frequency κE_0. (b) An off-resonant field causes **M** to precess in a cone at frequency $g = \sqrt{(\kappa E_0)^2 + (\Omega - \omega_0)^2}$.

in time. In fact, to observe optical nutation, we need a field which is sufficiently strong that $\kappa E_0 > \gamma, \gamma_{ab}$, since otherwise the motion of **M** would be damped out before it could precess even once about $\mathbf{\Omega}$.

The effect we are discussing is called optical nutation because it is the optical analog of the nmr phenomenon of transient nutation (Torrey, 1949). The word "nutation" was originally used to describe the oscillatory motion which is superimposed on the precession of a spinning top in a gravitational field (cf. Goldstein, 1950). **M** in our vector model has a very similar motion. We noted in Section 3.4.2 that the coordinate frame we are using for the vector model can be thought of as rotating at the optical field frequency Ω'. From this point of view, the precession of **M** about $\mathbf{\Omega}$ appears as a slow oscillation of **M** superimposed on its rapid precession about the III axis at frequency Ω'.

Let us now turn to a more quantitative treatment of optical nutation. Suppose that the sample is initially in thermal equilibrium and that all relaxation times are equal. Considering first the simplified situation in which we ignore steady-state terms, the behavior of the system is given by Eq. (3-85). For an optical field turned on at time t_1, this becomes

$$\mathbb{M}(t) = e^{-\gamma t'} e^{-\beta t'} \mathbb{M}(t_1) \tag{3-107}$$

where $t' = t - t_1$. Since $\mathbb{M}$ starts out in thermal equilibrium, $\mathbb{M}(t_1) = (0, 0, n_a - n_b)$, and since we have assumed $\gamma = \gamma_{ab}$, the quantity $\exp(-\beta t')$ is given by Eq. (3-93). Thus,

$$M(t) = \begin{pmatrix} u \\ v \\ w \end{pmatrix} = e^{-\gamma t'}(n_a - n_b) \begin{pmatrix} \dfrac{\delta \kappa E_0}{g^2}(1 - \cos gt') \\ \dfrac{\kappa E_0}{g}\sin gt' \\ 1 - \dfrac{\kappa^2 E_0{}^2}{g^2}(1 - \cos gt') \end{pmatrix} \tag{3-108}$$

Equation (3-108) shows quantitatively the behavior of the vector model discussed above. For example, consider the molecules which have $\delta = \Omega' - \omega_0 = 0$. They are exactly resonant with the optical field so $g = \sqrt{\delta^2 + \kappa^2 E_0{}^2} = \kappa E_0$ and Eq. (3-108) reduces to

$$\begin{pmatrix} u \\ v \\ w \end{pmatrix} = e^{-\gamma t'}(n_a - n_b) \begin{pmatrix} 0 \\ \sin \kappa E_0 t' \\ \cos \kappa E_0 t' \end{pmatrix} \tag{3-109}$$

This describes a circular motion of **M** in the II–III plane. Note that the frequency of this motion is κE_0. Thus, on resonance, it takes an amount of time $2\pi/\kappa E_0$ to make a transition from the lower state to the upper state and back down again. κE_0 is often referred to as the resonant "Rabi flopping frequency," after I. I. Rabi, who first derived results similar to Eqs. (3-108) and (3-109) to describe the behavior of nuclear spins in molecular beam magnetic resonance experiments (Rabi, 1937). In that case the nuclear spins actually "flop" from spin up to spin down or vice versa when making transitions. From the general case of (3-108), we see that the generalized Rabi flopping frequency is just g.

Equation (3-108) describes the behavior of any given velocity group. To find the polarization we must add up the contributions from all velocities as shown in Eq. (3-69). This gives us

$$P_s = -\frac{\mu_{ab}\kappa E_0}{\bar{u}\sqrt{\pi}} e^{-\gamma t'}(N_a - N_b) \int_{-\infty}^{\infty} e^{-v_z^2/\bar{u}^2} \frac{\sin gt'}{g} dv_z \qquad (3\text{-}110)$$

This equation follows from (3-108) and the fact that n_a and n_b have a velocity dependence as shown in Eq. (3-59). We need not consider P_c because it gives rise to the field $\mathscr{E}_s$ and in an optically thin sample $\mathscr{E}_s$ does not contribute appreciably to the transmitted intensity when we have a strong optical field E_0 present [see Eqs. (3-51)–(3-53)].

Ordinarily, the strength of the optical field is such that $\kappa E_0 \ll k\bar{u}$. This means that the function $(\sin gt')/g$ is large only over a narrow range of velocities compared to the Gaussian $\exp(-v_z^2/\bar{u}^2)$ for all except extremely short times t'. To a good approximation, then, one can pull the Gaussian factor out of the integral and evaluate it at the central peak of the $(\sin gt')/g$ function, i.e., at $\delta = 0$. The integral can then be done analytically to yield a zero-order Bessel function behavior for P_s;

$$P_s = -\frac{\mu_{ab}\kappa E_0}{\bar{u}\sqrt{\pi}} e^{-(\Omega-\omega_0)^2/k^2\bar{u}^2} e^{-\gamma t'}(N_a - N_b)\frac{\pi}{k}J_0(\kappa E_0 t') \qquad (3\text{-}111)$$

Having found the polarization, we may now calculate the optical intensity leaving the sample from Eqs. (3-50) and (3-53), which show that

$$I(t) = I_0 - \tfrac{1}{2}E_0\Omega L P_s(t - \tau_1) \qquad (3\text{-}112)$$

Recalling that τ_1 is just t_1 evaluated at the end of the sample, $z = L$ (cf. discussion in Section 3.3.1), we have, from (3-111),

$$I(t) = I_0 - \frac{\kappa^2 E_0^2\sqrt{\pi}\hbar\Omega L}{2k\bar{u}} e^{-(\Omega-\omega_0)^2/k^2\bar{u}^2}(N_b - N_a)e^{-\gamma(t-\tau_1)}J_0[\kappa E_0(t - \tau_1)]$$

$$(3\text{-}113)$$

This is a very interesting result. First, notice that even though each velocity group is oscillating at its own Rabi flopping frequency g [see Eq. (3-108)], when we add up all the contributions to find the polarization, we obtain a Bessel function which oscillates at κE_0, the Rabi flopping frequency for exact resonance! One could hardly ask for simpler behavior. In Figure 3-9 we show a plot of the signal predicted by (3-113) for a typical case in which $\kappa E_0 = 5\gamma$. The result closely resembles the optical nutation signals which one actually observes (cf. Figures 3-10 and 3-11).

The nutation signal is largest at short times $t - \tau_1$. Noting that both the exponential and J_0 approach unity for small $t - \tau_1$, we obtain for the peak value of the optical nutation signal,

$$\Delta I_{\text{peak}} = \frac{\kappa^2 E_0^2 \sqrt{\pi} \hbar \Omega L}{2k\bar{u}} (N_b - N_a) e^{-(\Omega - \omega_0)^2/k^2 \bar{u}^2} \tag{3-114}$$

This expression is exactly what one obtains for *unsaturated* steady-state absorption, i.e., absorption for the case of a small optical field such that $\kappa E_0 \ll \gamma$ [cf. Eq. (3-63) with the condition $\kappa E_0 \ll \gamma$ imposed]. While interesting, this result is not unexpected, since for very short times after the field is turned on, the level populations have not yet changed appreciably and one sees the full unsaturated absorption. The difference between steady-state absorption and ΔI_{peak} is that, in the latter case, E_0 is very large, so ΔI_{peak} is much larger than any steady-state absorption signal. This means that optical nutation can be used as a very sensitive method of searching for transitions in

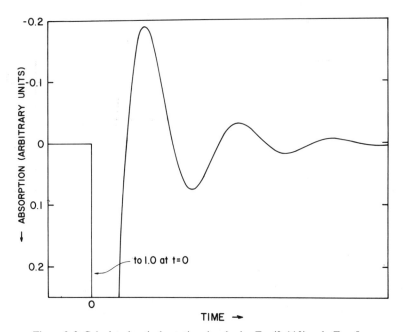

Figure 3-9. Calculated optical nutation signal using Eq. (3-113) and $\kappa E_0 = 5\gamma$.

low-pressure gases. By repeating the nutation experiment at a high repetition rate, even very weak absorption signals can usually be detected directly on an oscilloscope.

Our result (3-113) does have one peculiar feature. The maximum signal ΔI_{peak} seems to appear instantaneously when the field is turned on, since both J_0 and the exponential have maxima when $t - \tau_1 = 0$. This is physically absurd because our molecular transition has a limited bandwidth, given roughly by the Doppler width $k\bar{u}$, and no system with a finite bandwidth can respond instantaneously. Looking back, we see that the problem lies in Eq. (3-111), which was obtained by pulling the Gaussian factor out of the integral over v_z. This amounts to assuming that the Gaussian is infinitely wide. If we leave the Gaussian inside the integral in (3-110) and do the integration numerically, one finds that the absorption is zero at $t - \tau_1 = 0$ and rises to its maximum value of ΔI_{peak} in a time which is approximately $1/k\bar{u}$.

The simple Bessel function behavior of (3-113) is the response of molecules which were in levels a or b at time $t = t_1$. In addition, there will be a signal from molecules which enter state a or b at random times after $t = t_1$ due to collisions. This signal is described by the steady-state terms which involve $M(\infty)$. When these are included, we have

$$M(t) = e^{-\gamma t'}e^{-\beta t'}M(t_1) + (1 - e^{-\gamma t'}e^{-\beta t'})M(\infty) \qquad (3\text{-}115)$$

instead of Eq. (3-107). If we now repeat the derivation following Eq. (3-107) using Eq. (3-86) to evaluate $M(\infty)$, we find

$$I(t) = I_0 - \frac{\kappa^2 E_0^2 \sqrt{\pi}\hbar\Omega L}{2k\bar{u}} e^{-(\Omega-\omega_0)^2/k^2\bar{u}^2}(N_b - N_a)\left\{e^{-\gamma(t-\tau_1)}J_0[\kappa E_0(t-\tau_1)]\right.$$

$$\left. + \frac{\gamma}{\Gamma} - e^{-\gamma(t-\tau_1)}\frac{\gamma k}{\pi}\int_{-\infty}^{\infty}\left[\frac{\cos g(t-\tau_1)}{\delta^2+\Gamma^2} + \frac{\gamma}{g}\frac{\sin g(t-\tau_1)}{\delta^2+\Gamma^2}\right]dv_z\right\} \qquad (3\text{-}116)$$

where $\Gamma^2 = \kappa^2 E_0^2 + \gamma^2$.

In addition to the Bessel function obtained earlier, we have two terms due to the molecules which enter incoherently after $t = \tau_1$. The first term is a constant, γ/Γ, and is just the steady-state absorption given in Eq. (3-63). The second term involves an integral which must be evaluated numerically. It is an exponentially decaying term which is equal to γ/Γ at $t = \tau_1$. The difference between it and the previous term represents the growth of the steady-state absorption due to the incoherently entering molecules. Notice that both terms are very small compared to the Bessel function for strong optical fields, because they scale as $1/\kappa^2 E_0^2$. Hence, the simpler Eq. (3-113) will often give an adequate representation of the optical nutation signal.

If desired, our results could be generalized further by using the expression given in Appendix B for $\exp(-\beta t')$. This would allow us to relax the assumption that $\gamma_{ab} = \gamma$. However, having $\gamma_{ab} \neq \gamma$ only changes somewhat the decay of the nutation signal, with no significant change in oscillation frequency. Hence, this case is not of much interest, since we will see that the nutation signal is a poor choice to use as a tool for studying relaxation times.

Thus far, we have done a fairly complete analysis of optical nutation assuming a plane-wave optical field. In a real experiment, of course, one has a laser beam which has an intensity variation (beam profile) transverse to its direction of propagation. If one uses a large area detector which monitors the entire beam, it is clear that the signal one observes could be modified considerably from the plane-wave results we have been discussing. To see the effect of a laser beam profile on optical nutation, consider a collimated TEM$_{00}$ beam. We then have $E_0 \exp(-R^2/2w_0^2)$ as the optical field, where R is the radial distance from the center of the beam and w_0 is the laser beam spot radius (the radius at which the intensity has fallen off to $1/e$ of the beam center value). Assuming the optical field varies sufficiently slowly with R that it can be regarded as constant over a few wavelengths of light, the molecules at any given radius R will see what appears to be a plane wave. This will be the case for any laser beam of reasonable size, i.e., a few millimeters or larger in diameter. Now suppose that the entire laser beam is monitored by the detector in an experiment. The signal it sees will be given by

$$S = 2\pi \int_0^\infty I(R)R\,dR \qquad (3\text{-}117)$$

where $I(R)$ is the transmitted intensity given in Eqs. (3-113) or (3-116) provided that we replace E_0 in those equations by $E_0 \exp(-R^2/2w_0^2)$. Ignoring the steady-state terms and using (3-113), we have explicitly

$$S = \pi w_0^2 I_0 - \frac{\pi^{3/2}\hbar\Omega L}{k\bar{u}}(N_b - N_a)e^{-\gamma(t-\tau_1)}\int_0^\infty \kappa^2 E_0^2 e^{-R^2/w_0^2}$$

$$\times J_0[\kappa E_0 e^{-R^2/2w_0^2}(t-\tau_1)]R\,dR$$

$$= \pi w_0^2 \left\{ I_0 - \frac{\sqrt{\pi}\hbar\Omega L}{k\bar{u}}(N_b - N_a)e^{-\gamma(t-\tau_1)}\frac{\kappa E_0}{t-\tau_1}J_1[\kappa E_0(t-\tau_1)] \right\} \qquad (3\text{-}118)$$

For simplicity we have set $\Omega = \omega_0$ here, i.e., assumed the excitation is at Doppler line center. Equation (3-118) is a rather surprising result. It is similar to the plane-wave case, except that J_0 has been replaced by J_1 and we have an additional factor $1/(t-\tau_1)$. $J_1(x)$ looks essentially the same as $J_0(x)$, the main difference being a $\pi/2$ phase shift, and their oscillation "frequencies" are equal to better than 2%. Hence, the nutation frequency one observes when the entire beam is monitored does not differ significantly from the value one would predict using a plane-wave theory. The damping, however, is greater, owing to the $1/(t - \tau_1)$ factor. This additional damping is a result of the destructive interference of the various nutation frequencies which occur at different radii R.

To a large extent the beam profile integration, and hence the additional damping, may be avoided by using a small area detector which monitors just the central portion of the beam. However, we have nowhere taken into account the transverse motion of molecules across the beam or diffraction effects. Little is known about either of these phenomena, but clearly both will cause the nutation signal for even a point detector to be a mixture of slightly different nutation frequencies. Hence, one probably cannot ever quite obtain plane-wave results, although the use of short samples and large-diameter laser beams will certainly allow us to come close. In the remainder of the chapter we will therefore ignore beam profile effects and concentrate on the plane-wave case.

The discussion above points up the fact that optical nutation is not a very good way to measure the relaxation rate γ, since the damping of the nutation signal is due to a number of factors. In fact, at low pressures, the observed damping is almost independent of γ. We notice, however, that the optical nutation *frequency* has remained $\kappa E_0 = \mu_{ab}E_0/\hbar$ during all our manipulations. This suggests that by carefully measuring both E_0 and the nutation frequency, one could obtain μ_{ab}, the transition-dipole matrix element. We discuss this point further below after examining the optical nutation experiments that have been done to date.

The first observation of optical nutation was made by Hocker and Tang (1968), who observed nutation in SF_6 using a Q-switched CO_2 laser. The nature of the effect in the optical region had been previously discussed by Tang and Silverman (1966). A much clearer observation of optical nutation was reported in 1969 by Hocker and Tang. In that paper they briefly discussed the measurement of transition-dipole matrix elements using nutation and report an estimate of μ_{ab} in SF_6. Somewhat later, Brewer and Shoemaker (1971) reported observations of optical nutation for CH_3F and NH_2D in their original paper on Stark switching. Finally, this author has recently made very clean observations of optical nutation in NH_2D (Shoemaker and Van Stryland, 1976), which, for the first time, allow a quantitative comparison with theory. One of the traces is shown in Figure 3-10. The solid line is the experiment and the dots are a theoretical fit using Eqs. (3-116) and (3-117). A second trace at lower laser power is shown in Figure 3-11. In both cases the agreement between theory and experiment is excellent. Furthermore, both traces can be fit using the same value of μ_{ab}, thus giving an accurate measurement of the transition-dipole matrix element.

A few comments should be made regarding the observation of optical nutation using Stark switching. In our theoretical treatment we have

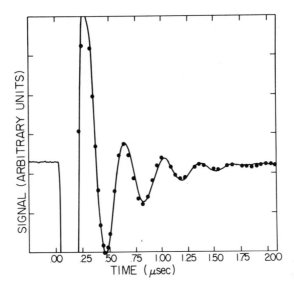

Figure 3-10. Observed and calculated optical nutation signals for the $(\nu_2, J, M) = (0_a, 4_{04}, \pm 4) \rightarrow$ $(1_s, 5_{14}, \pm 5)$ transition in NH_2D with an incident power of 2.18 W (from Shoemaker and Van Stryland, 1976).

assumed that the optical field is zero until time $t = t_1$, when it suddenly is turned on. With Stark switching the field is always on, but the molecules of interest are far off resonance, and hence unexcited, until $t = t_1$, when they are suddenly shifted into resonance and begin to interact with the field (cf. Section 3.5.2). Thus, the theoretical treatment is the same whether we are

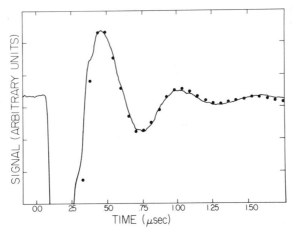

Figure 3-11. Observed and calculated optical nutation signals for the $(0_a, 4_{04}, \pm 4) \rightarrow (1_s, 5_{14}, \pm 5)$ transition in NH_2D with an incident power of 1.13 W (from Shoemaker and Van Stryland, 1976).

using Stark switching or a pulsed laser. The only difference is that in Stark switching there is a second group of molecules which are shifted out of resonance. These molecules emit radiation which shows up in the experiment as a heterodyne beat with the laser. This does not interfere significantly with the nutation experiment for two reasons. First, as we show in the next section, the emission is very short-lived and dies away during the first half-cycle of the optical nutation signal. In Figure 3-11, for example, one can see the tail end of the beat-frequency oscillation occurring between 0.25 and 0.3 μs. Second, one can make the beat completely disappear experimentally by making the Stark shift $\Delta\omega_0$ large enough that the heterodyne beat frequency (which is $\Delta\omega_0$ also) lies outside the bandwidth of the detector and hence is not seen.

We have noted several times that optical nutation can be used to measure μ_{ab}. This point is of more than just passing interest, because transition-dipole matrix elements are important quantities. They fundamentally determine the strength of the molecule-field interaction and as such enter into many different kinds of calculations. Furthermore, it is difficult to measure μ_{ab} using conventional techniques, especially for molecular transitions in the infrared. Optical nutation appears to be well suited to such measurements and, indeed, offers some unique advantages. The usual way one measures μ_{ab} in the infrared is to determine the absolute absorption for a transition over a wide range of sample pressures (see Gerry and Leonard, 1966, for example). This information, together with an accurate calculation of the number of molecules in the coupled levels yields the transition-dipole matrix element. While this method often works well, there are many situations where it does not. These include measurements on molecules which have dense spectra and thus overlapping lines at high pressure, isotopically substituted molecules where the isotopic composition is not accurately known, low-vapor-pressure materials, and hot-band transitions where the level populations may be difficult to determine. In these cases optical nutation is a very attractive alternative, since it requires only a single measurement of the nutation frequency made at low pressure together with a measurement of the optical field strength in the center of the beam (obtained by measuring the beam profile and total power). No knowledge of the level populations is needed.

3.7. Optical Free Induction Decay

3.7.1. Theory and Experiment

We now turn our attention to a second basic coherent transient effect, optical free induction decay (FID). Despite the strange name, optical FID is basically quite simple: It is just the radiation emitted by molecules after they have been excited by a pulse. As in the previous case of optical nutation, the name optical FID comes from nuclear magnetic resonance. In nmr, one can excite a sample of nuclear spins with a radio-frequency pulse. Following the pulse, the spins precess freely in the static external magnetic field H_0. This *free* precession *induces* an oscillating voltage in the pickup coil used as a detector, and the signal *decays* away in time as the precessing spins get out of phase with each other, owing to inhomogeneities in H_0. Hence, the name "free induction decay" for the nmr effect (Hahn, 1950a).

Optical FID is just the optical analog of FID in nmr. We excite molecules to a superposition of states a and b with a pulse of radiation, thus producing a phased array of oscillating dipole moments in the sample. When the pulse is turned off, the molecular dipoles continue to oscillate and radiate a field which is observed as the optical FID. Owing to their translational motion, the dipoles get out of phase with each other, and this causes the observed emission to die away.

The vector model picture of optical FID is very instructive. Consider a sample initially in thermal equilibrium which is excited by an intense pulse whose frequency is $\Omega = \omega_0$. In fact, for simplicity, let us assume that the pulse intensity is such that we have $\kappa E_0 \gg k\bar{u}$. We then have $\kappa E_0 \gg \delta$ for all molecular velocities v_z. This produces a driving vector $\boldsymbol{\Omega}$ which lies nearly along the I axis for all molecules, and they will all interact resonantly. As we showed previously in Figure 3-8a, the **M** vector for resonant molecules during a pulse precesses in a circle in the II–III plane. This time, however, we suppose that after turning on the intense pulse at t_1, we turn it off again at t_2, adjusting the time interval $t_2 - t_1$ such that $\kappa E_0(t_2 - t_1) = \pi/2$; i.e., we allow **M** to precess by 90° and then turn the field off. This situation is illustrated in Figure 3-12. For obvious reasons the pulse we have applied is called a $\pi/2$ or

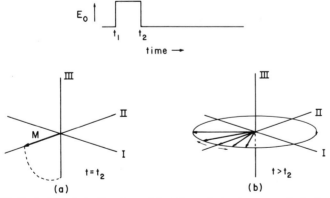

Figure 3-12. Optical free induction decay. (a) At the end of an intense optical pulse ($t = t_2$) the **M** vectors for all molecular velocities lie along the $-$II axis, producing a strong emission. (b) For $t > t_2$ the **M** vectors dephase as they precess at differing rates in the I–II plane.

$90°$ pulse. By making the pulse sufficiently intense, we have caused the **M** vectors for all molecules to behave identically; they all lie along the $-$II axis at time t_2. For times $t > t_2$, we have $\kappa E_0 = 0$, so the driving vector $\boldsymbol{\Omega}$ lies along the III axis and has the magnitude $|\boldsymbol{\Omega}| = |kv_z|$, since we assumed that $\Omega = \omega_0$. Each vector **M** will therefore precess in the I–II plane, but the **M** vectors for molecules of differing velocity will precess at different rates because their $\boldsymbol{\Omega}$'s are different. This causes the vectors to fan out in the I–II plane as shown in Figure 3-12b. In drawing this figure we have superimposed the **M** vectors for several velocity groups of molecules.* We recall from Section 3.4.1 that u and v are proportional to the in- and out-of-phase components of the oscillating dipole moment for some velocity group v_z. Hence, the direction of **M** in the I–II plane shows the phase of the induced dipole moment with respect to the optical field. Since all the molecules were excited by the same optical field, a diagram like Figure 3-12 also shows the phase of the various **M** vectors with respect to each other (as viewed from a common frame of reference). The relative phases are very important in an FID experiment, since they largely determine the observable behavior of the system. We can see this by noting that the polarization component P_s is found by adding up the v components of each **M** vector; i.e., we integrate v over all v_z as in Eqs. (3-69). At $t = t_2$, all the vectors are in phase with each other, P_s is a maximum, and the system will radiate strongly. As time progresses, however, the **M** vectors get out of phase with each other as shown in Figure 3-12b. This fanning out of the **M** vectors is called *Doppler dephasing*. When we add up the v components to get P_s, we now find a smaller value and the radiation is weaker. As the **M** vectors continue to fan out, P_s gets smaller and smaller and eventually becomes zero. Notice that the Doppler dephasing has nothing to do with collisional relaxation, as we have neglected collisions thus far. The Doppler dephasing results solely from the fact that we have a distribution of different molecular velocities.

We can gain further understanding by doing a simple quantitative calculation. As we did above, suppose we apply a short intense pulse such that $\kappa E_0 \gg k\bar{u}$ and adjust its length so that $\kappa E_0(t_2 - t_1) = \pi/2$. We will not assume that $\Omega = \omega_0$, however. Since the pulse is so short we will neglect steady-state terms as well as $\exp(-\gamma t)$ in calculating the molecular response

---

* More precisely, what we have done in drawing Figure 3-12 is to formulate the vector model for each velocity group in a coordinate frame which is (translationally) at rest with respect to the laboratory. In this frame the frequency of the optical field is Ω for all molecules, but the apparent resonance frequency for each velocity group is different and given by $\omega_0 + kv_z$. In our original introduction of the vector model (see Section 3.4.2) we used a moving coordinate frame in which the molecule was at rest. However, the vector model is identical in either frame because the motion of **M** depends only on δ, the detuning from resonance, not the frequency of the optical field. In the laboratory frame $\delta = \Omega - (\omega_0 + kv_z)$, and in the moving frame $\delta = \Omega' - \omega_0 = \Omega - kv_z - \omega_0$.

to the pulse. Hence,

$$M(t_2) = e_t^{-\beta_{21}}M(t_1) \tag{3-119}$$

where $t_{21} = t_2 - t_1$. Furthermore, since $\kappa E_0 \gg (\Omega' - \omega_0)$, we have $g \approx \kappa E_0$ in the expression for $\exp(-\beta t_{21})$. Assuming that the system is in thermal equilibrium at $t = t_1$, and using Eq. (3-93) we find that

$$M(t_2) = \begin{pmatrix} u \\ v \\ w \end{pmatrix} = (n_a - n_b) \begin{pmatrix} \dfrac{\delta}{\kappa E_0}(1 - \cos \kappa E_0 t_{21}) \\ \sin \kappa E_0 t_{21} \\ \cos \kappa E_0 t_{21} \end{pmatrix} \cong \begin{pmatrix} 0 \\ n_a - n_b \\ 0 \end{pmatrix} \tag{3-120}$$

We now use $M(t_2)$ as the initial condition for calculating the behavior of the system for times $t > t_2$. During this time period $\kappa E_0 = 0$, and one can use the solutions to the full set of population matrix equations given in Eqs. (3–72) to describe the system. We need only u and v, which are

$$u(t) = e^{-\gamma_{ab}(t-t_2)}(n_a - n_b) \sin \delta(t - t_2)$$
$$v(t) = e^{-\gamma_{ab}(t-t_2)}(n_a - n_b) \cos \delta(t - t_2) \tag{3-121}$$

It is now a simple matter to put in explicitly the v_z dependence of n_a, n_b, and δ, integrate over v_z to obtain P_c and P_s, and then use Eqs. (3-50) to find the field emitted by the sample. Although this calculation is straightforward and gives the correct result, it does not display some of the physics as clearly as one would like. Instead, we proceed in a slightly different manner by first calculating the average oscillating dipole moment, $\bar{\mu}_{ab}(z, v_z, t)$ for a velocity group of molecules. Using Eqs. (3-37) and (3-65), we find that

$$\bar{\mu}_{ab}(z, v_z, t) = \mu_{ab}[u(t) \cos(\Omega't - kz') - v(t) \sin(\Omega't - kz')]$$

$$= \mu_{ab}e^{-\gamma_{ab}(t-t_2)}(n_a - n_b)[\sin(\Omega' - \omega_0)(t - t_2) \cos(\Omega't - kz')$$

$$- \cos(\Omega' - \omega_0)(t - t_2) \sin(\Omega't - kz')]$$

$$= -\mu_{ab}e^{-\gamma_{ab}(t-t_2)}(n_a - n_b) \sin(\omega_0 t - kz' + \phi') \tag{3-122}$$

for times $t > t_2$. Here, $\phi' = (\Omega' - \omega_0)t_2$ is simply a fixed phase whose value depends on the exact time at which the excitation pulse ends. The third line of (3-122) follows immediately from a trigonometric addition formula.

We have here a very interesting result. It says that immediately after the end of the excitation pulse, every velocity group of molecules has an oscillating dipole moment which oscillates at ω_0, the natural resonance frequency of the molecules. The frequency Ω' of the excitation pulse is completely irrelevant. This result verifies a statement we have made several times previously—that when $\kappa E_0 = 0$, all molecules in a superposition of states a and b have an oscillating dipole moment whose frequency is ω_0.

The polarization in the sample can be found by simply integrating (3-122) over all velocities. We note that both z' and ϕ' have a velocity dependence since $z' = z - v_z t$ and $\phi' = (\Omega - k v_z - \omega_0) t_2$. Rearranging the argument of the sine function, we have

$$
\begin{aligned}
P(z, t) &= \frac{\mu_{ab}}{\bar{u}\sqrt{\pi}} e^{-\gamma_{ab}(t-t_2)} (N_b - N_a) \int_{-\infty}^{\infty} e^{-v_z^2/\bar{u}^2} \\
&\quad \times \sin\left[(\omega_0 + k v_2)(t - t_2) - kz + \Omega t_2\right] dv_z \\
&= \left[\mu_{ab}(N_b - N_a) e^{-\gamma_{ab}(t-t_2)} e^{-(1/4)k^2 \bar{u}^2 (t-t_2)^2}\right] \\
&\quad \times \sin\left[\omega_0(t - t_2) - kz + \Omega t_2\right]
\end{aligned} \tag{3-123}
$$

This result shows very clearly what is involved in Doppler dephasing. Although each oscillating dipole is oscillating at ω_0, the molecules are moving with velocity v_z. Hence, in the laboratory frame, their oscillation frequency appears to be $\omega_0 + k v_z$. When we add up all the different oscillation frequencies to obtain the polarization P, destructive interference causes P to decay away very rapidly. We call this decay Doppler dephasing because the molecules initially start off oscillating in phase at $t = t_2$, but then rapidly get out of phase with each other because they oscillate at different apparent frequencies due to the Doppler effect. The decay of P is nonexponential and is given by the factor $\exp\left[-\frac{1}{4}k^2\bar{u}^2(t-t_2)^2\right]$.

Before using (3-123) to calculate the emitted electromagnetic field, we need to rearrange the argument of the sine function, which looks somewhat peculiar in its present form. The key is to note that t_2 has a z dependence, as was discussed in Section 3.3.1. We have $t_2 = \tau_2 - (L - z)/c$, where τ_2 is the time at which the cutoff of the excitation pulse reaches the end of the sample, $z = L$. Putting this into (3-123), we find that the sine function becomes $\sin(\omega_0 t - k_\omega z + \phi)$, where $k_\omega = \omega_0/c$ is the proper wave vector for a traveling wave of frequency ω_0, and $\phi = (\Omega - \omega_0)(\tau_2 - L/c)$ is just a fixed phase angle. Thus, the term in brackets preceding the sine function in (3-123) is just the polarization amplitude for a plane wave of frequency ω_0. Denoting this amplitude by P_ω, we can now write down equations for the polarization and emitted field analogous to Eqs. (3-42),

$$
\begin{aligned}
P(z, t) &= P_\omega \sin(\omega_0 t - k_\omega z + \phi) \\
\mathscr{E}(z, t) &= \mathscr{E}_\omega \cos(\omega_0 t - k_\omega z + \phi)
\end{aligned} \tag{3-124}
$$

The emitted field is written as a cosine because we know that a plane-wave polarization will drive a field 90° out of phase with itself (see Appendix A).

P_ω and $\mathscr{E}_\omega$ are exactly analogous to P_s and $\mathscr{E}_c$ in Eqs. (3-42)–(3-50), so we can immediately write down the expression for $\mathscr{E}_\omega$ in terms of P_ω. It is

$$\mathscr{E}_\omega(L, t) = -\frac{\omega_0 L}{2\varepsilon_0 c} P_\omega(t - \tau_2) \tag{3-125}$$

Hence, the emitted field amplitude at the end of the sample is

$$\mathscr{E}_\omega(L, t) = -\frac{\mu_{ab}\omega_0 L}{2\varepsilon_0 c}(N_b - N_a)e^{-\gamma_{ab}(t-\tau_2)}e^{-(1/4)k^2\bar{u}^2(t-\tau_2)^2} \tag{3-126}$$

and the intensity one would detect is

$$I(t) = \frac{\mu_{ab}^2\omega_0^2 L^2}{8\varepsilon_0 c}(N_b - N_a)^2 e^{-2\gamma_{ab}(t-\tau_2)}e^{-(1/2)k^2\bar{u}^2(t-\tau_2)^2} \tag{3-127}$$

We see from this that the emission is a maximum at $t = \tau_2$ and that it has a rapid nonexponential decay due to Doppler dephasing. The one feature of the emitted intensity that is surprising at first glance is the $L^2(N_b - N_a)^2$ dependence. Ordinarily, one expects to see a linear dependence on length and number of molecules as in Eqs. (3-63) or (3-113). Because of the N^2 intensity dependence, a number of authors have been led to call the optical FID emission "superradiant". We discuss this point in detail in Section 3.7.2.

Another feature of the FID that is implicit in our treatment is the directionality of the emission. It occurs only in the forward direction. Of course, it could hardly be otherwise here, since we assumed plane-wave excitation. However, if we consider FID from a sample excited by a laser beam of finite size, we also find the same type of result. Nearly all the emission is concentrated within a very small angle of the forward direction, because only in the forward direction does one have a phased array of dipoles. A field propagating at even a slight angle will find that dipoles near the end of the sample are emitting out of phase, thus producing destructive interference in that direction.

Note also that because we assume an optically thin sample, we have $\mathscr{E}_c \ll E_0$. Hence, we may assume that the emission process itself will not significantly change the state of the system; i.e., the emission is not so intense that an appreciable fraction of the energy absorbed during excitation of the system is radiated away. In an optically thick sample, this process must be considered since the system can then radiate so strongly that the excitation is radiated away, leaving all the molecules down in the lower state. This phenomenon is known as radiation damping (Bloembergen and Pound, 1954).

In practice, one rarely has sufficiently intense pulses to satisfy $\kappa E_0 \gg \Omega' - \omega_0$ for all velocity groups of molecules. Instead, one usually has

$\kappa E_0 \ll k\bar{u}$, so that only molecules in a narrow velocity interval are excited resonantly. Also, there is no reason to restrict ourselves to pulses whose length is such that $\kappa E_0(t_2 - t_1) = \pi/2$. Optical FID in a Doppler-broadened medium can be observed for pulses of any length. It is convenient to characterize pulses by their "area," which we define as

$$\theta = \kappa \int_{-\infty}^{\infty} E_0(t)\, dt$$

$$= \kappa E_0 \,\Delta t \qquad \text{for rectangular pulses} \qquad (3\text{-}128)$$

Δt is the length of the pulse, e.g., $t_2 - t_1$. The concept of a pulse area is extremely useful in pulse-propagation problems such as self-induced transparency (McCall and Hahn, 1967). Note that the area θ is just the angle through which resonantly excited molecules precess during the pulse.

In the general case where θ is arbitrary and we do not require $\kappa E_0 \gg k\bar{u}$ for all molecules, the state of the system at the end of the pulse is given by Eq. (3-108) with $t' = t_2 - t_1$. This assumes that the system is in thermal equilibrium at $t = t_1$ and that the pulse is short enough that steady-state terms may be neglected. While some molecules are excited resonantly, we now also have a considerable amount of off-resonant excitation. This is illustrated in Figure 3-13, where we show a plot of u and v at time $t = t_2$ as a function of detuning from resonance. The graph is for the case of a $\pi/2$ excitation pulse $(t_2 - t_1 = \pi/2\kappa E_0)$, and the frequency axis is labeled in units of κE_0, giving curves which are valid for a 90° pulse of any intensity. We see that the excitation is nonnegligible even for detunings of as much as from 5 to $10\kappa E_0$.

Turning now to a quantitative calculation, we find that u and v following a short pulse of arbitrary area are

$$u(t) = e^{-\gamma_{ab}(t-t_2)}(n_a - n_b)\frac{\kappa E_0}{g}\left\{\frac{\delta}{g}[1 - \cos g(t_2 - t_1)]\cos \delta(t - t_2)\right.$$

$$\left. + \sin g(t_2 - t_1)\sin \delta(t - t_2)\right\}$$

$$\qquad (3\text{-}129)$$

$$v(t) = e^{-\gamma_{ab}(t-t_2)}(n_a - n_b)\frac{\kappa E_0}{g}\left\{\sin g(t_2 - t_1)\cos \delta(t - t_2)\right.$$

$$\left. - \frac{\delta}{g}[1 - \cos g(t_2 - t_1)]\sin \delta(t - t_2)\right\}$$

These expressions follow immediately from Eqs. (3-108) and (3-72). Let us now assume that the pulse intensity is such that $\kappa E_0 \ll k\bar{u}$. Then only a small fraction of molecules within the Doppler profile will be excited, and we can pull the Gaussian factor $\exp(-v_z^2/\bar{u}^2)$ outside the integrals over velocity

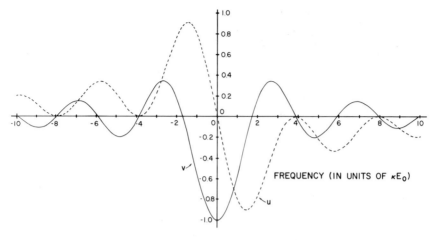

Figure 3-13. The **M** vector components u and v following a $\pi/2$ pulse. u and v are plotted versus detuning from exact resonance (or equivalently vs. molecular velocity).

when we calculate P_c and P_s from u and v. We then see immediately that P_c vanishes because $u(t)$ is an odd function of v_z. P_s is given by

$$
P_s = \left[\frac{\mu_{ab}\kappa E_0}{k\bar{u}\sqrt{\pi}} (N_b - N_a)e^{-(\Omega - \omega_0)^2/k^2\bar{u}^2} \right] e^{-\gamma_{ab}(t-t_2)} \int_{-\infty}^{\infty} \left\{ \frac{\sin g(t_2-t_1)}{g} \right.
$$

$$
\left. \times \cos \delta(t-t_2) - \frac{\delta}{g^2}[1 - \cos g(t_2-t_1)] \sin \delta(t-t_2) \right\} d\delta \tag{3-130}
$$

For convenience we have changed the variable of integration from v_z to $\delta = \Omega - \omega_0 - kv_z$. The first term in this integral can be done analytically, but as it stands, the second cannot. However, we note that the second term is zero when $\delta = 0$ and only becomes large when $\delta > \kappa E_0$. Thus, a reasonable approximation to the second term might be obtained by replacing $g = \sqrt{\delta^2 + \kappa^2 E_0^2}$ with δ. Using a half-angle formula, that term then becomes $2 \sin^2 [\delta(t_2-t_1)/2] \sin \delta(t-t_2)/\delta$, which is integrable. Thus,

$$
P_s \cong Ce^{-\gamma_{ab}(t-t_2)}\{\pi J_0[\kappa E_0\sqrt{(t_2-t_1)^2 - (t-t_2)^2}] - \pi\} \quad (\text{for } t-t_2 < t_2-t_1)
$$

$$
= 0 \quad (\text{for } t-t_2 > t_2-t_1) \tag{3-131}
$$

where C is the group of constants in braces in Eq. (3-130). This is a very interesting result. For times $t-t_2$ shorter than the pulse length, the polarization (and hence the emission) can be quite large, but at $t-t_2 = t_2-t_1$ the emission vanishes and stays zero for all later times. This is not a result of our approximations. By working in the Fourier transform domain, one can show rigorously that $P_s = 0$ for times $t - t_2$ longer than the pulse length (Hopf and Scully, 1969). Equation (3-131) is an approximation, however, and comparison with exact numerical calculations shows that the approximation we used is a good one only for small area pulses ($\theta \sim \pi/2$ or less).

One can rationalize the disappearance of the emission by noting that, crudely speaking, the bandwidth $\Delta\nu$ of molecules excited at t_2 is approximately $1/(t_2 - t_1)$, the inverse of the pulse length. Since P_s, the integral of $v(t)$ over δ, looks very much like the Fourier transform of $v(t_2)$, we might expect that it will disappear in a time $1/\Delta\nu \approx t_2 - t_1$.

It is now trivial to obtain the emitted intensity. One simply uses (3-50) to find $\mathscr{E}_c$ and then squares the result. We obtain

$$I(t) = \frac{\pi}{8\varepsilon_0 c}\left[\frac{\Omega L \mu_{ab} \kappa E_0}{k\bar{u}}(N_a - N_b)e^{-(\Omega-\omega_0)^2/k^2\bar{u}^2}\right]^2 e^{-2\gamma_{ab}(t-\tau_2)}$$

$$\times \{J_0[\kappa E_0\sqrt{(\tau_2-\tau_1)^2 - (t-\tau_2)^2}] - 1\}^2 \tag{3-132}$$

for times $t - \tau_2 < \tau_2 - \tau_1$. Notice that because of the J_0 Bessel function dependence, the emission can show considerable oscillatory structure. This will occur for large area excitation pulses which have $\kappa E_0(t_2 - t_1) \gg 1$. Furthermore, $I(t)$ can have a local maximum for times $t - \tau_2$ near $\tau_2 - \tau_1$, and hence the emission can show a small peak just before it vanishes. This phenomenon has been called an "edge echo" (Bloom, 1955). Both the sudden disappearance of the emission and edge echoes have actually been observed; first in nmr (Bloom, 1955), and more recently in the infrared (Brewer and Shoemaker, 1972).

Finally, we note that the frequency of the emission is Ω in the example we have just discussed. While it is still true that the oscillating dipoles oscillate at ω_0 following the pulse, we have only excited a small band of velocities whose laboratory-frame resonant frequencies $\omega_0 + kv_z$ lie in a narrow region centered on Ω. Hence, that is the emission frequency we observe.

So far we have only considered short pulses for which we could ignore steady-state terms. For longer pulses it is a simple matter to include the steady-state terms, but the Doppler integrals arising from them cannot be done analytically. However, everything becomes simple again if we consider very long pulses. In this case the system will have settled down to steady state by the end of the pulse and we have $\mathsf{M}(t_2) = \mathsf{M}(\infty)$. Hence, using Eqs. (3-72) and (3-86), we obtain

$$v(t) = \frac{\kappa E_0(n_a - n_b)}{\delta^2 + \Gamma^2}e^{-\gamma_{ab}(t-t_2)}[\gamma_{ab}\cos\delta(t-t_2) - \delta\sin\delta(t-t_2)]$$

$$\tag{3-133}$$

where $\Gamma^2 = (\gamma_{ab}/\gamma)\kappa^2 E_0^2 + \gamma_{ab}^2$.

We need not consider $u(t)$ if we assume as before that the width of excited molecules is much less than the Doppler width; i.e., $\Gamma \ll k\bar{u}$. This is the usual case for low-pressure gases. $u(t)$ is then an odd function of v_z, giving $P_c = 0$. The inequality $\Gamma \ll k\bar{u}$ also allows us to remove the Gaussian factor from the integral over v_z when we calculate P_s. The integral is then readily done, to yield

$$P_s = \frac{\mu_{ab}\kappa E_0}{k\bar{u}\sqrt{\pi}}(N_b - N_a)e^{-\gamma_{ab}(t-t_2)}e^{-(\Omega-\omega_0)^2/k^2\bar{u}^2}\left[\frac{\pi\gamma_{ab}}{\Gamma}e^{-\Gamma(t-t_2)} - \pi e^{-\Gamma(t-t_2)}\right]$$

(3-134)

and hence

$$\mathscr{E}_c = \left[\frac{\sqrt{\pi}\mu_{ab}\kappa E_0\Omega L}{2\varepsilon_0 ck\bar{u}}(N_b - N_a)e^{-(\Omega-\omega_0)^2/k^2\bar{u}^2}\right]\left(1 - \frac{\gamma_{ab}}{\Gamma}\right)e^{-(\gamma_{ab}+\Gamma)(t-\tau_2)}$$

(3-135)

If desired, the emitted intensity may be obtained from $I(t) = \frac{1}{2}c\varepsilon_0\mathscr{E}_c^2$. This simple result shows several interesting features. First, to see a large emission, we need $\Gamma \gg \gamma_{ab}$, which implies that $\kappa E_0 \gg \gamma_{ab}$. This means that one should use a fairly strong optical field to prepare the sample. Also, we notice that the decay of the emission goes as $\exp[-(\gamma_{ab}+\Gamma)(t-\tau_2)]$. This is exactly what we would expect. The γ_{ab} contribution comes from collisional decay, while the Γ contribution describes the Doppler dephasing. We have excited a bandwidth of molecular velocities characterized by Γ during the steady-state preparation, and these molecules then radiate at all laboratory-frame frequencies $\omega_0 + kv_z$ within this bandwidth. As they do so, they get out of phase with each other in a time which is approximately $1/\Gamma$.

One incorrect feature of Eq. (3-135) is that it predicts a maximum in the emission at the instant the excitation pulse is turned off, $t - \tau_2 = 0$. Actually, the reradiated field starts out at time $t - \tau_2 = 0$ with the small value $\mathscr{E}_c = C\gamma_{ab}/\Gamma$ [where C is the collection of constants in brackets in Eq. (3-135)] and then rises rapidly in a time $\sim 1/k\bar{u}$ to the value $\mathscr{E}_c = C(1 - \gamma_{ab}/\Gamma)$ predicted by Eq. (3-135). As one might suspect, the problem is caused by the fact that we pulled the Gaussian factor out of the integrals over v_z. A careful inspection shows that for very small times $t - \tau_2$, one cannot do this for the integral over v_z of the second term in Eq. (3-133). This point is discussed in detail by Foster *et al.* (1974).

One final feature of the FID in (3-135) is that most of the signal is due to off-resonant molecules. In a strong field, u is small in steady-state compared to v for slightly off-resonant molecules. As an example, the two quantities are shown in Figure 3-14 for the case of an optical field such that $\kappa E_0 = 5\gamma_{ab}$. An examination of Eqs. (3-72) and (3-133)–(3-135) shows that it is $u(t_2)$ (and hence the off-resonant molecules) which produce most of the field $\mathscr{E}_c$.

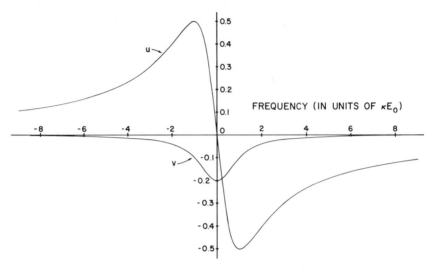

Figure 3-14. u and v as a function of detuning for the case of steady-state absorption in a strong optical field ($\kappa E_0 = 5\gamma_{ab}$).

Initially, these molecules give no contribution to $v(t)$ because $u(t_2)$ for molecules excited above resonance ($\delta > 0$) is just cancelled by $u(t_2)$ for molecules excited below resonance. For times $t > t_2$, however, these molecules give a large contribution as their Bloch vectors **M** begin to precess in the I–II plane.

We now turn to a discussion of the optical FID experiments that have been done to date. The first report of optical FID is a paper by Cheo and Wang in 1970. However, their SF_6 gas samples were optically thick ($\alpha L = 1.4$ and larger), and while free induction decay certainly occurs within such samples, the molecular emission from the front of the sample is so large that it excites molecules farther back. The result is that one simply observes a complete reshaping of the input pulse at the detector. Whether one wants to call this an observation of optical FID is a matter of semantics. It certainly cannot be described quantitatively by the theory we have been discussing. The first really clear observation of optical FID was made in 1972 by Brewer and Shoemaker using Stark switching. Free induction decays were observed using both short pulses and long pulse steady-state preparation and the signals compared with theory. Subsequently, Hall (1973) has observed optical FID in CH_4 using a frequency-switched He–Ne laser, and Brewer and Genack (1976) have observed FID in I_2 using a frequency-switched dye laser.

At this point we should stop briefly to consider the form of the signal one will observe in a Stark switching (or frequency-switching) FID experiment. Since the optical field is always present and only the resonance frequency is switched, this will require a slight modification of our previous theoretical results. Consider first the simplest case, where we have a resonant cw laser beam passing through a sample and we suddenly turn on a step-function electric field that shifts the molecular resonance frequency from ω_0 to $\omega_s = \omega_0 + \Delta\omega_0$. Molecules whose velocities v_z are such that $\Omega - kv_z - \omega_0 \approx 0$ are then shifted out of resonance, and a new group of molecules with velocities v_z' such that $\Omega - kv_z' - \omega_s \approx 0$ are shifted into resonance (see Figure 3-7). Provided that the shift $\Delta\omega_0$ is large enough, we may regard the two groups of molecules as being entirely distinct and treat them independently. The observed signal can then be found by simply summing all fields and squaring to obtain $I(t)$. Thus,

$$I(t) = \tfrac{1}{2}c\varepsilon_0(E_0 + \mathscr{E}_c^{v_z'} + \mathscr{E}_c^{v_z})^2$$

$$= \tfrac{1}{2}c\varepsilon_0(E_0^2 + 2\mathscr{E}_c^{v_z'}E_0 + 2\mathscr{E}_c^{v_z}E_0) \tag{3-136}$$

where we have used the fact that both reradiated fields $\mathscr{E}_c$ are small to neglect all terms of the form $\mathscr{E}_c\mathscr{E}_c$. The contribution $2\mathscr{E}_c^{v_z'}E_0$ from molecules that are switched into resonance is clearly just given by the expressions developed in the previous section on optical nutation [see Eqs. (3-113) and (3-116)] provided that we replace ω_0 by ω_s.

Finding the contribution $2\mathscr{E}_c^{v_z}E_0$ from the molecules switched out of resonance is also quite simple. In Stark switching the resonance frequency is changed to ω_s for times $t > t_2$, so instead of precessing in the I–II plane with a frequency $\delta = \Omega' - \omega_0$, the molecules precess at a frequency $\delta_s = \Omega' - (\omega_0 + \Delta\omega_0)$. Looking back at our derivations, we see that this changes the expression for $v(t)$ to

$$v(t) = e^{-\gamma_{ab}(t-t_2)}[v(t_2)\cos\delta_s(t-t_2) - u(t_2)\sin\delta_s(t-t_2)] \tag{3-137}$$

where $v(t_2)$ and $u(t_2)$ are identical to the explicit expressions given for them in Eq. (3-133). Strictly speaking, we should also write down the expression for $u(t)$. Unlike the pulsed optical field case, in Stark switching $u(t)$ gives rise to a nonvanishing polarization P_c. However, the resulting field $\mathscr{E}_s\sin(\Omega t - kz)$ is 90° out of phase with the laser field $E_0\cos(\Omega t - kz)$, and hence the heterodyne beat signal $2\mathscr{E}_sE_0$ is zero. Thus, we will simply ignore $u(t)$. Noting that $\delta_s = \delta - \Delta\omega_0$, we may write Eq. (3-137) as

$$v(t) = e^{-\gamma_{ab}(t-t_2)}\{[v(t_2)\cos\delta(t-t_2) - u(t_2)\sin\delta(t-t_2)]\cos\Delta\omega_0(t-t_2)$$

$$+ [v(t_2)\sin\delta(t-t_2) + u(t_2)\cos\delta(t-t_2)]\sin\Delta\omega_0(t-t_2)\} \tag{3-138}$$

Now in a Stark-switching experiment, we generally work under conditions where the bandwidth of any group of excited molecules is narrow compared to the Doppler width $k\bar{u}$. This means that we may pull the Gaussian factor $\exp(-v_z^2/\bar{u}^2)$ outside the integral when we integrate $v(t)$ over v_z to find the polarization. Furthermore, for excitation pulses of any length or intensity, one can readily verify that $u(t_2)$ is always an odd function of δ, and $v(t_2)$ an even function. Hence when we integrate $v(t)$ over v_z (or, equivalently, over δ), the terms multiplying $\sin \Delta\omega_0(t-t_2)$ in Eq. (3-138) will give zero. If we examine the remaining terms, we see that one has exactly the result obtained for a pulsed optical field, Eq. (3-133), except for an additional factor $\cos \Delta\omega_0(t-t_2)$. Hence, the emitted field $\mathscr{E}_c^{v_z}$ for molecules switched out of resonance in a Stark-switching experiment will just be given by $\mathscr{E}_c$ for the pulsed optical field case multiplied by $\cos \Delta\omega_0(t-\tau_2)$.

The physical reason for the appearance of a $\cos \Delta\omega_0 t$ term is quite simple. Had we turned the field off at $t=t_2$ instead of Stark switching, we would expect to see an emission centered at Ω, because the excited molecules have lab-frame natural resonance frequencies ω_0+kv_z centered at Ω. In Stark switching, however, we have switched the natural resonance frequency to $\omega_0+\Delta\omega_0$, and hence we expect to see emission at the frequency $\Omega+\Delta\omega_0$. This is exactly what one obtains when the $\mathscr{E}_c$ for a pulsed optical field is multiplied by the factor $\cos \Delta\omega_0(t-\tau_2)$.

An alternative way to find the emitted field in a Stark switching experiment is to calculate $\bar{\mu}_{ab}(z, v_z, t)$ and then integrate over all velocities to find the polarization and emitted field $\mathscr{E}_\omega$ as we did in Eqs. (3.122)–(3-126). For the case at hand we find that

$$\mathscr{E}_\omega(z, t) = -\frac{\Omega L}{2\varepsilon_0 c} P_s(t-\tau_2) \cos \left[(\Omega+\Delta\omega_0)t - \left(k+\frac{\Delta\omega_0}{c}\right)z + \phi \right] \quad (3\text{-}139)$$

where P_s is given by Eq. (3-134) and $\phi = -\Delta\omega_0(\tau_2 - L/c)$.

This field is completely equivalent to $\mathscr{E}_c \cos \Delta\omega_0(t-\tau_2)$, where $\mathscr{E}_c$ is given in Eq. (3-135), because $\langle \mathscr{E}_\omega E_0 \rangle_T = \mathscr{E}_c E_0 \cos \Delta\omega_0(t-\tau_2)$. The angle brackets $\langle \ \rangle_T$ denote a time average over several optical cycles. Note that we have a perfectly phased array of dipoles at the frequency $\Omega+\Delta\omega_0$, despite the fact that we excited the molecules at frequency Ω. This is a result of our assumption that the Stark-field pulse sweeps down the length of the sample at the speed of light. In doing so, it shifts the frequency of each dipole at just the right time to have it properly phased with respect to the rest of the dipoles.

Having found the reradiated fields for molecules switched into and out of resonance, we can put everything together to obtain

$$I(t) = I_0 - \frac{\kappa^2 E_0^2 \sqrt{\pi} \hbar \Omega L}{2k\bar{u}} (N_b - N_a) \{ e^{-(\Omega-\omega_s)^2 k^2 \bar{u}^2} e^{-\gamma(t-\tau_2)} J_0[\kappa E_0(t-\tau_2)]$$

$$-e^{-(\Omega-\omega_0)^2/k^2\bar{u}^2} \left(1 - \frac{\gamma_{ab}}{\Gamma}\right) e^{-(\gamma_{ab}+\Gamma)(t-\tau_2)} \cos \Delta\omega_0(t-\tau_2) \} \quad (3\text{-}140)$$

as the signal one expects to see in a Stark-switching experiment where the

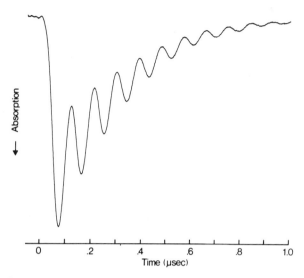

Figure 3-15. Optical free induction decay (FID) in NH_2D. The FID signal appears as a heterodyne beat with the laser, superimposed on optical nutation signal (from Brewer and Shoemaker, 1972).

Stark field is turned on at τ_2. We have chosen here the simpler form of the optical nutation signal, which neglects small steady-state terms. Figure 3-15 shows an experimental trace of optical FID in NH_2D, and we see that it does indeed look like the signal predicted by Eq. (3-140). In fact, one can make a good quantitative fit of the data using this equation (Hopf *et al.*, 1973). In Figure 3-15 we have lowered the laser power and expanded the time scale compared to the optical nutation traces shown in the previous section. Thus, we see only the first downward spike of the optical nutation trace. The optical FID appears as a heterodyne beat with the laser which is superimposed on the nutation background. By observing the amplitude of the beat as a function of time we have a direct measurement of the optical FID signal. Note that because we observe the effect as a heterodyne beat, we are really measuring the emitted field $\mathscr{E}_c$ rather than the emitted intensity.

In Section 3.6 we said that FID need not interfere with an optical nutation experiment, because the FID signal will die out during the first cycle of the optical nutation. Equation (3-140) demonstrates this point. At the end of the first half-cycle of the J_0 function, $\kappa E_0(t - \tau_2) \simeq 2$, and since $\Gamma > \kappa E_0$, the FID signal will already be reduced by a factor of more than $\exp(-2)$.

We have considered in detail the Stark-switching signal one will observe for a step-function Stark pulse. However, the same arguments hold for the FID signal coming from a short Stark pulse, and the polarization for the emitted field is simply given by Eq. (3-131) multiplied by $\cos \Delta\omega_0(t - \tau_2)$. The one feature that differs in this case is the optical nutation signal. The molecules giving rise to this signal see a resonant optical field that is interrupted for a short period of time and then turned back on. This produces an optical nutation signal whose amplitude is different from the signals discussed in Section 3.6. We discuss this signal in more detail in Section 3.9.2, where we show that it can be used to measure the level lifetimes γ_a and γ_b.

Some interesting signals can be obtained when one excites optical FID in a system containing more than two levels. R. G. Brewer and his coworkers have observed FID in one such system (Foster *et al.*, 1974), with the result shown in Fig. 3-16. Here an initially degenerate transition was excited using $\Delta M_J = 0$ selection rules, and then a Stark field was applied that shifts the various M_J sublevels and removes the degeneracy. Each $M_J \rightarrow M_J$ transition is thus excited independently and radiates an independent FID signal as described by Eq. (3-140). However, the FID amplitude and heterodyne beat frequency for each transition is different because μ_{ab} and $\Delta\omega_0$ have different values. The observed FID signal then has the form

$$S = \sum_{M_J = -J}^{J} A_{M_J} e^{-(\gamma_{ab} + \Gamma_{M_J})(t - \tau_2)} \cos \Delta\omega_{M_J}(t - \tau_2) \qquad (3\text{-}141)$$

where A_{M_J} and $\Delta\omega_{M_J}$ are the FID amplitude and Stark shift associated with a given $M_J \rightarrow M_J$ transition. For the CH_3F $J = 4 \rightarrow 5$ transition shown in Figure 3-16, the summation produces a set of periodic pulses. This is a result of the fact that both the upper and lower states have first-order Stark effects

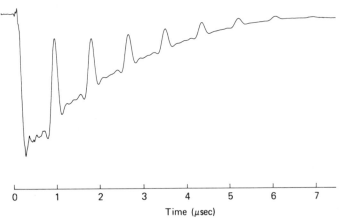

Figure 3-16. Optical FID from several $M_J \rightarrow M_J$ components in a $J = 4 \rightarrow 5$ transition of $^{13}CH_3F$. The total FID signal appears pulsed due to constructive and destructive interference between the individual FID emissions (from Foster *et al.*, 1974; reproduced by permission).

giving $\Delta\omega_{M_J} = CM_J$, where C is a constant. One can easily show that putting this expression in for $\Delta\omega_{M_J}$ in Eq. (3–141) will give a series of pulses that quantitatively fit the data shown in the figure. If one uses $\Delta M_J = \pm 1$ selection rules instead of $M_J = 0$, an exact analysis becomes extremely complicated, although one still expects to see a signal having the general form of Eq. (3-141).

On searching the literature for other observations of optical FID in gases, one finds that there is one other interesting FID experiment reported. In this experiment, by Yablonovitch and Goldhar in 1974, CO_2 gas was resonantly excited using essentially steady-state preparation. The excitation was then suddenly cut off using a gas breakdown cell, so that the sample emitted an optical FID signal. The unique feature of the experiment is that it was done in a medium that was very thick optically. One then sees no signal transmitted through the cell prior to cutoff of the excitation. Using the picture of the absorption process described in Section 3.3, we explain this zero transmitted signal as being due to destructive interference between the incident field and the reradiated field of the molecules. Therefore, when we cut off the incident field, only the reradiated field is left, and we suddenly see a pulse coming out of the cell after E_0 is cut off. This is a rather crude picture of the effect, but it contains the essence of the physics. A detailed, quantitative discussion is beyond the scope of this chapter as it involves a proper treatment of the resonant propagation effects in an optically thick medium. In addition to illustrating some interesting physics, this type of experiment shows promise as a useful technique for generating short laser pulses, since the pulse length can be tailored at will by varying the sample length or pressure.

As yet, optical FID has not found a good practical application other than the one just mentioned. Clearly, it is not of great use in measuring collisional relaxation times because the decay of the FID signal is generally dominated by Doppler dephasing. If one attempts to make the dephasing rate small by lowering the laser power, the signal also gets very small, as one can see from Eq. (3-135). Another possible application is to use FID as a means of obtaining high-resolution spectra. Suppose that we excite two or more closely spaced transitions and measure the resulting FID signal. The Fourier transform of this time-domain signal will give the frequency spectrum of the excited transitions. This technique works very well as a method of obtaining high-resolution spectra in nmr and in the microwave region. One disadvantage of the method in the optical region is that the line width of each spectral component will be $\gamma_{ab} + \Gamma$ (assuming steady-state preparation) rather than just the natural line width γ_{ab}. As we just mentioned, reducing the laser power to reduce Γ also reduces the signal, since it is proportional to $1 - \gamma_{ab}/\Gamma$. However, the same problem is present for competing techniques, such as Lamb dip spectroscopy, so this application may be quite useful in the future. Some preliminary experiments along these lines have already been done (Grossman *et al.*, 1977).

3.7.2. Superradiance

In discussing optical FID, we noted that the emitted intensity was proportional to the square of the initial population difference, $(N_b - N_a)^2 \equiv N^2$. If we look back at Eqs. (3-119)–(3-127), where we initially derived this N^2 dependence, it is easy to see why the result occurs. When we added up the oscillating dipole moments to obtain the macroscopic polarization, we found that at $t - t_2 = 0$ all the dipoles added up exactly in phase, to give a polarization that is simply N times as large as the average dipole moment induced in a single molecule. Hence, we obtain an emitted *field* that is proportional to N. None of this should be at all surprising. We have simply formed a phased array of oscillating dipoles by forcing each molecule to interact identically with a coherent optical field. It is true that this behavior is unusual in the optical region, but in the rf region, for example, phased systems of nuclear spins are very common. Note also that N^2 emission from a phased array of oscillating dipoles is a completely classical phenomenon. A phased array of N radio antennas will produce a directional radio signal whose intensity goes as N^2.

If, on the other hand, we consider the situation from a purely quantum-mechanical point of view, the N^2 emission looks somewhat more mysterious. In elementary quantum mechanics, one learns that there are two types of emission, stimulated and spontaneous. However, stimulated emission cannot be involved here since at $t = t_2$, where the N^2 emission is a maximum, the population difference $w(t_2)$ is zero [see Eq. (3-120)], and hence the net stimulated emission rate is zero also. Thus, it must be spontaneous emission that is occurring, despite the fact that ordinarily the spontaneous emission rate is proportional to N rather than N^2. Now the reason the intensity of spontaneous emission normally goes as N is that it is usually an incoherent process in which each molecule radiates a field whose phase is uncorrelated with the phase of the fields emitted by other molecules. However, there is no need for this to always be so. In his celebrated 1954 paper, R. H. Dicke pointed out that if one considers the entire gas sample and its radiated field as a single quantum-mechanical system, there exist states of the system in which the spontaneous emission rate is abnormally high and in fact can go as N^2. In these states the molecular wavefunctions are correlated (phased) in such a way as to make the spontaneous emission a coherent process in which the emitted field is just N times the field that would be emitted by a single molecule. Dicke further proposed that "for want of a better term, a gas which is radiating strongly because of coherence will be called super-radiant." This, then, is the origin of the term "superradiance."

As we have discussed it so far, the phenomenon of superradiance or coherent spontaneous emission is simply the explanation in quantum-mechanical language of a familiar classical process, emission by a phased array of dipoles. If this were all there were to the subject, superradiance would be in many respects a rather trivial phenomenon, since it is then just a new name for a well-known classical effect. There are, however, other aspects of superradiance that are not trivial at all. In particular, consider a situation in which we start off an experiment with a completely inverted transition, i.e., all molecules in the upper state. From a semiclassical point of view we predict that such a system will just sit there. It cannot radiate a field because ρ_{ab} is zero. If we now consider the fully quantum-mechanical calculation as Dicke did, though, we find a completely different behavior. At first the system just radiates spontaneously, as one might expect. However, given suitable molecular densities and level decay times, one finds that the first spontaneously emitted photons induce correlations in the other molecules so that they have an enhanced probability of emitting photons in the same direction as the first ones. In classical language we would say that a phased array of dipoles is beginning to develop. As this process continues, the entire sample becomes highly correlated and the emission rate starts to rise, at first slowly and then very quickly, until the sample is radiating in one direction at a rate proportional to N^2. The system then continues to radiate until all molecules have returned to the ground state. This effect has been observed experimentally (Skribanowitz *et al.*, 1973; Gross *et al.*, 1976) and has aroused considerable interest. Clearly, the superradiance that appears in such experiments is not a trivial phenomenon. Excellent quantum-mechanical treatments of superradiance can be found in Allen and Eberly (1975) and in Sargent *et al.* (1974).

This brings us to the problem of nomenclature. The term "super-radiance" has been taken to mean different things by different authors, and as a result there exists considerable confusion in the literature. One use that should be avoided is to describe high-gain lasers such as the nitrogen laser, which can be made to lase without mirrors, as superradiant. Although it is true that one can initially have a complete population inversion and that spontaneous emission provides the initial field to start the device operating, one also has continuous incoherent pumping of molecules to the upper state during the emission process. This prevents the system from developing the correlations that occur in superradiance, and one simply has spontaneous emission amplified by stimulated emission. For this case the output intensity is proportional to N rather than N^2, and the system stops lasing when only only half the molecules are in the ground state; i.e., $N_b - N_a = 0$. Thus, a mirrorless laser should not be thought of as a superradiant device.

Unfortunately, this is not the end of the confusion. Although Dicke's original definition of superradiance was very broad and would essentially include any N^2 emission process, many authors restrict the definition much more severely. As we saw above, N^2 emission can occur in two cases: a nontrivial case, where the system starts out completely inverted and develops phase correlations on its own, and a relatively trivial classical case, where the correlations are simply built in passively by a coherent excitation pulse. Many people in quantum optics now tend to refer only to the first case as being a manifestation of "superradiance" or "Dicke superradiance." However, agreement on this point is by no means universal. From a pedagogical point of view, the restricted definition has considerable merit. If one is doing a semiclassical treatment of the N^2 emission following a pulse, it is confusing to the student to refer to this as superradiance. The reason is that in virtually every theoretical paper written on superradiance, including Dicke's, the treatment is a fully quantized one using the full apparatus of quantum electrodynamics. On looking at this literature, the student tends to get the impression that the N^2 emission following a pulse is some complicated and mysterious many-body phenomenon, when in fact it can be understood by simple classical arguments. Thus, it would seem preferable not to refer to superradiance at all when discussing an N^2 emission such as optical FID.

3.8. Photon Echo

3.8.1. Two-Pulse Echoes

We now come to a most unusual phenomenon, the photon echo. To observe the echo, one applies to a sample two optical pulses, separated by a time interval Δt. Under the proper conditions, one then finds that after another time interval Δt beyond the second pulse, the sample suddenly emits a pulse of radiation, the photon echo. It is a dramatic effect, appearing as it does during a time interval when nothing is being done to the sample.

As with previous effects, the photon echo is an optical analog of an nmr transient effect, in this case of the spin echo discovered by Hahn (1950b). The reason for calling both phenomena echoes is obvious. We send two pulses into a sample, and a little while later it returns a pulse back to us. The photon echo has the distinction of being the first optical coherent transient effect to be observed. The experiment was done in 1964 by Kurnit *et al.*, who observed an echo in ruby and also named the effect "photon echo." We note in passing that although it is catchy and concise, the name "photon echo" is in one sense misleading: The concept of a photon does not enter into the experiment at all. Everything about the photon echo can be understood using semiclassical arguments only.

At first glance, the echo phenomenon looks quite mysterious, since it involves a burst of emission that suddenly seems to appear out of nothing. Actually, however, the explanation is fairly simple. The photon echo is nothing more than a reappearance of the optical free induction decay (FID) signal that follows the first pulse. We recall from the discussion of Section 3.7 that the decay of the FID signal is usually dominated by Doppler dephasing. This means that even after the FID signal has disappeared, each velocity group of excited molecules still has a large oscillating dipole moment. We see no radiation only because the dipoles of the different velocity groups are out of phase with each other, giving a net polarization of zero. If a way can be found to cause a rephasing of the dipoles, the FID emission will suddenly appear again. This is exactly what the second pulse in a photon echo experiment does.

Clearly the key to understanding in detail how a photon echo is produced is to understand how an optical pulse can induce a rephasing of the oscillating dipoles. This can be done very easily using the vector model. We consider the pulse sequence shown in Figure 3-17 and assume as we did in the previous section that $\kappa E_0 \gg k\bar{u}$. Since Ω is assumed to lie within the Doppler line width, this implies that $\kappa E_0 \gg \delta$ for all molecules and they will

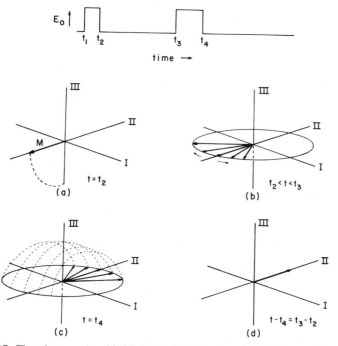

Figure 3-17. The photon echo. (a)–(d) depict the behavior of the **M** vectors for several molecular velocities at various times in the pulse sequence.

all respond resonantly to the pulses. Picking the first pulse area to be $\pi/2$, we find the **M** vectors for all molecules lying along the $-$II axis at $t = t_2$ as shown in Figure 3-17a. This initially produces a strong FID signal during the interval $t_2 \le t \le t_3$, but the signal dies away within a time given approximately by $1/k\bar{u}$ due to Doppler dephasing. The dephasing process is shown in Figure 3-17b. As we did previously, we have superimposed the **M** vectors for several different velocity groups on a common frame of reference so that the relative phases of the oscillating dipoles can readily be seen (see the footnote on p. 256). For simplicity, we also assume that $\Omega = \omega_0$ here, so that in the interval following the first pulse, each **M** vector dephases by an amount $\delta(t_3 - t_2) = kv_z(t_3 - t_2)$. Now at time t_3, an intense pulse is turned on again, but this time we choose the pulse area to be π. Again $\kappa E_0 \gg k\bar{u}$, so all molecules interact resonantly, and thus each vector **M** precesses by 180° in a cone about the I axis. This is illustrated in Figure 3-17c. The net result is that at $t = t_4$ we still have all the **M** vectors in the I–II plane, but their phases have been rearranged. Looking at the drawing we see that the vectors that dephased the farthest from the $-$II axis are moved ahead only a little, while those that dephased only a little are moved ahead a lot. Hence, the order of the vectors has been inverted, and they will all converge together along the $+$II axis when $t - t_4 = t_3 - t_2$, as shown in Figure 3-17d. Since the vectors are then all in phase, they will momentarily produce a strong emission signal before dephasing again.

We can be a little more precise by noticing that if we measure the relative phase of each **M** vector by an angle α between the vector and the $-$II axis, then the 180° pulse has the effect of changing α to $\pi - \alpha$. Since the vectors dephase in the I–II plane at a rate kv_z between pulses, we have $\alpha(t_3) = kv_z(t_3 - t_2)$ before the 180° pulse and $\alpha(t_4) = \pi - kv_z(t_3 - t_2)$ just afterward. Following the pulse, the vectors continue their dephasing motion so $\alpha(t) = \pi - kv_z(t_3 - t_2) + kv_z(t - t_4)$. Clearly, at $t - t_4 = t_3 - t_2$, we have $\alpha(t) = \pi$ for all molecules regardless of velocity, so all the dipoles are in phase with each other.

In an earlier section we noted that for a time after the original photon echo experiment in ruby, many people believed that echoes could not occur in gases. The reasoning was that echo formation requires a return of the oscillating dipoles to the phased array present at the end of the first pulse. However, since the dephasing is due to the translational motion of the molecules and this motion cannot be reversed, there is no way the molecules can get back to the proper spatial positions required for this phased array. Hence, there will be no echoes. Obviously, this conclusion is incorrect, since echoes are in fact observed in gases. The flaw in the argument is the false assumption that if molecules move from their original positions, they cannot be gotten back into a phased array. Consider molecules that move toward the front of the sample by a distance Δz between the first and second pulses.

These molecules will see the second pulse begin at a time that is $\Delta z/c$ earlier than a molecule that did not move, and hence following the pulse their **M** vectors will reach the +II axis $2\Delta z/c$ earlier than they would have if the molecules had simply dephased by some other mechanism without moving. But the moving molecules have also changed position by an amount $2\Delta z$ at this point, and therefore it takes the field emitted by the moving molecules an extra time $2\Delta z/c$ to reach the position of a nonmoving molecule. As a result, their field and the field emitted by a nonmoving molecule have exactly the same phase at the time of the echo. Hence we do have a phased array, despite the molecular motion. A very good discussion of this point can be found in Allen and Eberly (1975). Since we have properly taken the molecular motion into account in our formulation of the population matrix, we need not worry about this motion explicitly in any quantitative treatment of echo problems. The theory automatically takes care of it.

Although the vector model shows clearly how the second pulse causes rephasing of the dipoles, it does not give any information about the dependence of the echo on collisional relaxation times. This dependence is of great interest because it turns out that the photon echo is one of the most powerful tools we have to study relaxation processes. To examine this point, we consider a quantitative calculation. For simplicity we again assume that $\kappa E_0 \gg k\bar{u}$ during the pulses. Since the two pulses are very short, we can also ignore relaxation during the pulses. Furthermore, we let the first pulse be a $\pi/2$ pulse and the second pulse be a π pulse. The first half of the calculation is then identical to that done in the previous section on optical FID. $M(t_2)$ is given by Eq. (3-120), and u and v at time t_3 are given by Eq. (3-121) evaluated at $t = t_3$,

$$
\begin{aligned}
u(t_3) &= e^{-\gamma_{ab} t_{32}}(n_a - n_b) \sin \delta t_{32} \\
v(t_3) &= e^{-\gamma_{ab} t_{32}}(n_a - n_b) \cos \delta t_{32}
\end{aligned}
\tag{3-142}
$$

where $t_{32} = t_3 - t_2$. In general, we would also need to know $w(t_3)$ in order to calculate the effect of the second pulse, but for a π pulse, this is not necessary. The effect of the π pulse is calculated from

$$
M(t_4) = e^{-\beta t_{43}} M(t_3),
\tag{3-143}
$$

since we may ignore relaxation during the pulse. The exponential is given by Eq. (3-93), and since $\kappa E_0 \gg \delta = \Omega' - \omega_0$, we may set $g \approx \kappa E_0$ in that expression and neglect terms of order $\delta/\kappa E_0$ compared to unity. Thus,

$$
M(t_4) = \begin{pmatrix} u(t_4) \\ v(t_4) \\ w(t_4) \end{pmatrix} = \begin{pmatrix} 1 & 0 & 0 \\ 0 & \cos \kappa E_0 t_{43} & \sin \kappa E_0 t_{43} \\ 0 & -\sin \kappa E_0 t_{43} & \cos \kappa E_0 t_{43} \end{pmatrix} \begin{pmatrix} u(t_3) \\ v(t_3) \\ w(t_3) \end{pmatrix}
\tag{3-144}
$$

For the case at hand, $\kappa E_0 t_{43} = \pi$, so Eq. (3-144) becomes very simple. The pulse just changes the sign of v and w. After the pulse we again use the solutions to the full set of population matrix equations (3-72), to obtain

$$u(t) = e^{-\gamma_{ab}(t-t_4)} e^{-\gamma_{ab}t_{32}} (n_a - n_b)[\sin \delta t_{32} \cos \delta(t-t_4) - \cos \delta t_{32} \sin \delta(t-t_4)]$$

$$v(t) = e^{-\gamma_{ab}(t-t_4)} e^{-\gamma_{ab}t_{32}} (n_a - n_b)[-\sin \delta t_{32} \sin \delta(t-t_4) - \cos \delta t_{32} \cos \delta(t-t_4)]$$

$$(3\text{-}145)$$

These equations give u and v for times $t \geq t_4$. They may be rewritten in the simpler form

$$u(t) = -e^{-\gamma_{ab}(t-t_4+t_{32})} (n_a - n_b) \sin \delta[(t-t_4) - t_{32}]$$

$$v(t) = -e^{-\gamma_{ab}(t-t_4+t_{32})} (n_a - n_b) \cos \delta[(t-t_4) - t_{32}]$$

$$(3\text{-}146)$$

A glance at these equations shows that an echo is inevitable. To find the polarization, u and v must be integrated over v_z, i.e., over all values of δ, and the only way an integral over δ of something like $\cos \delta t$ can have a large value is for t to be zero or near zero. In Eq. (3-146) we see that the time argument multiplying δ is $(t-t_4) - (t_3-t_2)$, so the polarization will be large when $t-t_4 = t_3 - t_2$. The characteristic mark of an echo, then, is to have u and v components that oscillate at δ and have time arguments that go to zero at some time after the end of the second pulse. This is a very useful fact to know, since in more complicated situations we may have several terms in the expressions for u and v. The discussion above shows that we need only consider those terms with the proper time argument to find the echo signal.

If we look back to see how we obtained the time argument $t - t_4 - t_{32}$ in Eqs. (3-146), we see that it was the change in the sign of v by the second pulse that did it. Had we not applied the second pulse and changed the sign of v, the second terms on the right-hand side in Eqs. (3-145) would be positive, giving a time argument in Eqs. (3-146) of $(t-t_4) + t_{32}$ instead of $(t-t_4) - t_{32}$. This expression is always large for $t > t_4$ and gives no echo. It represents only the very small tail of the FID following the first pulse.

Having found u and v, we next calculate $\bar{\mu}_{ab}(z, v_z, t)$ as we did for the optical FID [see Eq. (3-122)]. We find for the oscillating dipole moment of any group of molecules,

$$\bar{\mu}_{ab}(z, v_z, t) = \mu_{ab} e^{-\gamma_{ab}(t-t_4+t_{32})} (n_a - n_b) \sin(\omega_0 t - kz' + \phi') \quad (3\text{-}147)$$

where $\phi' = \delta(t_4 + t_{32})$. If we compare this to Eq. (3-122), we see that our expression is identical to the expression obtained for FID except for a change in overall sign and the replacement of t_2 by $t_4 + t_{32}$ in the expression for ϕ'. Hence, we may just repeat the derivation given in Eqs. (3-122)–(3-127) and obtain immediately

$$I(t) = \frac{\mu_{ab}^2 \omega_0^2 L^2}{8\varepsilon_0 c}(N_b - N_a)^2 e^{-2\gamma_{ab}(t-\tau_4+\tau_{32})} e^{-(1/2)k^2\bar{u}^2[(t-\tau_4)-\tau_{32}]^2}$$

$$(3\text{-}148)$$

On comparing this with Eq. (3-127), we see that, ignoring the exponential decay due to collisions, the shape of the echo in time is simply that of two optical FIDs placed back to back, and its peak amplitude is identical to the peak amplitude of the FID signal. When collisional decay is included, the echo shape in a low-pressure gas is not significantly affected because $\gamma_{ab} \ll k\bar{u}$, and hence the echo width is very narrow compared to the collisional decay time.

The maximum echo amplitude occurs at $t - \tau_4 = \tau_{32}$ and is easily measured in an experiment because no pulses are present then. From Eq. (3-148) we find that

$$I_{\max} = \frac{\mu_{ab}^2 \omega_0^2 L^2}{8\varepsilon_0 c}(N_b - N_a)^2 e^{-4\gamma_{ab}\tau_{32}} \qquad (3\text{-}149)$$

This equation suggests a very nice way of measuring collisional relaxation times. One simply does a series of echo experiments in which only the pulse separation τ_{32} is varied, and measures $I_{\max}$ each time. If we then make a plot of ln $(I_{\max})$ versus τ_{32}, we will obtain a straight line whose slope is just $4\gamma_{ab}$. Note that we do not need to know the absolute of $I_{\max}$ to measure γ_{ab}. Just knowing the relative values of $I_{\max}$ for the different pulse separations τ_{32} is sufficient, because we only need the slope of the line.

In Eq. (3-149) we see that the decay of the echo peak depends only on γ_{ab}, regardless of the values of the level decay times γ_a and γ_b. However, one might legitimately wonder whether or not this result is only due to the fact that we used a second pulse whose area was exactly π. Looking back to Eq. (3-144), we see that for this case, and this case only, the population difference $w(t)$ does not enter the calculation. Thus, we should check to see whether the echo still decays as γ_{ab} when the second pulse is not exactly a π pulse. The calculation proceeds exactly as before, except that we will also need to know $w(t)$ in the interval between pulses. Since we want to be careful about relaxation times, we use the $w(t)$ obtained from the full set of population matrix equations given in Eqs. (3-74). We find it convenient to rewrite these equations as

$$w(t) = \tfrac{1}{2}(e^{-\gamma_a t'} + e^{-\gamma_b t'})w(t_2) + \tfrac{1}{2}(e^{-\gamma_a t'} - e^{-\gamma_b t'})s(t_2)$$
$$+ [(n_a - n_b) - (e^{-\gamma_a t'}n_a - e^{-\gamma_b t'}n_b)]$$
$$s(t) = \tfrac{1}{2}(e^{-\gamma_a t'} + e^{-\gamma_b t'})s(t_2) + \tfrac{1}{2}(e^{-\gamma_a t'} - e^{-\gamma_b t'})w(t_2)$$
$$+ [(n_a + n_b) - (e^{-\gamma_a t'}n_a + e^{-\gamma_b t'}n_b)] \qquad (3\text{-}150)$$

where $t' = t - t_2$ and s is the population sum, $\rho_{aa} + \rho_{bb}$. On examining these equations, we notice that the terms in brackets in both equations give a contribution to $w(t)$ and $s(t)$ that does not depend on $w(t_2)$ or $s(t_2)$. Hence, these terms are steady-state terms that give the contribution from molecules that enter state a or b *after* $t = t_2$. These molecules will only give an optical FID signal immediately following the second pulse and cannot contribute to the photon echo. Since we are interested only in the echo signal here, we ignore them and further rewrite Eqs. (3-150) as

$$w(t) = e^{-\gamma_{av}t'}[w(t_k)\cosh\gamma_d t' - s(t_k)\sinh\gamma_d t']$$
$$s(t) = e^{-\gamma_{av}t'}[s(t_k)\cosh\gamma_d t' - w(t_k)\sinh\gamma_d t']$$
$$(3\text{-}151)$$

where $\gamma_{av} = \frac{1}{2}(\gamma_a + \gamma_b)$, $\gamma_d = \frac{1}{2}(\gamma_a - \gamma_b)$, and $t_k = t_2$ for the case at hand. We now proceed to calculate the echo signal. At the end of the initial $\pi/2$ pulse, we have $w(t_2) = 0$ [see Eq. (3-120)]. $s(t_2)$ remains at its thermal equilibrium value $s = n_a + n_b$ during the short pulse because, according to Eqs. (3-67), the differential equation for s is not directly coupled to the optical field. At $t = t_3$, then,

$$w(t_3) = -e^{-\gamma_{av}t_{32}}(n_a + n_b)\sinh\gamma_d t_{32} \qquad (3\text{-}152)$$

while $u(t_3)$ and $v(t_3)$ are given by Eqs. (3-142). The effect of the second pulse can now be calculated using Eq. (3-144), but instead of letting $\kappa E_0 t_{43} = \pi$, we use $\kappa E_0 t_{43} = \theta_2$, where θ_2 is an arbitrary pulse area. This gives u and v at $t = t_4$, and these values can be put back into Eqs. (3-72) to find u and v for times $t \geq t_4$. The result, after a little trigonometry, is

$$u(t) = e^{-\gamma_{ab}(t - t_4 + t_{32})}(n_a - n_b)\Big[\tfrac{1}{2}(1 + \cos\theta_2)\sin\delta(t - t_4 + t_{32})$$

$$-\tfrac{1}{2}(1 - \cos\theta_2)\sin\delta(t - t_4 - t_{32})$$

$$-e^{-(\gamma_{av} - \gamma_{ab})t_{32}}\frac{n_a + n_b}{n_a - n_b}\sinh\gamma_d t_{32}\sin\delta(t - t_4)\Big] \qquad (3\text{-}153)$$

$$v(t) = e^{-\gamma_{ab}(t - t_4 + t_{32})}(n_a - n_b)\Big[\tfrac{1}{2}(1 + \cos\theta_2)\cos\delta(t - t_4 + t_{32})$$

$$-\tfrac{1}{2}(1 - \cos\theta_2)\cos\delta(t - t_4 - t_{32})$$

$$-e^{-(\gamma_{av} - \gamma_{ab})t_{32}}\frac{n_a + n_b}{n_a - n_b}\sinh\gamma_d t_{32}\cos\delta(t - t_4)\Big]$$

We can stop the calculation at this point because the relaxation-time behavior of the echo is already apparent. Although both $u(t)$ and $v(t)$ consist of three terms, we recall the argument following Eqs. (3-146), which tells us that terms oscillating at $\delta t'$ give a significant contribution only at $t' = 0$. This allows us to identify the various terms in $u(t)$ and $v(t)$ as follows: The first term has its maximum at $t \approx t_2 (t_4 \approx t_3$ for short pulses) and thus represents the tail of the FID signal following the first pulse. The third term has a maximum at $t = t_4$ and is therefore the FID signal that follows the second pulse. The second term is the echo because it has a maximum at $t - t_4 = t_3 - t_2$, and we notice that its decay depends on γ_{ab} only.

Furthermore, a little thought should convince you that our result is perfectly general. Even if we do not assume that $\kappa E_0 \gg k\bar{u}$ and let both pulse areas be arbitrary, we will still find that the echo decay rate is γ_{ab}. The reason is that an echo can arise only from terms that contain $\sin \delta t_{32}$ or $\cos \delta t_{32}$, whereas the only terms in $u(t)$ or $v(t)$ that decay at anything other than γ_{ab} arise from $w(t_3)$, and $w(t_3)$ never contains terms oscillating at $\sin \delta t_{32}$ or $\cos \delta t_{32}$. Thus, we conclude that the photon echo can be used to measure γ_{ab} regardless of what the level lifetimes γ_a and γ_b are. We need only ensure that the pulses are short compared to the interval between them, so that relaxation during the pulses can be neglected.

Thus far we have always assumed that κE_0 was so much greater than $\delta = \Omega' - \omega_0$ that we could simply set $\delta / \kappa E_0 = 0$ for all molecules. While this gives simple results, it is a rather gross approximation. A more realistic approach is to expand each element of $\exp(-\mathbb{B}t')$ in a power series in $\delta / \kappa E_0$ and keep only the leading terms. In particular, let us keep terms of order $\delta / \kappa E_0$ but ignore terms of order $\delta^2 / \kappa^2 E_0^2$ and higher. We then find that Eq. (3-93), the general expression for $\exp(-\mathbb{B}t')$, becomes

$$e^{-\mathbb{B}t'} = \begin{pmatrix} 1 & \dfrac{\delta}{\kappa E_0}\sin \kappa E_0 t' & \dfrac{\delta}{\kappa E_0}(1 - \cos \kappa E_0 t') \\[2ex] -\dfrac{\delta}{\kappa E_0}\sin \kappa E_0 t' & \cos \kappa E_0 t' & \sin \kappa E_0 t' \\[2ex] \dfrac{\delta}{\kappa E_0}(1 - \cos \kappa E_0 t') & -\sin \kappa E_0 t' & \cos \kappa E_0 t' \end{pmatrix} \qquad (3\text{-}154)$$

This result follows immediately from the fact that $g = \kappa E_0(1 + \frac{1}{2}\delta^2 / \kappa^2 E_0^2 + \cdots)$. Equation (3-154) should describe the response to an intense pulse having $\kappa E_0 \gg k\bar{u}$ with considerably more accuracy than an expression such as (3-144), which we used previously. However, Eq. (3-154) is also much more complicated and thus makes an efficient method for calculating echoes imperative. Since we know the echo will decay as γ_{ab}, we may set all relaxation times equal and write the echo calculation in the form

$$\mathbb{M}(t) = e^{-\gamma_{ab}(t - t_1)} e^{-\mathbb{B}'(t - t_4)} e^{-\mathbb{B}t_{43}} e^{-\mathbb{B}'t_{32}} e^{-\mathbb{B}t_{21}} \mathbb{M}(t_1)$$

$$= e^{-\gamma_{ab}(t - t_1)} \mathbb{R}\mathbb{M}(t_1) \qquad (3\text{-}155)$$

where $\exp(-\mathbb{B}t_{43})$ and $\exp(-\mathbb{B}t_{21})$ are given by Eq. (3-154) and $\exp[-\mathbb{B}'(t - t_4)]$ and $\exp(-\mathbb{B}'t_{32})$ are given by Eq. (3-93) with $\kappa E_0 = 0$. $\mathbb{R}$ is the product of these four matrices and represents the system response to the entire pulse sequence. A straightforward calculation of $\mathbb{R}$ would be a major task. However, we have a far simpler method at our disposal: the 2×2 matrix representations introduced in Eqs. (3-100)–(3-106). To use this calculational technique we need only find the Cayley–Klein parameters α and β which are appropriate for each time interval, form 2×2 matrices from them, and multiply the matrices together to find the α and β which describe the entire pulse sequence. It is then trivial to calculate $\mathbb{R}$ using Eq. (3-104). Note that because $\mathbb{M}(t_1) = (0, 0, n_a - n_b)$ and we only need to know $u(t)$ and $v(t)$, just two elements of $\mathbb{R}$ need to be calculated, $\operatorname{Re}(-2\alpha\beta)$ and $\operatorname{Im}(-2\alpha\beta)$.

Using the general expression for α and β given in Eq. (3-105), we see that to first order in $\delta/\kappa E_0$, we have

$$\alpha = \cos \kappa E_0 t'/2 - i\frac{\delta}{\kappa E_0} \sin \kappa E_0 t'/2$$

$$\beta = -i \sin \kappa E_0 t'/2 \qquad (3\text{-}156)$$

during the pulses. We take $t' = t_2 - t_1$ for the first pulse and $t' = t_4 - t_3$ for the second pulse. Because $\delta/\kappa E_0$ is small, we can rewrite the expression for α more simply as follows:

$$\alpha = \cos \kappa E_0 t'/2\left(1 - i\frac{\delta}{\kappa E_0} \tan \kappa E_0 t'/2\right)$$

$$\simeq e^{-i\delta q'} \cos \kappa E_0 t'/2 \qquad (3\text{-}157)$$

where $q' = (1/\kappa E_0) \tan \kappa E_0 t'/2$. The last step follows from the fact that $\delta q'$ is small, giving $\cos \delta q' \simeq 1$ and $\sin \delta q' \simeq \delta q'$. Between pulses and after the second pulse, $\kappa E_0 = 0$ and we have simply

$$\alpha = \cos \delta t'/2 - i \sin \delta t'/2 = e^{-i\delta t'/2}$$

$$\beta = 0 \qquad (3\text{-}158)$$

with $t' = t_3 - t_2$ and $t - t_4$, respectively. We can now calculate the 2×2 matrix $\mathbb{r}_T$ which describes the effect of the entire pulse sequence by multiplying together the 2×2 matrices for each time interval:

$$\mathbb{r}_T = \begin{pmatrix} \alpha & \beta \\ -\beta^* & \alpha^* \end{pmatrix} = \begin{pmatrix} e^{-i(1/2)\delta(t-t_4)} & 0 \\ 0 & e^{i(1/2)\delta(t-t_4)} \end{pmatrix}$$

$$\times \begin{pmatrix} e^{-i\delta q_2} \cos \frac{1}{2}\theta_2 & -i \sin \frac{1}{2}\theta_2 \\ -i \sin \frac{1}{2}\theta_2 & e^{i\delta q_2} \cos \frac{1}{2}\theta_2 \end{pmatrix}$$

$$\times \begin{pmatrix} e^{-i(1/2)\delta t_{32}} & 0 \\ 0 & e^{i(1/2)\delta t_{32}} \end{pmatrix} \begin{pmatrix} e^{-i\delta q_1} \cos \frac{1}{2}\theta_1 & -i \sin \frac{1}{2}\theta_1 \\ -i \sin \frac{1}{2}\theta_1 & e^{i\delta q_1} \cos \frac{1}{2}\theta_1 \end{pmatrix} \qquad (3\text{-}159)$$

This gives

$$\alpha = e^{-i(1/2)\delta(t-t_4)}\{e^{-i\delta[(1/2)t_{32}+q_1+q_2]}\cos\tfrac{1}{2}\theta_1\cos\tfrac{1}{2}\theta_2$$
$$-e^{i(1/2)\delta t_{32}}\sin\tfrac{1}{2}\theta_1\sin\tfrac{1}{2}\theta_2\} \qquad (3\text{-}160)$$

$$\beta = -ie^{-i(1/2)\delta(t-t_4)}\{e^{-i\delta[(1/2)t_{32}+q_2]}\sin\tfrac{1}{2}\theta_1\cos\tfrac{1}{2}\theta_2$$
$$+e^{i\delta[(1/2)t_{32}+q_1]}\cos\tfrac{1}{2}\theta_1\sin\tfrac{1}{2}\theta_2\}$$

as the elements of the matrix r_T. In the preceding equations we have denoted the pulse areas $\kappa E_0 t_{21}$ and $\kappa E_0 t_{43}$ by θ_1 and θ_2, respectively. We also have $q_1 = (1/\kappa E_0)\tan\tfrac{1}{2}\theta_1$ and $q_2 = (1/\kappa E_0)\tan\tfrac{1}{2}\theta_2$. Multiplying α and β together, we obtain three terms:

$$-2\alpha\beta = \tfrac{1}{2}ie^{-i\delta(t-t_4)}[e^{-i\delta(t_{32}+q_1+2q_2)}\sin\theta_1(1+\cos\theta_2)$$
$$+2e^{-i\delta q_2}\cos\theta_1\sin\theta_2 - e^{i\delta(t_{32}+q_1)}\sin\theta_1(1-\cos\theta_2)]$$
$$(3\text{-}161)$$

The real and imaginary parts of this expression give the necessary elements of $\mathbb{R}$, and on using (3-155), we find that

$$u(t) = e^{-\gamma_{ab}(t-t_1)}(n_a-n_b)\,\text{Re}\,(-2\alpha\beta)$$
$$v(t) = e^{-\gamma_{ab}(t-t_1)}(n_a-n_b)\,\text{Im}\,(-2\alpha\beta) \qquad (3\text{-}162)$$

Recalling the discussion following Eq. (3-153), we note that only terms in $2\alpha\beta$ which oscillate at $\delta(t-t_4-t_{32})$ will contribute to the echo. Thus, we only need consider the last term in Eq. (3-161). The first two terms represent FID signals following the first and second pulses, respectively. Using only the echo term in (3-162), we now form the oscillating dipole moment as we did in Eq. (3–122). Here we find that

$$\bar{\mu}_{ab}(z, v_z, t) = \mu_{ab}e^{-\gamma_{ab}(t-t_1)}(n_a-n_b)\tfrac{1}{2}\sin\theta_1(1-\cos\theta_2)\sin(\omega_0 t - kz' + \phi')$$
$$(3\text{-}163)$$

where $\phi' = \delta(t_4 + t_{32} + q_1)$. This has the same form as our earlier echo result (3-147), except for some constants and a change in the phase factor ϕ'. Hence one can immediately write down the answer for the echo intensity by looking at Eq. (3-148). We find that

$$I(t) = \frac{\mu_{ab}^2\omega_0^2 L^2}{8\varepsilon_0 c}(N_b-N_a)^2\tfrac{1}{4}\sin^2\theta_1(1-\cos\theta_2)^2 e^{-2\gamma_{ab}(t-\tau_1)}$$
$$\times e^{-(1/2)k^2\bar{u}^2[(t-t_4)-(\tau_{32}+q_1)]^2} \qquad (3\text{-}164)$$

for the photon echo produced by two intense pulses of area θ_1 and θ_2. We see that two pulses of any area will give an echo except for special cases such as $\theta_1 = \pi$ or $\theta_2 = 2\pi$. As expected, the maximum echo intensity comes from a $\pi/2$, π pulse sequence, but other sequences also give surprisingly large echoes. A $\pi/2$, $\pi/2$ pulse sequence, for example, produces a polarization at the echo peak which is half as large as that obtained for the $\pi/2$, π sequence. One can understand this result by a careful consideration of the vector model picture, where one finds that the **M** vectors arrange themselves into a figure 8 centered on the $+II$ axis at the time of the echo (Hahn, 1950b). A second interesting feature of the echo in Eq. (3-163) is the timing of the echo peak which occurs at $t - t_4 = (t_3 - t_2) + q_1$. Thus, there is an extra delay by an amount q_1 over the result obtained in Eq. (3-148). Looking back at the derivation, we see that this delay was a result of our taking into account off-resonant molecules to first order in $\delta/\kappa E_0$. One can understand this extra delay by noting that the FID following the first pulse behaves as if its maximum value had occurred at time $t = t_2 - q_1$, which is *before* the end of the pulse. This fact is readily verified by direct calculation and is also discussed in detail by Allen and Eberly (1975). Since the echo is just a reconstruction of the FID, q_1 reappears as an extra delay in the echo maximum. The size of the delay q_1 can be found from its definition, $q_1 = (1/\kappa E_0) \tan \frac{1}{2}\theta_1$. For small first pulse areas $\theta_1 < \pi/2$, $\frac{1}{2}\theta_1$ is sufficiently small that we may take $\tan \frac{1}{2}\theta_1 \simeq \frac{1}{2}\theta_1$. Then, since by definition $\theta_1 = \kappa E_0 t_{21}$, we have $q_1 = \frac{1}{2}t_{21}$. This gives an often-quoted result for the timing of echo maximum; that it occurs at $t - t_4 = t_3 - t_2 + \frac{1}{2}t_{21}$. This expression starts to break down for $\theta_1 \gtrsim \pi/2$. With $\theta_1 = \pi/2$, for example, the exact result is $t - t_4 = t_3 - t_2 + 2t_{21}/\pi$.

So far we have only considered the case in which the pulses were so intense that $\kappa E_0 \gg k\bar{u}$. While this situation can be obtained in some pulsed laser experiments, one more often has the reverse case of $\kappa E_0 \ll k\bar{u}$, especially if Stark or frequency switching is done. In this situation the effects of molecules excited off resonance are large and must be taken into account exactly. One feature that simplifies this problem is the fact that even here the collisional decay of the echo depends only on γ_{ab}. One can see this by noting that the relaxation-rate arguments given earlier depend only on the general form of the equations and not on any assumptions about resonant excitation. Thus, we need not worry about collisional relaxation and can use the same method of calculation that was used in the previous example. As before, we start with Eq. (3-155) and calculate the necessary elements of $\mathbb{R}$ from the Cayley–Klein parameters. In fact, the calculation is identical except that we use the general expression (3-105) for α and β during the pulses instead of Eq. (3-156). Since the calculation is simply a straightforward multiplication of 2×2 matrices, we just give the result. One finds that the part of $2\alpha\beta$ having the proper $t - t_4 - t_{32}$ time dependence is

$$-2\alpha\beta = \frac{\kappa^3 E_0^{\,3}}{2g^3} e^{-i\delta(t - t_4 - t_{32})} \left[\frac{\delta}{g}(1 - \cos gt_{21})(1 - \cos gt_{43}) \right.$$
$$\left. -i \sin gt_{21}(1 - \cos gt_{43}) \right] \qquad (3\text{-}165)$$

As before, $u(t)$ and $v(t)$ are given by Eq. (3-162). Since we are assuming $\kappa E_0 \ll k\bar{u}$, we may pull the Gaussian factors out of the integrals over v_z when the polarization components P_c and P_s are calculated. We then see immediately that $u(t)$ can be ignored since it is an odd function of δ and hence will give $P_c = 0$. The result for P_s is

$$P_s = C\left[\int_{-\infty}^{\infty} \frac{1}{g^3} \sin gt_{21}(1 - \cos gt_{43}) \cos \delta(t - t_4 - t_{32})\, d\delta\right.$$

$$\left. + \int_{-\infty}^{\infty} \frac{\delta}{g^4}(1 - \cos gt_{21})(1 - \cos gt_{43}) \sin \delta(t - t_4 - t_{32})\, d\delta\right] \qquad (3\text{-}166)$$

where

$$C = -\frac{\mu_{ab}\kappa^3 E_0^3}{2k\bar{u}\sqrt{\pi}}(N_b - N_a)e^{-\gamma_{ab}(t-t_1)}e^{-(\Omega-\omega_0)^2/k^2\bar{u}^2}$$

Unfortunately, this integral cannot be done analytically, so numerical integration is required to carry the problem any further.

When the integration is done numerically, one finds that the echo behaves in many respects as it did in the previous $\kappa E_0 \gg k\bar{u}$ case. As an example, Figure 3-18 shows a calculation of the echo amplitude for a $\pi/2$, π pulse sequence using Eq. (3-166). The echo is, of course, smaller than it would be if all molecules were excited resonantly, but it still shows the extra delay of $2t_{21}/\pi$ that we found earlier. The pulse-area dependence of the echo also shows the expected behavior. When the second pulse area is varied, for example, one finds that the echo amplitude is a maximum for $\theta_2 \approx \pi$, and so forth. There is no pulse area for which the echo completely disappears, however. This point has been investigated by Schenzle *et al.* (1976), who noted that at $t - t_4 = t_{32}$, the integrals in Eq. (3-166) simplify and can be expressed in terms of a power series. One can then easily study the second-pulse-area dependence of the echo. When it comes to the detailed shape and width of the echo, however, we find a rather different behavior than we had previously. For $\kappa E_0 \gg k\bar{u}$, we found that the echo was just a reconstruction of the first pulse FID. That is clearly not the case when $\kappa E_0 \ll k\bar{u}$, as one can see from Figure 3-18. The FID following the first pulse

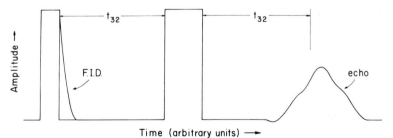

Figure 3-18. Free induction decay and photon echo amplitude for a $\pi/2$, π pulse sequence applied to a Doppler-broadened line ($\kappa E_0 \ll k\bar{u}$).

lasts for a time t_{21} as discussed in Section 3.7.1, whereas the echo is considerably wider than $2t_{21}$ and has a different shape as well. The reason for this behavior is that the echo shape and width depend on the bandwidth of molecules which have the proper excitation, and this bandwidth depends on the length of *both* the first and second pulses. In the simpler $\kappa E_0 \gg k\bar{u}$ case, both pulses excited the same bandwidth, i.e., the entire Doppler line. Note also that the echo is about a factor of 2 smaller than the FID signal.

Since one nearly always has $\kappa E_0 \ll k\bar{u}$ in a Stark switching (or frequency-switching) experiment, we should also examine how our theoretical treatment needs to be modified to describe the signal observed in this type of experiment. To see an echo, one simply applies two short electric field pulses across the sample. The echo will then arise from the molecules which are shifted into resonance during the Stark pulses. There is also a second group of molecules which is shifted out of resonance during the pulses. These molecules see a resonant field which is just interrupted briefly during the two pulses. Hence, they will give small optical nutation signals immediately after the first and second pulses. For reasonable pulse separations these signals will have died away by the time the echo occurs, however, so we can ignore this second group of molecules in calculating the echo signal, and consider only molecules which are switched *into* resonance during the pulses. To do the calculation, we let the natural resonance frequency of the molecules be ω_0 during the pulses, and be $\omega_0 + \Delta\omega_0$ otherwise. On looking at the derivation leading to Eq. (3-165), we see immediately that this only changes the exponential in the expression for $2\alpha\beta$ from $\exp[-i\delta(t - t_4 - t_{32})]$ to $\exp[-i\delta_s(t - t_4 - t_{32})]$, where $\delta_s = \Omega' - (\omega_0 + \Delta\omega_0)$. This means that the expression for P_s is identical to the P_s we obtained for the pulsed optical fields, Eq. (3-166), except that the $\cos\delta$ and $\sin\delta$ terms are now $\cos\delta_s$ and $\sin\delta_s$. We write this compactly as

$$P_s = C\left[\int_{-\infty}^{\infty} \mathscr{A} \cos\delta_s(t - t_4 - t_{32})\, d\delta + \int_{-\infty}^{\infty} \mathscr{B} \sin\delta_s(t - t_4 - t_{32})\, d\delta\right]$$

$$(3\text{-}167)$$

where $\mathscr{A}$ and $\mathscr{B}$ can be determined from (3-166). Noting that $\delta_s = \delta - \Delta\omega_0$, we can rewrite P_s as

$$P_s = C\left[\int_{-\infty}^{\infty} \mathscr{A} \cos\delta(t - t_4 - t_{32})\, d\delta + \int_{-\infty}^{\infty} \mathscr{B} \sin\delta(t - t_4 - t_{32})\, d\delta\right]$$

$$\times \cos\Delta\omega_0(t - t_4 - t_{32}) + C\left[\int_{-\infty}^{\infty} \mathscr{A} \sin\delta(t - t_4 - t_{32})\, d\delta\right.$$

$$\left. - \int_{-\infty}^{\infty} \mathscr{B} \cos\delta(t - t_4 - t_{32})\, d\delta\right] \sin\Delta\omega_0(t - t_4 - t_{32}) \qquad (3\text{-}168)$$

Now $\mathscr{A}$ is an even function of δ and $\mathscr{B}$ is an odd function of δ. Hence, the last two integrals in Eq. (3-168) vanish and we see that the polarization for a Stark switching experiment is just $P_s \cos \Delta\omega_0(t - t_4 - t_{32})$, where P_s is the polarization given in Eq. (3-166) for the pulsed-optical-field case. This is the same type of result that we found for optical FID.

The polarization gives rise to a field $\mathscr{E}_c$ which beats with the incident optical field E_c to give an echo signal

$$I(t) = I_0 - c\varepsilon_0 \mathscr{E}_c E_0$$

$$= I_0 - \frac{\Omega L}{2} E_0 P_s \cos \Delta\omega_0(t - \tau_4 - \tau_{32}) \tag{3-169}$$

where P_s is given by (3-166). Note that although $\mathscr{E}_s$ is nonzero when Stark switching is done, we need not consider it because $\mathscr{E}_s$ is out of phase with E_0 and thus the heterodyne beat $\mathscr{E}_s E_0$ vanishes.

Equation (3-169) predicts that the observed echo signal in a Stark switching experiment will be a heterodyne beat of frequency $\Delta\omega_0$ whose amplitude is proportional to the electric field $\mathscr{E}_c$ emitted by the molecules. Figure 3-19 shows the observed signal for a photon echo experiment in

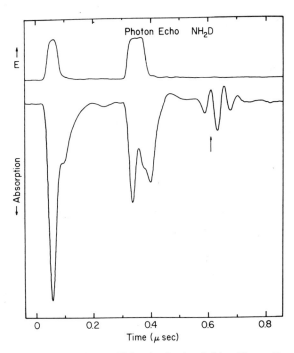

Figure 3-19. Observed photon echo in NH_2D using Stark switching. The small arrow below the echo indicates the time $t - t_4 = t_{32}$.

NH_2D, and we see that the echo does indeed behave as one expects. The small arrow below the echo indicates the time $t - \tau_4 = \tau_3 - \tau_2$. One can clearly see both the extra delay of the echo peak which is predicted by theory and the fact that the beat goes as $\cos \Delta\omega_0(t - \tau_4 - \tau_{32})$.

We now turn to a discussion of the photon echo experiments in gases that have been done to date. The first successful observations were made by Patel and Slusher (1968) some four years after the initial photon echo experiments in ruby (Kurnit *et al.*, 1964). They used two synchronized, Q-switched CO_2 lasers to generate the pulses and passed them through a 4.7-m SF_6 absorption cell to observe the echo. By measuring the echo decay as a function of pulse separation they were also able to measure γ_{ab} for SF_6. We note, however, that a later echo measurement by Meckley and Heer (1973) gives drastically different results for γ_{ab}. In both experiments the samples were optically thick ($\alpha L \simeq 2$), which complicates the interpretation. Shortly after the Patel–Slusher experiment, Bölger and Diels (1968) reported the observation of photon echoes in Cs vapor using a pulsed GaAs laser, and echoes have also been seen in SiF_4 (Nordstrom *et al.*, 1974).

In 1971, Brewer and Shoemaker reported the observation of photon echoes in $^{13}CH_3F$ and NH_2D using Stark switching. Subsequently, careful studies of the echo decay have been made in both molecules, but we defer a discussion of these experiments to Section 3.10, where velocity-changing collisions are discussed. These collisions turn out to be very important in the echo decay of $^{13}CH_3F$ and NH_2D. Because the echo is observed as a heterodyne beat in Stark switching experiments (cf. Figure 3-19), great sensitivity is available. This allows one to use optically thin samples and make careful, quantitative comparisons with theory. Recently, Brewer and Genack (1976) have also observed echoes for an electronic transition of I_2 using a frequency-switched dye laser. This technique has the same advantages as Stark switching and opens up the possibility of studying collisional effects in a wide variety of atomic, molecular, and even solid-state systems.

In pulsed laser experiments, one has an additional degree of freedom that has not been considered so far, and that is to vary the polarization of the second excitation pulse with respect to the first. To treat this case theoretically, one must relax the assumption of a fixed linear polarization made in Eq. (3-23) and retain the vector character of the transition dipole matrix element μ_{ab} so that the molecule-field interaction becomes $\boldsymbol{\kappa} \cdot \hat{e}E_0$ instead of just κE_0. Here $\boldsymbol{\kappa} = \mu_{ab}/\hbar$ and $\hat{e}$ is a unit vector describing the polarization of the electric field. This problem has been analyzed by Gordon *et al.* (1969), including the effects of the M_J degeneracy present in molecular transitions. They concluded that one could determine the rotational states involved in the transition, at least for small values of J, by studying the polarization dependence of the echo. However, one must be certain that only a single isolated transition is present, or erroneous conclusions will be obtained. For example, Gordon *et al.* applied their theory to the SF_6 transition overlapping the CO_2 P(20) line and concluded that it was a $J = 0 \to 1$ or $J = 1 \to 1$ transition. In fact, the SF_6 spectrum near P(20) consists of several lines, with J values ranging from $J = 55$ to $J = 60$ (McDowell *et al.*, 1976). Heer and Nordstrom (1975) later analyzed the polarization dependence of echoes in SF_6 and were able to determine which CO_2 lines were exciting Q-branch SF_6 transitions, and which were exciting P- or R-branch transitions.

Photon echoes have also been observed in three–level systems where one has two closely spaced upper levels a and c which can both be excited from a common lower level b. For such a system, one finds that the peak echo amplitude as a function of pulse delay is modulated at the frequency ω_{ac}, the splitting between the upper levels. It is tempting to think of the echo here as being simply the sum of the emitted fields from two independently excited transitions, $b \to a$ and $b \to c$, but this is incorrect. Instead, the problem of a coupled three-level system interacting with an optical field must be properly solved and the echo calculated from that solution. While this calculation is beyond the scope of this chapter, it has been done, first by Lambert *et al.* (1971), and later in more general form by Brewer and his coworkers (Schenzle *et al.*, 1976). Although the result is quite complicated, it simplifies considerably at time $t - \tau_4 = \tau_{32}$, and this special case is sufficient to determine the behavior of the echo peak as a function of the delay between pulses, τ_{32}. At $t - \tau_4 = \tau_{32}$, one finds the amplitude of the emitted field to be

$$\mathscr{E}_c(t - \tau_4 = \tau_{32}) = \frac{\Omega L}{4\varepsilon_0 c} e^{-\gamma(t - \tau_1)} (N_b - N_a) \sin \kappa_{av} E_0 \tau_{21} (1 - \cos \kappa_{av} E_0 \tau_{43})$$

$$\times \frac{1}{\mu_{av}^3} (\mu_{ab}^4 + \mu_{bc}^4 + 2\mu_{ab}^2 \mu_{bc}^2 \cos \omega_{ac} \tau_{32}) \tag{3-170}$$

where $\mu_{av} = \sqrt{\mu_{ab}^2 + \mu_{bc}^2}$ and $\kappa_{av} = \mu_{av}/\hbar$. To obtain this result one must assume that all relaxation times are equal and that the optical field during the pulses is so intense that $\kappa_{av} E_0 \gg k\bar{u}$. Equation (3-170) reduces to the two-level results we have discussed previously if $\mu_{bc} = 0$. Notice that it is the emitted *field* that is modulated at $\cos \omega_{ac} \tau_{32}$. Such a result cannot be obtained from two independently excited transitions. In fact, if we look at the terms in (3-170), we see that the two terms proportional to μ_{ab}^4 and μ_{bc}^4 have the form one would expect for two independent echoes arising from the $b \to a$ and $b \to c$ transitions. Hence, the modulated $\cos \omega_{ac} \tau_{32}$ term must arise from molecules which participate in both transitions simultaneously. The paper by Schenzle *et al.* (1976) presents an interesting qualitative explanation for the form of this term.

Modulated photon echoes have been observed in NH_2D by Shoemaker and Hopf (1974) and in ruby by Lambert (1973). The observations are in agreement with theory and allow one to measure very small splittings ω_{ac} directly.

The modulation of the photon echo amplitude is closely related to another phenomenon known as "quantum beats." In quantum beat experiments, atoms or molecules are excited to a superposition of two closely spaced levels a and c by a pulse. One then observes the spontaneous emission (fluorescence) emitted by the system and finds that the fluorescent intensity is modulated at the splitting ω_{ac}. This result can be qualitatively understood as being simply the beat between two emitted fields $E_1 \cos \omega_{ab}t$ and $E_2 \cos \omega_{cb}t$. Owing to the coherent pulse preparation, these fields start out in phase and hence produce an intensity $I(t) = \frac{1}{2}c\varepsilon_0(E_1{}^2 + E_2{}^2 + 2E_1E_2 \cos \omega_{ac}t)$. Quantum beat spectroscopy has been done on many systems using several different types of pulsed excitation (see Haroche et al., 1973, and references therein). In a sense, modulated photon echoes represent an extension of quantum beat spectroscopy, since modulated echoes can be observed in the infrared, where spontaneous emission from a single state is too weak to see. Note, however, that the physics of the two effects differs somewhat in that the modulation at ω_{ac} appears in the *field* in an echo experiment, while in a quantum beat experiment the ω_{ac} modulation is only present in the *intensity*.

We conclude this section on two-pulse echoes by noting that other variations of the two-pulse echo are also possible. Liao and Hartmann (1973), for example, have observed notched echoes and radiation-locked echoes. These are obtained when the first and second pulses, respectively, are made very long. Hartmann (1968) has also proposed a Raman echo experiment. For still other variations the reader is referred to the nmr literature. None of these echo schemes has yet found a useful application in the optical region, however.

3.8.2. Multiple-Pulse Echoes

In this section we consider what happens when three or more optical pulses are applied in a time short compared to molecular relaxation times. Although a variety of phenomena will appear, we will focus our attention primarily on two effects, the stimulated echo and the Carr–Purcell echo train, as these both have practical applications.

We begin by considering the general three-pulse problem. If one applies three pulses to a sample as shown in Figure 3-20, it is clear that an echo will be obtained between the second and third pulses provided that the second interval $t_{54} \equiv t_5 - t_4$ is greater than $t_{32} \equiv t_3 - t_2$. We might also expect to find one or two more echoes following the third pulse. As indicated in the figure, however, there are actually four echoes following the third pulse! This is a surprisingly large number. Let us see if we can discover how they all arise.

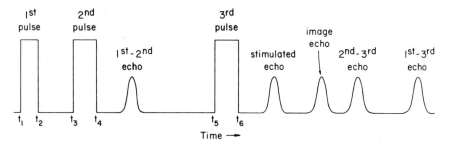

Figure 3-20. Schematic diagram showing photon echoes produced by a sequence of three pulses.

Two of the echoes are easily explained. The 1st–3rd pulse echo and the 2nd–3rd pulse echo are just ordinary two-pulse echoes which arise from the first and third and second and third pulses, respectively, Hence, there is nothing particularly interesting or unusual about these two echoes.

The image echo is the "echo of an echo" and is also easy to understand. At the peak of the 1st–2nd pulse echo, the oscillating dipoles are all back in phase, so the system looks just like it did at the end of the first pulse (see Figure 3-17). The only difference is that the dipoles are aligned along the $+II$ axis instead of the $-II$ axis. Thus, when a pulse is applied to the system following the original echo, another echo will occur in which the original echo is reformed. The motion of the **M** vectors during the formation of the image echo is analogous to the motion which occurs in the two-pulse echo sequence. In fact, one can see exactly what happens by holding Figure 3-17 upside down. Figure 3-17a then represents the system at time $t - t_4 = t_{32}$, Figure 3-17b at $t_5 > t > t_4$, Figure 3-17c at $t = t_6$, and Figure 3-17d at $t - t_6 = t_{54} - t_{32}$.

Although the image echo is no different in principle than an ordinary two-pulse echo, it does lead to some interesting possibilities. For example, we can apply a fourth pulse to re-form another echo from the image echo, then another pulse to re-form that echo, and so on *ad infinitum*. In general, applying more pulses would lead to a large number of echoes after each pulse, but under certain conditions we can arrange to have only the image echo appear. This situation is referred to as a Carr–Purcell echo train, and we discuss it in detail after considering the stimulated echo.

The stimulated echo is a phenomenon we have not encountered before, an echo which requires three pulses for its formation. To see how such an echo can arise, we first recall what the essential elements of the two-pulse echo were. In that case, we formed a set of oscillating dipoles with the first pulse, and found that they dephased as a result of the dipoles having different molecular velocities. The second pulse then reversed the sign of the v component of each **M** vector, and this act changed the phase of each oscillating dipole so that all the dipoles came back into phase at a time $\sim t_{32}$ after the end of the second pulse [see Eqs. (3-142)–(3-148)]. The stimulated echo is formed in the same manner, except that the reversal of the sign of v is accomplished in two steps. Instead of reversing the sign of v by a single 180° pulse, we reverse the sign by a sequence of two 90° pulses. The first of these pulses changes the v component into a w component along the III axis. The information remains stored there because dephasing takes place only in the I–II plane. At a later time we then change the w component back to a v component (but with reversed sign) by means of the final 90° pulse. Hence, the formation of the stimulated echo is really no different in principle than for a two-pulse echo. We have simply split the 180° second pulse into two 90° pieces and made use of the fact that w, the **M** vector component along the III axis, does not dephase. Figure 3-21 shows a vector model picture of stimulated echo formation. After the first 90° pulse the **M** vectors dephase in the usual manner, as shown in Figure 3-21a. The second 90° pulse then rotates each **M** vector into the I–III plane, as shown in Figure 3-21b. Following the second pulse the **M** vectors now continue to dephase in the

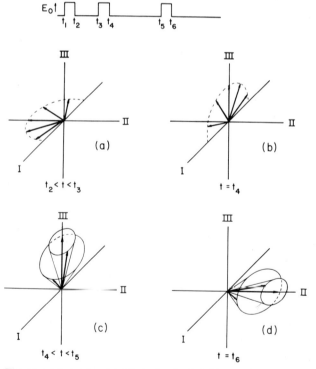

Figure 3-21. The stimulated echo. (a)–(d) depict the behavior of the **M** vectors for several molecular velocities at various times in the pulse sequence.

I–II plane. This causes the top of each **M** vector to rotate in a cone about the III axis, as shown in Figure 3-21c. The important point to notice here is that the angle between any **M** vector and the III axis remains constant, and hence $w(t)$ remains constant. This means that $w(t_5) = -v(t_3)$; i.e., we have stored information about the state of the system at $t = t_3$ along the III axis during the time interval t_{54}. The final pulse rotates $w(t_5)$ down into the I–II plane, as shown in Figure 3-21d. This results in an echo at a time t_{32} after the end of the third pulse. The configuration of the **M** vectors at the time of the echo is extremely difficult to draw because the exact positions of the **M** vectors on the cones shown in Figures 3-21c and d depend on the interval between the second and third pulses. When the mathematics are worked out, however, one finds that at $t = t_6 + t_{32}$ the **M** vectors have moved so as to add up more or less in phase and produce an echo. The reader is invited to trace out the motion of a few **M** vectors for himself and thereby verify that they do tend to lie near the +II axis at $t_6 + t_{32}$. Notice that the echo is not a perfect one, since the **M** vectors do not all end up on the +II axis. As a result, its amplitude is considerably less than the peak FID amplitude following the first pulse.

Let us now consider a very simple quantitative calculation of the stimulated echo. This will both reveal the relaxation-time dependence of the stimulated echo and show more clearly how information stored along the III axis is retrieved to produce a stimulated echo. We will assume that $\kappa E_0 \gg k\bar{u}$ so that all the molecules are excited resonantly, and let the three pulses be exactly $\pi/2$. As usual, we will also ignore relaxation during the pulses. At the start of the second pulse, $t = t_3$, we have the same situation that we had for the simple two-pulse echo treatment in Eq. (3-142):

$$u(t_3) = e^{-\gamma_{ab}t_{32}}(n_a - n_b) \sin \delta t_{32}$$

$$v(t_3) = e^{-\gamma_{ab}t_{32}}(n_a - n_b) \cos \delta t_{32} \qquad (3\text{-}171)$$

$$w(t_3) = 0$$

We are assuming here that $\gamma_a = \gamma_b \equiv \gamma$. In the more general case, $w(t_3)$ would be given by Eq. (3-152). The effect of a 90° pulse is simply to change w into v and v into $-w$, as can be seen from Eq. (3-144). Thus, at the end of the second pulse, we have

$$u(t_4) = e^{-\gamma_{ab}t_{32}}(n_a - n_b) \sin \delta t_{32}$$

$$v(t_4) = 0 \qquad (3\text{-}172)$$

$$w(t_4) = -e^{-\gamma_{ab}t_{32}}(n_a - n_b) \cos \delta t_{32}$$

We see that $-v(t_3)$ is now stored along the III axis as $w(t)$. The information that is stored here consists of the pattern $\cos \delta t_{32}$ in frequency space, as shown in Figure 3-22. During the time interval t_{54} the u and v components continue to dephase according to Eqs. (3-72) while w simply

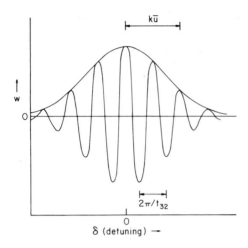

Figure 3-22. The population difference $w(t)$ during the time interval
between the second and third pulses,
plotted as a function of detuning from
resonance.

decays exponentially, as shown in Eq. (3-151). The third 90° pulse now
interchanges v and w so that we end up with

$$u(t_6) = e^{-\gamma_{ab}t_{54}} \cos \delta t_{54}[e^{-\gamma_{ab}t_{32}}(n_a - n_b) \sin \delta t_{32}]$$

$$v(t_6) = e^{-\gamma t_{54}}[-e^{-\gamma_{ab}t_{32}}(n_a - n_b) \cos \delta t_{32}]$$

(3-173)

at the end of the third pulse. Notice that the $\cos \delta t_{32}$ information has now
become a v component again. The net result of the last two pulses is that
$v(t_6) = \exp(-\gamma t_{54})[-v(t_3)]$. Thus, aside from an overall decay, the second
and third 90° pulses have simply changed the sign of v just as a single 180°
pulse does in a two-pulse echo. In fact, the terms in brackets in Eqs. (3-173)
are identical to the expressions one obtains for u and v following the 180°
pulse in a two-pulse echo treatment [see Eqs. (3-142)–(3-148)]. Following
the third pulse, u and v continue to dephase, so we obtain

$$u(t) = C[\cos \delta t_{32} \sin \delta(t - t_6) - A \cos \delta t_{54} \sin \delta t_{32} \cos \delta(t - t_6)]$$

$$v(t) = C[\cos \delta t_{32} \cos \delta(t - t_6) + A \cos \delta t_{54} \sin \delta t_{32} \sin \delta(t - t_6)]$$

(3-174)

where

$$C = -e^{-\gamma_{ab}(t - t_6 + t_{32})}(n_a - n_b)e^{-\gamma t_{54}}$$

and

$$A = e^{-(\gamma_{ab} - \gamma)t_{54}}$$

The next step is to expand the sin cos products appearing in Eqs. (3-174) so
that $u(t)$ and $v(t)$ can be written as a sum of terms each of which is only a
single sin or cos function. This may be done using product formulas of the
type $\sin \alpha \cos \beta = \frac{1}{2} \sin(\alpha + \beta) + \frac{1}{2} \sin(\alpha - \beta)$. We then make use of the fact

that terms oscillating at $\delta t'$ will contribute significantly to the polarization only near $t' = 0$ [see the discussion following Eq. (3-146)]. Since the stimulated echo occurs at time $t = t_6 + t_{32}$, we may pick out only those terms which oscillate at $\delta(t - t_6 - t_{32})$ and ignore the rest. On doing this, we find that only the first term in the expressions for $u(t)$ and $v(t)$ [Eq. (3-174)] contains the proper time dependence. Looking back we can see that these terms were exactly the ones which came from $v(t_3)$ and were stored along the III axis during the time interval t_{54}. We now have

$$u(t) = \tfrac{1}{2}C \sin \delta(t - t_6 - t_{32})$$
$$v(t) = \tfrac{1}{2}C \cos \delta(t - t_6 - t_{32})$$

(3-175)

as the parts of u and v which contribute to the stimulated echo. The rest of the terms contained in Eq. (3-174) contribute to other echoes or to FID signals following the pulses.

If we compare the result obtained in Eq. (3-175) with the simple two-pulse-echo result for $u(t)$ and $v(t)$ given in Eq. (3-146), we see that they are identical except for an extra factor of $\tfrac{1}{2} \exp(-\gamma t_{54})$ in the stimulated echo expression. Hence, the stimulated echo shape and intensity will be given by Eq. (3-148) multiplied by $\tfrac{1}{4}\exp(-2\gamma\tau_{54})$, and the peak intensity of the stimulated echo is

$$I_{\max} = \frac{1}{4}\frac{\mu_{ab}^2\omega_0^2 L^2}{8\varepsilon_0 c}(N_b - N_a)^2 e^{-4\gamma_{ab}\tau_{32}} e^{-2\gamma\tau_{54}}$$

(3-176)

This equation suggests how the stimulated echo may be usefully employed to study collisional relaxation. Although the collisional decay dependence of the echo involves both γ and γ_{ab}, we can study either one separately by simply varying one or the other of the pulse separations. For example, if we want to measure the decay rate γ (T_1 in nmr language), we hold the 1st–2nd pulse separation constant and observe the peak echo intensity as a function of the 2nd–3rd pulse separation. A plot of $\ln I_{\max}$ versus τ_{54} then yields directly a value for 2γ which is uncontaminated by γ_{ab}. This use of the stimulated echo to measure γ is the principal reason for our interest in it, since up till now we have not discussed any method which allows us to measure γ.

So far we have only looked at the relaxation time dependence of the stimulated echo for one special case: Three 90° pulses with $\kappa E_0 \gg k\bar{u}$ and $\gamma_a = \gamma_b$. However, it is easy to generalize this result. First, consider the case where we use pulses of arbitrary area and do not require $\kappa E_0 \gg k\bar{u}$. Although the algebra required to treat this situation will become very complex, only a few terms will actually contribute to the stimulated echo. As we noted earlier, these are the terms in $u(t)$ and $v(t)$ which oscillate at $\delta(t - t_6 - t_{32})$. One can now show that this information is sufficient to completely determine

Table 3-1   Terms Which Arise When Three Pulses Are Applied to
a Sample

Time dependence	Physical interpretation
$\exp\{-i\delta[(t-t_6+t_{54}+t_{32}]\}$	1st pulse FID
$\exp\{-i\delta[(t-t_6)+t_{54}]\}$	2nd pulse FID
$\exp\{-i\delta[(t-t_6)]\}$	3rd pulse FID
$\exp\{-i\delta[(t-t_6)+t_{32}]\}$	"Virtual echo"
$\exp\{-i\delta[(t-t_6)+t_{54}-t_{32}]\}$	1st–2nd pulse echo
$\exp\{-i\delta[(t-t_6)-t_{32}]\}$	Stimulated echo
$\exp\{-i\delta[(t-t_6)-t_{54}-t_{32}]\}$	1st–3rd pulse echo
$\exp\{-i\delta[(t-t_6)-t_{54}]\}$	2nd–3rd pulse echo
$\exp\{-i\delta[(t-t_6)-t_{54}+t_{32}]\}$	Image echo

the relaxation-time dependence of the stimulated echo. The key to the argument is to notice that between any two pulses (pulse i and pulse $i-1$), *every term* in $u(t_i)$ and $v(t_i)$ will oscillate at $\sin\delta(t_i-t_{i-1})$ or $\cos\delta(t_i-t_{i-1})$ and decay as $\exp[-\gamma_{ab}(t_i-t_{i-1})]$. This fact is evident from Eqs. (3-72). $w(t_i)$ and $s(t_i)$, on the other hand, *never* have terms oscillating at $\delta(t_i-t_{i-1})$ and always decay as $\exp[-\gamma(t_i-t_{i-1})]$, as can be seen from Eq. (3-151). The clear distinction between u,v terms and w,s terms is possible because the u,v and w,s equations of motion are completely uncoupled from each other between pulses [see Eqs. (3-67)]. A brief consideration of these facts should convince the reader that the stimulated echo can only arise from terms which were part of $u(t)$ and $v(t)$ during the intervals t_{32} and $t-t_6$ and were also part of $w(t)$ during the interval t_{54}. There is simply no other way to end up with a term of the form $\sin\delta(t-t_6-t_{32})$ or $\cos\delta(t-t_6-t_{32})$. $s(t)$ cannot be involved in the stimulated echo, because it is not coupled to u or v by the excitation pulses. Hence, we conclude that at the stimulated echo peak, $t-t_6=t_{32}$, the echo intensity will be proportional to $\exp(-4\gamma_{ab}t_{32})\exp(-2\gamma t_{54})$, just as we found in Eq. (3-176). This result is valid even if all the molecules do not interact resonantly and we use pulses of arbitrary area. We do assume, however, that the pulses are sufficiently short that relaxation during the pulses can be neglected.

We can also relax the assumption that $\gamma_a=\gamma_b$. This is a trivial generalization because γ_a and γ_b enter only into the equations for $w(t)$ and $s(t)$, and these equations only affect the stimulated echo through the behavior of $w(t)$ during the interval t_{54}. Using the full equation [Eq. (3-151)] for $w(t)$ and remembering that $s(t_4)$ does not have the right time dependence to contribute to the echo, we see that the most general result for the stimulated echo decay becomes

$$I_{max}\propto e^{-4\gamma_{ab}t_{32}}e^{-2\gamma_{av}t_{54}}\cosh^2\gamma_d t_{54}=\tfrac{1}{4}e^{-4\gamma_{ab}t_{32}}(e^{-\gamma_a t_{54}}+e^{-\gamma_b t_{54}})^2 \qquad (3\text{-}177)$$

In the simple stimulated echo treatment we did earlier, the pulse area dependence was not considered. We just assumed 90° pulses. To determine the area dependence it is most convenient to use the 2×2 matrix representation just as we did for two-pulse echoes. A useful by-product of this calculation will be the timing and pulse area dependence of the other three echoes which follow the third pulse. To this end we will take $t_{54} > t_{32}$ so that the 1st–2nd pulse echo and hence the image echo can occur.

To keep the algebra from becoming too cumbersome, we will denote the 2×2 matrix elements during the pulses by just α_i and β_i. The 2×2 matrices which describe the system between pulses will be written out explicitly but with $\exp(\pm i\frac{1}{2}\,\delta t')$ written as $\exp(\pm t')$. The matrix $\mathbf{r}_T$ which gives the total effect of the three pulses is then

$$
\mathbf{r}_T = \begin{pmatrix} e^{-(t-t_6)} & 0 \\ 0 & e^{t-t_6} \end{pmatrix} \begin{pmatrix} \alpha_3 & \beta_3 \\ -\beta_3^* & \alpha_3^* \end{pmatrix} \begin{pmatrix} e^{-t_{54}} & 0 \\ 0 & e^{t_{54}} \end{pmatrix} \begin{pmatrix} \alpha_2 & \beta_2 \\ -\beta_2^* & \alpha_2^* \end{pmatrix}
$$

$$
\times \begin{pmatrix} e^{-t_{32}} & 0 \\ 0 & e^{t_{32}} \end{pmatrix} \begin{pmatrix} \alpha_1 & \beta_1 \\ -\beta_1^* & \alpha_1^* \end{pmatrix} \tag{3-178}
$$

On multiplying out these matrices we obtain

$$
\alpha_T = e^{-(t-t_6)}[e^{-(t_{54}+t_{32})}\alpha_3\alpha_2\alpha_1 - e^{-(t_{54}-t_{32})}\alpha_3\beta_2\beta_1^*
$$
$$
- e^{(t_{54}-t_{32})}\beta_3\beta_2^*\alpha_1 - e^{(t_{54}+t_{32})}\beta_3\alpha_2^*\beta_1^*]
$$
$$
\beta_T = e^{-(t-t_6)}[e^{-(t_{54}+t_{32})}\alpha_3\alpha_2\beta_1 + e^{-(t_{54}-t_{32})}\alpha_3\beta_2\alpha_1^* \tag{3-179}
$$
$$
- e^{(t_{54}-t_{32})}\beta_3\beta_2\beta_1^* + e^{(t_{54}+t_{32})}\beta_3\alpha_2^*\alpha_1^*]
$$

To find $u(t)$ and $v(t)$, one must next calculate $-2\alpha_T\beta_T$. We obtain terms having nine distinct time dependences, each of which corresponds to a different physical effect, as shown in Table 3-1. The physical interpretation of each term can be found by noting the time at which the exponential goes to zero. For example, we see that the first entry in the table represents an FID following the first pulse, since it goes to zero at $t - t_6 = -t_{54} - t_{32}$, i.e., at the end of the first pulse. The first five terms displayed in the table are all negligible, because they represent coherences which occur at times $t < t_6$, and our calculation is for times $t > t_6$. We note in passing that the "virtual echo" term does not represent an actual echo which occurs before the third pulse. Between the second and third pulses only the ordinary two-pulse echo (labeled 1st–2nd pulse echo in the table) occurs. Following the third pulse, however, the virtual echo term appears, and the system acts exactly as if an echo had previously occurred at $t - t_6 = -t_{32}$ (Bloom, 1955).

With the aid of Table 3-1 it is easy to find $-2\alpha_T\beta_T$ for each of the four echoes which occur after $t = t_6$. In writing out these expressions explicitly we will assume that $\kappa E_0 \gg k\bar{u}$. Also, in order to properly describe the echo timing, we will take into account detuning to first order in $\delta/\kappa E_0$. α_i and β_i are then given by Eqs. (3-156) and (3-157) and we find that

$$-2\alpha_T\beta_T = -\frac{i}{4}[e^{-i\delta(t-t_6-t_{32}-q_1-q_2+q_3)}2\sin\theta_1\sin\theta_2\sin\theta_3$$

(stimulated echo)

$$-e^{-i\delta(t-t_6-t_{54}+t_{32}+q_1)}\sin\theta_1(1-\cos\theta_2)(1-\cos\theta_3)$$

(image echo)

$$+e^{-i\delta(t-t_6-t_{54}-q_2)}2\cos\theta_1\sin\theta_2(1-\cos\theta_3) \qquad (3\text{-}180)$$

(2nd–3rd pulse echo)

$$+e^{-i\delta(t-t_6-t_{54}-t_{32}-q_1-2q_2)}\sin\theta_1(1+\cos\theta_2)(1-\cos\theta_3)]$$

(1st–3rd pulse echo)

The real and imaginary parts of this expression give $u(t)$ and $v(t)$ as in Eq. (3-162). There is no need to carry the calculation further, as Eq. (3-180) already displays the timing and pulse area dependences of all four echoes. We see immediately that the stimulated echo is indeed largest when one uses three 90° pulses, as was assumed earlier. The other echoes all show the $\sin\theta_i(1-\cos\theta_i)$ area dependence that is characteristic of a two pulse echo. Finally, notice that the image echo term is reversed in sign from the others. This is a reflection of the fact that the image echo is formed along the $-$II axis instead of the $+$II axis.

Thus far, except for the discussion of relaxation-time dependence, we have assumed that the excitation pulses were so intense that $\kappa E_0 \gg k\bar{u}$. In Stark or frequency-switching experiments, this will not be the case, so we should also consider the opposite limit, $\kappa E_0 \ll k\bar{u}$. The only difference between this calculation and our previous treatment is that we now need to substitute the general formulas for α_i and β_i [see Eqs. (3-105)] into Eqs. (3-179). While this can easily be done, the resulting expressions are rather complicated and need to be numerically integrated over the Doppler profile to obtain the polarization in any event. Hence, we will not bother to write them down explicitly.

The results of the numerical calculations are worth noting, however. One finds that the timing of the echo peaks is still accurately given by Eq. (3-180), even though that equation was derived for the opposite limit, $\kappa E_0 \gg k\bar{u}$. The pulse area dependence of the echoes for $\kappa E_0 \ll k\bar{u}$, on the other hand, is not given by Eq. (3-180). This dependence, as well as the detailed echo shape, is a complicated function of all three pulse lengths and has not been fully explored as yet.

An example of a three-pulse echo experiment using Stark switching is shown in Figure 3-23. In the experiment a CO_2 laser was used to excite a single transition in $^{15}NH_3$ near 10 μm. The pulse areas were experimentally adjusted so that all the echoes could be seen clearly on a single trace. One

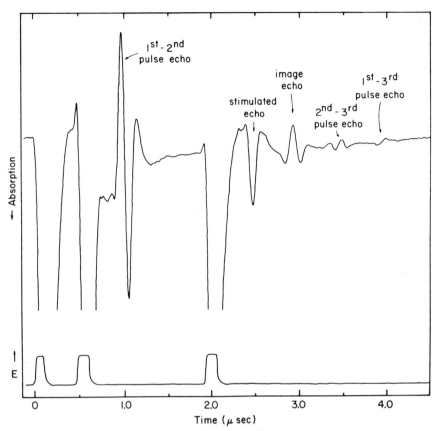

Figure 3-23. A three-pulse photon echo experiment in $^{15}NH_3$ using Stark switching. The lower trace shows the applied electric field pulses.

feature of the Stark switched echo experiment which stands out immediately is the fact that the echoes appear as heterodyne beats with the laser, and these echo beat signals have different phases with respect to each other. The phase of the echo heterodyne beat may be readily understood using arguments similar to those presented in the two-pulse echo case. We notice first that for any of the echoes, $-2\alpha_T\beta_T$ has the form

$$-2\alpha_T\beta_T = e^{-i\delta(t-t_6-t')}f(\alpha_1, \beta_1, \alpha_2, \beta_2, \alpha_3, \beta_3) \qquad (3\text{-}181)$$

where $t' = t_{32}$ for the stimulated echo, $t' = t_{54} - t_{32}$ for the image echo, etc. This result is for pulsed excitation where the optical field is zero between pulses. In the case of Stark switching, we switch the natural resonance frequency of the molecules from ω_0 to $\omega_0 + \Delta\omega_0$ between pulses. Since molecules which contribute to the echo are not resonant between pulses ($\delta \gg \kappa E_0$), we still have the same echo expression, Eq. (3-181), except that δ

in the exponential is changed to $\delta_s = \delta - \Delta\omega_0$. The polarization P_s is proportional to Im $(-2\alpha_T\beta_T)$, so it can be written in the form

$$P_s = C\left[\int_{-\infty}^{\infty} \mathscr{A}\cos\delta_s(t-t_6-t')\,d\delta + \int_{-\infty}^{\infty}\mathscr{B}\sin\delta_s(t-t_6-t')\,d\delta\right]$$

$$(3\text{-}182)$$

where $\mathscr{A} = (n_a - n_b)\,\text{Im}\,[f(\alpha_1, \ldots, \beta_3)]$ and $\mathscr{B} = -(n_a - n_b)\,\text{Re}\,[f(\alpha_1, \ldots, \beta_3)]$. Using the symmetry properties of α_i and β_i with respect to the detuning δ [e.g., Re (α_i) is an even function of δ, Im (α_i) is an odd function of δ], it is easy to show that Im $(-2\alpha_T\beta_T)$ is always an even function of δ, and hence that Re $[f(\alpha_1, \ldots, \beta_3)]$ must be an odd function and Im $[f(\alpha_1, \ldots, \beta_3)]$ an even function of δ. Equation (3-182) therefore has exactly the same form as the two-pulse result in Eq. (3-167), and we may repeat the argument given in Eqs. (3-167)–(3-169) to show that the echo heterodyne beat in a Stark- or frequency-switching experiment is simply given by $\cos \Delta\omega_0(t - t_6 - t')$.

This result enables us to understand what is observed in the three-pulse echo experiment of Fig. 3-23. With the exception of the image echo, all the echo beat signals have roughly the same phase. The differences that are present are due to the fact that the echoes have different widths and different delays [the q_i's in Eq. (3-180)] between $t - t_6 - t' = 0$ and the echo peak. The image echo, on the other hand, is $\sim 180°$ out of phase with the other echoes. This result arises because the image echo is formed along the $-$II axis and therefore gives rise to a polarization of the opposite sign from that produced by the echoes which are all formed along the $+$II axis. Mathematically, we can understand the $180°$ phase change by using Eqs. (3-69), (3-162), and (3-169) to obtain an expression for the transmitted intensity in terms of Im $(-2\alpha_T\beta_T)$. Ignoring relaxation, we find that

$$I(t) = I_0 - \left[\tfrac{1}{2}\mu_{ab}E_0\Omega L\int_{-\infty}^{\infty}(n_b - n_a)\,\text{Im}\,(-2\alpha_T\beta_T)\,dv_z\right]\cos\Delta\omega_0(t-t_6-t')$$

$$(3\text{-}183)$$

as the signal one expects to see in a Stark-switching experiment. Looking back to Eq. (3-180), we see that Im $(-2\alpha_T\beta_T)$ is positive for the image echo, but negative for all other echoes. The same is true if one looks at Im $(-2\alpha_T\beta_T)$ for the Doppler-broadened case where $\kappa E_0 \ll k\bar{u}$. Hence, the sample transmission at $t - t_6 - t' = 0$ will be negative for the image echo but positive for all other echoes; i.e., the image echo heterodyne beat is $180°$ out of phase with respect to the other echoes.

Very few three-pulse echo experiments have been done to date, although they have been observed in several systems, including $^{13}CH_3F$, $^{15}NH_3$, and ruby. As we noted earlier, the principal application for a three-pulse experiment would be the measurement of level decay rates using the stimulated echo. However, there are other ways to measure level decay rates, such as delayed nutation and adiabatic rapid passage, and these presently appear more attractive for such measurements. This probably accounts for the fact that three-pulse echoes have not yet been studied extensively.

We now turn our attention to the Carr–Purcell echo train, a phenomenon named after the two men who discovered it in nmr, H. Y. Carr and E. M. Purcell (Carr and Purcell, 1954). What they found was that there is a special pulse train which gives a very simple sequence of echoes. Furthermore, the echo decay is insensitive to the effects of molecular diffusion in the sample. In the optical region this same set of pulses can be used to produce echoes whose decay, unlike the two-pulse echo decay, is insensitive to the effects of velocity-changing collisions. We shall discuss the velocity-changing collision aspects of echoes in Section 3.10. For the moment we content ourselves with a discussion of how the Carr–Purcell echo sequence works and a demonstration of its dependence on the collisional relaxation times γ_{ij}.

Using the results of our previous treatment of three pulse echoes, the Carr–Purcell echo train can be easily understood. Consider the pulse train shown in Figure 3-24a. The first pulse is a 90° pulse, and all subsequent pulses are exactly 180° pulses. Furthermore, the spacing between the 180° pulses is $2(\tau + q_1)$, where τ is the spacing between the first two pulses and $q_1 = 1/\kappa E_0$. If the pulses are sufficiently intense that $\kappa E_0 \gg k\bar{u}$, one will observe a single photon echo between every pair of 180° pulses, as shown in Figure 3-24b. This simple behavior occurs because of the special pulse train we have chosen. Following the first two pulses we, of course, obtain a single two-pulse photon echo. Following the third pulse we would, in general, obtain four echoes, as given in Eq. (3-180). However, looking at that equation, we see that for $\theta_2 = \theta_3 = \pi$, all echoes except the image echo have zero amplitude and hence are absent. The only effect of the third pulse is therefore to re-form the original two-pulse echo. Since the fourth pulse is identical to the third pulse, it will also produce only an image echo of the previous echo. Similarly, the fifth pulse will re-create the echo following the fourth pulse, and so on.

We showed earlier that the image echo was formed in the same way that a two-pulse echo is formed. This also holds true for all the subsequent image echoes in this Carr–Purcell echo train. The only difference between one echo and another is that echoes are formed alternately along the $+II$ and $-II$ axes. We can show rigorously that all the image echoes are identical to the original echo (except for a sign) by considering the 2×2 matrices again. During the time interval after the first two pulses, α and β are given by Eq. (3-160). In particular, at the peak of the photon echo, which occurs at $t - t_4 = t_{32} + q_1$, α and β take the simple form

$$\alpha_p = -\frac{1}{\sqrt{2}} e^{-i(1/2)\delta q_1}, \qquad \beta_p = -\frac{i}{\sqrt{2}} e^{i(1/2)\delta q_1} \qquad (3\text{-}184)$$

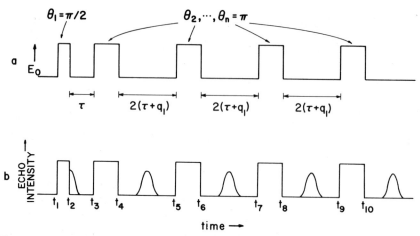

Figure 3-24. The Carr–Purcell echo experiment. (a) shows the pulse areas and spacing between pulses that are required. (b) shows the pulse train together with the echo signals one expects to observe.

for the 90–180° pulse sequence we are using. Now consider the time interval between the third and fourth pulses. The 2×2 matrix $\mathbf{r}$ which describes the evolution of the system in this interval is

$$\mathbf{r} = \begin{pmatrix} e^{-(1/2)\delta(t-t_6)} & 0 \\ 0 & e^{(1/2)\delta(t-t_6)} \end{pmatrix} \begin{pmatrix} 0 & -i \\ -i & 0 \end{pmatrix} \begin{pmatrix} e^{-(1/2)\delta(\tau+q_1)} & 0 \\ 0 & e^{(1/2)\delta(\tau+q_1)} \end{pmatrix}$$

$$\times \begin{pmatrix} \alpha_p & \beta_p \\ -\beta_p^* & \alpha_p^* \end{pmatrix}$$

$$= \begin{pmatrix} 0 & -ie^{-(1/2)\delta(t-t_6-\tau-q_1)} \\ -ie^{(1/2)\delta(t-t_6-\tau-q_1)} & 0 \end{pmatrix} \begin{pmatrix} \alpha_p & \beta_p \\ -\beta_p^* & \alpha_p^* \end{pmatrix} \qquad (3\text{-}185)$$

so

$$2\alpha\beta = 2\alpha_p^* \beta_p^* e^{-i\delta[(t-t_6)-(\tau+q_1)]} \qquad (3\text{-}186)$$

On looking at this result, it is clear that a single echo will occur at time $t - t_6 = \tau + q_1$ during the interval, since only then will the exponental factor vanish in $2\alpha\beta$. Furthermore, at the echo peak, $2\alpha\beta = 2\alpha_p^* \beta_p^* = -2\alpha_p\beta_p$, so the state of the system at $t - t_6 = \tau + q_1$ is identical to the state of the system at the peak of the first echo, except for a minus sign. Note also that the echo timing depends on the time interval between the previous echo and the third pulse as well as on the time elapsed after the third pulse. As we have discussed previously, this means that the collisional decay of the second echo must depend on γ_{ab} only.

Since, except for sign, the system at $t - t_6 = \tau + q_1$ is unchanged from its state at the first echo peak, we may now repeat the arguments above for the time interval following the fourth pulse to show that only a single echo decaying as γ_{ab} will occur in this interval. There is again a change of sign, so that at this echo peak $2\alpha\beta = 2\alpha_p\beta_p$. In fact, it should be clear that we can extend the argument to any number of pulses and show that in general, after n 180° pulses, the nth echo in the train will have $-2\alpha\beta = (-1)^n 2\alpha_p\beta_p = i(-1)^n$ at its peak. Using this general result it is trivial to calculate the peak echo amplitude for the nth echo. From Eqs. (3-162), (3-69), and (3-50), we see that

$$\mathscr{E}_{\max}(n\text{th echo}) = -\frac{\mu_{ab}\omega_0 L}{2\varepsilon_0 c}(N_b - N_a)[-e^{-2\gamma_{ab}(\tau + q_1)}]^n \qquad (3\text{-}187)$$

Thus, the collisional decay for the nth echo is just given by the collisional decay for the first echo raised to the nth power. As usual, we are neglecting decay during the pulses in the equation above. Clearly, the Carr–Purcell echo train provides a convenient way to measure γ_{ab}, since in a single experiment one obtains a whole train of evenly spaced echoes from which a value for γ_{ab} may be easily extracted. As we discuss in Section 3.10, however, an even more useful property of the Carr–Purcell echo train is the fact that it is insensitive to velocity-changing collisions.

The simple result obtained in Eq. (3-187) was for a train of 180° pulses spaced by $2(\tau + q_1)$. Let us now examine the effect of relaxing these stringent conditions on the pulse train. A change in the spacing between the pulses changes the Carr–Purcell echo train very little. It simply shifts around the positions of the echoes. For example, if we make the 180° pulse spacings larger than $2(\tau + q_1)$ by an amount Δ (but keep the 1st–2nd pulse spacing equal to τ), the first, third, fifth, etc., echoes will still occur at $\tau + q_1$ after their respective 180° pulses, while all the even-numbered echoes will simply be delayed to $\tau + q_1 + \Delta$ after their 180° pulses. Thus, the timing of the pulses is not critical. What is critical is that all the pulses after the first have an area of π. Deviations from this area cause other echoes to appear, as one can see from Eq. (3-180). In particular, a stimulated echo will occur after the third pulse at exactly the same time as the image echo. Furthermore, each image echo becomes smaller than the previous echo $[\frac{1}{2}(1 - \cos\theta_i) < 1$ for $\theta_i \neq \pi]$, thus causing an apparent decay of the echo train which is not due to collisional relaxation. The latter effect is not serious because the apparent decay is pressure-independent, and hence may be removed from the experimental data by plotting the total observed decay rate versus pressure and measuring only the slope of the resulting line. The appearance of the stimulated echo on top of the image echo is a problem, however, because it does not decay as γ_{ab}. This means that one cannot rely on the Carr–Purcell echo train to measure γ_{ab} if the pulse areas deviate appreciably from π.

The discussion so far has considered only very intense pulses, i.e., the $\kappa E_0 \gg k\bar{u}$ case, where, with sufficient care one can always adjust the pulse areas to π and avoid the problems mentioned above (although the transverse intensity profile of the beam may give trouble). If we consider Stark or frequency-switching experiments, however, where one generally has $\kappa E_0 \ll k\bar{u}$, the situation is rather different. We can still adjust the pulse areas to π, but for $\kappa E_0 \ll k\bar{u}$, not all the molecules respond resonantly, and these molecules see effective pulse "areas" of $gt_p = \sqrt{\delta^2 + \kappa^2 E_0^2} t_p$ rather than $\theta = \kappa E_0 t_p$ (t_p is the pulse width). Thus, one might legitimately wonder whether a simple Carr–Purcell echo train which decays as γ_{ab} can ever be obtained. To answer this question, one must turn to a computer calculation. We can begin by computing the relative amplitudes of the image and stimulated echoes following the third pulse. This can easily be done using Eqs. (3-105) and (3-179) to find $-2\alpha_T \beta_T$ and numerically integrating this quantity over all molecular velocities v_z. If we can find a pulse train for which the stimulated echo amplitude is negligible compared to the image echo, we may be able to produce a Carr–Purcell echo train which decays as γ_{ab} only. Unfortunately, when the calculation is done, no such pulse train can be found. A $\pi/2$, π, π pulse sequence, for example, gives a stimulated echo whose amplitude is nearly as large as the image echo. Thus, one is forced to conclude that in Stark- or frequency-switching experiments where $\kappa E_0 \ll k\bar{u}$, the Carr–Purcell echo train cannot be used to measure γ_{ab}. This fact does not appear to have been recognized in Carr–Purcell experiments done to date (Berman *et al.*, 1975).

Since the stimulated and image echoes appear in Stark-switching experiments as heterodyne beat signals which are 180° out of phase, one might think that a good Carr–Purcell echo train could not be obtained at all. This is not the case, however. Figure 3-25 shows a typical Carr–Purcell echo train obtained using Stark switching. We notice here that the echoes have large amplitudes, are uniform in shape, and decay exponentially (except for the first echo). How are we to understand this simple result? The key is again to look at the echo following the third pulse. It has the same phase as the first two-pulse echo, and, when relaxation is taken into account, a larger amplitude than the first echo. Since this clearly cannot be an image echo, it must be predominantly a stimulated echo. A computer calculation bears this out. For pulse sequences near $\pi/2$, $\pi/2$, $\pi/2$, the stimulated echo is much larger than the image echo and is also somewhat larger than the first two-pulse echo. (The reason a pulse train with areas near $\pi/2$, $\pi/2$, $\pi/2$ was used in the figure is that experimentally one obtains a clean Carr–Purcell echo train by adjusting the pulse lengths until the echo amplitudes are maximized.) Evidently none of the subsequent echoes in the train are image echoes either, since all the echoes have the same phase. The actual structure of later echoes in the train can be very complicated, however. For example, the echo following the fourth pulse has contributions from five different echoes: a stimulated echo from pulses 1, 2, and 4; a stimulated echo arising

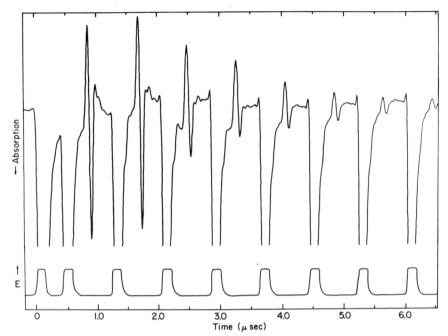

Figure 3-25. A Carr–Purcell echo experiment in NH_2D using Stark switching. The lower trace shows the applied electric field pulses.

from the first echo plus pulses 3 and 4; an image echo of the previous stimulated echo; an image echo of the previous image echo; and a two-pulse echo arising from pulses 1 and 3.

It should be obvious from the discussion above that there is little hope of obtaining unambiguous information about either the level decay rates or γ_{ab} from a Carr–Purcell experiment if $\kappa E_0 \ll k\bar{u}$. However, this does not mean that such experiments are pointless. If we can determine by other means that $\gamma_{ab} = \gamma_a = \gamma_b \equiv \gamma$, then Carr–Purcell echo experiments can be used to measure γ and to obtain a value which is free from the effects of velocity-changing collisions. Such experiments have been done (Schmidt *et al.*, 1973; Berman *et al.*, 1975) and will be discussed in Section 3.10.

3.9. Measurement of Level Decay Rates

In this section we discuss techniques which can be used to determine the level decay rates γ_a and γ_b. Using nmr language, we often refer to these methods as T_1 measurement techniques. However, it should be kept in mind that T_1, which is defined as $T_1 = 1/\gamma_a = 1/\gamma_b$, loses its meaning when $\gamma_a \neq \gamma_b$. The reason one is interested in the level decay rates should be obvious. These rates provide a measure of the collisional processes which change the level populations and thus can give useful information concerning gas-phase molecular interactions.

The discussion in this section will be confined to two T_1 measurement techniques: adiabatic rapid passage and delayed optical nutation. Both of these have been successfully used to measure level decay rates. Another possible measurement technique is the stimulated echo. However, we have already discussed this method in connection with multiple-pulse photon echoes, so it will not be considered further here.

3.9.1. Adiabatic Rapid Passage

Among the transient effects discussed in this chapter, adiabatic rapid passage (ARP) is unique in that the effect requires an optical field whose frequency is *swept*, not switched. The basic experiment is quite simple. One starts with a laser whose frequency is well below (above) the resonance frequency ω_0 of the transition. The laser frequency is then swept through the resonant frequency to a point well above (below) ω_0. If the proper intensity and sweep rate is used, one finds that the level populations have been completely inverted by this process; i.e., if all the molecules start out in the lower state, the sweep through resonance will leave the molecules all in the upper state. For this reason, ARP is sometimes also referred to as adiabatic inversion.

Although this is an interesting effect, it may not be immediately obvious how ARP can be used to measure level decay rates. The answer is that one just does two ARP experiments back to back with an adjustable delay τ_D between the two sweeps. The first sweep takes the system out of thermal equilibrium by placing the entire ground-state population into the excited state. This inverted population then decays back toward thermal equilibrium during the time delay τ_D and the amount of relaxation is determined by a measurement of the emission or absorption which is observed during the second ARP sweep. This determination is possible because the emission or absorption during ARP is proportional to the population difference which exists at the beginning of the sweep.

We now look more closely at adiabatic rapid passage in order to understand how the effects just described can occur. An obvious approach to understanding ARP is to use the familiar vector model. However, because the laser frequency (or resonance frequency) is swept during an ARP experiment and we have not considered this possibility previously in the chapter, it is necessary first to look back at the derivation of the vector model equations to see if they are valid for swept frequencies. The initial question that comes up is how one should write the optical field when the frequency is swept. We choose the following form:

$$E(z, t) = \hat{x}\{E_0(z, t) \cos [\Omega t - kz + \phi(z, t)]\} \qquad (3\text{-}188)$$

The frequency sweep has been introduced here in the form of a time-dependent phase factor $\phi(z, t)$. Since the frequency of the light is by definition the time rate of change of the total phase of the wave at fixed z, we have

$$\Omega_s = \Omega + \frac{\partial \phi}{\partial t} \qquad (3\text{-}189)$$

as the instantaneous frequency of the light. We will also place the requirement on $\phi(z, t)$ that it always satisfy the inequality $\partial\phi/\partial t \ll \Omega$; i.e., we require that the frequency sweep be much less than the optical frequency itself. This inequality justifies our having written the optical field in Eq. (3-188) as a quasi-monochromatic wave, and is always valid in any real experiment. We will often want to consider a linear frequency sweep for ARP experiments, and in this case we may write ϕ explicitly as $\frac{1}{2}v_s(t - t_1)^2$, where v_s is the frequency sweep rate in Hz/s. The instantaneous frequency is then $\Omega_s = \Omega + v_s(t - t_1)$.

Putting the optical field of Eq. (3-188) into the density matrix equations of motion, one obtains a set of population matrix equations for a two-level system which are virtually identical to Eqs. (3-28). The only difference is that the argument of the cosine function is now $\Omega't - kz' + \phi(z', t)$ instead of $\Omega't - kz'$. The next step is to make the rotating-wave approximation (RWA). Since $\partial\phi/\partial t \ll \Omega$, the optical field is still nearly monochromatic, and it is easy to convince oneself that the RWA is still valid. Hence, we set

$$\rho_{ab} \equiv \tilde{\rho}_{ab} e^{-i(\Omega't - kz' + \phi)} \qquad (3\text{-}190)$$

and substitute this form into the population matrix equations. Neglecting terms which oscillate at twice the optical frequency (the RWA), we find a set of equations which is identical to Eqs. (3-35) except that Ω' in the $\dot{\tilde{\rho}}_{ab}$ equation has been replaced by the instantaneous frequency $\Omega'_s = \Omega' + \dot{\phi}$. The next step is to rewrite these equations in terms of the real quantities u, v, w, and s exactly as was done in Section 3.4.1, to obtain

$$\dot{u} = \delta(t)v - \gamma_{ab}u$$

$$\dot{v} = -\delta(t)u + \kappa E_0 w - \gamma_{ab}v$$

$$\dot{w} = n_a\gamma_a - n_b\gamma_b - \kappa E_0 v - \tfrac{1}{2}(\gamma_a + \gamma_b)w - \tfrac{1}{2}(\gamma_a - \gamma_b)s \qquad (3\text{-}191)$$

$$\dot{s} = n_a\gamma_a + n_b\gamma_b - \tfrac{1}{2}(\gamma_a + \gamma_b)s - \tfrac{1}{2}(\gamma_a - \gamma_b)w$$

The only difference between these equations and Eqs. (3-67), where the frequency was not allowed to change, is that the detuning from resonance, $\delta(t) = \Omega' + \dot{\phi} - \omega_0$, is now a function of time instead of being a constant. Unfortunately, this simple change means that nearly all the solutions of the population matrix equations worked out in Section 3.4 are no longer valid, because we no longer have a set of differential equations with constant coefficients. The equations do have the same general form, however, so the vector model equation obtained by setting $\gamma_a = \gamma_b = \gamma_{ab} = 0$ is still $d\mathbf{M}/dt = \mathbf{M} \times \boldsymbol{\Omega}$ just as in Section 3.4, provided that the component of $\boldsymbol{\Omega}$ along the III axis is taken to be the instantaneous detuning $\delta(t)$ instead of the constant δ. It should also be noted that because of the way we have defined $\tilde{\rho}_{ab}$ in Eq. (3-190), the coordinate system for the present vector model can be regarded as one which is rotating at the instantaneous frequency of the optical field (see the discussion immediately preceding Section 3.4.3).

Having verified that the vector model is an appropriate description for swept optical fields, we may now make use of it to discuss ARP. Consider what happens to one velocity group of molecules during an ARP experiment. Initially, we have the molecules in thermal equilibrium and an optical field which is far below resonance. This means that the driving vector $\boldsymbol{\Omega}$ lies nearly along the $-$III axis and is almost parallel to $\mathbf{M}$. $\mathbf{M}$ therefore precesses in a very small cone about $\boldsymbol{\Omega}$, as shown in Figure 3-26a. This precession can be very rapid if the optical field is strong, because the precessional angular velocity is given by $|\boldsymbol{\Omega}| = g(t) = \sqrt{(\kappa E_0)^2 + (\Omega'_s - \omega_0)^2}$. Next suppose that we start to sweep the laser frequency toward resonance so that the angle θ between $\boldsymbol{\Omega}$ and the $-$III axis begins to increase. If the change in the direction of $\boldsymbol{\Omega}$ is very slow compared to the precession frequency of $\mathbf{M}$ about $\boldsymbol{\Omega}$, the precessional motion of $\mathbf{M}$ about $\boldsymbol{\Omega}$ will be relatively undisturbed and $\mathbf{M}$ will simply follow along with $\boldsymbol{\Omega}$ as θ changes while continuing to precess in a small cone about $\boldsymbol{\Omega}$ (see Figure 3-26b). We will later prove this statement rigorously, when a quantitative treatment is done. The fact that $\mathbf{M}$ follows along with $\boldsymbol{\Omega}$ is the key to understanding adiabatic rapid passage. We use a

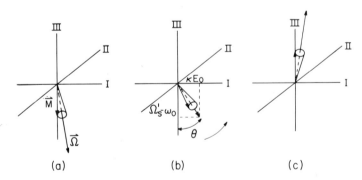

(a)   (b)   (c)

Figure 3-26. Vector model picture of adiabatic rapid passage. (a) Far below resonance $\mathbf{M}$ precesses in a small cone about $\boldsymbol{\Omega}$. (b) As the frequency is swept through resonance, the angle θ between $\boldsymbol{\Omega}$ and the $-$III axis increases. $\mathbf{M}$ continues to precess about $\boldsymbol{\Omega}$, following it along. (c) Far above resonance $\mathbf{M}$ ends up pointing nearly along the $+$III axis.

strong optical field and sweep its frequency slowly enough that the precession frequency is always much greater than $d\theta/dt$. **M** will then follow $\boldsymbol{\Omega}$ as we sweep through resonance, and if we stop the sweep far above resonance, **M** ends up pointing nearly along the +III axis, as shown in Figure 3-26c. Thus, we have inverted the level populations by sweeping through resonance.

The name adiabatic rapid passage comes, as usual, from nmr, where the analogous effect in spin systems has been observed for many years (Abragam, 1961). The word "adiabatic" is used in this name because the **M** vector follows smoothly along with $\boldsymbol{\Omega}$ as the optical field changes frequency. Notice, however, that the effect is not adiabatic in the sense that no transitions are made. In fact, every single molecule makes a transition! The phrase "rapid passage" comes from the second condition on the experiment—that the entire frequency sweep be done in a time τ_s that is short compared to the relaxation times $1/\gamma_a$, $1/\gamma_b$, or $1/\gamma_{ab}$. If the sweep were not rapid enough to meet this condition, the system would relax back toward thermal equilibrium during the sweep, and hence **M** would not necessarily remain aligned along $\boldsymbol{\Omega}$.

Mathematically, the two conditions one must meet to do an ARP experiment can be summarized by the inequalities

$$g(t) \gg \dot{\theta}, \qquad \gamma\tau_s \ll 1 \tag{3-192}$$

where τ_s is the total time required for the frequency sweep and $\gamma = \gamma_a$, γ_b, or γ_{ab}. These results can be put in a more useful form if we can express $d\theta/dt$ and τ_s in terms of $d\Omega'_s/dt$, the rate at which the frequency is swept. To do this we first look at the definition of $\boldsymbol{\Omega}$, which shows that

$$\theta = \tan^{-1}\left(\frac{\kappa E_0}{\omega_0 - \Omega'_s}\right) \tag{3-193}$$

if we write θ as a positive angle. Hence,

$$\dot{\theta} = \frac{\kappa E_0}{(\Omega'_s - \omega_0)^2 + \kappa^2 E_0^2}\frac{d\Omega'_s}{dt} \tag{3-194}$$

For a linear frequency sweep, $d\Omega'_s/dt$ is a constant and we see that $\dot{\theta}$ is largest near resonance, where $\Omega'_s - \omega_0 \approx 0$. Also, $g(t) = \sqrt{(\Omega'_s - \omega_0)^2 + \kappa^2 E_0^2}$ is smallest in this region. The limiting requirement on $g(t) \gg \dot{\theta}$ is therefore that it be satisfied at resonance, where it takes the form

$$\kappa E_0 \gg \frac{1}{\kappa E_0}\frac{d\Omega'_s}{dt} \tag{3-195}$$

Next, consider the total sweep time τ_s. For a given velocity group we want to sweep far enough that $\boldsymbol{\Omega}$ is aligned nearly along the III axis at the beginning and end of the sweep. Thus, we want an initial and final detuning which satisfies $|\Omega'_s - \omega_0| \gtrsim 5\kappa E_0$. Assuming a linear sweep rate again, this implies that $(d\Omega'_s/dt)\tau_s \gtrsim 10\kappa E_0$.

Although this expression gives the total sweep time needed for a single velocity group, one ordinarily works with Doppler-broadened transitions which contain many velocity groups having different apparent resonance frequencies. If, as is usually the case, we want to invert the entire transition and κE_0 is less than the Doppler width $k\bar{u}$, the condition on τ_s is that the total frequency sweep be on the order of $4k\bar{u}$, i.e., $(d\Omega'_s/dt)\tau_s \geqslant 4k\bar{u}$.

If the preceding expressions for τ_s are substituted into (3-192), we find that the condition $\gamma\tau_s \ll 1$ can be written as

$$\frac{1}{10\kappa E_0}\frac{d\Omega'_s}{dt} \gg \gamma, \qquad \kappa E_0 \gtrsim k\bar{u}$$

$$\frac{1}{4k\bar{u}}\frac{d\Omega'_s}{dt} \gg \gamma, \qquad \kappa E_0 \ll k\bar{u}$$

(3-196)

where $\gamma = \gamma_a,\ \gamma_b,$ or γ_{ab}.

Combining equations (3-195) and (3-196), we then have

$$(\kappa E_0)^2 \gg \frac{d\Omega'_s}{dt} \gg 10\kappa E_0\gamma;\ 4k\bar{u} \qquad (3\text{-}197)$$

This is the result we desire. It expresses in compact form the conditions on κE_0, $d\Omega'_s/dt$, and γ that must be met to obtain adiabatic rapid passage.

To get an idea of what these requirements mean in practice, let us put in some numbers. Suppose that we have a transition near $10\,\mu\text{m}$ in the infrared, and irradiate it with a 2-mm-diameter Gaussian laser beam of 20 W total power. If the transition dipole matrix element is 0.25 D, we then have $\kappa E_0 = 125$ MHz. Now $k\bar{u}$ is of the order of 500 MHz, while a typical relaxation rate for dipolar gases at low pressure (a few milliTorr) is $\gamma = 0.5$ MHz. Hence, we have

$$15{,}600 \gg \frac{d\Omega'_s}{dt} \gg 625;\ 1000$$

This shows that a frequency sweep rate $d\Omega'_s/dt$ of about 4000 MHz/μs will allow one to do an ARP experiment here. Note, however, that even with 20 W of power and a large transition-dipole matrix element, κE_0 is barely large enough. One would really like to have more power yet. This illustrates a general feature of ARP experiments: they require rather high power lasers.

Although the vector model provides a good intuitive picture of an ARP experiment, it would also be desirable to have a more rigorous mathematical treatment which, in particular, would enable one to calculate the absorption or emission signal one expects to observe during the frequency sweep. Our starting point is the set of equations for $\dot{u}$, $\dot{v}$, $\dot{w}$, and $\dot{s}$ given in Eq. (3-191). However, because the entire ARP sweep must be over in a time short compared to relaxation times ($\gamma\tau_s \ll 1$), we can ignore relaxation and set all the relaxation rates equal to zero in (3-191). This gives

$$\dot{u} = \delta(t)v$$

$$\dot{v} = -\delta(t)u + \kappa E_0 w \qquad (3\text{-}198)$$

$$\dot{w} = -\kappa E_0 v$$

These are simply the vector model equations. Because δ is a function of time, however, we cannot solve these equations in any simple and straightforward way. We could attempt a power-series solution, but this is unlikely to lead to a closed-form result, and would give little insight. Instead, we recall from our qualitative discussion of ARP the statement that the **M** vector follows along with the driving vector $\boldsymbol{\Omega}$, rotating around it in a very small cone. If this is the case, **M** is nearly always parallel to $\boldsymbol{\Omega}$, and we can then determine approximate solutions for u, v, and w by simple geometry (because u, v, and w must be proportional to the components of $\boldsymbol{\Omega}$).

Since the accuracy of the solutions obtained by such an approximation depends on how closely **M** remains aligned with $\boldsymbol{\Omega}$, let us first look at this point in more detail. What we would like to do is to prove that **M** always remains within a small angle of $\boldsymbol{\Omega}$. This proof is known in nmr as the adiabatic theorem (Abragam, 1961). To demonstrate it we transform Eqs. (3-198) into a doubly rotating coordinate frame in which the driving vector $\boldsymbol{\Omega}$ remains stationary along the III axis.* The new rotating frame is rotating at a velocity $\dot{\theta}$ with respect to the old frame, and the two frames are related by a simple rotation about the II axis through the angle θ (see Figure 3-26), i.e.,

$$\begin{pmatrix} u^r \\ v^r \\ w^r \end{pmatrix} = \begin{pmatrix} \cos\theta & 0 & \sin\theta \\ 0 & 1 & 0 \\ -\sin\theta & 0 & \cos\theta \end{pmatrix} \begin{pmatrix} u \\ v \\ w \end{pmatrix} \qquad (3\text{-}199)$$

The reverse transformation going from u^r, v^r, w^r to u, v, w has the same form but with θ replaced by $-\theta$. We can obtain differential equations for u^r, v^r, w^r by differentiating Eq. (3-199). Noting that from the definition of θ, we have

$$\cos\theta = \frac{-\delta(t)}{g(t)}, \qquad \sin\theta = \frac{\kappa E_0}{g(t)} \qquad (3\text{-}200)$$

where $g(t) = [\delta(t)^2 + \kappa^2 E_0^2]^{1/2}$, and using these equations along with Eqs. (3-198) and (3-199), we find that

$$\dot{u}^r = -g(t)v^r + \dot{\theta}w^r$$

$$\dot{v}^r = g(t)u^r \qquad (3\text{-}201)$$

$$\dot{w}^r = -\dot{\theta}u^r$$

---

* The new coordinate frame is referred to as "doubly rotating" because our original coordinate frame is usually regarded as one which is already rotating around the III axis at an angular velocity Ω'_s (see the discussion immediately preceding Section 3.4.3).

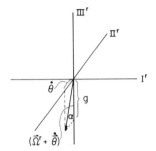

Figure 3-27. In a doubly rotating coordinate system where Ω remains fixed along the III$'$ axis, the effective driving vector is $(\Omega' + \dot{\theta})$. This vector has components $\dot{\theta}$ and g, as shown, and thus is offset from the III$'$ axis by an angle α.

If we put these equations in vector notation, we see that they can be written as

$$\dot{\mathbf{M}}' = \mathbf{M}' \times (\Omega' + \dot{\boldsymbol{\theta}}) \tag{3-202}$$

where $\dot{\boldsymbol{\theta}} = (0, -\dot{\theta}, 0)$. $\Omega' = (0, 0, -g(t))$ is just the driving vector Ω expressed in the new rotating coordinate system.

Equation (3-202) gives us the information we were looking for. Suppose that we initially start off with a fixed-frequency field far below resonance so that $\theta \simeq 0$ and $\dot{\theta} = 0$. $\mathbf{M}'$ and Ω' are then parallel, so nothing happens. Next we start to sweep the frequency so that $\dot{\theta} \neq 0$. This produces a component of the effective driving field $(\Omega' + \dot{\boldsymbol{\theta}})$ along the $-$II$'$ axis (see Figure 3-27), and hence an angle $\alpha = \tan^{-1}(\dot{\theta}/g)$ between $\mathbf{M}'$ and $(\Omega' + \dot{\boldsymbol{\theta}})$. $\mathbf{M}'$ will therefore precess about $(\Omega' + \dot{\boldsymbol{\theta}})$ in a cone of half-angle α. If we choose a sweep rate such that $\dot{\theta} =$ constant, this precession will continue undisturbed until the end of the sweep when we are far above resonance and $\theta \rightarrow \pi$. Thus, as long as we keep $\dot{\theta} \ll g(t)$, the angle α will be very small, and $\mathbf{M}'$ always remains very close to the $-$III$'$ axis (hence, $\mathbf{M}$ is always aligned nearly along Ω). This result completes the proof of the adiabatic theorem.

Having shown that $\mathbf{M}$ is always nearly parallel to Ω if $\dot{\theta} \ll g(t)$, we would now like to use this information to determine u, v, and w. However, if we assume that $\mathbf{M}$ is parallel to Ω, we find $u = |\mathbf{M}| \sin \theta$, $v = 0$, $w = -|\mathbf{M}| \cos \theta$. This is clearly an unacceptable result. Among other things it violates conservation of energy because the ground-state population ends up in the excited state during ARP, and our result predicts there is no absorption, since $v = 0$. What went wrong?

Looking back at the discussion of the adiabatic theorem, we note that while $\mathbf{M}$ does stay close to the direction of Ω, i.e., $\mathbf{M}'$ remains close to the $-$III$'$ axis, the actual behavior of $\mathbf{M}'$ is a precession about the effective driving vector $(\Omega' + \dot{\boldsymbol{\theta}})$, not the $-$III$'$ axis. Thus, it is more accurate to ignore the rapid, oscillatory precessional motion and say that $\mathbf{M}'$ is effectively parallel to the direction of $(\Omega' + \dot{\boldsymbol{\theta}})$ than it is to say that $\mathbf{M}$ is parallel to Ω. This small change makes a big difference in the final result. Taking $\mathbf{M}'$

parallel to $(\boldsymbol{\Omega}' + \dot{\boldsymbol{\theta}})$, we have

$$u' = 0$$
$$v' = -|\mathbf{M}| \sin \alpha \qquad (3\text{-}203)$$
$$w' = -|\mathbf{M}| \cos \alpha$$

where α is the angle between the effective driving vector and the $-\text{III}'$ axis as shown in Figure 3-27. By definition we have

$$\sin \alpha = \frac{\dot{\theta}}{\sqrt{g(t)^2 + \dot{\theta}^2}} \approx \frac{\dot{\theta}}{g(t)}, \qquad \cos \alpha = \frac{g(t)}{\sqrt{g(t)^2 + \dot{\theta}}} \approx 1 \qquad (3\text{-}204)$$

where the last equality in each expression follows from the fact that $\dot{\theta} \ll g(t)$. The next step is to transform back out of the doubly rotating frame by a rotation of $\mathbf{M}' = (u', v', w')$ through the angle $-\theta$. Using Eq. (3-204), this gives

$$u = |\mathbf{M}| \sin \theta$$
$$v = -|\mathbf{M}| \frac{\dot{\theta}}{g(t)} \qquad (3\text{-}205)$$
$$w = -|\mathbf{M}| \cos \theta$$

Note that while u and w are unchanged from what one obtains taking $\mathbf{M}$ parallel to $\boldsymbol{\Omega}$, we now have $v \neq 0$. The final step is to obtain an expression for $|\mathbf{M}|$. This is easily done by noting that

$$\frac{d}{dt} |\mathbf{M}| = \frac{1}{2|\mathbf{M}|} \frac{d}{dt} (u^2 + v^2 + w^2)$$
$$= \frac{1}{2|\mathbf{M}|} (2u\dot{u} + 2v\dot{v} + 2w\dot{w})$$
$$= 0 \qquad (3\text{-}206)$$

The last line follows from Eqs. (3-198) and shows that $|\mathbf{M}|$ is simply a constant. The value of this constant depends on the initial conditions at the beginning of the ARP sweep. For the purposes of this section we are interested in two types of initial conditions. The first is the situation where the system is initially in thermal equilibrium and we sweep upward starting from far below resonance. Thus far, we have always assumed that this was the case, and we then have that $|\mathbf{M}| = n_b - n_a$. There is also a second situation of interest, though, and that is the one where we have previously inverted the system with an ARP sweep, allowed the system to relax for a time, and then made a second ARP sweep. This is the technique used to make T_1 measurements. Here the $\mathbf{M}$ vector for all molecules still initially lies along the III axis, but its magnitude is now given by $|\mathbf{M}| = |\rho_{aa}(t_2) - \rho_{bb}(t_2)|$, where t_2 is the time at which the second ARP sweep is begun. We can write out this expression for $|\mathbf{M}|$ explicitly by noting that the off-resonant field is effectively absent between ARP sweeps, so that the density matrix solutions of Eq. (3-74) apply. If the delay between ARP sweeps is denoted by τ_D, then

$$|\mathbf{M}| = |\rho_{aa}(t_2) - \rho_{bb}(t_2)| = |(n_a - n_b)[1 - (e^{-\gamma_a \tau_D} + e^{-\gamma_b \tau_D})]| \quad (3\text{-}207)$$

This result follows from Eqs. (3-74) and the fact that at the end of the first ARP sweep we have $\rho_{aa}(t) = n_b$, $\rho_{bb}(t) = n_a$. The expression for $|\mathbf{M}|$ can now be substituted into Eqs. (3-205), but we need to be very careful about signs. The overall sign of the expressions for u, v, w depends on whether or not we sweep from above or below resonance and whether $\mathbf{M}$ lies initially along the +III or −III axis. A little thought will convince one that the following expressions are correct:

$$u(t) = \mp(n_a - n_b)[1 - (e^{-\gamma_a \tau_D} + e^{-\gamma_b \tau_D})]\sin\theta$$

$$v(t) = (n_a - n_b)[1 - (e^{-\gamma_a \tau_D} + e^{-\gamma_b \tau_D})]\dot\theta/g(t) \quad (3\text{-}208)$$

$$w(t) = (n_a - n_b)[1 - (e^{-\gamma_a \tau_D} + e^{-\gamma_b \tau_D})]\cos\theta$$

In these equations θ is always defined to be the positive angle such that $\theta \approx 0$ at the beginning of the sweep. The upper sign is used when the sweep starts from below resonance, and $\sin\theta$, $\cos\theta$ are then given by Eq. (3-200). The lower sign is used when the sweep starts from above resonance. In this case, $\cos\theta$ is changed to $\cos\theta = \delta(t)/g(t)$. Equations (3-208) are for a second ARP sweep which follows the first one after a time delay τ_D. The other situation where we start with a system in initially thermal equilibrium can be obtained from these equations by simply letting $\tau_D \to \infty$.

The reader may have noticed that two of the ARP sweeps being discussed here represent a possibility that we have not considered heretofore. Specifically, if we sweep from below resonance with $\mathbf{M}$ initially along the +III axis or sweep from above resonance with $\mathbf{M}$ initially along the −III axis, $\mathbf{M}$ and the driving field vector $\boldsymbol{\Omega}$ are nearly antiparallel at the beginning of the experiment. In this situation ARP still occurs just as it does when $\mathbf{M}$ and $\boldsymbol{\Omega}$ are initially parallel, except that $\mathbf{M}$ precesses in a small cone about an axis antiparallel to $\boldsymbol{\Omega}$. This is easily seen from a consideration of the vector model equation $d\mathbf{M}/dt = \mathbf{M} \times \boldsymbol{\Omega}$.

The absorption signal can now be readily calculated from the preceding expression for v provided that the equations relating v to $I(t)$ which were developed in Section 3.3 are still valid. The validity of these equations does need to be examined, however, because we are considering a swept frequency in this section. In particular, the optical field is now given by Eq. (3-188) instead of Eq. (3-23). This means that our definitions of $\mathbf{P}$ and $\mathscr{E}$ in Eq. (3-42) must have $\Omega t - kz + \phi(z, t)$ instead of $\Omega t - kz$ arguments in the sine and cosine functions. On substituting the new definitions into the wave equation (3-44), we find some terms which were not present previously.

These may be eliminated by noting the following facts. First, the instantaneous frequency and wave vector are given by $\Omega_s = \Omega + \dot{\phi}$ and $k_s = k - \partial\phi/\partial z$; hence, $k_s = \Omega_s/c$. Second, we have $\partial^2\phi/\partial z^2 - (1/c^2)\partial^2\phi/\partial t^2 = 0$, because the phase $\phi(z, t)$ is specified as part of the incident optical field. In an optically thin medium, this field obeys the homogeneous wave equation, and the equality for ϕ above follows directly when Eq. (3-188) is substituted into this wave equation. Finally, we need the fact that $\partial^2\phi/\partial t^2 \ll \Omega_s^2$, because the phase is assumed to be slowly varying compared to the optical frequency. The reader is invited to substitute the new definitions of $\mathbf{P}$ and $\mathscr{E}$ into the wave equations for himself and then use the relations just given to obtain the same amplitude wave equations (3-44) that we had for constant-frequency fields. The only difference is that k and Ω are now k_s and Ω_s. The derivation of the reduced wave equations then follows without change. The solutions, Eqs. (3-50), are also identical because Ω_s is a function of $t - t_1$ only, just like P_s and P_c [see the discussion preceding Eq. (3-49)].

Now we can calculate the absorption or emission signal for an ARP experiment. The relation between P_s and v is still given by Eq. (3-69). Combining this relation with Eqs. (3-50) and (3-53) gives

$$I(t) = I_0 + \tfrac{1}{2}\Omega_s L E_0 \mu_{ab} \int_{-\infty}^{\infty} v(t)\, dv_z$$

$$= I_0 - \frac{\mu_{ab}(N_b - N_a)\Omega_s L E_0}{2k\bar{u}\sqrt{\pi}}[1 - (e^{-\gamma_a \tau_D} + e^{-\gamma_b \tau_D})]\int_{-\infty}^{\infty} e^{-(kv_z/k\bar{u})^2} \frac{\dot{\theta}}{g(t)}\, d(kv_z)$$

$$(3\text{-}209)$$

The second equality follows when the expression for v given by Eq. (3-208) is used. Now

$$\frac{\dot{\theta}}{g(t)} = \frac{\kappa E_0}{[(\Omega_s - \omega_0 - kv_z)^2 + \kappa^2 E_0^2]^{3/2}} \frac{d\Omega_s}{dt} \qquad (3\text{-}210)$$

and if $\kappa E_0 \ll k\bar{u}$ this is a narrow function of kv_z compared to the Gaussian $\exp(-kv_z/k\bar{u})^2$. We may then pull the Gaussian factor outside the integral and evaluate it at the peak of $\dot{\theta}/g(t)$, i.e., at $kv_z = \Omega_s - \omega_0$. The integration can now be done, giving $2/\kappa^2 E_0^2$, so the transmitted intensity is

$$I(t) = I_0 - \frac{\hbar\Omega_s L(N_b - N_a)}{k\bar{u}\sqrt{\pi}} \frac{d\Omega_s}{dt}[1 - (e^{-\gamma_a \tau_D} + e^{-\gamma_b \tau_D})]e^{-(\Omega_s - \omega_0)^2/k^2\bar{u}^2}$$

$$(3\text{-}211)$$

This is the result we have been looking for. It gives the absorption or emission signal that one expects to see during an ARP sweep which follows a previous adiabatic inversion by a time delay τ_D. The signal for an ARP sweep with the system initially in thermal equilibrium is given by Eq. (3-211) with $\tau_D = \infty$.

Very often the frequency sweep in an ARP experiment is nearly linear ($d\Omega_s/dt$ = constant), and then we see from Eq. (3-211) that the observed signal reproduces in the time domain the Gaussian lineshape. The peak signal in this situation comes at $\Omega_s = \omega_0$, and at that point

$$\Delta I_{\mathrm{pk}}(\tau_D) = C(e^{-\gamma_a \tau_D} + e^{-\gamma_b \tau_D} - 1) \qquad (3\text{-}212)$$

where C is a constant. Furthermore, C is nothing other than $-\Delta I_{\mathrm{pk}}$ at $\tau_D = \infty$, so we can rewrite Eq. (3-212) as

$$\ln\left[\Delta I_{\mathrm{pk}}(\tau_D) - \Delta I_{\mathrm{pk}}(\infty)\right] = \ln\left[-\Delta I_{\mathrm{pk}}(\infty)\right] + \ln\left(e^{-\gamma_a \tau_D} + e^{-\gamma_b \tau_D}\right) \quad (3\text{-}213)$$

If $\gamma_a = \gamma_b$, a plot of $\ln\left[\Delta I_{\mathrm{pk}}(\tau_D) - \Delta I_{\mathrm{pk}}(\infty)\right]$ versus τ_D will yield a straight line, and the level lifetime (T_1) can be determined directly from the slope of the line. If $\gamma_a \neq \gamma_b$, however, the plot will give a curved line and a more sophisticated method of analysis will be needed to determine γ_a and γ_b.

Although Eq. (3-211) is only an approximate result, it should be quite accurate as long as the ARP conditions given in Eq. (3-197) are satisfied. The principal approximation was to ignore the oscillatory component of v due to the rapid precession of **M** about Ω, and take only its average value. If we had only a single velocity group of molecules present, this would mean that the absorption signal would have an oscillation superimposed on it. However, we actually have many velocity groups present, each of which has a different precessional frequency. Thus, when we sum over all these groups to get the absorption signal, the oscillation will be completely averaged away except for an initial transient at the very beginning of the sweep.

The theory presented in this section can be extended in several ways. Until now we have assumed that all the relaxation times were much longer than the total sweep time ($\gamma\tau_s \ll 1$), so that relaxation during the sweep could be ignored. Actually, however, ARP (in the sense that **M** remains approximately aligned along Ω during the sweep) will still take place even if $\gamma\tau_s \ll 1$ is not well satisfied. All that is really required is that the relaxation rates γ satisfy the inequality $\gamma \ll g(t)$; i.e., relaxation is slow compared to the precession frequency of **M** about Ω. This relation could hold even when the relaxation times are shorter than the total sweep time. The system would not end up totally inverted, of course, because relaxation occurs during the sweep, but a form of ARP will still take place. An analytical treatment of this situation is complicated because damping must be included in the equations of motion, but it has been worked out by Lehmberg and Reintjes (1975). The interested reader is referred to their work for further details.

A second direction in which ARP theory can be extended is to relax somewhat the condition $\dot{\theta} \ll g(t)$. As we have seen earlier, increasing $\dot{\theta}$ relative to $g(t)$ causes the angle between **M** and Ω to increase, so we cannot relax this condition too far without losing the entire concept of an adiabatic rapid passage. However, it would be interesting to know at what point the frequency sweep rate becomes so large relative to $g(t)$ that we are no longer able to effectively invert the system by a sweep through resonance. Although this would appear to be a very complicated problem, Horwitz (1975) has shown that a surprisingly simple treatment exists if we neglect relaxation and assume a linear frequency sweep. Working directly with the probability amplitudes c_i [see Eq. (3-8)], Horwitz utilizes a clever transformation to put the equations of motion into the form of a confluent hypergeometric equation. The solutions of this equation, while well known, are not themselves especially simple in form. However, if we do not ask for the detailed motion of the system during the sweep but only for its final state at the end of the sweep, an extremely simple answer is obtained. One finds that w_{end}, the population difference at the end of the sweep, is given by

$$w_{\text{end}} = w_{\text{start}} \left\{ 2 \exp\left[-\frac{\pi}{4} \kappa^2 E_0^2 / (d\Omega_s/dt) \right] - 1 \right\} \qquad (3\text{-}214)$$

Recalling from Eq. (3-195) that the ARP condition $\dot{\theta} \ll g(t)$ becomes $d\Omega_s/dt \ll \kappa^2 E_0^2$ for a linear frequency sweep, we see that in the limit of $\dot{\theta}$ negligible compared to $g(t)$, the exponential in Eq. (3-214) goes to zero and we have complete inversion, $w_{\text{end}} = -w_{\text{start}}$. As $\kappa^2 E_0^2$ becomes smaller compared to $d\Omega_s/dt$, however, we can use Eq. (3-214) to quantitatively determine the departure from complete inversion. For example, when $\kappa^2 E_0^2 = 3(d\Omega_s/dt)$, we see that $w_{\text{end}} = -0.81 w_{\text{start}}$. While one might not say that this situation satisfies the ARP condition $\kappa^2 E_0^2 \gg d\Omega_s/dt$ very well, we nonetheless can do a fairly good job of inverting the system with this sweep.

Before leaving our theoretical discussion of ARP, it is worth noting that there exists a substantial body of closely related literature dealing with the interaction between molecules and intense near-resonant light pulses. The subject is commonly known as "adiabatic following" because the **M** vector follows along with Ω just as in ARP. The principal difference is that in adiabatic following the field is usually assumed not to go through resonance or even closely approach it. In an optically thin medium such as we are concerned with in this chapter, this situation would lead to no effects of great interest. However, in an optically thick medium, some very unusual and interesting effects can occur when the conditions for adiabatic following are satisfied. These include resonantly enhanced self-focusing and defocusing, pulse self-steepening, optical shock formation, and spectral broadening due to self-phase modulation. The bulk of the experimental work on these effects as well as a considerable amount of theory has been done by D. Grischkowsky and his colleagues (see Grischkowsky, 1975). Excellent theoretical contributions have also been made by M. D. Crisp (1973).

We now turn to a discussion of the adiabatic-rapid-passage experiments done to date. The first ARP experiments were, of course, done in nmr many years ago (see Abragam, 1961). In the optical region, however, the first ARP experiment was not attempted until 1969, when Treacy and De Maria reported the observation of ARP in both SF_6 and NH_3. They used a special Q-switched CO_2 laser which produced 1-μs pulses as the source. It was found that these pulses were chirped with a frequency sweep of ~60 MHz and hence could be used as is for an ARP experiment. To observe ARP, the pulsed laser beam was split into two beams, only one of which passed through the sample. The beams were then monitored by two identical detectors, with the difference between detector outputs giving the absorption or emission produced by the sample. Evidence that ARP was actually occurring was provided by the fact that when a sequence of pulses was passed through the sample in a short time compared to the relaxation time, the first pulse showed absorption by the sample, while the second pulse gave an emission signal. No quantitative results were reported, however. There is also some question about whether ARP was really observed, at least in NH_3, because later spectroscopic work showed that there are no NH_3 transitions close enough to a CO_2 laser line to allow one to sweep the CO_2 laser frequency through a resonance without using a Stark field.

The first really clean ARP experiments in which quantitative results were obtained was done by M. M. T. Loy in 1974 on NH_3 gas. Loy utilized an ingenious combination of pulsed laser and Stark-switching techniques to observe ARP. To obtain the high laser intensity needed for ARP, he used a pulsed CO_2 laser giving a 5-μs-long pulse with a peak intensity of about 300 W/cm^2. The output of this laser did not have an appreciable frequency sweep, however, so Loy passed the pulse through a Stark cell and produced ARP by sweeping a transition frequency through resonance with an electric field applied across the cell. The pulse duration was sufficiently long that two ARP experiments, with an adjustable time delay τ_D between them, could be performed during a single pulse. Figure 3-28a shows the CO_2 laser pulse (with ARP signal superimposed on it) in the upper trace and the Stark voltage pulse in the lower trace. The frequency sweep occurs during the rise and fall times of the Stark pulse, and these times were adjusted so as to produce the proper sweep rates. By changing the Stark pulse length, the time delay τ_D can be varied. The Stark pulse amplitude was 2000 V/cm, giving a frequency sweep of ~800 MHz for the NH_3 $(\nu,J,K,M) = (0_s,5,5,\pm5) \to (1_a,5,5,\pm5)$ transition that was used.

By plotting the log of the ARP signal versus τ_D as discussed following Eq. (3-213), Loy found that the data could be fit rather well by a straight line (over the measurable range of delay times 0.2 to 1.2 μs). This would seem to indicate simple exponential decay. Repeating this for several different pressures, the level decay rate was found to be $\gamma = (0.75 \pm 0.035P)$ MHz, where P is the NH_3 pressure in milliTorr. The pressure-independent part of γ is due to transit-time effects.

Figure 3-28. Adiabatic rapid passage in NH_3. (a) Upper oscilloscope trace shows part of the CO_2 laser pulse with the ARP signal (small blips near the peak of the pulse) superimposed. Lower trace shows the Stark pulse applied to the plates of the sample cell. (b) ARP signals for several delay times τ_D, shown on an expanded time scale (from Loy, 1974; reproduced by permission).

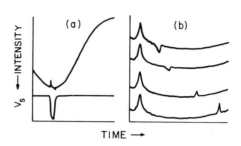

The result above bears further discussion, because it illustrates very clearly a pitfall which is likely to be encountered when attempting to measure level decay rates. From the experimental data as they stand, one is tempted to conclude that the upper- and lower-state level decay rates are equal. However, as Loy realized, this conclusion is not necessarily correct. If ln (ARP signal) gives a straight line from $\tau_D = 0$ to $\tau_D = \infty$, the decay is a single exponential. Unfortunately, the limited signal-to-noise ratio in an experiment confines one to a finite range of delay times, and one often cannot tell the difference between a single exponential and a double exponential over this range. The data can look like a straight line in either case, so additional information is needed. In Loy's experiment, additional information was available in the form of a microwave measurement of the ground-state relaxation time, which shows that $\gamma_b = (0.75 \pm 0.176P)$ MHz, a collisional relaxation rate which is some five times larger than the rate measured in the ARP experiment. We can understand this result mathematically by analyzing Loy's data using Eq. (3-213) with the value of γ_b given above substituted into it. This analysis shows that over the range of τ_D used in the ARP experiment, the measurement is sensitive primarily to the upper-state decay rate γ_a, and, in fact, γ_a is approximately equal to the rate γ measured in the experiment. The physical explanation for the large difference between γ_a and γ_b is that in the excited state the ν_2 vibrational motion (the NH_3 inversion mode) is so rapid that it averages away most of the dipole–dipole interaction which takes place during a collision (Loy, 1974). This gives a very small excited-state decay rate.

Since Loy's ARP experiment in 1974, there has been one other noteworthy ARP experiment. This was an attempt by Javan and his colleagues to do an ARP experiment *within* a Doppler-broadened line (Hamadani *et al.*, 1975). They had a very interesting idea. In previous ARP experiments, one had to sweep over the entire Doppler line in a short time compared to the relaxation time. This requires a high sweep rate and hence very high laser powers. It was this fact that forced Loy to use a pulsed CO_2 laser. If one only had to sweep over a few homogeneous linewidths instead of the entire Doppler linewidth, however, the sweep rates could be much lower, and lower-power cw lasers could be used. This is a tremendous experimental advantage.

Although it is simple enough to sweep over a few homogeneous widths and invert that part of the Doppler line, the relevant question is how to detect and/or make use of this inversion. One cannot simply make a second ARP sweep back over that same portion of the line, because very large transient effects occur at the beginning and end of each ARP sweep (the laser frequency is always within the Doppler linewidth) and completely mask the absorption or emission signal occurring during ARP. Instead, to detect the ARP, Hamadani *et al.* used two optical fields. The experimental arrangement is essentially that used for Lamb dip experiments. One cw laser beam is passed through the sample and its frequency is swept by moving one of the laser mirrors with a PZT. The second beam, a weak probe beam, is produced by splitting off a small portion of the first beam before it enters the cell. This probe beam is then sent through the cell in the *opposite* direction from the strong beam and is monitored by a detector. Consider now what happens when the strong beam frequency is initially somewhat below the Doppler line center frequency and we sweep through the line center to a point equally far above the line center. As the sweep is made, this beam successively inverts the population of resonant velocity groups, starting with $v_z = -v_0$ and continuing through $v_z = 0$ to $v_z = +v_0$. What does the probe beam see during this time? Since it is very weak, it simply gives a small absorption signal which is proportional to the population difference. It starts out resonant with molecules of velocity $+v_0$ because it is propagating in the $-z$ direction. During the first half of the sweep it passes through resonance with molecules having $v_z = +v_0$ to $v_z = 0$ and is just looking at molecules in thermal equilibrium. At $v_z = 0$, however, it starts to see molecules which have been inverted by the strong field, and thus the absorption signal suddenly changes into an emission signal. During the second half of the sweep, the probe beam is monitoring velocity groups between $v_z = 0$ and $v_z = -v_0$ which have previously been inverted by the strong beam. The signal during this period is the one of interest because it is proportional to the population difference for the previously inverted molecules. These have relaxed during the delay between the passage of the strong beam and the probe beam through their resonant frequency, so the probe beam signal measures the relaxation rate of the population difference.

This technique appears to be a feasible one for measuring level decay rates, although in practice it does have certain disadvantages and complications. The signal, for example, can be quite small, because we are looking for a small change in intensity of a weak probe beam. Also, since the total frequency sweep generally covers a substantial portion of the Doppler curve, both the laser power and the steady-state absorption vary during the sweep. This can complicate the data analysis. In the experiment of Hama-dani *et al.*, there was the added complication that a transition degenerate in M_J was being observed and some of the $M_J \rightarrow M'_J$ components did not have large-enough martix elements to satisfy the ARP conditions. As a result, they were able to obtain only semiquantitative measurements. It is hoped that further refinements in the technique will allow more accurate measurements to be made.

Before leaving the subject of ARP, it may be worthwhile to briefly mention two proposals for ARP experiments which have recently been made. The first, by Nebenzahl and Szöke (1974), is to use the momentum change $\hbar\Omega/c$ that is produced in an atom during ARP to do isotope separation. In the proposed apparatus, each resonant atom makes a number of successive adiabatic rapid passages, and all together these can produce a substantial momentum change. The experiment would be done inside a laser cavity, which allows one to recirculate the "used photons" and hence has the potential for being quite energy-efficient. A second very interesting future experiment has been described by Grischkowsky and Loy (1975). They show that in a two-photon transition, it is possible to use the ac Stark effect (also known as the dynamic or optical Stark effect) to obtain a "self-induced adiabatic rapid passage." This effect has no analog in the one-photon transitions that we have been considering. In essence, the experiment consists of sending two oppositely directed laser beams through a sample with frequencies such that the sum frequency is nearly resonant with a two-photon transition. This two-photon transition appears as a narrow Doppler-free resonance. Furthermore, because of the ac Stark effect, the frequency of this resonance depends on the laser *intensities*. Hence, laser pulses with the proper time-dependent intensity can, in fact, produce ARP by simultaneously being absorbed and causing the two-photon transition to sweep itself through resonance with the lasers. This looks like a very attractive experiment which hopefully will be done in the near future.

3.9.2. Delayed Optical Nutation

We now turn our attention to a second very successful method for measuring level decay rates, delayed optical nutation. As the name implies, the phenomenon of optical nutation is used in this type of experiment. The use is not direct, however, in the sense of using the decay of the optical nutation signal to measure decay rates. As discussed in Section 3.6, this is not a viable technique. Instead, the approach is similar in philosophy to the ARP method discussed in the previous section. We begin by noting that the peak height of the initial absorption spike in an optical nutation experiment (i.e., the peak value during the first half-cycle of nutation) is just directly proportional to the initial population difference $(N_b - N_a)$ [see Eq. (3-114)]. This suggests that if we disturb a system from thermal equilibrium somehow, let it relax for a period of time, and then suddenly shift back into resonance to produce an optical nutation signal, the height of the initial spike might provide a measure of the population difference existing at that time. We will prove shortly that this is, in fact, the case.

Given that the initial nutation signal can be used to monitor population differences at whatever time we choose, there remains the question of how we want to initially disturb the system from thermal equilibrium. Since these experiments have all been done using a Stark- or frequency-switching apparatus, the method chosen has been one that is very easy to implement on this type of apparatus—pulsed excitation. With pulsed excitation the delayed nutation experiment becomes very simple. We switch some velocity group into resonance, thus exciting it. We stay in resonance just long enough that the molecules experience a 180° pulse, and then switch back out of resonance. After waiting a delay time τ_D, during which the excited molecules begin to relax back toward equilibrium, we again switch into resonance and observe the height of the initial absorption spike which occurs. By repeating the experiment for different delay times τ_D, we can measure the decay rates of the level populations.

The pulse sequence used for the delayed nutation experiment just described is shown in Figure 3-29a. A second type of pulse sequence has also been used for these experiments, and is shown in Figure 3-29b. Here the velocity group of interest is excited using steady-state absorption, i.e., using a very long pulse. We then switch out of resonance for a delay time τ_D and back again as before. This type of pulse sequence (sometimes called one-pulse delayed nutation) does not drive the system from equilibrium as far as the two-pulse sequence of Figure 3-29a, because at best the population difference can only be driven to zero; it cannot be inverted. However, it has

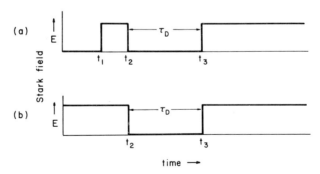

Figure 3-29. Pulse sequences and timing used for delayed nutation experiments. (a) Two-pulse delayed nutation. (b) One-pulse delayed nutation.

the advantage of being very simple to implement experimentally since no fancy pulse sequences are needed; an ordinary pulse generator will do.

Consider next a quantitative treatment of delayed nutation. Looking for the moment only at the short pulse excitation case shown in Figure 3-29a, we see that up to time $t = t_3$, the pulse sequence looks exactly like that used in an optical free-induction decay (FID) experiment. Hence, $u(t_3)$ and $v(t_3)$ are just given by the FID equations (3-129) evaluated at $t = t_3$. $w(t_3)$ can be obtained as follows. At the end of the first pulse, $w(t_2)$ can be found from Eq. (3-108); while $s(t_2)$ is just $n_a + n_b$, since s remains constant during a short pulse. Taking the sum and difference of these results, we find that

$$\rho_{aa}(t_2) = n_a + \tfrac{1}{2}(n_b - n_a)\frac{\kappa^2 E_0^2}{g^2}[1 - \cos g(t_2 - t_1)]$$

$$\rho_{bb}(t_2) = n_b - \tfrac{1}{2}(n_b - n_a)\frac{\kappa^2 E_0^2}{g^2}[1 - \cos g(t_2 - t_1)]$$

(3-215)

Equations (3-215) can now be substituted into Eqs. (3-74), which describe the behavior of the populations when the optical field is either zero or far off resonance. From these equations we obtain

$$w(t_3) = \rho_{aa}(t_3) - \rho_{bb}(t_3)$$

$$= (n_b - n_a)\left\{\frac{\kappa^2 E_0^2}{2g^2}[1 - \cos g(t_2 - t_1)](e^{-\gamma_a \tau_D} + e^{-\gamma_b \tau_D}) - 1\right\} \qquad (3\text{-}216)$$

where $\tau_D = t_3 - t_2$.

When we shift back into resonance at $t = t_3$, the behavior of the system is found from $\mathbb{M}(t) = \exp[-\mathbb{B}(t - t_3)]\mathbb{M}(t_3)$, where the exponential matrix is given by Eq. (3-93). As usual, we need only calculate $v(t)$, because the absorption signal when the laser field is on depends only on P_s [see Eqs. (3-50)–(3-53)]. Writing out $v(t)$ explicitly, we have

$$v(t) = -\frac{\delta}{g}\sin g(t - t_3)u(t_3) + \cos g(t - t_3)v(t_3) + \frac{\kappa E_0}{g}\sin g(t - t_3)w(t_3)$$

(3-217)

The next step is to integrate $v(t)$ over all velocities v_z to obtain P_s [see Eq. (3-69)]. The resulting integrals are quite complicated and, unfortunately, cannot be done analytically. An investigation of these integrals by numerical integration does reveal a great simplification, however, In the usual Stark switching situation, where we have κE_0 much less than the Doppler width $k\bar{u}$, the Gaussian factor $\exp(-v_z^2/\bar{u}^2)$ can be pulled outside the integrals, and we then find that

$$\int_{-\infty}^{\infty}\left[-\frac{\delta}{g}\sin g(t - t_3)u(t_3) + \cos g(t - t_3)v(t_3)\right]dv_z = 0 \qquad (3\text{-}218)$$

as long as $\tau_D > t_2 - t_1$; i.e., the delay between pulses is larger than the first pulse width. Hence,

$$P_s = -\mu_{ab}\kappa E_0 \int_{-\infty}^{\infty} \frac{\sin g(t-t_3)}{g} w(t_3) \, dv_z \qquad (3\text{-}219)$$

when $\tau_D > t_2 - t_1$.

The result shown in Eq. (3-218) is crucial if one wants to use delayed nutation to measure level decay rates, because $u(t_3)$ and $v(t_3)$ decay at the rate γ_{ab}, not at the level decay rates γ_a and γ_b. Hence, it would be desirable to understand why it occurs. Mathematically, it is not readily apparent why the integral vanishes, although a reformulation of the problem in a Fourier transform domain might do the trick. Physically, however, it is not too hard to see why one should expect a null result for the integral. Referring back to the discussion of optical free induction decay in Section 3.7, we note that following the initial pulse a large polarization P_s is present [see Eqs. (3-128)–(3-132)]. This is a result of the fact that the pulse causes the **M** vectors for the various excited velocity groups to have components in the I–III plane which are initially clustered about the II axis. During the delay between pulses the vectors dephase in the I–II plane, and at $\tau_D = t_2 - t_1$ the polarization vanishes and remains zero thereafter. What happens is that the amount of dephasing, $\delta\tau_D$, becomes so large that **M** vectors for molecules in closely neighboring velocity groups develop large phase differences in the I–II plane. In particular, phase differences on the order of π develop between velocity groups whose resonant frequency difference is small compared to the total bandwidth of excited molecules. This means that for every molecule whose **M** vector gives a positive contribution to P_s, there is another molecule of slightly differing velocity, which was excited nearly identically but whose **M** vector phase differs by π. The contributions from these pairs of molecules simply cancel each other, and we are left with no net polarization. Now consider what happens when the pulse comes on again at t_3 and the **M** vectors start to precess about the effective field Ω. As **M** vectors for the pairs of molecules just mentioned precess, they are excited nearly identically again, and this means that the contributions to $v(t)$ from their u and v components existing at t_3 will still just cancel each other. Hence, the integral in Eq. (3-218) gives zero.

Since the $u(t_3)$ and $v(t_3)$ terms do not contribute to the polarization, we are left with the simple expression of Eq. (3-219) for P_s. The transmitted intensity can now be found immediately using Eqs. (3-50) and (3-53). We have

$$I(t) = I_0 + \tfrac{1}{2}\kappa^2 E_0^2 \hbar\Omega L \int_{-\infty}^{\infty} \frac{1}{g} \sin g(t-t_3) w(t_3) \, dv_z \qquad (3\text{-}220)$$

On substituting in the explicit expression for $w(t_3)$ given in Eq. (3-216), this becomes

$$I(t) = I_0 - \frac{\kappa^2 E_0^2 \hbar \Omega L \sqrt{\pi}}{2k\bar{u}}(N_b - N_a)e^{-(\Omega-\omega_0)^2/k^2\bar{u}^2}$$

$$\times \{J_0[\kappa E_0(t-t_3)] - f(t-t_3)(e^{-\gamma_a \tau_D} + e^{-\gamma_b \tau_D})\} \qquad (3\text{-}221)$$

where

$$f(t-t_3) = \frac{\kappa^2 E_0^2}{2\pi} \int_{-\infty}^{\infty} \frac{1}{g^3} \sin g(t-t_3)[1 - \cos g(t_2-t_1)]\, d(kv_z)$$

We have assumed here that $\kappa E_0 \ll k\bar{u}$ so that the Gaussian factor $\exp(-v_z^2/\bar{u}^2)$ could be pulled outside the integral. The integral denoted by $f(t-t_3)$ has been evaluated by Schenzle et al. (1976), who found it to be a sum of three generalized hypergeometric series.

In a delayed nutation experiment our interest is in the peak value of the signal produced following t_3, not in its detailed shape. Let us denote the time at which this peak occurs as $t - t_3 = t_{pk}$. The signal we want to measure is then the peak intensity change, $I_0 - I(t_{pk})$, as a function of the delay time between pulses, τ_D. Calling this quantity $S(\tau_D)$, we have

$$S(\tau_D) = I_0 - I(t_{pk})$$

$$= C[J_0(\kappa E_0 t_{pk}) - f(t_{pk})(e^{-\gamma_a \tau_D} + e^{-\gamma_b \tau_D})] \qquad (3\text{-}222)$$

where C is the collection of constants appearing in Eq. (3-221). Noting that $S(\infty)$ is just $CJ_0(\kappa E_0 t_{pk})$, we can rewrite this as

$$S(\infty) - S(\tau_D) = Cf(t_{pk})(e^{-\gamma_a \tau_D} + e^{-\gamma_b \tau_D}) \qquad (3\text{-}223)$$

This expression is the most convenient one to use for data analysis. We simply look at how much the nutation peak following t_3 is reduced in height as compared to the peak signal value obtained for an infinite delay. This is illustrated in Figure 3-30. As indicated in the figure, $S(\infty)$ is usually obtained from a measurement of the peak absorption during the first pulse. This is a perfectly legitimate way to measure $S(\infty)$, provided that the first pulse length is longer than the rise time needed to reach the peak signal and provided that the time between repetitions of the delayed nutation experiment is long compared to the molecular relaxation times. Obviously, this measurement of $S(\infty)$ can be checked by making sure that $S(\infty) - S(\tau_D) \to 0$ for large delays τ_D. By plotting the logarithm of $S(\infty) - S(\tau_D)$ versus τ_D, γ_a and γ_b can be determined just as for ARP [see the discussion following Eq. (3-213)].

One point that has been glossed over in the treatment just presented is a discussion of the exact time t_{pk} at which the peak signal $S(\tau_D)$ occurs. This is

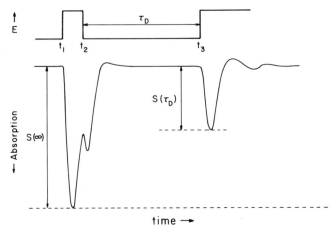

Figure 3-30. Two-pulse delayed nutation. The signal of interest is the difference $S(\infty) - S(\tau_D)$ as a function of the delay time τ_D.

not a trivial point, because an evaluation of Eq. (3-221) as it stands predicts that delayed nutation cannot be used to measure decay rates! To see how this can be, note that the signal in Eq. (3-221) is the difference of the two terms. The first term, $J_0[\kappa E_0(t - t_3)]$, reaches its peak value of unity almost immediately (a few nanoseconds after $t = t_3 = 0$), as we saw in Section 3.6. The second term, however, is virtually zero for small $t - t_3$ and does not reach a maximum until $\kappa E_0(t - t_3) \approx \pi/2$ (typically 50–100 ns). Further-more, even for very short delays τ_D and optimum first pulse areas of $\kappa E_0(t_2 - t_1) \approx 0.8\pi$, the second term has a maximum value of only about 0.5. Thus, Eq. (3-221) predicts that the peak signal will occur at very short $t - t_3$ and be almost solely due to the first term, $J_0[\kappa E_0(t - t_3)]$.

Fortunately, the researchers (Schmidt *et al.*, 1973) who did the first delayed nutation experiments were at that time unaware of the theoretical treatment presented above. They simply did the experiment and found that it worked very well indeed! The reason the technique works is quite simple. In the delayed nutation experiments to date, the detectors have not been ultrafast devices. Instead, they have been bandwidth-limited to about 8 MHz, giving a detector rise time which is typically 50 ns. Hence, the observed signal is not given by Eq. (3-221) but rather that expression convoluted with the detector's response function. This convolution reduces the height of the fast-rise-time J_0 term and also causes the peak signal to occur at $t - t_3 \geq 50$ ns, i.e., at times for which the second term has reached a large value. As a result, the experiment does indeed work and can be used to measure level decay rates.

There is one other complication present in a delayed nutation experi-ment which should be examined. These experiments have always been done using Stark or frequency switching *inside* a Doppler-broadened line. This means that there will be a second velocity group of molecules which is resonant during the delay time τ_D and then is suddenly switched out of resonance at $t = t_3$. These molecules will emit a free induction decay signal which occurs at the same time as the delayed nutation signal that one wants to observe. This FID signal appears as a heterodyne beat with the laser, and the first cycle of this beat could contribute to the peak signal observed just after t_3.

Fortunately, the FID is not usually a problem. First, one often can make the Stark or frequency shift large enough that the FID signal occurs at a frequency which is outside the bandwidth of the detector. In this case, the FID signal will be very small. Second, the initial FID signal as a function of τ_D is, to a first approximation, independent of collisional decay rates. This can be seen by examining the expressions that were derived in Section 3.7. Specifically, we note that the FID signal of interest is proportional to the polarization P_s evaluated at short times after the end of the excitation pulse (which has length τ_D). For short τ_D we then have $P_s = C[1 - J_0(\kappa E_0 \tau_D)]$ from Eq. (3-131), and for very long τ_D we have $P_s = C$ from Eq. (3-134). The latter expression is obtained by using the fact that $\kappa E_0 \gg \gamma_{ab}$ in a delayed nutation experiment. C is a constant factor which is independent of collisional decay rates. From these two limiting cases it is clear that the FID signal amplitude (the initial amplitude of the heterodyne beat signal) is an oscillatory function of τ_D and damps out to a constant initial amplitude for long τ_D. Thus, even if the first cycle of the FID does contribute to the delayed nutation measurement, we will merely see a small oscillation superimposed on the delayed nutation signal. Such oscillations have indeed been observed in this author's laboratory. The important point to note is that they do not affect the pressure-dependent decay of the signal, and thus do not prevent accurate measurement of γ_a and γ_b.

Thus far we have examined in some detail the two-pulse delayed nutation experiment shown in Figure 3-29a. The analysis of the one-pulse delayed nutation experiment shown in Figure 3-29b is very similar, the only difference being the fact that the initial excitation pulse is infinitely long in the one-pulse experiment. To treat this case we may use the steady-state results of Eq. (3-27), which give

$$u(t_2) - iv(t_2) = (n_a - n_b)\frac{\kappa E_0}{\delta^2 + \Gamma^2}(\delta - i\gamma_{ab})$$

$$\rho_{aa}(t_2) = n_a + \tfrac{1}{2}(n_b - n_a)\frac{\gamma_{ab}}{\gamma_a}\frac{\kappa^2 E_0^2}{\delta^2 + \Gamma^2} \qquad (3\text{-}224)$$

$$\rho_{bb}(t_2) = n_b - \tfrac{1}{2}(n_b - n_a)\frac{\gamma_{ab}}{\gamma_b}\frac{\kappa^2 E_0^2}{\delta^2 + \Gamma^2}$$

where

$$\Gamma^2 = \frac{1}{2}\left(\frac{1}{\gamma_a}+\frac{1}{\gamma_b}\right)\gamma_{ab}\kappa^2 E_0^2 + \gamma_{ab}^2$$

The behavior of the system during the delay τ_D is again given by Eqs. (3-72) and (3-74), and then Eq. (3-217) can be used to find the quantity of interest, $v(t)$, for $t > t_3$. As in the two-pulse experiment, however, $u(t_3)$ and $v(t_3)$ do not contribute significantly to the polarization P_s provided that the delay τ_D is longer than the time required for the FID signal following t_2 to disappear. In the present case this dephasing time is $(\gamma_{ab}+\Gamma)^{-1}$ [see the discussion following Eq. (3-135)], and one always uses delay times longer than this in an actual experiment. Thus, the $u(t_3)$ and $v(t_3)$ contributions are negligible, and Eq. (3-219) may be used to obtain P_s. The transmitted intensity can then be found from Eq. (3-220). Assuming that $\kappa E_0 \ll k\bar{u}$, we have

$$I(t) = I_0 - \frac{\kappa^2 E_0^2 \hbar \Omega L \sqrt{\pi}}{2k\bar{u}}(N_b - N_a)e^{-(\Omega-\omega_0)^2/k^2\bar{u}^2}$$

$$\times \left\{J_0[\kappa E_0(t-t_3)] - h(t-t_3)\left(\frac{\gamma_{ab}}{\gamma_a}e^{-\gamma_a\tau_D} + \frac{\gamma_{ab}}{\gamma_b}e^{-\gamma_b\tau_D}\right)\right\} \quad (3\text{-}225)$$

where

$$h(t-t_3) = \frac{\kappa^2 E_0^2}{2\pi}\int_{-\infty}^{\infty}\frac{\sin g(t-t_3)}{g(\delta^2+\Gamma^2)}d(kv_z)$$

This equation gives the signal one expects to see in a one-pulse delayed nutation experiment. It differs from the two-pulse result of Eq. (3-221) in several respects. The integral $h(t-t_3)$, which must be evaluated numerically, has a peak value which is nearly a factor of 2 smaller than the peak value attainable for $f(r-t_3)$. This means that the signal in a one-pulse delayed nutation experiment is smaller than in a two-pulse experiment. The temporal behavior of $h(t-t_3)$ and $f(t-t_3)$ is virtually identical, however, so the comments made earlier regarding the timing of t_{pk} apply to both experiments. A more important difference between one- and two-pulse delayed nutation is the somewhat altered relaxation behavior one expects to observe in the two experiments. Defining $S(\tau_D)$ and C as in Eq. (3-222), we have

$$S(\infty) - S(\tau_D) = Ch(t_{pk})\left(\frac{\gamma_{ab}}{\gamma_a}e^{-\gamma_a\tau_D} + \frac{\gamma_{ab}}{\gamma_b}e^{-\gamma_b\tau_D}\right) \quad (3\text{-}226)$$

for one-pulse delayed nutation. Comparing this to Eq. (3-223) we see that when $\gamma_a \neq \gamma_b$, the one-pulse experiment is more sensitive to the slower of the two decay rates, whereas the two-pulse experiment weights both decay rates equally. It may be possible to take advantage of this difference in making a measurement of γ_a and γ_b. A final difference between one- and two-pulse delayed nutation has to do with measuring $S(\infty)$. In the two-pulse case we mentioned that this could be conveniently done by measuring the height of the excitation pulse nutation signal at t_1. In the one-pulse experiment no such signal exists, and one must simply measure $S(\tau_D)$ for very long delay times to obtain $S(\infty)$.

We now turn to a consideration of the delayed nutation experiments which have been done to date. The technique was first developed by R. G. Brewer and his coworkers (Schmidt *et al.*, 1973). They used a cw CO_2 laser as the source and studied two-pulse delayed nutation in $^{13}CH_3F$ using Stark switching. In a subsequent paper (Berman *et al.*, 1975) further details of the experiment were given and the one-pulse delayed nutation technique was introduced. A level decay rate for the $^{13}CH_3F$ $(\nu_3, J, K) = (0,4,3) \rightarrow (1,5,3)$ transition was measured and found to be $\gamma = (0.080 + 0.089P)$ MHz, where P is the pressure in milliTorr. By doing other experiments involving photon echoes and coherent Raman beats they also concluded that in the $^{13}CH_3F$ transitions studied, $T_1 = T_2$, i.e., $\gamma_a = \gamma_b = \gamma_{ab}$. These experiments will be discussed in Section 3.10.

A second study made by Brewer and Genack (1976) used a frequency-switched dye laser to obtain level decay rates for an electronic transition in $^{127}I_2$ [$(\nu, J) = (2,59) \rightarrow (15,60)$ in the $X^1\Sigma_g^+ \rightarrow B^3\Pi_{0^+u}$ band]. For this experiment the analysis of delayed nutation presented in this section is incomplete because appreciable spontaneous emission can occur from the excited state to the ground state. As mentioned in Section 3.2, the pumping terms in the density matrix equations must be modified to treat this situation. A proper theoretical treatment can be found in Schenzle and Brewer (1976). We have generally ignored spontaneous emission because it is unimportant in the infrared, where most of the experiments have been done to date. In the visible and uv, however, it often cannot be ignored. In the I_2 transition that was studied, for example, spontaneous emission between the levels was estimated to be about 10% of the total decay rate. The level decay rates for I_2 were found to be roughly equal with a value of $\gamma = (0.71 + 0.029P)$ MHz. This result is about 10 times larger than the I_2 collisional decay rates found in fluorescence measurements. The discrepancy results from the fact that fluorescence is generally insensitive to vibrational- or rotational-state-changing collisions within the upper (excited electronic)-state manifold.

Delayed nutation measurements have also been done in this author's laboratory on NH_2D and $^{15}NH_3$ (Van Stryland, 1976). Figure 3-31 shows a two-pulse delayed nutation experiment on NH_2D using Stark switching and a CO_2 laser source. In this figure the nutation signals observed for a number of different delay times τ_D have been superimposed to show the relaxation behavior of the signal. The decay of $S(\infty) - S(\tau_D)$ toward zero with increasing τ_D is clearly evident. The results of the NH_2D measurements are

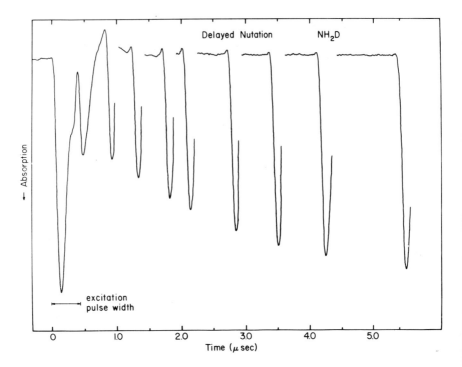

Figure 3-31. Two-pulse delayed nutation in NH_2D. The signals for a succession of experiments having different delay times have been superimposed in order to show the relaxation behavior. The excitation Stark pulse for each experiment is identical, having a width as marked in the figure.

interesting. We found that $\gamma_a \neq \gamma_b$ and were able to determine both relaxation rates from the delayed nutation experiments. This had to be done somewhat indirectly, because the decay of the signal in any one experiment looks like a single exponential over the limited range of τ_D accessible. However, if the observed decay rates are plotted as a function of pressure, we find a *curved* line and a least-squares fit to this line, taking into account the range of τ_D used in determining each point, yields the two decay rates. For NH_2D $[(\nu_2,J,M) = (0_a,4_{04},\pm4) \rightarrow (1_a,5_{05},\pm5)]$ we find that $\gamma_a = (0.07 + 0.063P)$ MHz and $\gamma_b = (0.07 + 0.12P)$ MHz. Note that the upper- and lower-state decay rates differ by about a factor of 2 (ignoring the 0.07-MHz contribution from transit-time effects). This is roughly what one would expect in light of Loy's ARP measurements on NH_3. He found that the upper-state decay rate in NH_3 was a factor of 5 slower than the ground state and interpreted this as being due to an averaging away of the dipole–dipole interaction by the NH_3 inversion. Our results support this interpretation, since in NH_2D the inversion frequency in $\nu_2 = 1$ is about a factor of 2 slower than in NH_3. Thus, the averaging out of the dipole–dipole interaction should not be as complete, giving an upper-state decay in NH_2D that is relatively faster than in NH_3, just as observed.

3.10. Velocity-Changing Collisions

3.10.1. Introduction

In this section we discuss a type of collisional decay channel which has previously been neglected, velocity-changing collisions (VC collisions). What is meant by this term is a collision which alters only the colliding molecule's velocity, leaving its internal quantum state completely unchanged. In a hard collision where the molecule changes its quantum state, the molecular velocity will also change, of course, but such collisions simply result in a loss of population from the energy levels of interest. We do *not* refer to such a collision as a VC collision.

Although we have not taken VC collisions into account previously, it would seem at first glance to be straightforward to include such effects in the theoretical framework. One merely adds in additional pumping and decay terms to the equations of motion to account for the additional collisional relaxation channel. Thus, although the actual solution of these equations may be complicated (because VC collisions will couple the equations of motion for differing molecular velocity groups), it may seem that the formal introduction of VC collisions, at least, does not appear to present a problem. However, Berman and Lamb (1970) have pointed out that this is not the case. Unless we put specific restrictions on the types of collisional interactions which can occur in the gas, a simple addition of VC collisions into the theory we have developed thus far is fundamentally wrong.

The source of the trouble lies in our initial assumption that the translational motion of the molecules could be treated classically. Specifically, we assumed that a density matrix ρ' could be associated with each molecule, and that each element of ρ' was a function of the classical position and velocity of the molecule. We then summed over molecules having the same velocity v_z to obtain a population matrix ρ, whose elements refer to all molecules which are at position $\mathbf{R}$ at time t and have velocity v_z. Following Berman and Lamb, let us now examine carefully what can happen during a collision to make this simple picture break down. Suppose that we have some molecule A with two energy levels a and b such that the intermolecular potential between A and a collision partner M is different depending on whether the molecule A is in state a or state b. This is just the type of situation which normally gives rise to unequal level decay rates, $\gamma_a \neq \gamma_b$.

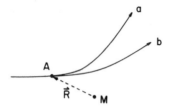

Figure 3-32. Two possible trajectories for the collision of a molecule A with a perturber molecule M. If A enters the collision in state a, it will interact with M through a potential $U_a(\mathbf{R})$ and follow trajectory a. If A enters in state b, it will interact through $U_b(\mathbf{R})$ and follow trajectory b. If A enters the collision in a superposition of states, this classical picture is no longer well defined.

Clearly, the classical trajectory of A during the collision will depend on which state molecule A is in, as shown in Figure 3-32. If A is in level a at the start of the collision, the interaction potential is some function $U_a(\mathbf{R})$ and will cause the molecule to move along the classical trajectory marked a. Associated with this trajectory is a corresponding velocity change $\Delta\mathbf{v}_a$. If, on the other hand, the molecule begins the collision in level b, a different intermolecular potential $U_b(\mathbf{R})$ will be present, causing the molecule to change its velocity by $\Delta\mathbf{v}_b$ and move along trajectory b. Now suppose that we irradiate the transition $b \rightarrow a$ with a laser in order to observe some coherent transient effect. The laser will cause some molecules to develop a nonzero off-diagonal density matrix element ρ'_{ab}, and as a result these molecules will undergo collisions while they are in a coherent superposition of levels a and b. The problem is now evident. If molecule A in Figure 3-32 is in a superposition of states, what is its trajectory after the collision? There is no classical answer to this question, and one is therefore forced to abandon the classical picture in which molecules have a well-defined velocity and trajectory following a collision! Instead the collision must be regarded as a quantum mechanical scattering process in which the center-of-mass motion of the molecule is quantized. Of course, if the collision is hard enough to cause the molecule to change its state, we do not care what its velocity is after the collision because it has left the active levels. The trouble occurs for weaker "phase-interrupting collisions," where the state of the molecule is unchanged and only the phase of ρ'_{ab} (i.e., the phase of the oscillating dipole moment) is changed. In this case one simply cannot say that collisions may result in both a change in the dipole's phase *and* a change in velocity. A classical treatment of the translational motion in effect assumes this, whereas the proper quantum-mechanical treatment shows that the phase-interrupting and velocity-changing aspects of a collision are so completely entangled that it is impossible to regard them as separate processes (Berman, 1975). Thus, in general it is just wrong to treat phase-interrupting and velocity-changing collisions as separate and distinct processes as one must do when translation is treated classically. Despite this, it is still tempting to do the calculations classically because the quantum-mechanical ones are so hard. Unfortunately, calculations using this approach can give predictions which do not agree with experiment (Berman, 1975).

In light of the discussion above, the prospects for a simple treatment of velocity changing collisions may not look very bright. Note, however, that the problems arise only when both velocity-changing and phase-interrupting collisions can occur. For many (perhaps most) transitions, this is not the case and one finds that one of these processes will completely dominate the other. It will be well worth our while to examine the circumstances under which this can happen, because a much simpler theoretical treatment can then be used. First, we note that the argument used to show the necessity for a quantum treatment of translation is applicable only when the collisional interaction in the upper and lower states is different. If the trajectories a and b in Figure 3-32 differ by only a negligible amount, a classical treatment of the translation will be perfectly adequate. More precisely, what we require for a classical picture to hold is that $k \, \Delta v$, the change in Doppler shift produced by the collision, be so nearly the same for the two trajectories that

$$|\mathbf{k} \cdot (\Delta \mathbf{v}_a - \Delta \mathbf{v}_b) T| \ll 1 \qquad (3\text{-}227)$$

Here $\mathbf{k}$ is the wave vector for the incident radiation and T is the lifetime of the molecule in either of the two levels a or b. If this relation holds, there will be no difference between the two trajectories as far as any interaction with the radiation is concerned. In particular, the resonant frequency $\omega_0 + kv_z$ will not depend on which state the molecule was in during the collision.

There is one further implication for transitions which have a nearly equal collisional interaction for states a and b. Not only are we able to treat VC collisions classically, but we also expect that phase-interrupting collisions (PI collisions) will be unimportant. The basis for this expectation is the following simple physical picture of how PI collisions occur. During a PI collision the colliding molecules approach each other fairly closely, interacting primarily through the electromagnetic field that each molecule produces at the location of the other. These fields cause the molecular energy levels to shift during the collision, and, if the upper and lower levels of a transition shift by different amounts, the resonance frequency will change also. The result is that a molecule with $\rho'_{ab} \neq 0$ will have the frequency of its oscillating dipole moment changed during the collision. After the collision this frequency will return to its original value but with an altered phase, because of the fact that the oscillation frequency has been shifted for a finite amount of time.

According to this argument, transitions which have similar collisional interactions in the upper and lower states will not exhibit PI collisions, because both the upper and lower levels will experience similar energy shifts, leading to little change in resonant frequency and hence little change in the phase of ρ'_{ab}. A very large number of transitions satisfy the criterion of having nearly equal upper- and lower-state interactions. This includes most transitions which occur within the same electronic state, i.e., fine-structure transitions in atoms and rotational and rotation–vibration transitions in molecules. In practice this means that one does not expect to see PI collisions for infrared transitions, and the experiments done to date bear this out. The only clear observations of PI collisions have been made on electronic transitions which generally lie in the visible and uv.

In molecules there is a further consideration which tends to suppress PI collisions—the relative ease with which state-changing collisions can occur. Any collision hard enough to appreciably affect the phase of ρ'_{ab} tends to also be strong enough to change the rotational state of the molecule. We can illustrate this with a simple calculation. In a typical infrared transition, ρ'_{ab} oscillates at 5×10^{13} Hz, while a collision typically lasts about 10^{-12} s. Therefore, only about 50 cycles of the ρ'_{ab} oscillation occurs during the collision, and in order to change the phase of ρ'_{ab} appreciably, the resonant frequency will need to be shifted by about 1%, or 5×10^{11} Hz. This shift is considerably larger than most rotational energy-level spacings, and any collision producing such a large shift will almost certainly change the rotational state. As a result, PI collisions are not observed in molecules even when one might expect to see them. An example is NH_2D, where PI collisions do not seem to be observed (Van Stryland, 1976) even though the level decay rates and collisional interactions are different in the upper and lower states, as we saw in Section 3.9.2.

We have just seen that for a very large class of transitions, VC collisions can be treated classically, and PI collisions will not occur. There is also a second type of limiting case where the general theory requiring quantized translation can be simplified. This is the situation where the collisional interactions in the upper and lower states are so different that PI collisions dominate the decay of ρ_{ab} and velocity-changing collisions are not observable. One can see how this result occurs by again referring to Figure 3-32. Suppose that $U_a(\mathbf{R})$ and $U_b(\mathbf{R})$ differ so much that trajectories a and b go off in substantially different directions. In this case quantum mechanics tells us that $\rho'_{ab} = 0$ after the collision. The simplest way to understand this is to notice that the collision partner M is doing exactly what the state selector magnet does in a Stern–Gerlach experiment. It deflects molecule A into one of two different directions, depending on which state A is in. If A enters the collision in a superposition of states, the collision will force it into one state or the other, provided that the trajectories a and b are spatially separated by a large enough distance. In particular, this separation must be large enough to prevent any spatial overlap of the quantum-mechanical wave packets which describe the state a and state b trajectories of the molecule. A collision satisfying the criterion above will destroy ρ'_{ab} and leave the molecule in either state a or state b. We choose to call such a collision a PI collision despite the fact that here the dipole is destroyed rather than simply having its phase altered. This generalization is useful experimentally because in both cases the collision only contributes to γ_{ab} without affecting the level decay rates γ_a or γ_b. For experiments which monitor ρ_{ab}, it should also be clear that the effects of the velocity change occurring in such a collision will be unobservable because ρ'_{ab} simply disappears.

From the preceding discussion, we can deduce the circumstances under which we expect to see PI collisions dominate the observed decay. We need a state-dependent collisional interaction which is sufficiently selective to force $\rho'_{ab} = 0$ any time we have a collision hard enough to change the molecular velocity. This condition is probably not rigorously met for any real transition, because there will always be collisions at large impact parameters where the collisional interaction for both states is weak. Nonetheless, it appears to be reasonably well satisfied for many atomic transitions and possibly for some molecular electronic transitions as well (Brewer and Genack, 1976). When PI collisions do dominate, $\gamma_{ab} \gg \gamma_a, \gamma_b$ and the decay of ρ_{ab} is simply an exponential, as we have assumed throughout the earlier sections of this chapter (Berman, 1975). In particular, the two-pulse echo experiment will show single exponential decay and allow measurement of γ_{ab}.

The decay of ρ_{aa} and ρ_{bb} is another story, however. These quantities are unaffected by PI collisions and therefore will be sensitive to velocity-changing collisions. In a delayed nutation experiment, for example, where only part of the Doppler line is excited, the collisional decay will not necessarily be that given in Section 3.9.2. Excitation of only a portion of the line disturbs the velocity distribution from thermal equilibrium, and during the delay between pulses this distribution will start to relax back to equilibrium via velocity-changing collisions. In molecules this relaxation is generally very slow compared to γ_a and γ_b and can be neglected. We discuss this point in more detail below. Relaxation of the velocity distribution for atoms is not generally as slow, however, primarily because collisions with large velocity changes can occur without causing a change in the state of the atom. In this case the relaxation behavior for delayed nutation and stimulated echo experiments will have to be modified. Since our interest in this chapter is mainly in infrared molecular transitions, we will not pursue this subject any farther.

At this point it may be worthwhile to stop briefly and recap what we have discovered so far in this section. We found that a general treatment of collisions can be very complicated when the possibility of velocity changes is included. Fortunately, most transitions can be adequately described using one of two limiting forms of the general theory. In one case, generally applicable to atoms, one has a strongly state-dependent collisional interaction. This causes phase-interrupting collisions to dominate the decay of ρ_{ab} while velocity-changing collisions are not observed. In the opposite limit, the collision is not strongly state-dependent, a situation which is typical of molecular infrared transitions. Here phase-interrupting collisions are negligible and velocity-changing collisions can be treated classically.

Throughout the rest of Section 3.10 we will be concerned exclusively with the latter case. In the remainder of this subsection, we introduce classical VC collisions quantitatively into the theory. This is followed by a discussion of the general approach to solving the equations of motion, and finally by a treatment of VC collision effects in specific experiments. As we shall see, these effects are generally observable only in photon echo experiments.

The inclusion of VC collisions in the population matrix equations of motion can be done in a rather simple and intuitive manner, provided one keeps in mind that the ultimate justification for such an approach lies in the fact that the result agrees with that obtained from a rigorous (but unfortunately very complicated) theoretical development. We begin by noting that the original population matrix equations of motion, Eqs. (3-20), which we obtained in Section 3.2.1, can be written in the form

$$\frac{\partial \rho_{ij}}{\partial t} = \left(\frac{\partial \rho_{ij}}{\partial t}\right)_{\text{no coll}} + \left(\frac{\partial \rho_{ij}}{\partial t}\right)_{\text{coll}} \tag{3-228}$$

where $(\dot{\rho}_{ij})_{\text{no coll}}$ is the time rate of change of ρ_{ij} that would occur in the absence of any collisions (or other pumping and decay mechanisms), and

$$\left(\frac{\partial \rho_{ij}}{\partial t}\right)_{\text{coll}} = \lambda_i(v_z)\delta_{ij} - \gamma_{ij}\rho_{ij} \tag{3-229}$$

is the rate of change if ρ_{ij} is due to collisions. This separation of $\dot{\rho}_{ij}$ into collisional and noncollisional contributions can always be made, provided that we assume binary collisions and the impact approximation (Berman, 1972). Notice that Eq. (3-229) has an extremely simple and direct physical meaning. $\lambda_i(v_z)$ is just the rate (molecules/s/cm^3) at which molecules with velocity v_z enter level i due to collisions. Thus, the collisional contribution to $\dot{\rho}_{ii}$ is the collisional pumping rate at which molecules in velocity group v_z enter level i minus the collisional decay rate, $\gamma_{ii}\rho_{ii}$, at which they leave level i. We now want to include the effects of VC collisions into the population matrix equations. In the limit of state-independent collisional interactions that we are considering, we found that VC collisions could be treated classically as a distinct, independent collisional process which results in nothing other than a change of velocity. In light of the present discussion, it therefore seems quite clear that VC collisions should be included by simply adding another pumping and decay rate into $(\dot{\rho}_{ij})_{\text{coll}}$. On doing this, we have

$$\left[\frac{\partial}{\partial t}\rho_{ij}(z, v_z, t)\right]_{\text{coll}} = \lambda_i\delta_{ij} + \int_{-\infty}^{\infty} W(v_z' \to v_z)\rho_{ij}(z, v_z', t)\, dv_z' - (\Gamma_{vc} + \gamma_{ij})\rho_{ij} \tag{3-230}$$

Here we have expressed the VC collision pump rate in terms of $W(v'_z \to v_z)$, the probability per unit time that molecules in velocity group v'_z will undergo a VC collision which takes them from v'_z to the new velocity v_z. The total pump rate for ρ_{ij} has exactly the form one would expect. It is the probability of a $v'_z \to v_z$ collision weighted by the value of ρ_{ij} for each group v'_z and integrated over all velocities v'_z. $W(v'_z \to v_z)$ is usually called a collision kernel, because all detailed information about the velocity-changing collisions are contained in it. Note that there is one important difference between the VC collision pump rate and the state-changing collision pump rate $\lambda_i \delta_{ij}$. For off-diagonal elements, $\lambda_i \delta_{ij}$ is zero, whereas the VC pump rate is not. This is because state-changing collisions are assumed to be incoherent; i.e., they destroy any off-diagonal density matrix elements that may have been present in a molecule before the collision. VC collisions, on the other hand, do not affect the value of any density matrix elements associated with a molecule; they only change its velocity. Hence, the off-diagonal population matrix element ρ_{ab} will be pumped by VC collisions. Equation (3-230) can also be derived rigorously by doing a fully quantum-mechanical treatment of the collision process and taking the appropriate classical limit for the case of a state-independent collisional interaction (Berman, 1972).

The decay rate due to VC collisions, Γ_{vc}, can also be expressed in terms of a probability for VC collisions. If $W(v_z \to v'_z)$ is the probability per unit time for a VC collision going from v_z to v'_z, we clearly have

$$\Gamma_{vc} = \int_{-\infty}^{\infty} W(v_z \to v'_z)\, dv'_z \qquad (3\text{-}231)$$

The probability $W(v_z \to v'_z)$ can be related to the probability $W(v'_z \to v_z)$ which appears in Eq. (3-230) by using the principle of detailed balance. This principle says that in thermal equilibrium, the number of molecules going from velocity group v'_z to v_z via a VC collision must be equal to the number going from v_z to v'_z. Hence,

$$W(v'_z \to v_z) n_i(v'_z) = W(v_z \to v'_z) n_i(v_z) \qquad (3\text{-}232)$$

where $n_i(v_z)$ is the thermal equilibrium population of level i having velocity v_z.

Having formally introduced VC collisions into the theory, we can now write down the population matrix equations of interest for a two-level system. Unfortunately, these equations will be very complicated to solve because the $\dot{\rho}_{ab}(v_z)$ equation, for example, is coupled not only to the $\dot{\rho}_{aa}(v_z)$ and $\dot{\rho}_{bb}(v_z)$ equations but also to the $\dot{\rho}_{ab}(v'_z)$ equations for all other velocity groups v'_z. However, we can simplify matters somewhat by restricting our attention to the case where the optical field is either zero or far off resonance, so that the coupling between $\dot{\rho}_{ab}$ and $\dot{\rho}_{aa}$ or $\dot{\rho}_{bb}$ disappears. This zero-field case will be sufficient to treat nearly all situations of interest, because the experiments one uses to measure relaxation times (such as photon echoes and delayed nutation) consist of short pulses separated by delays where the optical field is zero or off resonance. We treated this type of situation previously by ignoring relaxation during the short pulses and using zero-field solutions in between. The same treatment can be used when VC collisions are present by just changing the zero-field solutions appropriately. We will not consider the effect of VC collisions on phenomena such as optical nutation where the full set of coupled population matrix equations must be solved. VC collision effects are expected to be small for these phenomena and are unlikely to provide much useful information.

The zero-field equations can be written down immediately using Eqs. (3-28) with $\kappa E_0 = 0$, and Eqs. (3-228) and (3-230). We find that

$$\dot{\rho}_{aa}(v_z) = n_a \gamma_a + \int W(v'_z \to v_z)\rho_{aa}(v'_z)\, dv'_z - (\Gamma_{vc} + \gamma_a)\rho_{aa}(v_z)$$

$$\dot{\rho}_{bb}(v_z) = n_b \gamma_b + \int W(v'_z \to v_z)\rho_{bb}(v'_z)\, dv'_z - (\Gamma_{vc} + \gamma_b)\rho_{bb}(v_z) \qquad (3\text{-}233)$$

$$\dot{\rho}_{ab}(v_z) = -i\omega_0\rho_{ab}(v_z) + \int W(v'_z \to v_z)\rho_{ab}(v'_z)\, dv'_z - (\Gamma_{vc} + \gamma_{ab})\rho_{ab}(v_z)$$

As we did in Section 3.2.1, it is useful to transform the $\dot{\rho}_{ab}$ equation into a form involving the slowly varying amplitude $\tilde{\rho}_{ab}$ defined in Eq. (3-34). Since $\Omega' t - kz' = \Omega t - kz$, this transformation is identical for all velocities v_z, and we find that

$$\dot{\tilde{\rho}}_{ab}(v_z) = i\delta(v_z)\tilde{\rho}_{ab}(v_z) + \int W(v'_z \to v_z)\tilde{\rho}_{ab}(v'_z)\, dv'_z - (\Gamma_{vc} + \gamma_{ab})\tilde{\rho}_{ab}(v_z)$$
$$(3\text{-}234)$$

where $\delta(v_z) = \Omega - \omega_0 - kv_z$. Equations (3-233) and (3-234) are the basic equations we need to solve to determine the effects of VC collisions on transient-effect experiments.

3.10.2. Brownian Motion and Velocity-Changing Collisions

Looking at Eqs. (3-233) and (3-234), we note that each equation contains an unknown collision kernel $W(v'_z \to v_z)$. Clearly, we must specify this kernel in order to solve the equations explicitly. The question is, how to go about doing this. It would be desirable to calculate the kernel from first principles. However, intermolecular potentials are sufficiently complicated that there is little hope of using them as a starting point to obtain a mathematically tractable kernel. Instead, we take a phenomenological approach which relies on making a few reasonable assumptions to obtain a functional form for the kernel. Experimental results and numerical calculations using known intermolecular potentials can then be used to check its validity.

The question before us is: What do we know about VC collisions that might tell us something about the form of the collision kernel? The first thing that comes to mind is the fact that VC collisions are *weak* collisions; i.e., we do not expect a VC collision to result in a large change in velocity, especially for dipolar molecules. This implies that $W(v'_z \to v_z)$ is a highly localized function centered nearly on the original velocity v'_z. The velocity that the molecule had before the collison tends to persist and $W(v'_z \to v_z)$ will be very small for large changes $v'_z - v_z$. Now VC collisions are by no means the only phenomena in physics where a weak collision model is suitable. The Brownian motion problem, for example, is another case where this type of collision model is applicable. In fact, there is a very close analogy between Brownian motion and VC collisions. In Brownian motion, a heavy particle is suspended in a fluid and moves about randomly as a result of collisions with the much lighter particles making up the fluid. Because the fluid particles are so light, any single collision will change the massive particle's velocity only a little and the probability that a collision will take it from $\mathbf{v}'$ to $\mathbf{v}$ is a highly localized function centered somewhere near $\mathbf{v}'$, just as in the VC collision case.

We can exploit the analogy between Brownian motion and VC collisions by making use of the existing Brownian motion literature. In particular, we can use a beautiful paper by Keilson and Storer (1952), who derive an explicit collision kernel using essentially one assumption. Their reasoning, adapted to the case of VC collisions, is described below.

Clearly, we must begin by assuming something about the properties of $W(v'_z \to v_z)$. One very reasonable starting point is to assume that the *functional form* of $W(v'_z \to v_z)$ is independent of the initial velocity v'_z. Since, by definition, $W(v'_z \to v_z)$ depends only on the initial and final velocities v'_z and v_z, our assumption amounts to saying that W depends only on the change in velocity $v_z - v'_z$, not the absolute velocities. Thus, we are tempted to write W as a function of the single variable $\Delta v_z = v_z - v'_z$. This is a bit too restrictive, however, because W turns out to be an even function. If we just use Δv_z as the variable, we are implying that a collision is equally likely to cause an increase or a decrease in v'_z. This is incorrect, because a molecule moving along with velocity v'_z is obviously more likely to have this velocity component decreased by the collision than increased. Keilson and Storer chose to take this into account by writing W as a function of the variable $v_z - \alpha v'_z$ rather than $v_z - v'_z$. Here α is a constant whose value is somewhat less than unity. As a result of including α, we will find a probability of going from $v'_z \to v_z$ which is centered about $\alpha v'_z$ instead of v'_z. Hence, on the average, a collision will cause a decrease in v'_z.

We now have our starting point. We assume that $W(v_z' \to v_z)$ can be written as $W(v_z - \alpha v_z')$. Having done this, Keilson and Storer then show that it is the *only* assumption that one needs to make. If W is a function of $v_z - \alpha v_z'$ only, the principle of detailed balance [Eq. (3-232)] can be used to completely determine the form of the collision kernel! We will not repeat their derivation here. We simply give the result, which is that W must have the form $W(v_z' \to v_z) = A_0 \exp\left[-(v_z - \alpha v_z')^2/\Delta u^2\right]$, where $\Delta u^2 \equiv (1 - \alpha^2)\bar{u}^2$ and A_0 is a constant. On substituting this functional form into Eq. (3-231), which defines Γ_{vc}, we find that the VC collisional decay rate is a constant, $\Gamma_{vc} = A_0 \Delta u \sqrt{\pi}$. Note that the prediction of a velocity independent Γ_{vc} follows rigorously from our initial assumption. We can now write out the collision kernel explicitly. It is*

$$W(v_z' \to v_z) = \frac{\Gamma_{vc}}{\Delta u \sqrt{\pi}} e^{-(v_z - \alpha v_z')^2/\Delta u^2} \tag{3-235}$$

We see here that the Keilson–Storer kernel has a very simple form. It is just a Gaussian centered at $\alpha v_z'$ with width $2\,\Delta u$. In molecules, one expects $k\,\Delta u$ to be a very small fraction of the Doppler width $k\bar{u}$, and this is verified experimentally. Hence, α is very close to unity and we may write

$$\Delta u^2 \equiv (1 - \alpha^2)\bar{u}^2 \simeq 2(1 - \alpha)\bar{u}^2 \tag{3-236}$$

Strictly speaking, Δu is $\sqrt{2}$ times the rms change in velocity per collision, as one can readily verify from the definition of the kernel, Eq. (3-235). For the sake of brevity, however, we will often refer loosely to Δu as being the average change in velocity per collision.

It is quite gratifying to have obtained a simple analytic collision kernel from essentially a single assumption about the velocity independence of the form of the collision kernel. The validity of the kernel, however, will rest on its agreement with experiment. Unfortunately, there is not as yet sufficient evidence to decide how accurate the Keilson–Storer kernel is. We will show below that the predicted photon echo decay is in good agreement with experiments done to date. Such experiments do not provide a very stringent test of the kernel, though, largely because only the integral of $W(v_z' \to v_z)$ enters the equations of motion. As a result, nearly any smooth, sharply peaked function one uses for $W(v_z' \to v_z)$ will give similar results. This is reassuring in one sense, because it means that the general form of results obtained with the Keilson–Storer kernel are likely to be correct even if the kernel itself is not quite accurate. In the rest of the section, we will therefore use the Keilson–Storer kernel to examine the role of VC collisions on transient-effect experiments.

---

*In their original paper, Keilson and Storer derived a three-dimensional collision kernel $W(\mathbf{v}' \to \mathbf{v})$. The result in the text is easily obtained from this kernel by integrating $W(\mathbf{v}' \to \mathbf{v})$ over all final velocity components v_x and v_y.

We begin by looking at the $\dot{\rho}_{aa}$ and $\dot{\rho}_{bb}$ equations, Eqs. (3-233), which describe the behavior of the level populations in a transient experiment. Before starting the actual analysis, it may be worthwhile to consider qualitatively what type of behavior to expect. Our primary interest is in experiments where we excite the system, generally with a pulse, and then watch it relax back to equilibrium. Thus, an experiment begins with the excitation of some velocity bandwidth Δd of molecules within the Doppler line. Now, we noted earlier that the average velocity change Δu in a VC collision was very small. In fact, in experiments done to date, Δu is very small compared to the bandwidth of excited molecules. This tells us what to expect regarding the relaxation of the level populations. If $\Delta u \ll \Delta d$, a single VC collision simply cannot cause an appreciable amount of relaxation. The initial pulse has, crudely speaking, burned a large hole into the Doppler line. VC collisions will try to fill this hole, but because Δu is very small, it takes many collisions to produce velocity changes comparable to the hole width and thus fill it. This means that in general, state-changing collisions will completely swamp out VC collision effects in a transient experiment except in the special case of photon echoes.

With this simple physical picture in mind, we can now examine the decay of ρ_{aa} and ρ_{bb} quantitatively. To begin with we note that the equations for ρ_{aa} and ρ_{bb} have the same form, so we need consider only one of them, say ρ_{aa}. We can simplify this equation by defining a quantity $d_{aa}(z, v_z, t)$ which represents the deviation of the level population having velocity v_z from thermal equilibrium:

$$d_{aa} = \rho_{aa} - n_a \tag{3-237}$$

On substituting this definition into Eq. (3-233) and then using Eqs. (3-231) and (3-232), we find that

$$\dot{d}_{aa}(v_z) = \int W(v_z' \to v_z) d_{aa}(v_z') \, dv_z' - (\Gamma_{vc} + \gamma_a) d_{aa}(v_z) \tag{3-238}$$

Since we want to concentrate on VC collisions here, let us also eliminate the effects of γ_a by setting

$$d_{aa} = \tilde{d}_{aa} e^{-\gamma_a t} \tag{3-239}$$

Substituting this into (3-238) and multiplying by $\exp(\gamma_a t)$, we find that

$$\dot{\tilde{d}}_{aa}(v_z) = \int W(v_z' \to v_z) \tilde{d}_{aa}(v_z') \, dv_z' - \Gamma_{vc} \tilde{d}_{aa}(v_z) \tag{3-240}$$

This equation contains only the effects of VC collisions on level populations. Equation (3-240) has the form of a linearized Boltzmann or transport equation and is a well-known equation in the Brownian motion literature. Keilson and Storer (1952) present a solution of this equation using the kernel of Eq. (3-235). However, the solution is quite complicated and takes the form of an infinite series. A much simpler treatment can be obtained by noting that, as discussed earlier, the velocity width of the kernel (Δu) is very narrow compared to the bandwidth of molecules one excites with the initial pulse. This means that $\tilde{d}_{aa}(v'_z)$ is a slowly varying function of velocity compared to $W(v'_z \to v_z)$ and hence it is useful to expand $\tilde{d}_{aa}(v'_z)$ in Eq. (3-240) in a power series about $v'_z = v_z$ [i.e., near the peak of $W(v'_z \to v_z)$]. Hence,

$$\dot{\tilde{d}}_{aa}(v_z) = \int W(v'_z \to v_z) \sum_{n=0}^{\infty} \frac{1}{n!} \frac{\partial^n \tilde{d}_{aa}(v_z)}{\partial v_z{}^n} (v'_z - v_z)^n \, dv'_z - \Gamma_{\text{vc}} \tilde{d}_{aa}(v_z) \tag{3-241}$$

Notice that we no longer have an infinite set of coupled differential equations. The power-series expansion has eliminated the coupling. Because $W(v'_z \to v_z)$ is nonzero over only a small velocity bandwidth compared to the total bandwidth of excited molecules, the first few terms in the expansion should be sufficient to give a good approximation of the integral. On putting the collision kernel, Eq. (3-235), into Eq. (3-241) and keeping the first three terms, we find that

$$\dot{\tilde{d}}_{aa} = \Gamma_{\text{eff}} \tilde{d}_{aa} + \Gamma_{\text{eff}} v_z \frac{\partial \tilde{d}_{aa}}{\partial v_z} + D \frac{\partial^2 \tilde{d}_{aa}}{\partial v_z{}^2} \tag{3-242}$$

where $\Gamma_{\text{eff}} = \Gamma_{\text{vc}}(1-\alpha)$ and $D = \Gamma_{\text{vc}} \Delta u^2/4$. In obtaining this equation we have used the fact that $\Delta u \ll \bar{u}$ and hence that α is very nearly 1 to simplify Γ_{eff} and D. Equation (3-242) is a very well known type of equation called a Fokker–Planck equation. This equation describes systems which approach equilibrium via a diffusive process. The effective relaxation rate is Γ_{eff} and D is a diffusion constant.

Berman (1974) has examined the validity of retaining only the first three terms of the series in some detail. He finds that if the width of the collision kernel Δu and the bandwidth of excited molecules Δd satisfy

$$\frac{\Delta u^2}{\Delta d^2} \ll 1 \tag{3-243}$$

then the Fokker–Planck equation (3-242) is a good approximation to the linearized Boltzmann equation (3-240). Now the outstanding feature of systems described by a Fokker–Planck equation is that they relax toward equilibrium on a time scale that is very long compared to the collision times; i.e., it takes a large number of collisions for the system to relax an appreciable amount. One can most easily see this by examining the solution of Eq. (3-242). Suppose that at time $t = 0$ we have excited some bandwidth of molecules Δd within the Doppler line. For simplicity, let us assume that excitation pulse is Gaussian so that the distribution of excited molecules is a

Gaussian function of frequency (rather than the usual $(\sin v_z/v_z)^2$ distribution that would be obtained by exciting the system with a rectangular pulse of radiation). At $t = 0$ we then have

$$\tilde{d}_{aa}(v_z, 0) = \frac{C}{\Delta d \sqrt{\pi}} e^{-(v_z - v_0)^2/\Delta d^2} \tag{3-244}$$

where v_0 is the velocity group excited exactly on resonance and C is a constant. We want to solve Eq. (3-242) subject to this initial condition. The solution of the Fokker–Planck equation, while not trivial, is well known and can be found in many articles (see Berman, 1974). Therefore, we will not give the details of the solution but simply state the result, which is

$$\tilde{d}_{aa}(v_z, t) = \frac{C}{s \sqrt{\pi}} \exp\left[-(v_z - e^{-\Gamma_{\mathrm{eff}}t}v_0)^2/s^2\right] \tag{3-245}$$

where

$$s^2 = [\bar{u}^2(1 - e^{-2\Gamma_{\mathrm{eff}}t}) + \Delta d^2 e^{-2\Gamma_{\mathrm{eff}}t}]$$

We see here that the velocity distribution excited by the pulse decays to a thermal equilibrium distribution in a very simple manner. At all times the distribution is a Gaussian, centered at $\exp(-\Gamma_{\mathrm{eff}}t)v_0$ and having width s. The width is Δd at $t = 0$ and increases at a rate $2\Gamma_{\mathrm{eff}}$ toward the equilibrium width $\bar{u}$.

This result quantitatively verifies our earlier statements that it takes many collisions to cause appreciable relaxation. The VC collision rate is Γ_{vc}, whereas the rate at which the excited velocity distribution relaxes toward equilibrium is $\Gamma_{\mathrm{eff}} = \Gamma_{\mathrm{vc}}(1 - \alpha)$. Because $1 - \alpha = \Delta u^2/2\bar{u}^2$ is typically very small (10^{-4} or less), it can take hundreds of VC collisions to produce an appreciable change in $\tilde{d}_{aa}$.

Having obtained a solution for $\tilde{d}_{aa}$, we could now go back and write out the full solutions for $\rho_{aa}(t)$ and $\rho_{bb}(t)$, including the effects of VC collisions. However, the experiments to date show that the VC collision rate Γ_{vc} is of the same order of magnitude as the level decay rates γ_a and γ_b. This means that Γ_{eff} is completely negligible compared to γ_a and γ_b and may be ignored. Thus, we conclude that in a transient-effect experiment, *the observable decay rate of ρ_{aa} or ρ_{bb} will be unaffected by VC collisions*. In particular, this means that VC collision effects will not appear in experiments which measure level decay rates (except for stimulated echo experiments). The theory already given in Section 3.9 provides a complete description of the observable level decays.

The conclusion that VC collisions may be neglected depends, of course, on the validity of the assumptions we made in the course of the analysis. Looking back, we see that essentially two assumptions were made. The first is that VC collisions are sufficiently weak that the average velocity change per collision (Δu) is very small compared to both the thermal velocity $\bar{u}$ and the velocity bandwidth of molecules excited by a pulse, Δd. The second assumption is that Γ_{vc} is not orders of magnitude larger than the level decay rates. Both assumptions can be verified by experiment, in particular by photon echo experiments, as discussed in the next section. One final point to note is that the conclusions we have drawn do not depend on the fact that we used the Keilson–Storer collision kernel. Its use allowed us to obtain convenient analytic results, but any sharply peaked kernel with a narrow width would give similar behavior. Only the details would change.

3.10.3. Photon Echoes and Velocity-Changing Collision Measurements

Much of the discussion in the preceding section is based on the assumption that the VC collision rate is not orders of magnitude greater than other collisional decay rates and that the average change in velocity per collision was very small. This leads to the conclusion that experiments such as delayed optical nutation are insensitive to VC collision effects and cannot easily be used to verify our assumptions. How, then, can we measure Γ_{vc} and Δu? The answer is to use photon echo experiments. In this section we discuss quantitatively the effect of VC collisions on two-pulse photon echo experiments and then go on to show how the results of these experiments may be combined with other results to obtain values for Γ_{vc} and Δu.

Since we have just shown in the preceding section that the effective relaxation rate $\Gamma_{eff} = \Gamma_{vc}(1 - \alpha)$ due to VC collisions is extremely small compared to γ_a or γ_b (and hence to γ_{ab}), the assertion that photon echoes are sensitive to VC collisions may be surprising. However, if we recall how a photon echo is formed (see discussion in Section 3.8.1), it is easy to see how VC collisions can destroy the echo with a decay rate Γ_{vc} rather than Γ_{eff} and thus significantly affect the observed echo decay. In an echo experiment, oscillating dipoles induced in the molecules by the initial pulse dephase by an amount $kv_z\tau$ in the interval τ between the first and second pulses. Ideally, the second pulse then changes the amount of dephasing to $\pi - kv_z\tau$, so that at a time t after the second pulse, each molecule has dephased by $\pi - kv_z\tau + kv_zt$. Thus, at time $t = \tau$ after the second pulse, all the molecules will have the same phase and an echo will occur. Now suppose that VC collisions occur during the experiment. For simplicity, consider a molecule which undergoes a collision immediately after the second pulse, changing its velocity from v_z to $v_z + \Delta v_z$. The relative phase for this molecule at time t is $\pi - kv_z\tau + k(v_z + \Delta v_z)t$, so that at $t = \tau$ its phase is $\pi - k\,\Delta v_z\tau$. If the delay τ is large enough that $k\,\Delta v_z \gtrsim 1$, then the single VC collision will have destroyed that molecule's contribution to the echo since it is not properly phased. Now although the average change in velocity per collision Δu is very small (typically less than 1% of the average thermal velocity), the phase change $k \, \Delta u\tau$ is by no means small. For example, at $10 \, \mu m$, $k \, \Delta u\tau > 1$ for delay times τ larger than $2 \, \mu s$ with a Δu of only $100 \, \text{cm/s}$. Since delay times typically range from 0.5 to $30 \, \mu s$, $k \, \Delta u\tau > 1$ is easily obtained in practice.

From the discussion above one can see qualitatively what to expect for the decay of the photon echo. For very short delay times such that $k \, \Delta u\tau \ll 1$, VC collisions will affect the echo very little and the observed decay rate will just be γ_{ab}. For very long delay times such that $k \, \Delta u\tau \gg 1$, every VC collision will produce a large random phase change and cause the echo to decay just as a phase-changing collision would. In this regime the observed decay rate will be $\gamma_{ab} + \Gamma_{vc}$.

By being a little more careful, we can even predict how VC collisions will begin to affect the decay rate for $k \, \Delta u\tau < 1$ but not so small as to be negligible. First, we note that a molecule whose phase is $\pi - k \, \Delta v_z \tau$ rather than π at the time of the echo has an **M** vector which makes an angle $k \, \Delta v_z \tau$ with respect to the $+\text{II}$ axis rather than being aligned on the axis (see discussion in Section 3.8.1). Its contribution to the echo is thus $\cos k \, \Delta v_z \tau$ instead of unity. Next consider what the average value of $k \, \Delta v_z \tau$ for a VC collision is. The rms change in velocity if $\Delta u/\sqrt{2}$, so we might use $k \, \Delta u\tau/\sqrt{2}$ for short τ. However, collisions can occur at any time between the first pulse and the echo, while only collisions that occur near the second pulse will produce a dephasing $k \, \Delta v_z \tau$. Collisions occurring near the echo or the first pulse will give very little dephasing. Thus, a better guess for the average dephasing would be $\frac{1}{2} k \, \Delta u\tau/\sqrt{2}$. Now consider what the size of the echo will be. We get a contribution of unity from molecules which have not collided plus a contribution of $\cos (k \, \Delta u\tau/2\sqrt{2})$ from those which have made a VC collision. The fraction of molecules which have not collided is $\exp [-2(\Gamma_{vc} + \gamma_{ab})\tau]$ and the fraction which have undergone a VC collision but not a state- or phase-changing collision is $\exp (-\gamma_{ab}\tau)[1 - \exp (-2\Gamma_{vc}\tau)]$. Hence, the peak echo signal will be proportional to

$$S = e^{-2\gamma_{ab}\tau}[e^{-2\Gamma_{vc}\tau} + (1 - e^{-2\Gamma_{vc}\tau}) \cos (k \, \Delta u\tau/2\sqrt{2})] \quad (3\text{-}246)$$

Now if τ is sufficiently small, we will be in a single collision regime where most molecules have not collided even once, i.e., $2\Gamma_{vc}\tau < 1$. In experiments done to date, $\Gamma_{vc} < k \, \Delta u$, so this condition is easily satisfied when $k \, \Delta u\tau < 1$. Hence, we may expand both the cosine and the Γ_{vc} exponential at short times, to obtain

$$k \, \Delta u\tau < 1: \quad S \approx e^{-2\gamma_{ab}\tau}[1 - 2\Gamma_{vc}\tau + 2\Gamma_{vc}\tau(1 - k^2 \, \Delta u^2\tau^2/16)]$$

$$= e^{-2\gamma_{ab}\tau}(1 - \Gamma_{vc}k^2 \, \Delta u^2\tau^3/16)$$

$$= e^{-2\gamma_{ab}\tau}e^{-\Gamma_{vc}\tau^3 k^2 \, \Delta u^2/8} \quad (3\text{-}247)$$

Here only the first two terms in all expansions have been retained. Notice that VC collisions show up as an $\exp(-\tau^3)$ dependence of the echo decay. For extremely short times, the τ^3 term is negligible, but for somewhat longer times it produces an increasingly large deviation from the simple γ_{ab} decay rate. This is a result of the increasing effectiveness of the VC collisions in destroying the perfect rephasing of the dipoles at the echo peak. Aside from a small change in the numerical factor $\frac{1}{8}$, Eq. (3-247) agrees exactly with the prediction obtained from a rigorous quantitative treatment and is also in agreement with experiment.

In the opposite limit where $k\,\Delta u\tau \gg 1$, it is also easy to see why there is no net contribution to the echo from molecules which have made a VC collision. Although each molecule which has experienced a VC collision will make a contribution $\cos(k\,\Delta v_z\tau/2)$ to the echo, just as in the short time limit, the fact that $k\,\Delta u\tau \gg 1$ means that the cosine function will be positive for some molecules (those experiencing a small Δv_z, for example), but negative for other molecules experiencing a somewhat larger Δv_z. Contributions to the echo from these two sets of molecules cancel each other, and we are left with only contributions from molecules which have experienced no collisions. Thus,

$$k\,\Delta u\tau \gg 1: \quad S \simeq {}^{-2\gamma_{ab}\tau}e^{-2\Gamma_{vc}\tau} \tag{3-248}$$

Aside from a constant numerical factor, this result is again in agreement with the predictions of a rigorous theory and is also in agreement with experiment. It is worth noting that both the short and long time-limiting forms have been obtained with no assumptions as to the specific form of the collision kernel. We only assumed that $W(v_z' \to v_z)$ was a sharply peaked function of width Δu; i.e., it is a Brownian motion or weak collision type of kernel.

The $\exp(-\tau^3)$ dependence for short times is reminiscent of the decay behavior that is observed in nmr (Hahn, 1950). There, spin echo experiments on nonviscous liquids also yield an echo amplitude which goes as $\exp(-\tau^3)$. This behavior has a somewhat different origin in the nmr case, however. During a spin echo experiment all molecules experience a large number of collisions. These cause the molecules to diffuse through the liquid and experience a large number of different local magnetic fields as they do so. Thus, the dephasing rate for each molecule has a small, fluctuating component which produces an imperfect rephasing of the spins at the time of the echo. One can readily show that this leads to an $\exp(-\tau^3)$ behavior so long as $\eta(\tau)\tau \ll 1$, where $\eta(\tau)$ represents the average total phase deviation produced by the fluctuating component during time τ. In nmr $\eta(\tau)$ is so small that $\eta(\tau)\tau$ remains much less than 1 for all observable delay times τ. Thus, the difference between the nmr and optical cases is clear. In nmr the $\exp(-\tau^3)$ behavior results from spins which experience many collisions, with each collision producing an extremely small change in the dephasing rate. In the optical case, this behavior arises from molecules which experience a *single* VC collision that produces a substantial change in the dephasing rate.

We consider next a quantitative treatment of VC collisions in two-pulse echo experiments. Since the theoretical development of this subject is largely due to the pioneering research of R. G. Brewer and P. R. Berman along with their coworkers (Schmidt *et al.*, 1973; Berman *et al.*, 1975), we shall draw heavily on their work in our treatment of the echo problem. We note in passing that an alternative formulation of VC collision effects in photon echoes has been presented by Heer (1974a,b). In these papers the time dependence of the echo decay is interpreted in terms of a differential forward-scattering cross section rather than a collision kernel. The same general form for the echo decay is predicted by this theory, and we will not pursue it further in this chapter.

To begin the quantitative treatment, we note that relaxation can be ignored during the short excitation pulses in an echo experiment. Hence, the only difference from our previous treatment of photon echoes lies in the zero-optical field solutions. In particular we need only consider the zero-field equation for $\tilde{\rho}_{ab}$, because the two-pulse echo arises only from terms present in $u(t)$ and $v(t)$ during the zero-field intervals [recall that $2\tilde{\rho}_{ab}(t) = u(t) - iv(t)$]. This is the reason why the echo decays purely as γ_{ab} when VC collisions are neglected. We want, therefore, to solve the $\dot{\tilde{\rho}}_{ab}$ equation, (3-234), for an arbitrary time interval $t - t_k$ with arbitrary initial conditions. With this solution in hand, it will then be fairly straightforward to obtain the echo behavior.

The first step in solving the $\dot{\tilde{\rho}}_{ab}$ equation is to simplify it by setting

$$\dot{\tilde{\rho}}_{ab}(v_z, t) \equiv e^{[i\delta(v_z) - \gamma_{ab}](t - t_r)} f(v_z, t - t_r) \qquad (3\text{-}249)$$

where $\delta(v_z) = \Omega - \omega_0 - kv_z$. t_r is some point in time which we choose as a reference time. If the zero-field time interval under consideration begins at time $t = t_k$, then we normally choose $t_r = t_k$; i.e., we choose the beginning of the time interval as the reference point. Mathematically, however, there is no need to do this, and on one occasion it will be convenient to pick a $t_r \neq t_k$.

When Eq. (3-249) is substituted into Eq. (3-234), we find that

$$\frac{\partial}{\partial t} f(v_z, t - t_r) = \int W(v_z' \to v_z) e^{ik(v_z - v_z')(t - t_r)} f(v_z', t - t_r) \, dv_z' - \Gamma_{vc} f(v_z, t - t_r)$$

$$(3\text{-}250)$$

The function $f(v_z, t - t_r)$ contains only information about the effects of velocity-changing collisions, as can be seen by setting $W(v_z' \to v_z) = \Gamma_{vc} = 0$. We then have $\partial f / \partial t = 0$, so that f is just a constant and $\tilde{\rho}_{ab}(t)$ in Eq. (3-249) becomes the standard zero-field solution used in previous sections.

Although the form of Eq. (3-250) resembles the linearized Boltzmann equation (3-240), the presence of the complex exponential factor in the integral causes the solutions to be rather different. Obviously, the equation is not an easy one to solve. $f(v_z, t - t_r)$ is a complex function and its value at any time is coupled to the values of $f(v_z', t - t_r)$ for all neighboring velocity groups v_z'. However, Berman *et al.* (1975) have shown that a very good approximate solution may be obtained by writing $f(v_z, t - t_r)$ as the product of a real amplitude $C_0 \exp[y(t - t_r)]$ and a phase $x(t - t_r)v_z$. Thus, we try a solution of the form

$$f(v_z, t - t_r) = C_0 e^{y(t - t_r)} e^{ix(t - t_r)v_z} \qquad (3\text{-}251)$$

Here C_0 is a constant and the only velocity dependence of f is assumed to be the multiplicative factor v_z in the second exponential. Substituting this form for f into Eq. (3-250), we obtain

$$i\dot{x}(t')v_z + \dot{y}(t') = \int W(v_z' \to v_z) e^{i[kt' - x(t')](v_z - v_z')} \, dv_z' - \Gamma_{vc} \qquad (3\text{-}252)$$

where $t' = t - t_r$. Notice that because neither $x(t')$ nor $y(t')$ depends on v_z, this substitution has uncoupled the equation from the equations of motion for all other velocity groups. To make further analytic progress we must put in a specific form for the collision kernel. Using the Keilson–Storer kernel of Eq. (3-235), the integration over v_z' in Eq. (3-252) can be done in a straightforward manner, to yield

$$i\dot{x}(t')v_z + \dot{y}(t') = \Gamma_{vc} e^{-[kt' - x(t')]^2 \Delta u^2/4} e^{-i[kt' - x(t')](1-\alpha)v_z} - \Gamma_{vc} \qquad (3\text{-}253)$$

On examining this differential equation, we notice that the first term on the right-hand side is negligible unless the exponent in the first exponential is a fairly small number. This, however, implies that the argument of the complex exponential is extremely small. To see this, note that $1 - \alpha = \Delta u^2/2\bar{u}^2$ and v_z is at most on the order of $\bar{u}$. Hence, $(1 - \alpha)v_z \simeq \Delta u^2/2\bar{u}$. Now if the first exponent is less than, say, about 4, we have $[kt' - x(t')] \Delta u/2 < 2$, so that

$$[kt' - x(t')](1 - \alpha)v_z \simeq [kt' - x(t')]\frac{\Delta u}{2}\frac{\Delta u}{\bar{u}} \ll 1$$

The last inequality follows from the fact that $\Delta u/\bar{u}$ is a very small number, typically 10^{-3} or less.

The argument above shows that we can expand the complex exponential in Eq. (3-253) and keep only the first two terms, to obtain

$$i\dot{x}(t')v_z + \dot{y}(t') = -\Gamma_{vc} + \Gamma_{vc} e^{-[kt' - x(t')]^2 \Delta u^2/4}\{1 - i[kt' - x(t')](1 - \alpha)v_z\} \qquad (3\text{-}254)$$

This form of the equation will be valid for all values of $kt' - x(t')$, because whenever this quantity is large enough to invalidate the expansion, the exponential in front will make the entire second term negligible. From Eq. (3-254) we immediately obtain two equations,

$$\dot{x}(t') = -\Gamma_{vc}e^{-[kt'-x(t')]^2 \Delta u^2/4}[kt' - x(t')](1 - \alpha)$$
$$\dot{y}(t') = \Gamma_{vc}[e^{-[kt'-x(t')]^2 \Delta u^2/4} - 1]$$

(3-255)

These equations are not analytically soluble as they stand. We still need another approximation. Looking back at the definitions of $x(t')$ and $v(t')$, we note that $x(t')$ represents a phase shift of $\tilde{\rho}_{ab}$ produced by the VC collisions. Intuition suggests that this shift will be extremely small, and thus it may be profitable to assume that

$$|x(t')| \ll |kt'|$$

(3-256)

If we can verify this inequality after finding $x(t')$, the assumption is a valid one. Using this assumption in Eq. (3-255), we can immediately write down solutions for the equations. We have

$$x(t - t_r) = x(t_k - t_r) - \Gamma_{vc}(1 - \alpha) \int_{t_k}^{t} e^{-k^2 \Delta u^2 (t''-t_r)^2/4} k(t'' - t_r)\, dt''$$

$$y(t - t_r) = y(t_k - t_r) - \Gamma_{vc}(t - t_k) + \Gamma_{vc} \int_{t_k}^{t} e^{-k^2 \Delta u^2 (t''-t_r)^2/4}\, dt''$$

(3-257)

Note that the integrations extend from the beginning of the zero-field interval, $t'' = t_k$, to an indefinite time $t'' = t$. The constants $x(t_k - t_r)$ and $y(t_k - t_r)$ are determined from the initial conditions.

We can summarize the solution of the $\dot{\rho}_{ab}$ equation by combining Eqs. (3-249) and (3-251), which show that

$$\tilde{\rho}_{ab}(v_z, t) = C_0 e^{(i\delta - \gamma_{ab})(t-t_r)} e^{y(t-t_r)} e^{ix(t-t_r)v_z}$$

(3-258)

with initial condition

$$\tilde{\rho}_{ab}(v_z, t_k) = C_0 e^{(i\delta - \gamma_{ab})(t_k-t_r)} e^{y(t_k-t_r)} e^{ix(t_k-t_r)v_z}$$

(3-259)

The functions $x(t - t_r)$ and $y(t - t_r)$ are given by Eq. (3-257) provided we can show that $|x(t - t_r)| \ll |k(t - t_r)|$.

We are now ready to do a calculation of the photon echo. Since we are only interested in learning about the relaxation behavior here, we consider only the simplest echo case, where one has such intense pulses that $\kappa E_0 \gg k\bar{u}$ and we use a $\pi/2$, π pulse sequence. Other more complicated cases must exhibit the same relaxation behavior because all two-pulse echoes arise only from terms in $\tilde{\rho}_{ab}$ during the delays between pulses. These zero-field $\tilde{\rho}_{ab}$ solutions are the same regardless of what happens during the pulses. In particular, the relaxation behavior for $kE_0 \ll k\bar{u}$ with arbitrary pulse areas is identical to that for the simpler case we are about to consider. The detailed algebra for the $kE_0 \ll k\bar{u}$ case can be found in Berman *et al.* (1975).

The pulse sequence for the two-pulse echo is shown in Figure 3-17. At time $t = t_1$ the system is in thermal equilibrium. The $\pi/2$ pulse then gives $u(t_2) = 0$, $v(t_2) = n_a - n_b$, so $\tilde{\rho}_{ab}(t_2) = \frac{1}{2}[u(t_2) - iv(t_2)] = -i(n_a - n_b)/2$. This is the initial condition we need for the zero-field interval $t_3 - t_2$. If we now choose the reference time t_r to be the beginning of the interval, we have $t_r = t_k = t_2$, and putting the initial condition into Eq. (3-259) gives $x(0) = y(0) = 0$ and $C_0 = -i(n_a - n_b)/2$. The solutions for $x(t - t_2)$, $y(t - t_2)$ in Eq. (3-257) now become

$$x(t - t_2) = -\frac{2\Gamma_{vc}(1 - \alpha)}{k\,\Delta u^2}[1 - e^{-k^2\Delta u^2(t - t_2)^2/4}]$$

$$y(t - t_2) = -\Gamma_{vc}(t - t_2) + \frac{2\Gamma_{vc}}{k\,\Delta u}\int_0^{k\Delta u(t - t_2)/2} e^{-\eta^2}\,d\eta$$

(3-260)

The integral in the equation for y cannot be done analytically. However, by expressing it as we have, in terms of the dummy variable $\eta = k\,\Delta u(t - t_2)/2$, the integral becomes a tabulated function, $\frac{1}{2}\sqrt{\pi}\,\text{erf}\,[k\,\Delta u(t - t_2)/2]$, where erf (x) is the error function or probability integral. The solutions given in Eq. (3-260) are valid if $|x(t - t_k)| \ll |k(t - t_2)|$. It is easy to verify this inequality by noting that for short times $k\,\Delta u(t - t_2) < 1$, the exponential in $x(t - t_2)$ can be expanded to give $|x(t - t_2)| \approx \frac{1}{2}\Gamma_{vc}(1 - \alpha)k(t - t_2)^2$. This is clearly much less than $k(t - t_2)$, because $\Gamma_{vc}(1 - \alpha)(t - t_2) \ll 1$ for all delay times $t - t_2$ of experimental interest. This point was discussed in Section 3.10.2, where we found that the population distribution relaxes toward equilibrium at the extremely slow rate $\Gamma_{vc}(1 - \alpha)$. For times $k\,\Delta u(t - t_2) > 1$, the exponential in $x(t - t_2)$ can be neglected and the inequality written as

$$\frac{2\Gamma_{vc}(1 - \alpha)(t - t_2)}{k^2\,\Delta u^2(t - t_2)^2} \ll 1$$

which is also clearly valid for $k\,\Delta u(t - t_2) > 1$.

We now have a solution for the delay between pulses. At $t = t_3$ this solution gives

$$\tilde{\rho}_{ab}(t_3) = -\frac{1}{2}i(n_a - n_b)e^{i\delta t_{32}}e^{-\gamma_{ab}t_{32}}e^{y(t_{32})}e^{ix(t_{32})v_z}$$

(3-261)

where $t_{32} = t_3 - t_2$ is the delay between pulses and $x(t_{32})$, $y(t_{32})$ are given by Eq. (3-260) evaluated at $t = t_3$. The π pulse which begins at t_3 now changes $\tilde{\rho}_{ab}(t_3)$ into $\tilde{\rho}_{ab}^*(t_3)$. This is easily seen from Eq. (3-144), which shows that a π pulse just changes the sign of $v(t)$, leaving $u(t)$ unaffected. Thus, at the end of the pulse we have

$$\tilde{\rho}_{ab}(t_4) = \tilde{\rho}_{ab}^*(t_3)$$

(3-262)

For the interval $t > t_4$, it would now seem logical to proceed as in the previous interval and set $t_r = t_k = t_4$. The trouble with doing this is that the factor $\exp(-i\delta t_{32})$ in the initial condition $\tilde{\rho}_{ab}(t_4)$ forces us to have $x(0) = kt_{32} - x(t_{32})$ as the initial condition on x. The constant term kt_{32} in the solution for $x(t - t_4)$ then makes the inequality $|x(t - t_4)| \ll |k(t - t_4)|$ invalid, and our approximate solutions, Eqs. (3-257), are incorrect. To get around this problem we choose the reference time t_r to be $t_4 + t_{32}$; i.e., the reference time is at the time of the echo. This removes the offending term, as we will now see. We set $t_k = t_4$, $t_r = t_4 + t_{32}$. Substituting Eqs. (3-261) and (3-262) into (3-259), the initial condition becomes

$$\tfrac{1}{2}i(n_a - n_b)e^{-i\delta t_{32}}e^{-\gamma_{ab}t_{32}}e^{y(t_{32})}e^{-ix(t_{32})v_z} = C_0 e^{-i\delta t_{32}}e^{\gamma_{ab}t_{32}}e^{y(-t_{32})}e^{ix(-t_{32})v_z}$$

$$(3\text{-}263)$$

The $\exp(-i\delta t_{32})$ factors cancel, and the initial condition on x now is $x(t_k - t_r = -t_{32}) = -x(t_{32})$. Picking $C_0 = \tfrac{1}{2}i(n_a - n_b)e^{-2\gamma_{ab}t_{32}}$, we also have $y(t_k - t_r = -t_{32}) = y(t_{32})$. With these initial conditions put into Eq. (3-257), the integrals can easily be done and we find, after some rearrangement,

$$x(t - t_4 - t_{32}) = -2x(t_{32}) - \frac{2\Gamma_{vc}(1 - \alpha)}{k\,\Delta u^2}[1 - e^{-k^2\,\Delta u^2(t - t_4 - t_{32})^2/4}]$$

$$(3\text{-}264)$$

$$y(t - t_4 - t_{32}) = 2y(t_{32}) - \Gamma_{vc}(t - t_4 - t_{32}) + \int_0^{k\,\Delta u(t - t_4 - t_{32})/2} e^{-\eta^2}\,d\eta$$

We must now show that $|x(t - t_4 - t_{32})| \ll |k(t - t_4 - t_{32})|$. The form of this inequality is identical to the form we obtained for the $t_3 - t_2$ interval, except that we now have a very small constant term $2x(t_{32})$ present in the expression for $x(t - t_4 - t_{32})$. This term will not affect the validity of the inequality for large values of $|t - t_4 - t_{32}|$, but right near the time of the echo, where $t - t_4 - t_{32} \to 0$, the inequality will no longer hold. This means that for a very short period of time just before the echo, the solutions shown in Eq. (3-264) will not be valid. This time period (given roughly by $4\Gamma_{vc}(1 - \alpha)/k^2\,\Delta u^2$) is so short, however, that the error incurred by using the solutions for all times is negligible. If desired, this point can be verified by comparing Eqs. (3-264) with a numerical integration of Eqs. (3-255).

Using these results in Eq. (3-258), we can write down the complete solution to the echo problem. For $t > t_4$, we have

$$\tilde{\rho}_{ab}(v_z, t) = \tfrac{1}{2}i(n_a - n_b)e^{-\gamma_{ab}(t - t_4 + t_{32})}e^{i(\Omega - \omega_0)(t - t_4 - t_{32})}$$

$$\times e^{y(t - t_4 - t_{32})}e^{-i[k(t - t_4 - t_{32}) - x(t - t_4 - t_{32})]v_z}$$

$$(3\text{-}265)$$

Using $2\tilde{\rho}_{ab} = u - iv$ and ignoring the tiny phase shift $x(t - t_4 - t_{32})$, we then find that

$$v(t) = -(n_a - n_b)e^{-\gamma_{ab}(t - t_4 + t_{32})}e^{y(t - t_4 - t_{32})}\cos\delta(t - t_4 - t_{32}) \quad (3\text{-}266)$$

where $y(t-t_4-t_{32})$ is given by Eq. (3-264). Notice that, aside from an altered decay rate, $v(t)$ obtained here has exactly the same form that we obtained when VC collisions were neglected [Eq. (3-146)].

For relaxation measurements, the quantity of interest is the echo signal at the echo peak as a function of delay time t_{32}. We are particularly interested in the signal for Stark-switching experiments, because the only VC collision measurements to date have been obtained using this technique. The echo signal in this case is proportional to the emitted electric field $\mathcal{E}_c$. This field is easily found, using Eqs. (3-69) and (3-50), to be

$$\mathcal{E}_c(\text{echo peak}) = \frac{\mu_{ab}\Omega L}{2\varepsilon_0 c}(N_b - N_a)$$

$$\times \exp\left(-2\gamma_{ab}\tau - 2\Gamma_{vc}\tau + \frac{4\Gamma_{vc}}{k\,\Delta u}\int_0^{k\,\Delta u\tau/2} e^{-\eta^2}\,d\eta\right) \quad (3\text{-}267)$$

where τ denotes the delay between pulses (t_{32}) and we have, for simplicity, evaluated $\mathcal{E}_c$ at $t-t_4=t_{32}$, which is very near the time of the echo peak. This equation is the final result which we wish to compare with experiment.

It is interesting to note that for short times $k\,\Delta u\tau < 1$, the error-function integral in Eq. (3-267) can be written as a power-series expansion whose first two terms are $\frac{1}{2}k\,\Delta u\tau - \frac{1}{24}(k\,\Delta u\tau)^3$. Also, for $k\,\Delta u\tau \gg 1$, the integral becomes $\sqrt{\pi}/2$. Thus, we find that the short and long time limits of $\mathcal{E}_c$ are

$$k\,\Delta u\tau < 1: \quad \mathcal{E}_c(\text{echo peak}) = Ce^{-2\gamma_{ab}\tau}e^{-\Gamma_{vc}k^2\,\Delta u^2\tau^3/6} \quad (3\text{-}268)$$

$$k\,\Delta u\tau \gg 1: \quad \mathcal{E}_c(\text{echo peak}) = Ce^{-2\gamma_{ab}\tau}e^{-2\Gamma_{vc}\tau}e^{2\sqrt{\pi}\Gamma_{vc}/k\,\Delta u} \quad (3\text{-}269)$$

where C is a constant. Note that for very short times $k\,\Delta u\tau \ll 1$, the τ^3 term in Eq. (3-268) is negligible compared to the $-2\gamma_{ab}\tau$ term. It is gratifying to see that the form of these results agrees with Eqs. (3-247) and (3-248), which were deduced from general considerations independent of the form of the collision kernel. Berman et al. (1975) have also shown that the τ^3 dependence of Eq. (3-268) plus higher-order corrections can be obtained by expanding the exponential in Eq. (3-252). This approach yields an infinite series solution for $\mathcal{E}_c$ whose delay time dependence is independent of the choice of collision kernel.

We have just shown in considerable detail exactly how two-pulse echo experiments are sensitive to VC collisions. Clearly, it would be very desirable to use this sensitivity to obtain quantitative information about VC collisional processes. However, it is usually hard to do this using only the two-pulse echo experiment. There are three unknown quantities, γ_{ab}, Γ_{vc}, and Δu, and it is very difficult to reliably extract values for all three from a single echo decay curve which can be measured only over a limited range of delay times. One good way to obtain additional information is to do multiple-pulse echo experiments. With this in mind, let us now briefly examine the characteristics of these experiments.

Consider first the Carr–Purcell echo experiment. Here we have a $\pi/2$, π pulse sequence separated by a delay τ, followed by additional π pulses with delays of $\sim 2\tau$ between them. Echoes then occur halfway between each π pulse, as shown in Figure 3-24. If the pulses are very intense, with $\kappa E_0 \gg k\bar{u}$, the relaxation behavior of this experiment is trivial to obtain. The amplitude of the nth echo in the Carr–Purcell(C–P) train is simply given by the two-pulse echo decay result raised to the nth power. This is discussed in detail for a case where VC collisions are neglected in Section 3.8.2 [see Eq. (3-187)]. However, the argument is unaffected when VC collisions are included. The basic reason for the result is that, at the peak of each echo, the dipoles are in phase. Thus, the relaxation behavior between any one echo and the next is identical to that between the initial $\pi/2$ pulse and the first echo. The result that

$$\mathscr{E}_c(n\text{th C-P echo peak}) = [\mathscr{E}_c \text{ (two-pulse echo peak)}]^n$$

then follows immediately.

The C–P echo train is useful because it is relatively insensitive to VC collisions. To see this, we recall the qualitative discussion of how VC collisions affect the echo, which is presented at the beginning of this subsection. The central point of the discussion is the fact that VC collisions reduce the echo size by causing molecules to dephase at one rate prior to the π pulse, and then rephase at a *different* rate afterward. As a result, the colliding molecules end up out of phase by an average amount $k\,\Delta u\tau/2\sqrt{2}$ at the time of the echo. In a C–P experiment the same thing happens. A VC collision occurring after the ith echo in the train causes the affected molecule to dephase at one rate and rephase at a different rate, so that it is out of phase by an average amount $k\,\Delta u\tau_{cp}/2\sqrt{2}$ at the time of echo $i+1$ (where $2\tau_{cp}$ is the spacing between π pulses). The point, however, is that the molecule gets no further out of phase than this at the time of subsequent echoes $i+2$, $i+3$, etc.! The reason is that although the molecule is slightly out of phase at echo $i+1$, it dephases and rephases at the same rate between echo $i+1$ and $i+2$, and therefore gets no further out of phase; similarly for all subsequent echoes. We see from this discussion that by keeping the π pulses close together so that τ_{cp} is small we can keep the average dephasing small, even for C–P echoes which occur at very long delays after the start of an experiment. This means that the decay of the echoes in a C–P echo train will be much less affected by VC collisions than will a two-pulse echo. As a quantitative illustration, suppose that we do a two-pulse echo experiment with a delay time τ such that $k\,\Delta u\tau < 1$. According to Eq. (3-268), the echo in this case will occur at 2τ and be reduced by an amount

$\exp(-\Gamma_{vc}k^2 \Delta u^2 \tau^3/6)$ due to VC collisions. Now suppose we do a C–P experiment with closely spaced pulses such that $\tau_{cp} = \tau/n$, where n is an integer. The nth echo in this train will also occur after a total delay of 2τ, but its amplitude is found from Eq. (3-268) raised to the nth power; i.e., it is reduced by only $[\exp(-\Gamma_{vc}k^2 \Delta u^2 \tau_{cp}^3/6)]^n = \exp(-\Gamma_{vc}k^2 \Delta u^2 \tau^3/6n^2)$, owing to VC collisions. If n is a reasonably large number, say 3 or more, the effect of VC collisions on the C–P experiment is clearly negligible compared to the effect on a two-pulse experiment. Thus, the C–P echo train decays at a rate which is essentially just γ_{ab}, allowing us to measure γ_{ab} independent of VC collisions.

While the theory just presented tells us how to measure γ_{ab} when we have intense pulses with $\kappa E_0 \gg k\bar{u}$, one is in trouble when a Stark- or frequency-switching experiment with $\kappa E_0 \ll k\bar{u}$ is done. It is shown in Section 3.8.2 that the C–P echo train in this case is dominated by contributions from stimulated echoes, with only a fraction of the echo signal coming from image echoes of the type just discussed. To understand what is measured here, we clearly must also understand the effects of VC collisions on stimulated echoes. Fortunately, within limits, it is fairly easy to do this. In a stimulated echo experiment one has a delay t_{32} between the first two pulses, followed by a delay t_{54} between the second and third pulse. In Section 3.8.2 we showed that the stimulated echo arises solely from terms present in ρ_{aa} and ρ_{bb} during the delay t_{54}, and solely from terms present in $\tilde{\rho}_{ab}$ during the delay t_{32} and the interval (also of length t_{32}) between the third pulse and the echo. Thus, the echo decays as $\tilde{\rho}_{ab}$ during the two intervals t_{32}, and as ρ_{aa}, ρ_{bb} during the delay t_{54}. Since we found the effect of VC collisions on ρ_{aa} and ρ_{bb} to be negligible in Section 3.10.2 (over the time scale of a transient experiment), we might expect to have no decay due to VC collisions during t_{54}. This conclusion is, in fact, correct, but only under the conditions that $k\,\Delta ut_{32} \ll \pi$. To see how this condition arises, we recall that during the interval t_{54}, the information about the stimulated echo is stored along the III axis as a pattern $\cos \delta t_{32}$ in frequency space, i.e., $\rho_{aa}(v_z) - \rho_{bb}(v_z) \propto \cos(\Omega - \omega_0 - kv_z)t_{32}$ [see Eqs. (3-171)–(3-176) and Figure 3-22]. Anything which destroys this arrangement will destroy the echo. For example, a VC collision which changes the velocity by $\Delta v_z = \pi/kt_{32}$ can take a molecule out of a velocity group where $\rho_{aa} - \rho_{bb}$ is a maximum and leave it in a group where $\rho_{aa} - \rho_{bb}$ is a minimum. Such collisions effectively fill in the holes in the $\cos \delta t_{32}$ pattern and destroy the echo. Hence, we conclude that if the rms velocity change Δu is on the order of π/kt_3, single VC collisions can cause the stimulated echo to decay significantly. The reason we found no VC collision effects in Section 3.10.2 is that we assumed that $\rho_{aa}(v_z)$ did not vary significantly over a range of velocities Δu. Since $\rho_{aa}(v_z) - \rho_{bb}(v_z) \propto \cos \delta t_{32}$ is a periodic function of v_z with period $2\pi/kt_{32}$, this assumption is violated unless $\Delta u \ll \pi/kt_{32}$.

When the interval between the first two pulses is sufficiently short that $k \Delta u t_{32} \ll \pi$, single VC collisions cannot appreciably alter the $\cos \delta t_{32}$ pattern and we can neglect them during the interval t_{54}. In this case the stimulated echo decay goes as

$$k \Delta u t_{32} \ll \pi: \quad \mathscr{E}_c(\text{stim. echo peak}) \propto e^{-2\gamma_{ab}t_{32}} e^{-\Gamma_{vc}k^2 \Delta u^2 t_{32}^3/6}(e^{-\gamma_a t_{54}} + e^{-\gamma_b t_{54}})$$

$$(3\text{-}270)$$

Here we have used Eq. (3-268) to describe the decay during the intervals t_{32}, and Eq. (3-177) for the interval t_{54}. The stimulated echo decay for the case where $k \Delta u t_{32}$ is large has not yet been worked out in detail.

With this background, we now return to a consideration of the C–P echo train when $\kappa E_0 \ll k\bar{u}$. Because the pulses are closely spaced in these experiments, the condition $k \Delta u t_{32} \ll \pi$ is satisfied, and Eq. (3-270) can be used to describe most of the stimulated echo contributions which dominate the C–P echo train. Thus, in general, we expect VC collisions to have little effect on the C–P echo train [the t^3 term in Eq. (3-270) is very small for $k \Delta u t_{32} \ll \pi$]. In the experiments done to date, this expectation appears to be verified as the observed C–P echo trains show only a simple exponential decay. We note, however, that the neglect of VC collisions in C–P experiments cannot be rigorously true since stimulated echoes arising from widely separated initial pulses can contribute to a given C–P echo as well as two-pulse echoes from widely separated pulses (see Section 3.8.2). Apparently, the contributions from these echoes are quite small.

We are now ready to consider the actual transient-effect experiments on VC collisions which have been done. These are unfortunately very few in number. There have been two studies made on $^{13}CH_3F$ (Berman *et al.*, 1975; Grossman *et al.*, 1977) and one on NH_2D (Van Stryland, 1976). The 1975 paper by Berman *et al.* is by far the most complete study, so we will concentrate initially on it. A CO_2 laser and Stark switching was used in the experiments to examine the effect of VC collisions on the $(\nu_3, J, K) = (0,4,3) \rightarrow (1,5,3)$ transition in $^{13}CH_3F$. A combination of several different techniques were used in this study, including delayed nutation, Carr–Purcell echoes, and two-pulse echoes. The decay curves as a function of time at 0.32-mTorr pressure are shown in Figure 3-33. A great deal of information is available in this figure. First, we see that the two-pulse echo decay curve is in excellent agreement with the theoretical prediction of Eq. (3-267). This gives us some confidence in using the theoretical model to evaluate VC collision parameters. To actually extract values for Δu and Γ_{vc}, it is most convenient to make use of the long and short time-limiting forms given in Eqs. (3-268) and (3-269). First, we note that at *very* short times the decay of the logarithm of the echo amplitude A is a straight line whose equation is $A = \ln C - 2\gamma_{ab}\tau$ (where C is a constant), while at long times it is a straight line whose equation is $A = \ln C - 2(\gamma_{ab} + \Gamma_{vc})\tau + 2\sqrt{\pi}\Gamma_{vc}/k \Delta u$.

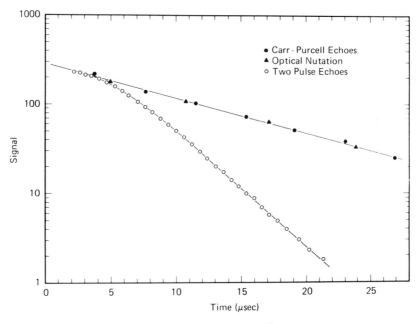

Figure 3-33. Decay curves of optical coherent transients in $^{13}CH_3F$. The CO_2 laser intensity is about 350 mW/cm² in a 1.3-cm-diameter beam, and the Stark pulses are 40 V/cm plus an 80-V/cm dc bias. The time axis is the total elapsed time in each experiment; for example, $t = 2\tau$ in the two-pulse echo experiment (from Schmidt *et al.*, 1973; reproduced by permission).

If we extend these straight lines until they intersect, the intersection point will occur at $2\tau = 2\sqrt{\pi}/k\,\Delta u$. In the figure this point occurs at $2\tau = 6.0\ \mu s$ (the time axis in the figure is the total elapsed time = twice the delay time τ), so we obtain $\Delta u = 91$ cm/s. This is indeed a very small number compared to the thermal velocity of $\bar{u} = 3.75 \times 10^4$ cm/s and to the velocity bandwidth of molecules excited by the echo pulses, as we have assumed throughout our theoretical analysis. Γ_{vc} is also easily obtained from the fact that the difference in slope of the long and very short time limits is just Γ_{vc}. By obtaining Γ_{vc} in this way for a number of different pressures, we can obtain the pressure-dependent part of Γ_{vc}, which is the value of interest.

The trouble with obtaining both Δu and Γ_{vc} in the manner described above is that there are very few data points in the echo decay curve with which to determine the short time-limit slope $-2\gamma_{ab}$. Thus, in practice one wants to obtain this slope from other experiments, such as Carr–Purcell or delayed nutation measurements. Now, delayed nutation signals decay as $\exp(-\gamma_a t) + \exp(-\gamma_b t)$, while Carr–Purcell echoes decay as some complicated mixture of γ_a, γ_b, and γ_{ab}. Thus, it is unclear in general how to make use of these experiments. In the case of $^{13}CH_3F$, however, we are fortunate enough to have a fairly simple situation. First, the decay rate of the delayed nutation signal agrees well with the lower-state decay rate γ_b obtained by Jetter *et al.* (1973) using microwave techniques. This means that $\gamma_a = \gamma_b \equiv \gamma$. We will therefore have $\gamma_{ab} = \gamma$, provided that phase-changing collisions are not significant. All available evidence suggests that this is the case. Experimentally, we see in Figure 3-33 that the Carr–Purcell decay and the delayed nutation decay both coincide and show the same slope. This seems to indicate that $\gamma_{ab} = \gamma$, unless the C–P echo decay has a completely negligible contribution from γ_{ab}. We note in passing that the identical decay rates for C–P echoes and delayed nutation does prove our earlier assertion that the C–P echoes are insensitive to VC collisions. There is also a strong theoretical reason for believing that phase-changing collisions are negligible—the fact that the upper and lower states are very similar. The principal difference between the two states is the fact that in the upper state the C–F stretching vibrational motion is excited. However, this excitation appears to affect the molecular properties very little. The dipole moment, for example, differs by less than 3% in the two states (Shoemaker *et al.*, 1974). Also, we have $\gamma_a = \gamma_b$. Thus, the two states should behave similarly during a collision, and it would be very surprising if phase-changing collisions were significant.

Assuming that the C–P echo or delayed nutation signals decay as $\exp(-2\gamma_{ab}\tau)$, we can use the value of γ_{ab} to help obtain Γ_{vc} and Δu. In addition to improving the accuracy of the techniques mentioned earlier for extracting Γ_{vc} and Δu, we can also now make use of the τ^3-dependent region of the two-pulse echo decay. This is done by adding $2\gamma_{ab}\tau$ to each point of the ln (echo amplitude) vs. time decay curve. If the resulting curve is plotted vs. $(2\tau)^3$ we find a straight line for short times (up to about 7 μs in $^{13}CH_3F$). The reason for this result is obvious if one simply looks at Eq. (3-268). The slope of the line gives $\Gamma_{vc}k^2 \Delta u^2/48$. Yet another technique is to note that the straight-line behavior of ln (echo amplitude) observed at long delay times can be extrapolated back to $\tau = 0$ to obtain a zero time intercept of $\ln C + 2\sqrt{\pi}\Gamma_{vc}/k \Delta u$. On finding ln C from the $\tau = 0$ intercept of the C–P echo decay, we can obtain $\Gamma_{vc}/\Delta u$. If the values of Γ_{vc} and Δu obtained by all these methods are self-consistent, one can have some confidence in the results.

One unexpected complication that arises in the data analysis is a phenomenon which Berman *et al.* (1975) refer to as an intensity-dependent dephasing. It is observed only in the two-pulse echo decay and consists of a contribution to the VC collisional decay rate Γ_{vc} which is pressure-*independent* and proportional to the laser intensity. The origin of this effect is not understood as yet, but it can be effectively eliminated by making measurements at various laser intensities and extrapolating to zero intensity. The details are described in Berman *et al.* (1975).

When the $^{13}CH_3F$ data were analyzed, the following results were found:

$$\Delta u = 85 \text{ cm/s}$$

$$\gamma_{ab} = \gamma_a = \gamma_b = (0.089P)\,\text{MHz} \qquad \sigma = 500 \text{ Å}^2$$

$$\Gamma_{vc} = (0.074P)\,\text{MHz}, \qquad \sigma_{vc} = 430 \text{ Å}^2$$

with P in mTorr. Here only the pressure-dependent contributions to γ_{ab} and Γ_{vc} are given, and the corresponding cross sections are calculated using the relation $\sigma = \gamma/N\bar{u}\sqrt{2}$ where $\gamma = \gamma_{ab}$ or Γ_{vc}. $N = N_A P/RT$ is the number of molecules/cm^3. We note that the VC collision cross section is very large, nearly equal to the total inelastic collision cross section. Heer (1974b) has presented a calculation indicating that this is to be expected for molecules having substantial dipole moments.

Brewer and his coworkers have performed another very interesting study of $^{13}CH_3F$ (Grossman *et al.*, 1977). In this work they looked at the velocity dependence of the echo decay for a single $M_J \rightarrow M_J$ component of the $J = 4 \rightarrow 5$ $^{13}CH_3F$ transition discussed above. This was done by noting that while the total echo signal in a Stark switching experiment is a sum of contributions from all the allowed $M_J \rightarrow M_J$ components in the transition, each component has a different Stark effect and therefore a different heterodyne beat frequency with the laser. This allows one to separate out the individual contributions by Fourier-transforming the echo signal obtained for a given time delay τ. On doing this for many different delays, the echo decay of each $M_J \rightarrow M_J$ component can be individually measured. Furthermore, by applying a dc bias field to the Stark cell, one can shift the center frequency ω_0 of a given component with respect to the laser and thus select for study any desired velocity group within the Doppler line width of an $M_J \rightarrow M_J$ component. By studying the echo decay for a number of different velocity groups, some interesting results were obtained. The inelastic or state changing decay rate $\gamma_a = \gamma_b$ was found to be independent of velocity. This is not surprising because one expects the collision cross section for molecules which interact via a dipole–dipole interaction to be inversely proportional to velocity. The number of collisions per second is directly proportional to velocity, and therefore the velocity dependence cancels in the observed decay rate (see Anderson, 1949, for example). The surprising result is that the VC collision rate Γ_{vc} *does* depend on velocity with a dependence which is characteristic of a $1/r^6$ interaction potential (second-order dipole–dipole or van der Waals' interaction). This result was not expected and is not well understood at present. The velocity dependence is not large (Γ_{vc} varies by about 35% over the Doppler line width), but it does imply that the Keilson–Storer kernel is not completely accurate, since the K–S kernel predicts a velocity-independent Γ_{vc}. Another result found in this work is that the decay rates γ and Γ_{vc} were identical for all $M_J \to M_J$ components.

Experiments have also been performed in this author's laboratory on VC collision effects in NH_2D (Van Stryland, 1976). As in the previous CH_3F work, delayed nutation, Carr–Purcell echo, and two-pulse echo experiments were done using a CO_2 laser and Stark switching. Unfortunately, in the case of NH_2D, the upper- and lower-state interactions are quite different, producing $\gamma_a \neq \gamma_b$ (see Section 3.9.2), and giving us no reason to believe *a priori* that the Carr–Purcell echo train will decay as γ_{ab}. Some additional information is needed to extract VC collision information from the two-pulse echo data. In NH_2D this information is available from pressure-broadening measurements made by Plant and Abrams (1976). By measuring the unsaturated absorption profile over a wide range of pressures, they found that $\gamma_{ab} = (0.126P)$ MHz, where P is the pressure in mTorr. This result is in excellent agreement with the C–P echo decay measurements, which give $(0.124P)$ MHz, and also is within experimental error of the value for $(\gamma_a + \gamma_b)/2$ obtained from delayed nutation measurements. These data indicate that phase-changing collisions are not important in this transition and allow us to analyze the data in a manner similar to that done for CH_3F. The results are

$$\Delta u = 400 \text{ cm/s}$$

$$\gamma_{ab} = (0.126\,P)\,\text{MHz}, \qquad \sigma = 530 \text{ Å}^2$$

$$\Gamma_{vc} = (0.11P)\,\text{MHz}, \qquad \sigma_{vc} = 430 \text{ Å}^2$$

with P in mTorr. The relaxation rate quoted here for γ_{ab} is from Plant and Abrams (1976). These results apply to both the NH_2D $(\nu_2, J, M) = (0_a, 4_{04}, \pm4) \to (1_a, 5_{05}, \pm5)$ and $(0_a, 4_{04}, \pm4) \to (1_s, 5_{14}, \pm5)$ transitions, which exhibit identical relaxation behavior.

It is interesting to note that the collision cross sections obtained here are nearly identical to those for CH_3F. Δu is larger than in CH_3F but is still far smaller than $\bar{u}$ or the bandwidth of molecules excited by microsecond or shorter pulses. We expect that the numbers obtained for NH_2D and CH_3F are typical of the behavior to be expected for dipolar molecules. Little is known about VC collisions in nonpolar molecules.

There is one other type of experiment which has not been discussed yet but which is somewhat relevant to VC collision measurements. This is the coherent Raman beat experiment. The decay behavior of this transient effect is essentially independent of velocity-changing collisions, even those in which the velocity change is comparable to the thermal velocity. Thus, by showing that the coherent Raman beat decay is identical to a level decay rate such as γ_a obtained from delayed nutation measurements, one can completely rule out any possibility that VC collisions with a large Δv_z are occurring and affecting the observed level decay rates.

Coherent Raman beats is the name given a long-lived beat signal which occurs when a three-level system with a common lower (upper) level and two closely spaced upper (lower) levels are excited in a Stark switching experiment. A thorough discussion of the three-level dynamics involved here is beyond the scope of this chapter, so we content ourselves with a simplified, qualitative discussion that will indicate what the effect is and why it is independent of VC collisions. More detailed discussions can be found in Shoemaker and Brewer (1972) and Brewer and Hahn (1973). To see how the effect occurs, consider the situation illustrated in Figure 3-34a. We suppose that initially ($t < 0$) the two upper levels are degenerate and that a laser is resonant with both the $3 \to 1$ and $3 \to 2$ transitions. This causes the system to be in a coherent superposition of all three levels at $t = 0$. We then apply a Stark field which splits levels 1 and 2 by $h\omega_{12}$. Although the previously excited molecules are now no longer in resonance, the intense laser field is still present and causes these molecules to make two-photon transitions between levels 1 and 2, as shown schematically in Figure 3-34b. These transitions are near-resonant Raman transitions in which laser energy at frequency Ω' (in the rest-frame coordinate system of the molecule) is absorbed and a shifted frequency $\omega'_e = \Omega' \pm \omega_{12}$ is radiated. Because of the coherent preparation for $t < 0$ and because the two-photon transitions are nearly resonant with an intermediate state (level 3), the Raman-shifted radiation is quite intense. Furthermore, in the forward direction (the direction of laser beam propagation), the emitting molecules are all in phase with each other, just as in the one-photon effect of optical FID (see Section 3.7.1). As a result, the emitted intensity at ω'_e is highly directional and increases as N^2, where N is the number of emitting molecules. If a detector is placed behind the sample cell, the molecular emission will beat with the laser to produce a heterodyne beat signal of frequency ω_{12} at the detector output. This is the signal that is observed, as shown in Figure 3-35. The large initial

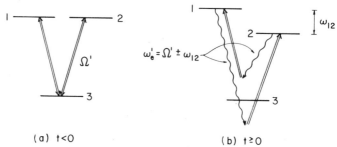

(a) $t < 0$ (b) $t \geq 0$

Figure 3-34. Energy-level diagram for coherent Raman beats. (a) For $t < 0$ the laser is resonant and excites the $3 \to 1$ and $3 \to 2$ transitions. (b) At $t = 0$ a Stark field splits levels 1 and 2, shifting the laser out of resonance. Two-photon transitions then occur in which radiation is emitted at frequency ω'_e.

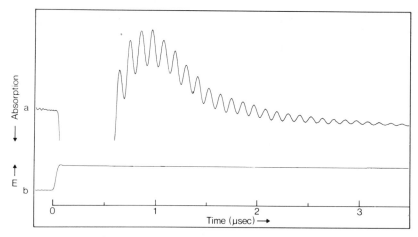

Figure 3-35. Coherent Raman beats in $^{13}CH_3F$ (from Shoemaker and Brewer, 1972). (a) Optical response showing the Raman beats. (b) Applied Stark field.

transient in this figure is an optical nutation signal from molecules which are shifted *into* resonance by the Stark field at $t = 0$.

It is interesting to note that in the original paper reporting the phenomenon of coherent Raman beats (Shoemaker and Brewer, 1972), the effect was initially referred to as "two-photon superradiance," because of the directional N^2 emission characteristics. For the reasons discussed in Section 3.7.2, however, it was felt soon afterward that this was not a very appropriate name, and the effect was renamed coherent Raman beats, which is considerably more accurate and descriptive.

Unlike the situation in FID or photon echo experiments, Doppler dephasing plays no role in the decay of the coherent Raman beat signal. The reason for this is easy to see. In the molecule's rest frame the emitted frequency is $\omega'_e = \Omega' \pm \omega_{12}$. This follows from conservation of energy. In the laboratory frame, however, the emitted frequency is

$$\omega_e = \omega'_e + k_e v_z$$
$$= (\Omega - k v_z) \pm \omega_{12} + k_e v_z$$
$$\approx \Omega \pm \omega_{12}$$

Hence, the emission frequency is the same no matter what the velocity v_z is. The last line in this equation follows from the fact that ω_{12} is typically only a few MHz. Thus, $k \simeq k_e$. What has happened is that the Doppler downshift of the absorbed laser power at frequency Ω is almost exactly canceled by the Doppler upshift of the emitted radiation. This is the characteristic of Raman scattering processes observed in the forward direction.

Since ω_e is identical for all molecular velocities, it is clear that the decay of the coherent Raman beat signal is unaffected by VC collisions, at least to first order. There is a weak velocity dependence of the Raman signal due to the fact that the Raman transition probability depends on the detuning of level 3 from exact resonance, but this is generally a negligible effect. If one finds that γ_1 or γ_2 obtained from delayed nutation measurements is the same as the Raman beat decay (which depends on γ_{12}, the decay of the off-diagonal element ρ_{12}), it is safe to conclude that no VC collisions with large Δv_z are contributing to γ_1 or γ_2. In $^{13}CH_3F$, Berman *et al.* (1975) found $\gamma_1 = \gamma_2 = \gamma_{12}$, which proves that no large Δv_z collisions were occurring in that system. For further details of this experiment and a discussion of some of the complications which can arise, see Shoemaker and Brewer (1972), Brewer and Hahn (1973), and Berman *et al.* (1975).

We have examined the subject of VC collisions and their effect on coherent transient experiments in considerable detail in this section. However, this area of research is clearly not yet fully developed, and a great deal remains to be done both experimentally and theoretically. On the theoretical side, perhaps the most obvious needs are for a model to explain the velocity dependence of VC collisions, for an explanation of the intensity-dependent dephasing phenomenon, and for a detailed understanding of what is measured in a C–P echo train. Experimentally, only a small beginning has been made. Thorough studies of collisional decay rates, including the velocity dependence, clearly need to be made in a number of different molecular gases. By doing so, one might hope to gain some understanding of the relation between molecular structure and collisional relaxation. In addition, it would be highly desirable to develop new experimental methods, particularly those which can be used to provide a good test of the collision kernel. As this research continues, we will undoubtedly gain new insights into both molecular collision phenomena and the interaction of radiation and matter. Also, if past history is any guide, a number of exciting surprises surely lie in store for us.

Appendix A. Justification of the Reduced Wave Equation

In the text we obtained a set of wave equations (3-44) which describe the reradiated field amplitudes $\mathscr{E}_c$ and $\mathscr{E}_s$ produced by a polarization whose in- and out-of-phase components have amplitudes P_c and P_s. Specifically, we found that

$$\frac{\partial^2 \mathscr{E}_c}{\partial z^2} - \frac{1}{c^2}\frac{\partial^2 \mathscr{E}_c}{\partial t^2} - 2k\left(\frac{\partial \mathscr{E}_s}{\partial z} + \frac{1}{c}\frac{\partial \mathscr{E}_s}{\partial t}\right) = \mu_0\frac{\partial^2 P_c}{\partial t^2} + 2\Omega\mu_0\frac{\partial P_s}{\partial t} - \Omega^2\mu_0 P_c$$

$$\frac{\partial^2 \mathscr{E}_s}{\partial z^2} - \frac{1}{c^2}\frac{\partial^2 \mathscr{E}_s}{\partial t^2} + 2k\left(\frac{\partial \mathscr{E}_c}{\partial z} + \frac{1}{c}\frac{\partial \mathscr{E}_c}{\partial t}\right) = \mu_0\frac{\partial^2 P_s}{\partial t^2} - 2\Omega\mu_0\frac{\partial P_c}{\partial t} - \Omega^2\mu_0 P_s$$

$$(3\text{-}A1)$$

We would now like to show that these equations may be reduced to a set of soluble first-order differential equations if the amplitudes $\mathscr{E}_c$, $\mathscr{E}_s$, P_c, and P_s are all slowly varying. To do this we note that the second derivative terms may be neglected in (3-A1) if the following inequalities are satisfied:

$$\left|\frac{\partial^2 \mathscr{E}_c}{\partial z^2}\right| \ll 2k \left|\frac{\partial \mathscr{E}_s}{\partial z}\right| \tag{3-A2a}$$

$$\left|\frac{\partial^2 \mathscr{E}_s}{\partial t^2}\right| \ll 2\Omega \left|\frac{\partial \mathscr{E}_c}{\partial t}\right| \tag{3-A2b}$$

$$\left|\frac{\partial^2 P_c}{\partial t^2}\right| \ll 2\Omega \left|\frac{\partial P_s}{\partial t}\right| \tag{3-A2c}$$

plus a similar set with $\mathscr{E}_c$ and $\mathscr{E}_s$, P_c and P_s interchanged. Consider first the third inequality, (3-A2c). For a sufficiently small interval Δt, $\partial P_s / \partial t \simeq \Delta P_s / \Delta t$. Picking $\Delta t = 1/\Omega$, i.e., one optical period for the light wave, the inequality becomes

$$\left|\frac{\Delta(\Delta P_c)}{1/\Omega^2}\right| \ll 2\Omega \left|\frac{\Delta P_s}{1/\Omega}\right| \tag{3-A3}$$

which is just

$$|\Delta(\Delta P_c)| \ll 2|\Delta P_s|$$

Now P_s and P_c are not independent quantities because they are coupled through the population matrix equations of motion [see Eqs. (3-67) and (3-69)]. Thus, any time there is a change in P_s, there will be a corresponding change in P_c and in general we find that ΔP_c is of the same order of magnitude as ΔP_s.* This being the case, we may replace ΔP_s by ΔP_c in (3-A3), to obtain

$$|\Delta(\Delta P_c)| \ll 2|\Delta P_c| \tag{3-A4}$$

Stated in words, (3-A4) requires that the change in ΔP_c during one optical period must be much less than ΔP_c itself. But this must be true, since a large fractional change in ΔP_c during one optical period would imply that P_c had Fourier components near Ω, contrary to our initial hypothesis of slowly varying amplitudes. Hence, the third inequality in (3-A2) is valid.

The same argument may be repeated for (3-A2b) provided that we can show that $\Delta \mathscr{E}_c$ is of the same order of magnitude as $\Delta \mathscr{E}_s$. This is hard to do rigorously. Intuitively, however, it is clear that if the in- and out-of-phase polarization components P_c and P_s are coupled, then the in- and out-of-phase field components $\mathscr{E}_c$ and $\mathscr{E}_s$ radiated by P_c and P_s must be coupled

---

*The one exception is the case of exact resonance, where $\Omega' - \omega_0 = 0$. The coupling between P_c and P_s then vanishes and our argument must be modified, as discussed at the end of this appendix.

also. Hence, we expect that $\Delta\mathcal{E}_c$ and $\Delta\mathcal{E}_s$ during one optical period will have similar magnitudes, since ΔP_s and ΔP_c do, and (3-A2b) can be justified. In a similar manner, (3-A2a) may be justified by considering the spatial change in $\mathcal{E}_c$ and $\mathcal{E}_s$ over one optical wavelength.

Having made all the inequalities in (3-A2) at least plausible, we drop all second derivative terms in Eq. (3-A1). In addition, we will drop the first derivative terms in P_s and P_c. This can be done if we have

$$\left|\frac{\partial P_s}{\partial t}\right| \ll \frac{\Omega}{2} P_c \tag{3-A5}$$

and similarly for P_c and P_s interchanged. But this inequality holds also for essentially the same reasons that we used to justify Eqs. (3-A2); i.e., P_c and P_s cannot change much during one optical period if they are slowly varying amplitudes.

On using Eqs. (3-A2) and (3-A5) we find that the wave equations (3-A1) become

$$\frac{\partial\mathcal{E}_c}{\partial z} + \frac{1}{c}\frac{\partial\mathcal{E}_c}{\partial t} = -\frac{\Omega}{2\varepsilon_0 c}P_s$$

$$\frac{\partial\mathcal{E}_s}{\partial z} + \frac{1}{c}\frac{\partial\mathcal{E}_s}{\partial t} = \frac{\Omega}{2\varepsilon_0 c}P_c \tag{3-A6}$$

These are the desired reduced wave equations which are given in the text as Eqs. (3-45). We note in passing that a similar set of reduced wave equations may be obtained for the case of plane waves in an optically thick medium, except that there the arguments justifying the reduction become somewhat more subtle, especially where the dropping of the first derivatives of **P** are concerned (Hopf, 1974).

There is some interesting physics hidden in these wave equations. Notice, for example, that the polarization component P_s, which is 90° out of phase with the driving field $E_0 \cos{(\Omega t - kz)}$ produces a reradiated field $\mathcal{E}_c$ which is in phase with the driving field. On the surface this does not appear to make sense physically, since an oscillating dipole clearly must radiate a field having the same phase as the dipole's oscillation. However, we assumed a plane-wave solution for the wave equation, which means that we are looking at the radiation from many dipoles at different points in space, not just one dipole. A useful point of view is to envision the medium as being made up of many thin sheets of molecules, with each sheet perpendicular to the propagation direction of the light. One can then show that the net field from each sheet of dipoles is shifted by 90° from the oscillation phase of any dipole in the sheet as Eq. (3-A6) indicates. Good discussions of the radiation from a dipole sheet may be found in Sargent *et al.* (1974) and in Feynman *et al.* (1963). In the reduced wave equations the phase shift is automatically built into the equations when the assumption of plane-wave solutions is made.

One final note concerning the reduced wave-equations. When a transition is excited exactly on line center, one finds that $P_c = 0$ and the reduced wave equations then predict that $\mathscr{E}_s = 0$. On the surface this appears to be a problem since we assumed that $|\partial^2 \mathscr{E}_c / \partial t^2| \ll 2k |\partial \mathscr{E}_s / \partial t|$ to obtain those equations. Although this inequality cannot be true if $\mathscr{E}_s = 0$, our reduced wave equations are still a good approximation. To see this note that for $P_c = 0$, Eqs. (3-A1) become

$$2k\left(\frac{\partial \mathscr{E}_s}{\partial z} + \frac{1}{c}\frac{\partial \mathscr{E}_s}{\partial t}\right) = \frac{\partial^2 \mathscr{E}_c}{\partial z^2} - \frac{1}{c^2}\frac{\partial^2 \mathscr{E}_c}{\partial t^2} - 2\Omega\mu_0\frac{\partial P_s}{\partial t}$$

$$2k\left(\frac{\partial \mathscr{E}_c}{\partial z} + \frac{1}{c}\frac{\partial \mathscr{E}_c}{\partial t}\right) = -\Omega^2\mu_0 P_s \tag{3-A7}$$

To obtain the second equation here we have neglected $\partial^2 P_s / \partial t^2$ compared to $\Omega^2 P_s$ and dropped the second derivatives of $\mathscr{E}_s$ compared to first derivatives of $\mathscr{E}_c$. The second equation is then already one of the reduced wave equations and can be solved to find $\mathscr{E}_c$. The first equation also looks like a reduced wave equation except that instead of having zero, we have

$$\frac{\partial^2 \mathscr{E}_c}{\partial z^2} - \frac{1}{c^2}\frac{\partial^2 \mathscr{E}_c}{\partial t^2} - 2\Omega\mu_0\frac{\partial P_s}{\partial t}$$

for the driving term. However, because all amplitudes are slowly varying, these derivatives are very small quantities, and thus the error made in neglecting them will not be serious. The difference is that instead of obtaining an extremely small value for $\mathscr{E}_s$ (because the driving terms are very small), we find that $\mathscr{E}_s = 0$ using the reduced wave equations.

Appendix B. Matrix Formulation of the Bloch Equations

3.B1. Matrix Solution of the Optical Bloch Equations

In the first part of this appendix we show that the matrix form of the optical Bloch equations can be solved just as if it were a simple scalar equation. The equation of interest is

$$\dot{\mathbb{M}} = -(\gamma_{ab} + \mathbb{B})\mathbb{M} + \mathbb{A} \tag{3-B1}$$

In order to work with this equation we first need to know what a function which has a matrix for its argument means. This is simply a matter of definition. $f(\mathbb{B})$, where f is some function and $\mathbb{B}$ is a square matrix, is defined to be the matrix obtained from a power-series expansion of f in powers of $\mathbb{B}$. Thus, $\exp(\mathbb{B})$ means that

$$e^{\mathbb{B}} \equiv 1 + \mathbb{B} + \frac{1}{2!}\mathbb{B}^2 + \frac{1}{3!}\mathbb{B}^3 + \cdots \tag{3-B2}$$

Here $\mathbb{B}^n$ means the matrix multiplication of $\mathbb{B}$ with itself n times, for example, $\mathbb{B}^3 = \mathbb{B}\mathbb{B}\mathbb{B}$. Obviously, the power-series expansion of f must exist and be convergent; otherwise, the definition is meaningless. Clearly, $f(\mathbb{B})$ is itself a square matrix whose elements can in principle be calculated from the elements of $\mathbb{B}$.

The exponential function $\exp(\mathbb{B}t)$ is the only matrix function we will need, so we confine our attention to it. Using (3-B2), we can now demonstrate several useful properties of this function. First,

$$\mathbb{B}e^{\mathbb{B}t} = e^{\mathbb{B}t}\mathbb{B} \tag{3-B3}$$

This follows immediately from (3-B2) and the fact that a scalar such as t commutes with any matrix. Note that matrix multiplication is not commutative in general, however. Next we see that

$$\frac{d}{dt}e^{\mathbb{B}t} = \mathbb{B} + \mathbb{B}^2 t + \frac{1}{2!}\mathbb{B}^3 t^2 + \cdots = \mathbb{B}e^{\mathbb{B}t} \tag{3-B4}$$

provided that $\mathbb{B}$ is not itself a function of time. Also,

$$\mathbb{B}\int e^{\mathbb{B}t} = \mathbb{B}\left(t + \frac{1}{2!}\mathbb{B}t^2 + \frac{1}{3!}\mathbb{B}^2 t^3 + \cdots\right) + \mathbb{B}C'$$

$$= e^{\mathbb{B}t} - 1 + \mathbb{B}C' \tag{3-B5}$$

where C' is a matrix whose elements are the constants of integration. Thus,

$$\int e^{\mathbb{B}t} = \mathbb{B}^{-1}e^{\mathbb{B}t} + C \tag{3-B6}$$

where $\mathbb{B}^{-1}$ denotes the inverse of $\mathbb{B}$ and $C = C' - \mathbb{B}^{-1}$. Finally, we note that

$$e^{\mathbb{B}t}e^{-\mathbb{B}t} = \left(1 + \mathbb{B}t + \frac{1}{2!}\mathbb{B}^2 t^2 + \cdots\right)\left(1 - \mathbb{B}t + \frac{1}{2!}\mathbb{B}^2 t^2 - \cdots\right)$$

$$= 1 \tag{3-B7}$$

so

$$e^{-\mathbb{B}t} = (e^{\mathbb{B}t})^{-1} \tag{3-B8}$$

Taken together, Eqs. (3-B3)–(3-B8) allow us to manipulate the matrix differential equation for M essentially as if it were a scalar equation. When this is done, we find that

$$\mathsf{M}(t) = e^{-(\gamma_{ab} + \mathbb{B})t'}\mathsf{M}(t_k) + [1 - e^{-(\gamma_{ab} + \mathbb{B})t'}](\gamma_{ab} + \mathbb{B})^{-1}\mathsf{A} \tag{3-B9}$$

where $t' = t - t_k$, as described in Section 3.4.3. Finally, we note that the matrix exponentials in (3-B9) can be simplified because

$$e^{-\gamma_{ab}t'}e^{-\mathbb{B}t} = (1 - \gamma_{ab}t + \tfrac{1}{2}\gamma_{ab}^2 t^2 - \cdots)(1 - \mathbb{B}t + \tfrac{1}{2}\mathbb{B}^2 t^2 - \cdots)$$

$$= 1 - (\gamma_{ab} + \mathbb{B})t + \tfrac{1}{2}(\gamma_{ab} + \mathbb{B}^2 t^2 - \cdots)$$

$$= e^{-(\gamma_{ab} + \mathbb{B})t} \tag{3-B10}$$

3.B2. Evaluation of exp $(-\mathbb{B}t')$

All that we have discovered about the matrix exp $(-\mathbb{B}t')$ thus far is its power-series expansion, and several algebraic properties which follow from that expansion. Since it is not a closed form, the power series does not look particularly promising as a method of evaluating the exponential explicitly. However, in this section we show that the expansion can be rewritten as a polynomial with no approximations! To accomplish this we follow the original presentation of Jaynes (1955). Notice first that any $n \times n$ square matrix such as $\mathbb{B}$ has a set of eigenvalues which may be found from the secular determinant

$$|\mathbb{B} - \lambda 1| = 0 \tag{3-B11}$$

To actually find the eigenvalues, one expands the determinant, obtaining a polynomial in λ, which we write as

$$\sum_{k=0}^{n} C_k \lambda^k = 0 \tag{3-B12}$$

The various roots λ of this polynomial are the eigenvalues of the matrix. Now there is a well-known theorem in linear algebra, called the Cayley–Hamilton theorem, which states that given a square matrix $\mathbb{B}$ having eigenvalues given by Eq. (3-B12), the equation

$$\sum_{k=0}^{n} C_k \mathbb{B}^k = 0 \tag{3-B13}$$

also holds with the C_k's in (3-B12) and (3-B13) being identical. The proof of this result is not trivial but may be found in almost any linear algebra textbook (cf. Hohn, 1958).

To apply this theorem, we expand the secular determinant for $\mathbb{B}$ to find the polynomial corresponding to (3-B12). The result is

$$-\lambda^3 + \Delta\gamma\lambda^2 - g^2\lambda + \Delta\gamma\delta^2 = 0 \tag{3-B14}$$

where $g^2 = \delta^2 + \kappa^2 E_0^2$. Thus, from the Cayley–Hamilton theorem,

$$\mathbb{B}^3 - \Delta\gamma\mathbb{B}^2 + g^2\mathbb{B} - \delta^2\Delta\gamma 1 = 0 \tag{3-B15}$$

This result can now be used to truncate the power-series expansion of the exponential,

$$e^{-\mathbb{B}t'} = \mathbb{1} - \mathbb{B}t' + \frac{1}{2!}\mathbb{B}^2(t')^2 - \frac{1}{3!}\mathbb{B}^3(t')^3 + \cdots \qquad (3\text{-}B16)$$

To do this we notice that any given term in (3-B16) of order $\mathbb{B}^3$ or higher may be removed by applying Eq. (3-B15). For example, the $\mathbb{B}^3$ term may be removed by multiplying Eq. (3-B15) by $(1/3!)(t')^3$ and adding the result to (3-B16). If we now consider starting with the term of order $\mathbb{B}^n$, n an arbitrarily large number, we can eliminate it as just described, then repeat the procedure for the term of order $\mathbb{B}^{n-1}$, and so on. The end result of these eliminations will be an expression

$$e^{-\mathbb{B}t'} = a_0\mathbb{1} + a_1\mathbb{B} + a_2\mathbb{B}^2 \qquad (3\text{-}B17)$$

where a_0, a_1, and a_2 are scalar functions of $t' = t - t_k$. Of course, it is not feasible to actually carry out the successive eliminations to find a_0, a_1, and a_2. To determine these, we differentiate (3-B17) using (3-B4) to rewrite the left-hand side. This gives us

$$-a_0\mathbb{B} - a_1\mathbb{B}^2 - a_2\mathbb{B}^3 = \dot{a}_0\mathbb{1} + \dot{a}_1\mathbb{B} + \dot{a}_2\mathbb{B}^2 \qquad (3\text{-}B18)$$

or, using (3-B15) to eliminate the $\mathbb{B}^3$ term,

$$\dot{a}_0\mathbb{1} + \dot{a}_1\mathbb{B} + \dot{a}_2\mathbb{B}^2 = -(\delta^2 \Delta\gamma a_2)\mathbb{1} + (g^2 a_2 - a_0)\mathbb{B} - (\Delta\gamma a_2 + a_1)\mathbb{B}^2 \qquad (3\text{-}B19)$$

This equation must hold for all values of the matrix $\mathbb{B}$, so we may equate coefficients to obtain

$$\dot{a}_0 = -\delta^2 \Delta\gamma a_2$$
$$\dot{a}_1 = g^2 a_2 - a_0 \qquad (3\text{-}B20)$$
$$\dot{a}_2 = -\Delta\gamma a_2 - a_1$$

These equations may now be solved for the a's to give an explicit expression for $\exp(-\mathbb{B}t')$. They are reproduced as Eqs. (3-90).

3.B3. Solution of the Bloch Equations for Strong Optical Fields

A general solution to the optical Bloch equations would be obtained if Eqs. (3-B20) could be solved. Unfortunately, this cannot be done, as the solutions involve the roots of a cubic equation. This may be seen by differentiating the $\dot{a}_2$ equation twice to obtain an equation for a_2 only:

$$\dddot{a}_2 + \Delta\gamma\ddot{a}_2 + g^2\dot{a}_2 + \delta^2 \Delta\gamma a_2 = 0 \qquad (3\text{-}B21)$$

A homogeneous linear differential equation such as this will have solutions of the form $\exp(\lambda t')$. On substituting in a solution of this form we find an equation for λ:

$$\lambda^3 + \Delta\gamma\lambda^2 + g^2\lambda + \delta^2\,\Delta\gamma = 0 \tag{3-B22}$$

An explicit expression for the roots of this cubic is too cumbersome to be of any use, so we must look for special cases where the equation simplifies. The two obvious situations where this will happen are when δ or $\Delta\gamma$ are zero. $\Delta\gamma = 0$ is the case discussed in detail in the text, while $\delta = 0$ is just not very useful. However, there is also one other useful situation where an approximate solution may be found. This is the high-power case, in which $|\kappa E_0| \gg |\Delta\gamma|$, first discussed by Torrey in 1949 in connection with nmr. To obtain the solution we need to extract one real root of (3-B22). This is done by rewriting (3-B22) as

$$\lambda(\lambda^2 + g^2) + \Delta\gamma(\lambda^2 + \delta^2) = 0 \tag{3-B23}$$

Thus,

$$\lambda = -\Delta\gamma\frac{\lambda^2 + \delta^2}{\lambda^2 + g^2}$$

or

$$\lambda = -\Delta\gamma\left(1 - \frac{g^2 - \delta^2}{g^2 + \lambda^2}\right) \tag{3-B24}$$

Now a cubic equation must have one real root, and if λ is real, we have

$$0 \le \frac{g^2 - \delta^2}{g^2 + \lambda^2} \le 1$$

It then follows from (3-B24) that $|\lambda| \le |\Delta\gamma|$. Now suppose that the optical field is sufficiently large that $|\Delta\gamma| \ll \kappa E_0$. Then $|\Delta\gamma| \ll g = (\delta^2 + \kappa^2 E_0^2)^{1/2}$, and hence $|\lambda| \ll g$. This means that λ^2 is negligible compared to g^2, and Eq. (3-B24) becomes

$$\lambda = -\Delta\gamma\left(1 - \frac{g^2 - \delta^2}{g^2}\right) = -\frac{\delta^2\,\Delta\gamma}{g^2} \tag{3-B25}$$

Thus we have extracted an approximate root of (3-B22) provided that $|\Delta\gamma| \ll g$. Since $\lambda = -\delta^2\,\Delta\gamma/g^2$ is a root of Eq. (3-B22), that equation may be rewritten as

$$\left(\lambda + \frac{\delta^2\,\Delta\gamma}{g^2}\right)(\lambda^2 + b\lambda + c) = 0 \tag{3-B26}$$

b and c can be determined by comparing this equation with Eq. (3-B22) and equating coefficients of equal powers of λ to obtain

$$b + \delta^2 \Delta\gamma/g^2 = \Delta\gamma$$
$$c + b\delta^2 \Delta\gamma/g^2 = g^2 \qquad \text{(3-B27)}$$
$$c\delta^2 \Delta\gamma/g^2 = \delta^2 \Delta\gamma$$

This is a consistent set of equations for b and c provided that terms of order $\Delta\gamma^2/g^2$ are neglected compared to unity. We find $b = \Delta\gamma(1 - \delta^2/g^2)$ and $c = g^2$. The quadratic $\lambda^2 + b\lambda + c = 0$ may next be solved to find the remaining two roots of λ, which are

$$\lambda = -\frac{\Delta\gamma}{2}(1 - \delta^2/g^2) \pm ig \qquad \text{(3-B28)}$$

if we again ignore $\Delta\gamma^2/g^2$ compared to unity. We have now found a set of approximate roots for the cubic equation (3-B22) and thus can write down the general solution for a_2:

$$a_2 = Ae^{-\eta t'} + Be^{-\rho t'} \sin gt' + Ce^{-\rho t'} \cos gt' \qquad \text{(3-B29)}$$

where

$$\eta = \frac{\delta^2 \Delta\gamma}{g^2}$$

$$\rho = \frac{\Delta\gamma}{2}[1 - (\delta^2/g^2)]$$

The constants A, B, and C must be determined from the initial conditions. As discussed in the text following Eqs. (3-91), we have $a_0(0) = 1$, $a_1(0) = a_2(0) = 0$ at $t' = 0$. Using these relations and Eqs. (3-B20), we find that $\dot{a}_2(0) = 0$ and $\ddot{a}_2(0) = 1$. We may now obtain a set of three equations for A, B, and C by using (3-B29) and its first and second derivatives, all evaluated at $t' = 0$. This gives us

$$0 = A + C$$
$$0 = -\eta A + gB - \rho C \qquad \text{(3-B30)}$$
$$1 = \eta^2 A - 2\rho gB + (\rho^2 - g^2)C$$

which are readily solved to yield $A = 1/g^2$, $B = (\eta - \rho)/g^3$, $C = -1/g^2$. We have neglected $\Delta\gamma^2/g^2$ compared to 1 to obtain A, B, and C in the form shown. Thus,

$$a_2 = \frac{1}{g^2}\left[e^{-\eta t'} + e^{-\rho t'}\left(\frac{\eta - \rho}{g}\sin gt' - \cos gt'\right)\right] \qquad \text{(3-B31)}$$

The coefficient a_1 is now readily obtained by noting that according to Eqs. (3-B20), $a_1 = -\Delta\gamma a_2 - \dot{a}_2$. This gives us

$$a_1 = \frac{1}{g^2}[-2\rho e^{-\eta t'} - e^{-\rho t'}(g \sin gt' - 2\rho \cos gt')] \qquad (3\text{-}B32)$$

Finally, we also have from Eqs. (3-B20) that $a_0 = g^2 a_2 - \dot{a}_1$, so

$$a_0 = e^{-\eta t'} + \frac{\eta}{g} e^{-\rho t'} \sin gt' \qquad (3\text{-}B33)$$

In the last two expressions for a_1 and a_0 we have again neglected terms of order $\Delta\gamma^2/g^2$ compared to unity. We now have expressions for a_0, a_1, and a_2 in the high-power case, where $|\kappa E_0| \gg |\Delta\gamma|$. These coefficients go into Eq. (3-B17), which written out explicitly is

$$e^{-\mathbf{B}t'} = \begin{pmatrix} a_0 - a_2\delta^2 & -a_1\delta & a_2\delta\kappa E_0 \\ a_1\delta & a_0 - a_2 g^2 & -a_1\kappa E_0 - a_2\,\Delta\gamma\kappa E_0 \\ a_2\delta\kappa E_0 & a_1\kappa E_0 + a_2\,\Delta\gamma\kappa E_0 & a_0 - a_1\,\Delta\gamma - a_2\kappa^2 E_0^2 \end{pmatrix}$$

$$(3\text{-}B34)$$

Given an arbitrary $\mathbf{M}(t_\kappa)$, $\mathbf{M}(t)$ for a later time can be found by simply multiplying $\exp(-\gamma_{ab}t')\mathbf{M}(t_\kappa)$ by $\exp(-\mathbf{B}t')$ if steady-state terms are ignored. Thus, we have found a complete solution (to first order in $\Delta\gamma/g$) for the optical Bloch equations when a strong optical field is present. Although complicated, the solution is relatively easy to handle on a computer.

References

Abragam, A., 1961, *The Principles of Nuclear Magnetism*, Oxford University Press, Inc., New York.

Allen, L., and Eberly, J. H., 1975, *Optical Resonance and Two-Level Atoms*, John Wiley & Sons, Inc., New York.

Anderson, P. W., 1949, *Phys. Rev.* **76**:647.

Berman, P. R., 1972, *Phys. Rev.* **A5**:927.

Berman, P. R., 1974, *Phys. Rev.* **A9**:2170.

Berman, P. R., 1975, *Appl. Phys.* **6**:283.

Berman, P. R., and Lamb, W. E., Jr., 1970, *Phys. Rev.* **A2**:2435.

Berman, P. R., Levy, J. M., and Brewer, R. G., 1975, *Phys. Rev.* **A11**:1668.

Bloembergen, N., and Pound, R. V., 1954, *Phys. Rev.* **95**:8.

Bloom, A. L., 1955, *Phys. Rev.* **98**:1105.

Bölger, B., and Diels, J. C., 1968, *Phys. Lett.* **28A**:401.

Bordé, C. J., Hall, J. L., Kunasz, C. V., and Hummer, D. G., 1976, *Phys. Rev.* **A14**:236.

Brewer, R. G., and Genack, A. Z., 1976, *Phys. Rev. Lett.* **36**:959.

Brewer, R. G., and Hahn, E. L., 1973, *Phys. Rev.* **A8**:464.

Brewer, R. G., and Shoemaker, R. L., 1971, *Phys. Rev. Lett.* **27**:631.

Brewer, R. G., and Shoemaker, R. L., 1972, *Phys. Rev.* **A6**: 2001.

Carr, H. Y., and Purcell, E. M., 1954, *Phys. Rev.* **94**:630.

Cheo, P. K., and Wang, C. H., 1970, *Phys. Rev.* **A1**:225.

Corben, H. C., and Stehle, P., 1960, *Classical Mechanics*, 2nd ed., John Wiley & Sons, Inc., New York.

Crisp, M. D., 1969, *Opt. Commun.* **1**:59.

Crisp, M. D., 1973, *Phys. Rev.* **A8**:2128.

Dicke, R. H., 1954, *Phys. Rev.* **93**:99.

Feynman, R. P., Vernon, F. L., and Hellwarth, R. W., 1957, *J. Appl. Phys.* **28**:49.

Feynman, R. P., Leighton, R. B., and Sands, M., 1963, *The Feynman Lectures on Physics*, Addison-Wesley Publishing Company, Inc., Reading, Mass.

Fiutak, J., 1963, *Can. J. Phys.* **41**:12.

Foster, K. L., Stenholm, S., and Brewer, R. G., 1974, *Phys. Rev.* **A10**:2318.

Gerry, E. T., and Leonard, D. A., 1966, *Appl. Phys. Lett.* **8**:227.

Goldstein, H., 1950, *Classical Mechanics*, Addison-Wesley Publishing Company, Inc., Reading, Mass.

Gordon, J. P., Wang, C. H., Patel, C. K. N., Slusher, R. E., and Tomlinson, W. J., 1969, *Phys. Rev.* **179**:294.

Grischkowsky, D., 1975, Adiabatic Following, in: *Laser Applications to Optics and Spectroscopy, Physics of Quantum Electronics* (S. F. Jacobs *et al.*, eds.), Vol. 2, pp. 437–452, Addison-Wesley Publishing Company, Inc., Reading, Mass.

Grischkowsky, D., and Loy, M. M. T., 1975, *Phys. Rev.* **A12**:1117.

Gross, M., Fabre, C., Pillet, P., and Haroche, S., 1976, *Phys. Rev. Lett.* **36**:1035.

Grossman, S. B., Schenzle, A., and Brewer, R. G., 1977, *Phys. Rev. Lett.* **38**:275.

Hahn, E. L., 1950a, *Phys. Rev.* **77**:297.

Hahn, E. L., 1950b, *Phys. Rev.* **80**:580.

Hall, J. L., 1971, The lineshape problem in laser-saturated molecular absorption, in: *Lectures in Theoretical Physics* (K. T. Mahanthappa and W. E. Brittin, eds.), Vol. 12A, pp. 161–210, Gordon and Breach Science Publishers, Inc., New York.

Hall, J. L., 1973, Saturated absorption spectroscopy with applications to the 3.39 μm methane transition, in: *Atomic Physics 3* (S. J. Smith and G. K. Walters, eds.), pp. 615–646, Plenum Publishing Corporation, New York.

Hall, J. L., 1975, Colloques Internationaux du C.N.R.S., No. 217. Reprinted in: *Laser Applications to Optics and Spectroscopy, Physics of Quantum Electronics* (S. F. Jacobs *et al.*, eds.), Vol. 2, pp. 401–421, Addison-Wesley Publishing Company, Inc., Reading, Mass.

Hamadani, S. M., Mattick, A. T., Kurnit, N. A., and Javan, A., 1975, *Appl. Phys. Lett.* **27**:21.

Haroche, S., Paisner, J. A., and Schawlow, A. L., 1973, *Phys. Rev. Lett.* **30**:948.

Hartmann, S. R., 1968, *IEEE J. Quantum Electron.* **QE4**:802.

Heer, C. V., 1974a, *Phys. Lett.* **49A**:213.

Heer, C. V., 1974b, *Phys. Rev.* **A10**:2112.

Heer, C. V., and Nordstrom, R. J., 1975, *Phys. Rev.* **A11**:536.

Hocker, G. B., and Tang, C. L., 1968, *Phys. Rev. Lett.* **21**:591.

Hocker, G. B., and Tang, C. L., 1969, *Phys. Rev.* **184**:356.

Hohn, F. E., 1958, *Elementary Matrix Algebra*, Macmillan Publishing Co., Inc., New York.

Hopf, F. A., 1974, Amplifier theory, in: High Energy Lasers and Their Applications, *Physics of Quantum Electronics* (S. F. Jacobs *et al.*, eds.), Vol. 1, pp. 77–176, Addison-Wesley Publishing Company, Inc., Reading, Mass.

Hopf, F. A., and Scully, M. O., 1969, *Phys. Rev.* **179**:399.

Hopf, F. A., Shea, R. F., and Scully, M. O., 1973, *Phys. Rev.* **A7**:2105.

Horwitz, P., 1975, *Appl. Phys. Lett.* **26**:306.

Jackson, J. D., 1962, *Classical Electrodynamics*, John Wiley & Sons, Inc., New York.

Jaynes, E. T., 1955, *Phys. Rev.* **98**:1099.

Jetter, H., Pearson, E. F., Norris, C. L., McGurk, J. C., and Flygare, W. H., 1973, *J. Chem. Phys.* **59**:1796.

Keilson, J., and Storer, J. E., 1952, *Quart. of Appl. Math.* **10**:243.

Kiefer, J. E., Nussmeier, T. A., and Goodwin, F. E., 1972, *IEEE J. Quantum Electron.* **QE8**:173.

Kurnit, N. A., Abella, I. D., and Hartmann, S. R., 1964, *Phys. Rev. Lett.* **13**:567.

Lamb, W. E., Jr., 1976, Notes on the use of density matrices in quantum electronics, in: *Laser Photochemistry, Tunable Lasers and Other Topics, Physics of Quantum Electronics* (S. F. Jacobs *et al.*, eds.), Vol. 4, pp. 403–420, Addison-Wesley Publishing Company, Inc., Reading, Mass.

Lambert, L. Q., 1973, *Phys. Rev.* **B7**:1834.

Lambert, L. Q., Compaan, A., and Abella, I. D., 1971, *Phys. Rev.* **A4**:2022.

Lehmberg, R. H., and Reintjes, J., 1975, *Phys. Rev.* **A12**:2574.

Liao, P. F., and Hartmann, S. R., 1973, *Phys. Lett.* **A44**:361.

Liu, W., and Marcus, R. A., 1975, *J. Chem. Phys.* **63**:272.

Loy, M. M. T., 1974, *Phys. Rev. Lett.* **32**:814.

Macomber, J. D., 1976, *The Dynamics of Spectroscopic Transitions*, John Wiley & Sons, Inc., New York.

Mattick, A. T., Sanchez, A., Kurnit, N. A., and Javan, A., 1973, *Appl. Phys. Lett.* **23**:675.

McCall, S. L., and Hahn, E. L., 1967, *Phys. Rev. Lett.* **18**:908.

McDowell, R. S., Galbraith, H. W., Krohn, B. J., and Cantrell, C. D., 1976, *Opt. Comm.* **17**:178.

Meckley, J. R., and Heer, C. V., 1973, *Phys. Lett.* **A46**:41.

Mollow, B. R., 1969, *Phys. Rev.* **188**:1969.

Nebenzahl, I., and Szöke, A., 1974, *Appl. Phys. Lett.* **25**:327.

Nordstrom, R. J., Gutman, W. M., and Heer, C. V., 1974, *Phys. Lett.* **A50**:25.

Patel, C. K. N., and Slusher, R. E., 1968, *Phys. Rev. Lett.* **20**:1087.

Plant, T. K., and Abrams, R. L., 1976, *J. Appl. Phys.* **47**:4006.

Rabi, I. I., 1937, *Phys. Rev.* **51**:652.

Sargent, M., Scully, M. O., and Lamb, W. E., Jr., 1974, *Laser Physics*, Addison-Wesley Publishing Company, Inc., Reading, Mass.

Schenzle, A., and Brewer, R. G., 1976, *Phys. Rev.* **A14**:1756.

Schenzle, A., Grossman, S., and Brewer, R. G., 1976, *Phys. Rev.* **A13**:1891.

Schmidt, J., Berman, P. R., and Brewer, R. G., 1973, *Phys. Rev. Lett.* **31**:1103.

Schuda, F., Stroud, C. R., Jr., and Hercher, M., 1974, *J. Phys.* **B7**:L198.

Scully, M. O., Stephen, M. J., and Burnham, D. C., 1968, *Phys. Rev.* **171**:213.

Shoemaker, R. L., and Brewer, R. G., 1972, *Phys. Rev. Lett.* **28**:1430.

Shoemaker, R. L., and Hopf, F. A., 1974, *Phys. Rev. Lett.* **33**:1527.

Shoemaker, R. L., and Van Stryland, E. W., 1976, *J. Chem. Phys.* **64**: 1733.

Shoemaker, R. L., Stenholm, S., and Brewer, R. G., 1974, *Phys. Rev.* **A10**:2037.

Skribanowitz, N., Herman, I. P., MacGillivray, J. C., and Feld, M. S., 1973, *Phys. Rev. Lett.* **30**:309.

Slichter, C. P., 1963, *Principles of Magnetic Resonance*, Harper & Row, Publishers, New York.

Tang, C. L., and Silverman, B. D., 1966, Dynamic effects on the propagation of intense light pulses in optical media, in: *Physics of Quantum Electronics* (P. Kelley *et al.*, eds.), pp. 280–293, McGraw-Hill Book Company, New York.

Torrey, H. C., 1949, *Phys. Rev.* **76**:1059.

Treacy, E. B., and De Maria, A. J., 1969, *Phys. Lett.* **29A**:369.

Van Stryland, E. W., 1976, Ph.D. thesis, University of Arizona.

Yablonovitch, E., and Goldhar, J., 1974, *Appl. Phys. Lett.* **25**:580.

Yariv, A., 1964, *Proc. IEEE* **52**:719.

4

Coherent Spectroscopy in Electronically Excited States

Charles B. Harris and William G. Breiland

4.1. Introduction

This chapter deals with one of the new generation of coherence experiments that have been developing in the past few years. In this instance the coherence techniques are applied to levels within electronically excited states, and the coherence is monitored by a double-resonance method. Although the techniques are applicable to many traditional magnetic resonance experiments, they also form the basis for investigating molecular dynamics in spectroscopy and promise to reveal a great deal about the time evolution of excited states and the nature of time-dependent interactions between molecular states in solids. The specific development given here will be confined to coherence experiments in the molecular excited triplet states in zero field; however, it is not too difficult to visualize how the methods could be applied to a wide variety of other systems and problems, some of which are outlined in other chapters of this volume. It is important to stress that the generality of the approach makes it useful to any double-resonance experiment in a multilevel system regardless of whether the levels are associated with spin systems or electronic states, or both, as is the case with optically detected magnetic resonance. For this reason we will develop the theory rather completely.

Charles B. Harris • Department of Chemistry, and Materials and Molecular Research Division of Lawrence Berkeley Laboratory, University of California, Berkeley, California   *William G. Breiland* • Department of Chemistry, University of Illinois, Urbana, Illinois

Our philosophy, after a brief introduction, is to divide the material into three major categories; a theoretical section, an experimental section, and an applications section. A thorough discussion will be presented with the hope that each section might be self-contained.

4.1.1. Historical Development

The history of optically detected magnetic resonance (ODMR) originates from the combination of the insightful early work of Lewis and Kasha (1944, 1945) on the nature of the triplet states, from the first ESR studies of organic triplet states by Hutchison and Mangum (1958, 1961), and from the pioneering work of Geschwind *et al.* (1965), who demonstrated that *optically detected* magnetic resonance was possible in *solids* by reporting the optically detected ESR of the metastable (^2E) state of Cr^{3+} in Al_2O_3. At this point several investigators attempted to do similar experiments in organic triplet states, the first successful experiments being reported by Sharnoff (1967) and Kwiram (1967).

Optically detected magnetic resonance not only increased the number of measurable triplet-state ESR systems by orders of magnitude, but it later developed into a powerful spectroscopic tool. Schmidt and van der Waals, in 1968, established in a *zero-field* double-resonance ESR experiment that, owing to symmetry considerations, intersystem crossing occurs unequally in a highly preferential manner to the individual zero-field spin sublevels and results in a highly non-Boltzmann distribution between the sublevels at sufficiently low temperatures. These observations prompted a variety of new experiments along the lines of magnetic resonance and also led to the subsequent development of phosphorescence microwave double resonance (PMDR) by Tinti *et al.* (1969). In PMDR one uses a high-resolution spectrometer to monitor the phosphorescence from individual sites or from individual vibronic transitions, while one applies a microwave field to couple two of the three triplet sublevels. In this fashion, PMDR can be used to monitor and assign spectral lines originating from selective spin sublevels. It became readily apparent that a very informative spectroscopic probe had been discovered. The consequence of utilizing the sensitivity and selectivity of optically detected magnetic resonance led in short succession to many varied experiments aimed at understanding many new properties of triplet states. The relative rates of intersystem crossing and radiative decay from the triplet state were found by a variety of methods (Schmidt *et al.*, 1969; El-Sayed and Olmsted, 1971; Winscom and Maki, 1971; Harris and Hoover, 1972). The zero-field esr absorption spectra were explained in terms of the dipole–dipole, nuclear quadrupole, and nuclear hyperfine interactions (Schmidt and van der Waals, 1968). Optically detected

ENDOR (Harris *et al.*, 1969; Chan *et al.*, 1969) and EEDOR (Kuan *et al.*, 1970) were reported, and a number of other effects, such as level anticrossing (Veeman and van der Waals, 1970), transferred hyperfine structure from guest to host molecules (Fayer *et al.*, 1970), and PMDR studies on triplet excitons (Francis and Harris, 1971a, b), were observed. All these studies served to demonstrate the viability of PMDR techniques as valuable tools in probing both excited-state structure through the zero-field parameters and the dynamics of intra- and intermolecular energy transfer in organic excited states. More detailed accounts and specific references on the development of this field may be obtained from a number of excellent review articles (El-Sayed, 1971, 1974; Kwiram, 1972; Dennison, 1973).

In all these experiments no consideration of coherent coupling of the radiation field with the spin sublevels was made nor was it necessary for the interpretation of most of the experiments. It soon became apparent, however, that coherence effects would provide a wealth of data and yield a variety of new and unique approaches to triplet-state dynamics.

4.1.2. Recent Advances

Within just a few years, many properties of organic triplet-state ESR have become fairly well understood through the use of zero-field optically detected magnetic resonance, and attention has been directed toward further applications of magnetic resonance concepts to the excited-state problem. The ramifications of coherent coupling with a strong microwave field on the excited state were first considered by Harris in 1971. Harris pointed out that the coherent components could not be observed directly in the phosphorescence except in the special case in which the polarization of emission to some ground-state level is the *same* for both triplet sublevels, in which case a quantum beat phenomenon would occur as an amplitude modulation of the phosphorescence intensity at the Larmor frequency of the transition. Initial attempts were made to observe the coherent terms in excited triplet states utilizing conventional detection of the coherent component. In 1972 Schmidt observed the first triplet-state spin echo (see also Hahn, 1950), and later Carr–Purcell echo trains (Carr and Purcell, 1954) were observed (van't Hof *et al.*, 1973), utilizing conventional ESR techniques. Although these experiments are valuable in many situations because of the favorable non-Boltzmann population distribution in many triplet states, they sacrifice the sensitivity of optical detection.

The major important difference between conventional and double-resonance methods lies in the fact that the double-resonance technique effectively monitors changes in the populations of the energy levels coupled by the oscillating field and does not measure the oscillating dipole moment as does conventional detection. As a result, coherent information contained in the off-diagonal matrix elements of the time-dependent density matrix describing the ensemble cannot in general be observed *directly*. Thus, one is apparently faced with the dilemma that to take advantages of the increased sensitivity of double-resonance techniques one must relinquish those measurements of relaxation phenomena that are based on coherence effects.

However, it was realized by Breiland *et al.* in 1973 that this difficulty could be circumvented by the proper application of a probe pulse, which effectively samples the "degree" of coherence at any time. From this it becomes obvious that coherence phenomena may be observed in any multilevel system using any observable that can monitor any property of the population of one of the levels associated with the coherence, provided, of course, that the observable used can follow the ensemble on a time scale compatible with the relaxation effects that one wishes to measure. Indeed, most of the conventional methods of multiple-pulse nmr can be applied to excited-triplet-state optically detected magnetic resonance. Moreover, some experiments, such as spin locking (Redfield, 1955; Solomon, 1959b) and adiabatic demagnetization in the rotating frame (Slichter and Holton, 1961; Anderson and Hartmann, 1962), when optically detected in triplet states (Harris *et al.*, 1973; Brenner *et al.*, 1974) provide *unique* ways to study a wide variety of problems in excited states, including energy transfer dynamics and vibrational relaxation.

In an attempt to provide continuity to the historical development of coherence in a two-level system and to cast these methods as an extension of this development, we begin in the next section with the use of standard methods of density matrix formulation and apologize in advance for the initial repetition of what is well known to many workers in a wide variety of fields. We hope, however, that simple and straightforward examples will aid the uninitiated reader in adopting some of these methods in his or her own particular discipline or speciality.

4.2. Theoretical Considerations

4.2.1. General Aspects of Coherence in Excited States

Coherent coupling in excited-state systems has as its basis the very general geometrical model (Feynman *et al.*, 1957) that has been treated in Chapters 2 and 3 of this volume. Two alterations must be made, however, in order to deal with optically detected coherence experiments. First, the pulse sequences must be amended with a final "probe pulse" that allows coherence to be monitored by means of an observable that measures only the relative populations in the two-level system, as is done in double-resonance techniques. Second, it is desirable to separate the relaxation terms, T_1 and T_2, from the kinetic parameters that describe the excited-state feeding and decay processes. In doing this separation, one is faced with a problem in which the number of two-level systems in the ensemble is no longer constant in time, and the theory must be arranged to account for this fact. Fortunately, neither one of these alterations is so drastic that pulse experiments cannot be visualized in terms of a modified geometrical model, but, as is the case in all specific applications, care must be exercised in certain situations where canonical pulse techniques are applied to excited-triplet-state systems.

The first part of Section 4.2 is a rather general description that treats these alterations in detail. The second part applies the models that are developed in the first part to pulse experiments in excited-triplet-state systems, and various ramifications, pitfalls, and applications are discussed.

We wish to concern ourselves with the model system illustrated in Figure 4-1. In this model two levels are chosen from the excited-state manifold and are connected by a strong, coherent driving field. The geometrical picture is utilized to describe the time evolution of the two-level ensemble, but, since they are of considerable theoretical and physical importance, we wish to include explicitly the incoherent feeding *rates*, F_a and F_b, that arise, for example, in intersystem crossing, and the decay *rate constants*, k_a and k_b, that could arise from both radiative and radiationless decay to other states. In general, the specific feeding and decay processes are quite complicated and involve many levels, so the intractable multilevel problem is greatly simplified by adopting these feeding and decay parameters, and by focusing attention on only the two levels of interest; the rest of

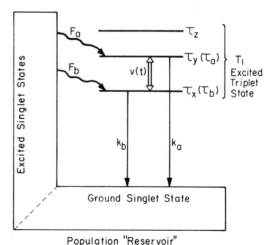

Figure 4-1. Pictorial representation of the model system presented in the discussion. F_a and F_b are constant feeding rates from an excited-singlet-state reservoir, and k_a and k_b are the total decay rates to the ground singlet state. τ_x, τ_y, and τ_z are the triplet spin sublevels in the zero-field basis. τ_a and τ_b correspond to the two spin sublevels coupled by a coherent driving field $V(t)$.

the multilevel system is considered to be a "reservoir" in the thermodynamic sense, to act as a source and sink for excited-state population.

If the total population in the two-level system fluctuates in time, the three parameters r_1, r_2, and r_3 are not sufficient to describe the density matrix completely and the geometrical model breaks down at this point. The four parameters ρ_{aa}, ρ_{bb}, r_1, and r_2 are sufficient, however, and one may obtain a modified geometrical model in which ρ_{aa} is represented by a vector along the $+z$ axis and ρ_{bb} is similarly along the $-z$ axis. At any instant of time the standard geometrical model is valid and r_3 is given by $\rho_{aa} - \rho_{bb}$, as usual. Utilizing this concept, the processes of feeding and decay are simply visualized as follows. First, we consider the total number of molecules present as an ensemble of two-level systems. If a given two-level system is excited, i.e., if the state of a molecule is $|a\rangle$ or $|b\rangle$ or any linear combination of states $|a\rangle$ and $|b\rangle$, it is considered to contribute a "pure state" density matrix vector to the ensemble average. If the state of the molecule is something other than $|a\rangle$ or $|b\rangle$, i.e., it is in the reservoir, it contributes nothing. The final vector describing the two-level ensemble at any instant of time is thus the vector sum of all the molecules that are in some superposition of $|a\rangle$ and $|b\rangle$. An incoherent feeding process implies that population suddenly appears in one of the two eigenstates, $|a\rangle$ or $|b\rangle$. This will appear as if an individual vector contribution suddenly appears along the $+$ or $-z$ axis, respectively. Decay is easily pictured as a vector that suddenly vanishes from the ensemble contribution. The time evolution of these contributions is governed on a macroscopic scale by the kinetic parameters F_a, F_b, k_a, and k_b, and is visualized as the vector sum of tiny vectors that randomly appear along $\pm z$ and are immediately driven by the EM field that may be present. It is assumed that the driving field does not affect any states in the reservoir.

The simplest illustration of this concept is to consider feeding and decay into a single state, say $|a\rangle$. Referring to Figure 4-2, we choose the initial situation to be one in which all the population is in the reservoir, resulting in no vector contributions. As population is fed into $|a\rangle$ at a constant rate, individual vector contributions will add and cause a buildup along the $+z$ axis. As this process is occurring, some of the contributions present in the sum will vanish as a result of decay from $|a\rangle$, and in the absence of a driving field, a steady-state situation will eventually occur in which the vector sum lies along the $+z$ axis (cf. Figure 4-2c). An identical situation applies for state $|b\rangle$, but now the contributions lie along the $-z$ axis. The total population difference is determined by the sum of these two contributions. If a driving field is now suddenly turned on, the vectors present at that instant of time will evolve according to the torque equation with the additional condition that each contribution has a probability of vanishing that is related to its position in the geometrical space. For example, a vector along $+z$ has a probability of decay proportional to k_a, since it is in $|a\rangle$, and similarly a

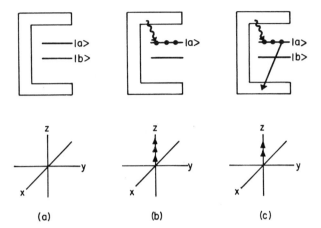

Figure 4-2. Buildup of a steady-state signal from feeding and decay processes. (a) All population is initially in the reservoir. (b) Constant feeding causes the buildup of vector contributions in the $|a\rangle$ state. (c) Competition between feeding and decay finally results in a fixed steady-state population that is represented as a vector sum along the $+z$ axis of the rotating frame.

probability of k_b if it should lie directly along the $-z$ axis and thus be in a pure $|b\rangle$ state. For intermediate positions the probability is a weighted average determined by its $+$ and $-z$ components, plus the probability of decay for the in-plane components that is determined by the interaction Hamiltonian and expressed roughly by $1/T_2$. The important thing to note is that the rather complicated situation for intermediate positions is still simple to understand qualitatively, and is subject to exact analytical verification for quantitative studies, as we shall show shortly.

What about population that subsequently enters the system while the two levels are being driven with the EM field? This is easily visualized by breaking up the time evolution into tiny segments. During each segment population is allowed to enter as we have just described and then undergoes nutation according to the torque equation. It is then added to the contributions already present, and the process is repeated with the next time segment. In this fashion the entire time evolution of a coherently driven excited-state system is easily visualized.

As an example of the use of this modified picture, consider the buildup of a steady-state on-resonance ESR signal. In this situation a driving field is applied on resonance in the presence of feeding and decay. Figure 4-3 illustrates the mental picture that may be employed to visualize how the signal is generated. If conventional detection is used, the size of the y component will represent the measured signal. Note that in the absence of a

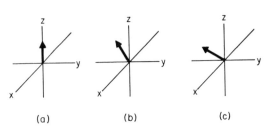

Figure 4-3. Evolution of an on-resonance steady-state absorption signal (ω_{eff} applied along x). (a) Initial population represented as a vector along the $+z$ direction. (b) The driving field is applied along x, causing the vector to precess in the zy plane. (c) At low powers the vector approaches the steady-state value given by an orientation in the zy plane. (d) The vector is composed of a series of smaller vectors, each oriented in a different direction of the zy plane depending on the time at which the states were created. (e) Under high power a saturated signal results in which the single contributions are driven very rapidly around the effective field, resulting in the uniform disk of vector contributions having a vector sum zero.

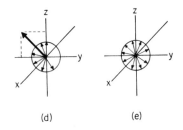

driving field, a steady-state z component results that is larger than the z component in the presence of the driving field. The difference between these components would yield the double-resonance signal. Note also that the signal can be seen to "saturate" under high driving fields, for then the vector contributions begin to undergo more than one nutation before they vanish. For very high fields, a uniform disk is formed whose resultant is zero. This causes the conventionally detected signal to disappear and the double-resonance signal to reach a maximum.

4.2.2. Equation of Motion for the Model System

Once a realistic model has been developed, the next step is to include all the relevant parameters in an equation of motion for the density matrix. The most general form would necessarily include the effects of a coherent driving field of arbitrary strength, feeding to the individual states $|a\rangle$ and $|b\rangle$, decay from the states, and finally, relaxation in terms of the appropriate parameters T_1 and T_2. This is easily done in a stepwise fashion by successive additions of the appropriate terms following the development of Breiland *et al.* (1976).

4.2.2.1. Basic Torque Equation in the Rotating Frame

Without feeding, decay, or relaxation, the situation reduces to the straightforward two-level problem that has been discussed in earlier chapters and thoroughly treated in such treatises as Abragam (1961). As is generally the case, it is more convenient to work with an equation of motion that is expressed in rotating frame coordinates. Briefly, the development is as follows. The Hamiltonian is given by a time-independent term,

$$\mathcal{H} = (\hbar\omega_0/2)\sigma_3 \tag{4-1}$$

that determines the two-level energy spacing, plus a time-dependent driving field term,

$$V(t) = \frac{\hbar\omega_1}{2}\exp\left[(-i\omega t\sigma_3)/2\right]\sigma_1\exp\left[(i\omega t\sigma_3)/2\right] \tag{4-2}$$

The rotating field approximation has been made and σ_i are the Pauli spin matrices. Rotating-frame coordinates are obtained by a unitary transformation on the density matrix,

$$\rho'(t) = U^{-1}\rho(t)U \tag{4-3}$$

where the unitary operator is given by

$$U \equiv \exp\left[(-i\sigma_3\omega t)/2\right] \tag{4-4}$$

This results in an equation of motion for $\rho'(t)$ which is particularly simple since the rotating frame Hamiltonian is time-independent.

$$\frac{d\rho'}{dt} = \frac{-i}{\hbar}[\mathcal{H}', \rho'] \tag{4-5}$$

$$\mathcal{H}' = [\hbar(\omega_0 - \omega)/2]\sigma_3 + (\hbar\omega_1/2)\sigma_1 \tag{4-6}$$

From this point on, primed notation will be dropped, since it will always be assumed that one is dealing with the rotating frame coordinates.

This equation of motion, (4-5), may be expressed in terms of the modified r-vector components and forms the basis for further additions.

$$\frac{dr_1}{dt} = -(\omega_0 - \omega)r_2$$

$$\frac{dr_2}{dt} = (\omega_0 - \omega)r_1 - \omega_1(\rho_{aa} - \rho_{bb})$$

$$\frac{d\rho_{aa}}{dt} = \omega_1 r_2/2 \tag{4-7}$$

$$\frac{d\rho_{bb}}{dt} = -\omega_1 r_2/2$$

Of course, these reduce to the rotating frame torque equation, where $r_3 = \rho_{aa} - \rho_{bb}$, and the equation of motion describes how the two-level system evolves under the influence of the coherent driving field.

4.2.2.2. Addition of Feeding and Decay Terms

Incoherent feeding to the eigenstates $|a\rangle$ and $|b\rangle$ is described by including the feeding rates, F_a and F_b (Figure 4-1), into the equations of motion for ρ_{aa} and ρ_{bb}, respectively. It is also possible to include *coherent* feeding by adding terms to the r_1 and r_2 equations of motion (Brock *et al.*, 1977), but since this occurs only under special conditions for most excited-state zero-field experiments, we shall delay consideration of it until the applications section.

Decay is slightly more complicated, since it can occur from both eigenstates and superposition states. Physically, the *probability* for being in state $|a\rangle$ or $|b\rangle$ decays exponentially with rate constants k_a and k_b, respectively.

$$\frac{d\rho_{aa}}{dt} = -k_a \rho_{aa}$$

$$\frac{d\rho_{bb}}{dt} = -k_b \rho_{bb} \tag{4-8}$$

This can be expressed mathematically by assuming that the *amplitude* for being in the eigenstate decays as

$$\frac{da}{dt} = \frac{-k_a}{2} a$$

$$\frac{db}{dt} = \frac{-k_b}{2} b \tag{4-9}$$

which yields (4-8) when the definitions for ρ_{aa} and ρ_{bb} are used. From this it follows that the r_1 and r_2 components will decay as the *average* of the decay constants

$$\frac{dr_1}{dt} = \frac{-(k_a + k_b)}{2}(ab^* + ba^*) = -k_{av}r_1$$

$$\frac{dr_2}{dt} = -k_{av}r_2 \tag{4-10}$$

which is reasonable, since the superposition state may be viewed as "undecided" about which eigenstate will eventually exhibit spontaneous emission.

Now the effects of feeding and decay may be added explicitly to Eq. (4-7):

$$\frac{dr_1}{dt} = -(\omega_0 - \omega)r_2 - k_{av}r_1$$

$$\frac{dr_2}{dt} = (\omega_0 - \omega)r_1 - \omega_1(\rho_{aa} - \rho_{bb}) - k_{av}r_2$$

$$\frac{d\rho_{aa}}{dt} = \omega_1 r_2/2 - k_a\rho_{aa} + F_a$$

$$\frac{d\rho_{bb}}{dt} = -\omega_1 r_2/2 - k_b\rho_{bb} + F_b$$

(4-11)

The various terms may be compared to the rotating frame Bloch equations, and the following similarities noted: (1) k_{av} plays the same role as the transverse relaxation T_2, and (2) the combination of feeding and decay will appear as if population is transferred incoherently from $|a\rangle$ to $|b\rangle$ despite the fact that no explicit mechanism has been provided for this. Specifically, the apparent incoherent transfer of population from state $|a\rangle$ to state $|b\rangle$ can occur by decay from state $|a\rangle$ to the reservoir and by subsequent feeding to the sate $|b\rangle$. In this sense an effective T_1 process is occurring, although it is important to note that this process is by no means a "thermalization," i.e., a relaxation to a Boltzmann population distribution, but rather it is a T_1 process to a "spin temperature" that is determined solely by the relative values of the kinetic parameters, and thus could result in any conceivable population distribution.

The similarities with T_1 and T_2 lead one to define *effective* relaxation times

$$t_1 \equiv k_{av}/k_a k_b, \qquad t_2 \equiv 1/k_{av} \qquad (4\text{-}12)$$

These are useful in expressing the steady-state solutions of Eq. (4-11) found by setting all time derivatives to zero. These solutions provide analytical expressions for the slow-passage continuous-wave (cw) line shapes to be expected from a standard spectroscopic analysis of the two-level system. First, noting that the steady-state populations in the absence of a driving field are given by

$$\rho^0_{aa} = F_a/k_a$$
$$\rho^0_{bb} = F_b/k_b$$
$$r_3^{\,0} = \rho^0_{aa} - \rho^0_{bb}$$

(4-13)

the steady-state solutions are

$$r_1^{ss} = \frac{r_3^0 \omega_1 \Delta \omega t_2^2}{1 + \Delta \omega^2 t_2^2 + \omega_1^2 t_1 t_2}$$

$$r_2^{ss} = \frac{-r_3^0 \omega_1 t_2}{1 + \Delta \omega^2 t_2^2 + \omega_1^2 t_1 t_2}$$

$$\rho_{aa}^{ss} = \frac{\rho_{aa}^0 (1 + \Delta \omega^2 t_2^2) + \omega_1^2 t_1 t_2^2 [(F_a + F_b)/2]}{1 + \Delta \omega^2 t_2^2 + \omega_1^2 t_1 t_2}$$

$$\rho_{bb}^{ss} = \frac{\rho_{bb}^0 (1 + \Delta \omega^2 t_2^2) + \omega_1^2 t_1 t_2^2 [(F_a + F_b)/2]}{1 + \Delta \omega^2 t_2^2 + \omega_1^2 t_1 t_2}$$

(4-14)

where $\Delta \omega \equiv \omega_0 - \omega$.

These forms are, of course, similar to those obtained in magnetic resonance and represent the usual Lorentz lineshapes with a "power-broadening" term, $\omega_1^2 t_1 t_2$, in the denominator.

4.2.2.3. Exact Solutions, Including Feeding and Decay

Equations (4-7) and (4-5) are equivalent statements, Eq. (4-5) being a more compact form. Since $\mathcal{H}'$ in Eq. (4-5) is time-independent, the density matrix has a simple matrix solution which amounts to a similarity transformation on $\rho(t=0)$ with time-evolution operators.

$$\rho'(t) = S^\dagger \rho'(0) S \tag{4-15}$$

The evolution operator is given by

$$S = \exp(i \mathcal{H}' t / \hbar) \tag{4-16}$$

and has an explicit matrix form in the $|a\rangle, |b\rangle$ basis given by

$$S = \begin{bmatrix} \cos \dfrac{\omega_{\text{eff}} t}{2} + \dfrac{i \Delta \omega}{\omega_{\text{eff}}} \sin \dfrac{\omega_{\text{eff}} t}{2} & \dfrac{i \omega_1}{\omega_{\text{eff}}} \sin \dfrac{\omega_{\text{eff}} t}{2} \\[3mm] \dfrac{i \omega_1}{\omega_{\text{eff}}} \sin \dfrac{\omega_{\text{eff}} t}{2} & \cos \dfrac{\omega_{\text{eff}} t}{2} - \dfrac{i \Delta \omega}{\omega_{\text{eff}}} \sin \dfrac{\omega_{\text{eff}} t}{2} \end{bmatrix} \tag{4-17}$$

$$\omega_{\text{eff}} \equiv \sqrt{\omega_1^2 + \Delta \omega^2}$$

Equation (4-11) may also be written in a more compact form,

$$i\hbar \frac{d\rho}{dt} = [\mathcal{H}, \rho] - [K, \rho]_+ + F \tag{4-18}$$

where the decay matrix, K, is given by

$$K \equiv \frac{i\hbar}{2}\begin{bmatrix} k_a & 0 \\ 0 & k_b \end{bmatrix} \tag{4-19a}$$

and the feeding matrix, F, by

$$F \equiv i\hbar \begin{bmatrix} F_a & 0 \\ 0 & F_b \end{bmatrix} \tag{4-19b}$$

The $[\ ]_+$ indicates anticommutation. The solution to Eq. (4-18) has a simple form,

$$\rho(t) = Q^+(\rho(0) - \rho_{ss})Q + \rho_{ss} \tag{4-20}$$

where the evolution operator, Q, is defined by

$$Q \equiv \exp\left[i(\mathcal{H} + K)t/\hbar\right] \tag{4-21}$$

and the steady-state values of the density matrix ρ_{ss} are obtained from Eq. (4-14). Note that Eq. (4-20) collapses easily to (4-15) when K and F are zero.

It is important to realize that the matrix Q is not unitary, since it contains the imaginary matrix K in the argument of the exponent. This implies that the operations that are indicated in Eq. (4-21) will not result in a similarity transformation. However, this is to be expected, since the decay process implies that the trace of the density matrix is no longer preserved. An explicit form for Q is useful for pulsed driving field applications and is readily obtained from Putzer's method:

$$Q = \exp\left(-k_{av}t/2\right) \begin{bmatrix} \cos \Omega t/2 - \dfrac{k_D - i\Delta\omega}{\Omega} \sin \Omega t/2 & \dfrac{i\omega_1}{\Omega} \sin \Omega t/2 \\[2em] \dfrac{i\omega_1}{\Omega} \sin \Omega t/2 & \cos \Omega t/2 + \dfrac{k_D - i\Delta\omega}{\Omega} \sin \Omega t/2 \end{bmatrix}$$

$$k_D = \frac{k_a - k_b}{2}$$

$$\Omega = [\omega_1{}^2 + (\Delta\omega + ik_D)^2]^{1/2} \tag{4-22}$$

Special cases of Eq. (4-20) will be dealt with later in reference to zero-field triplet-state coherence experiments, but we wish now to continue with the last step in the successive additions by including relaxation in the equation of motion.

4.2.2.4. Addition of Relaxation Terms

Relaxation is readily incorporated into Eq. (4-11), to yield the desired equation of motion for the model system depicted in Figure 4-1. We shall use the modified form of the Bloch equations (Bloch, 1946) first suggested by Redfield in 1955, in which both relaxation terms T_2 and T_{2e} are included. The term T_2 is a "normal" transverse relaxation and is associated with the r_2 component. For coherent components that lie along the r_1 direction, which is the direction of the applied driving field, relaxation is no longer an energy-conserving process and can result in a term that is dependent on the strength of the applied driving field, hence the different term T_{2e} (usually referred to as $T_{1\rho}$). Also, since the eigenstate populations are separated for purposes of feeding and decay, it is convenient to separate the spin–lattice relaxation into its component parts. We define W_a to be a rate constant associated with spin–lattice transitions out of $|a\rangle$ and into $|b\rangle$, and W_b to be the similar term for transitions from $|b\rangle$ to $|a\rangle$. The canonical T_1 term is given by

$$\frac{1}{T_1} = W_a + W_b \tag{4-23}$$

The spin–lattice rates will usually be related by a Boltzmann factor

$$\frac{W_b}{W_a} = \exp\left(-\hbar\omega_0/kT\right) \tag{4-24}$$

As was the case for incoherent feeding and decay, the spin–lattice terms are inserted as decay parameters in the "K matrix" of Eq. (4-19):

$$K \equiv \frac{i\hbar}{2}\begin{bmatrix} k_a + W_a & 0 \\ 0 & k_b + W_b \end{bmatrix} \tag{4-25a}$$

and as additional feeding terms in the "F matrix,"

$$F \equiv i\hbar\begin{bmatrix} F_a + W_b\rho_{bb} & 0 \\ 0 & F_b + W_a\rho_{aa} \end{bmatrix} \tag{4-25b}$$

With these definitions the final equation of motion is expressed by

$$\frac{dr_1}{dt} = -\Delta\omega r_2 - [k_{av} + (W_a + W_b)/2 + 1/T_{2e}]r_1$$

$$\frac{dr_2}{dt} = \Delta\omega r_1 - [k_{av} + (W_a + W_b)/2 + 1/T_2]r_2 - \omega_1(\rho_{aa} - \rho_{bb})$$

$$\tag{4-26}$$

$$\frac{d\rho_{aa}}{dt} = \omega_1 r_2/2 - (k_a + W_a)\rho_{aa} + W_b\rho_{bb} + F_a$$

$$\frac{d\rho_{bb}}{dt} = -\omega_1 r_2/2 - (k_b + W_b)\rho_{bb} + W_a\rho_{aa} + F_b$$

As before, the steady-state solutions are easily obtained and may be expressed in forms similar to magnetic resonance expressions provided that the proper effective relaxation parameters are defined.

$$r_1^{ss} = \frac{r_3^0 \omega_1 \Delta\omega T T_e}{1 + \Delta\omega^2 T T_e + \omega_1^2 T\tau}$$

$$r_2^{ss} = \frac{-r_3^0 \omega_1 T}{1 + \Delta\omega^2 T T_e + \omega_1^2 T\tau}$$

$$\rho_{aa}^{ss} = \frac{\rho_{aa}^0 (1 + \Delta\omega^2 T T_e) + (\omega_1^2 T\tau / k_{av})[(F_a + F_b)/2]}{1 + \Delta\omega^2 T T_e + \omega_1^2 T\tau}$$

$$\rho_{bb}^{ss} = \frac{\rho_{bb}^0 (1 + \omega^2 T T_e) + (\omega_1^2 T\tau / k_{av})[(F_a + F_b/2]}{1 + \Delta\omega^2 T T_e + \omega_1^2 T\tau}$$

(4-27)

The definitions

$$\frac{1}{T} \equiv k_{av} + \frac{W_a + W_b}{2} + \frac{1}{T_2}$$

$$\frac{1}{T_e} \equiv k_{av} + \frac{W_a + W_b}{2} + \frac{1}{T_{2e}}$$

$$\tau \equiv k_{av}/(k_a k_b + k_a W_b + k_b W_a)$$

$$\rho_{aa}^0 = \frac{F_a(k_b + W_b) + F_b W_b}{k_a k_b + k_a W_b + k_b W_a}$$

$$\rho_{bb}^0 = \frac{F_b(k_a + W_a) + F_a W_a}{k_a k_b + k_a W_b + k_b W_a}$$

$$r_3^0 = \rho_{aa}^0 - \rho_{bb}^0$$

(4-28)

are to be used in Eq. (4-27).

4.2.2.5. Exact Solutions, Including Feeding, Decay, and Relaxation

Equation (4-26) may be put into a more compact form for solution by adopting the Liouville form of the equation of motion that expresses the elements of the 2×2 density matrix as a four-vector. Equation (4-26) is then written as

$$\frac{d\rho}{dt} = L\rho + F$$

(4-29)

The explicit forms for L and F are given by

$$
L = \begin{bmatrix}
-(k_a + W_a) & \dfrac{i\omega_1}{2} & \dfrac{-i\omega_1}{2} & W_b \\[2ex]
\dfrac{i\omega_1}{2} & \dfrac{-1}{2}\left(\dfrac{1}{T}+\dfrac{1}{T_e}\right)-i\,\Delta\omega & \dfrac{1}{2}\left(\dfrac{1}{T}-\dfrac{1}{T_e}\right) & \dfrac{-i\omega_1}{2} \\[2ex]
\dfrac{-i\omega_1}{2} & \dfrac{1}{2}\left(\dfrac{1}{T}-\dfrac{1}{T_e}\right) & \dfrac{-1}{2}\left(\dfrac{1}{T}+\dfrac{1}{T_e}\right)+i\,\Delta\omega & \dfrac{i\omega_1}{2} \\[2ex]
W_a & \dfrac{-i\omega_1}{2} & \dfrac{i\omega_1}{2} & -(k_b + W_b)
\end{bmatrix}
$$

$$(4\text{-}30)$$

$$
F = \begin{bmatrix} F_a \\ 0 \\ 0 \\ F_b \end{bmatrix}, \qquad
\rho = \begin{bmatrix} \rho_{aa} \\ \rho_{ab} \\ \rho_{ba} \\ \rho_{bb} \end{bmatrix}
\qquad (4\text{-}31)
$$

Equation (4-29) has a simple solution,

$$\rho(t) = \exp\{Lt\}[\rho(0) - \rho_{ss}] + \rho_{ss} \qquad (4\text{-}32)$$

where ρ_{ss} is obtained from Eq. (4-27). Despite the fact that L is not Hermitian, it is a straightforward matter to determine the exponential matrix with Putzer's method (1966), and, since this involves the solution of a quartic equation with real coefficients, exact closed-form solutions are possible. One unfortunate aspect of this general solution, however, is that it still remains rather tedious to solve even in the simplest of special cases, and often it is more convenient to work with Eq. (4-20) for investigating experiments on a qualitative basis, since special cases collapse to simple forms.

4.2.2.6. Discussion

At this point it is useful to look at the equation of motion, Eq. (4-26), to see how the explicit inclusion of the kinetic parameters might alter the interpretation of the modified Bloch equations. First, it should be noted that by setting all the kinetic parameters to zero, which amounts to a fixed population in the two-state system, the modified Bloch equations are recovered immediately if one recalls that T_1 is defined by Eq. (4-23). Equivalently, if the relaxation processes are very much faster than feeding or decay, it is obvious that standard coherence experiments can be performed on excited-state systems, since the equations of motion are identical for fixed and variable total populations in this case. On the other hand, even if there is no relaxation in the sense of transverse or spin–lattice times, the excited-state system will have *effective* relaxation times that are determined solely by the kinetic parameters. These considerations become important when one is trying to separate "true" relaxation mechanisms that are related to time-dependent *inter*molecular interactions from the kinetic parameters that depend largely on electronic *intra*molecular interactions. The problems arise, for example, in a spin locking, experiment in which the power-dependent transverse relaxation time, T_{2e}, could approach or exceed the decay lifetime contribution, $1/k_{av}$. A similar situation exists for the temperature-independent decay, k_a or k_b, versus the temperature-dependent spin–lattice relaxation rates, W_a and W_b. In this case the apparent lifetimes of the states will appear to depend on temperature, and accurate spin–lattice times will be difficult to determine since the "true" lifetimes cannot be determined independently. The parameters must then be fit simultaneously to a temperature-dependent study. Aside from these complications, however, there are no major difficulties in understanding Eqs. (4-26), since they are still basically the Bloch equations.

The fact that we have assumed that no feeding occurs to the coherent components implies that conditions can be arranged to effectively isolate an ensemble of excited-state systems at a given point in time. This may be accomplished by the spin-locking sequence illustrated in Figure 4-4, and can be utilized to monitor dynamic situations without interference from succeeding feeding processes. This feature will be utilized in the application section at the end of this chapter, although it should be noted that some examples require "spin memory" of the total conservation of **r**. This is particularly true when the triplet-state energy is exchanged between closely lying states of the same molecule or aggregate. For the present, however, we will restrict our attention to incoherent feeding.

At first it might well be assumed that the excited-state lifetime places a limit on the time that a coherent component may be maintained since it is not fed, but this is certainly not true. Steady-state solutions for the coherent components are nonzero in the presence of a driving field, since the coherent field continually drives newly fed population into the in-plane components where it may decay, only to be replenished by new population, exactly in the manner that the T_1 and T_2 processes work in fixed-population systems.

4.2.2.7. Inhomogeneous Relaxation and Expected Line Shapes

Another type of relaxation that must be considered in any two-level system is the inhomogeneous broadening. In this case the solutions to Eq. (4-26) must be summed over an appropriate distribution function that reflects the inhomogeneous line shape, usually Lorentzian or Gaussian in

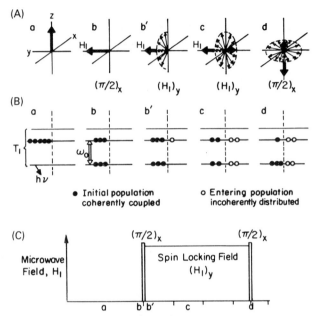

Figure 4-4. (A) Diagrammatical representation of the rotating-frame pseudo-magnetization for a system in which intersystem crossing occurs only to one level. (a) Pseudo-magnetization before the application of microwaves; (b) after an initial $\pi/2$ pulse is applied along the rotating-frame x axis. (b') The field is shifted 90° to the y axis, spin locking the initial pseudo-magnetization. The partial disk indicates that the incoming population is being driven in a plane perpendicular to the spin-locked population. (c) A component of pseudo-magnetization developing along the negative y axis due to longitudinal relaxation along the spin-locking field is indicated by the double-headed arrow. The disk represents an entering population after several hundred nanoseconds, incoherently distributed and precessing about the spin-locking field. (d) After the application of the final $\pi/2$ pulse along the x direction, the remaining spin-lock population is restored to the z axis. (B) Illustration of the spin sublevel population change in the laboratory frame under the application of the spin-locking pulse sequence. (a) All population is in the middle spin sublevel. (b) Microwaves couple the middle and bottom sublevels for a time which produces a $\pi/2$ pulse. The sublevel populations are equalized. (b') The system is spin-locked. The coherent population is equally distributed between the two spin sublevels, but an incoherent population begins to enter. (c) The coherent population decreases and the incoherent population becomes equally distributed. (d) The final $\pi/2$ pulse increases the bottom sublevel population. No changes occur that are due to the incoherent population. (C) Representation of the microwave pulse sequence.

form. Many systems have Lorentzian shapes characterized by the inhomogeneous relaxation time, T_2^*, and have a distribution about some center frequency, $\bar{\omega}_0$, given by the normalized distribution,

$$g(\omega_0) = \frac{T_2^*}{\pi} \frac{1}{1 + (\omega_0 - \bar{\omega}_0)^2 T_2^{*2}} \tag{4-33}$$

Integration with the distribution given in (4-33) is useful in describing the steady-state solutions to the general equation of motion, for they give analytical solutions for slow-passage line shapes to be expected experimentally from real two-level excited-state systems. This yields, in the notation of Eq. (4-28),

$$\langle r_1^{ss} \rangle = r_3{}^0 \omega_1 \overline{\Delta\omega}/D$$

$$\langle \rho_{xx}^{ss} \rangle = \frac{1}{D} \left[\rho_x^0 \left(\frac{1}{TT_e} + \frac{1}{T_2^{*2}} + \frac{1}{T_2^*} \frac{2 + \omega_1{}^2 T\tau}{\sqrt{(TT_e)(1 + \omega_1{}^2 T\tau)}} + \overline{\Delta\omega}^2 \right) \right.$$

$$\left. + \frac{\omega_1{}^2 T\tau}{k_{av}} \frac{F_a + F_b}{2} \left[\frac{1}{TT_e} + \frac{1}{T_2^*} \frac{1}{\sqrt{(TT_e)(1 + \omega_1{}^2 T\tau)}} \right] \right]$$

$$(4\text{-}34)$$

$$D = \overline{\Delta\omega}^2 + \{ (1/T_2^*) + [(1 + \omega_1 T\tau)/TT_e]^{1/2} \}^2$$

$$\overline{\Delta\omega} \equiv \bar{\omega}_0 - \omega$$

The subscript x indicates either a or b. Physically, $\overline{\Delta\omega}$ indicates how far one is off resonance from the inhomogeneously broadened line center. Equations (4-34) are valid for all driving field strengths and reflect the rather complicated saturation characteristics of the r-vector components. For sufficiently low driving field strengths, however, the components assume simple Lorentz forms,

$$\langle r_1^{ss} \rangle_{low} = \frac{r_3{}^0 \omega_1 \overline{\Delta\omega} \Gamma^2}{1 + \overline{\Delta\omega}^2 \Gamma^2}$$

$$\langle r_2^{ss} \rangle_{low} = \frac{-r_3{}^0 \omega_1 \Gamma}{1 + \overline{\Delta\omega}^2 \Gamma^2}$$

$$(4\text{-}35)$$

$$\langle \rho_{xx}^{ss} \rangle_{low} = \rho_x{}^0 + \frac{\omega_1{}^2 \Gamma\tau (F_a + F_b)}{(1 + \overline{\Delta\omega}^2 \Gamma^2)(k_a + k_b)}$$

$$\langle r_3^{ss} \rangle_{low} = r_3{}^0 \left(1 - \frac{\omega_1{}^2 \Gamma\tau}{1 + \overline{\Delta\omega}^2 \Gamma^2} \right)$$

$$\frac{1}{\Gamma} \equiv \frac{1}{T} + \frac{1}{T_2^*}$$

For low fields $T_e \to T$, and this has been used in the form for Γ.

4.2.3. Relationship Between the Geometrical Model and Double-Resonance Observables

4.2.3.1. Density Matrix and Dipole Emission

In order to utilize a double-resonance technique to measure coherence properties, one must be aware of the fact that the double-resonance signal cannot usually measure the coherence directly. This is due to the fact that the dipole radiation to the ground state reflects the populations in the levels $|a\rangle$ and $|b\rangle$, corresponding to ρ_{aa} and ρ_{bb}, respectively, and will not under normal conditions exhibit any effects due to the loss of coherence, represented by the in-plane r_1 and r_2 components. It is worthwhile to restate this in quantitative, analytical form in order to clarify the problem that one faces with double-resonance coherence experiments, and also to provide a convincing solution to this problem.

The model system pictured in Figure 4-1 has been expanded in Figure 4-5 to include the explicit forms for dipole emission to the ground state. The electric-dipole matrix elements, μ_{ag} and μ_{bg}, are generic notations for the large number of possible emission channels from an excited state to the ground-state manifold. PMDR generally monitors a single emission channel.

The expectation value for dipole emission from the excited state to the ground state is determined easily by constructing a projection operator for dipole emission to each level g_n,

$$\hat{\mu}_n \equiv |\mu|g_n\rangle\langle g_n|\mu| \tag{4-36}$$

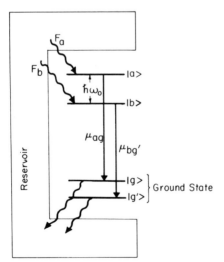

Figure 4-5. Optically detected double resonance. Transitions between $\langle a|$ and $\langle b|$ are observed as a change in the dipolar emission to the ground-state manifold. μ_{ag} denotes the matrix elements $\langle a|\mu|g\rangle$, where μ is the electric-dipole operator. Only two of the large number of possible transition channels to the ground state have been shown.

and determining the expectation value in the usual way:

$$\langle \hat{\mu}_n \rangle = \text{Tr} \, [\rho(t)\hat{\mu}_n]. \tag{4-37}$$

Working in the $|a\rangle$, $|b\rangle$ basis, the explicit forms for ρ and $\hat{\mu}_n$ are

$$\rho = \begin{bmatrix} \rho_{aa} & \dfrac{r_1 - ir_2}{2} \\[2mm] \dfrac{r_1 + ir_2}{2} & \rho_{bb} \end{bmatrix} \tag{4-38}$$

$$\mu_n = \begin{bmatrix} |\mu_{an}|^2 & \mu_{an}\mu_{bn}^* \\[2mm] \mu_{an}^*\mu_{bn} & |\mu_{bn}|^2 \end{bmatrix} \tag{4-39}$$

and the dipole emission for each spectral line would be of the form

$$\langle \hat{\mu}_n \rangle = \rho_{aa}k_{an} + \rho_{bb}k_{bn} + \text{Re} \, [\mu_{an}\mu_{bn}^*(r_1 - ir_2)] \tag{4-40}$$

where the spectral decay rate constants k_{xn} have been substituted for the probabilities $|\mu_{xn}|^2$. This equation is the same equation as derived by Harris (1971) from a similar construct. Equation (4-40) reveals that the radiative contribution to each spectral line in the emission spectrum will consist of the usual radiation from each of the two levels $|a\rangle$ and $|b\rangle$, which is proportional to the respective populations ρ_{aa} and ρ_{bb}, plus a "quantum beat" or coherent term that is nonzero only if the matrix elements μ_{an} and μ_{bn} are simultaneously nonzero and have the same polarization direction. In the laboratory frame of reference, the coherent components are fluctuating at a frequency ω, which is close to ω_0, the separation frequency. Thus, if any coherence is present in the two-level system, it could manifest itself as a modulation of the emitted light at the frequency ω_0 only when the polarization of the moments μ_{an} and μ_{bn} are the same. Specifically, the laboratory-frame emission would result in

$$\langle \hat{\mu}_n \rangle = \rho_{aa}k_{an} + \rho_{bb}k_{bn} + \mu_{an}\mu_{bn}(r_1 \cos \omega t + r_2 \sin \omega t) \tag{4-41}$$

where r_1 and r_2 are the rotating frame components and the dipole matrix elements are assumed real. For many organic molecules, the symmetry is sufficiently low in most cases to preclude this quantum beat possibility. One usually encounters a situation in which emission is possible to the same ground-state level only if the polarizations are different, with the result that the two-state coherence is not manifest in the emission.

4.2.3.2. Probe Pulse Method

There is a straightforward way to circumvent the problem discussed above. The method (cf. Breiland *et al.*, 1975) about to be described has the advantage of being quite general for any two-level system and is useful for virtually all symmetry conditions. It thus opens up a wide variety of possibilities for coherence experiments that utilize an observable that is proportional to the two-level populations rather than measuring the coherence directly in the r_1 or r_2 terms. Since photons from excited-state species are being detected, a large enhancement in sensitivity is effected. Indeed, the sensitivity can be as high as optical spectroscopy itself, and for triplet states, experiments can be performed with as few as 10^4 photons.

A hypothetical experiment serves best to illustrate the probe pulse technique clearly. Suppose that at time t_0 during an experiment a coherent component has been created and its decay is to be monitored. A conventional detection technique could do this directly, since the observable used is directly proportional to the coherence, represented as the in-plane y component in Figure 4-6. Under usual conditions, Eq. (4-41) demonstrates that the dipole emission cannot monitor this decay. However, the loss of coherence may be mapped out point by point in time by applying a $\pi/2$ pulse that rotates the coherent component vector to the z axis, where it will manifest itself as a *change* in the dipole emission. To find the size of the coherent component at some other time, the entire experiment must be repeated with

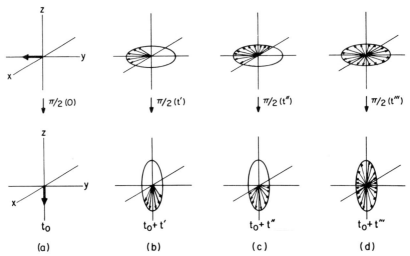

Figure 4-6. Use of a $\pi/2$ probe pulse to monitor the coherent state. At t_0 a coherent component is created along the y axis by a $\pi/2$ pulse applied along x. This corresponds to a r_2 component in the Feynman *et al.* (1957) representation. Conventional detection monitors the dephasing processes continuously on a real-time basis, illustrated for several times in the series along the top half of the figure. In the bottom half a $\pi/2$ probe pulse has been applied at times t_0, t', t'', t''' in four separate experiments. The z component now measures the size of what was r_2 along y, and this is seen as a change in the emission from the two-level system because of a change of the net projection along z.

the probe pulse applied at another time. In this fashion the decay $r_2(t_0 + t)$ is mapped out as a function of the *parameter t* rather than as a real time process. Figure 4-6 demonstrates this mapping process. The probe pulse technique essentially acts as a boxcar integrator, sampling the size of the coherent component at a point along the decay.

For a quantitative understanding, assume that the density matrix is represented at the time $t_0 + t$ by

$$\rho(t_0 + t) = \begin{bmatrix} \rho_{aa} & \dfrac{r_1 - ir_2}{2} \\ \dfrac{r_1 + ir_2}{2} & \rho_{bb} \end{bmatrix} \qquad (4\text{-}42)$$

As illustrated in Figure 4-7, either component r_1 or r_2 can be monitored by applying the probe pulse with the appropriate phase. r_1 can be measured by a $\pi/2$ ($\phi = 90°$) pulse, while r_2 can be measured by a $\pi/2$ ($\phi = 0°$) pulse. We apply a $\pi/2$ pulse to monitor this decay, and, using Eqs. (4-15) and (4-17), obtain for $\Delta\omega = 0$, $\omega_1 t = \pi/2$,

$$\rho = \begin{bmatrix} \sqrt{2}/2 & -i\sqrt{2}/2 \\ -i\sqrt{2}/2 & \sqrt{2}/2 \end{bmatrix} \begin{bmatrix} \rho_{aa} & \dfrac{r_1 - ir_2}{2} \\ \dfrac{r_1 + ir_2}{2} & \rho_{bb} \end{bmatrix} \begin{bmatrix} \sqrt{2}/2 & i\sqrt{2}/2 \\ i\sqrt{2}/2 & \sqrt{2}/2 \end{bmatrix} \qquad (4\text{-}43)$$

yielding

$$\rho = \frac{1}{2} \begin{bmatrix} \rho_{aa} + \rho_{bb} + r_2 & i(\rho_{aa} - \rho_{bb}) + r_1 \\ -i(\rho_{aa} - \rho_{bb}) + r_1 & \rho_{aa} + \rho_{bb} - r_2 \end{bmatrix}$$

The dipole radiation now reflects the size of r_2 as a parametric modulation of the emission

$$\langle \hat{\mu}_n(t_0 + t) \rangle = \frac{\rho_{aa} + \rho_{bb}}{2}(k_{an} + k_{bn}) + \frac{r_2(t)}{2}(k_{an} - k_{bn})$$

So as r_2 decays with time, the magnitude of dipole emission resulting after the probe pulse will reflect this decay as a function of the probe pulse delay parameter, t.

Special cases of Eq. (4-43) are worth noting. If at t_0 there are no Z components, i.e., $\rho_{aa} = \rho_{bb}$, the *change* in dipole emission will measure $r_2(t)$ directly. Specifically, before the probe pulse, the emission is given by

$$\langle \hat{\mu}_n \rangle = \rho_{aa} k_{na} + \rho_{bb} k_{nb} = \rho_{aa}(k_{na} + k_{nb}) \qquad (4\text{-}44)$$

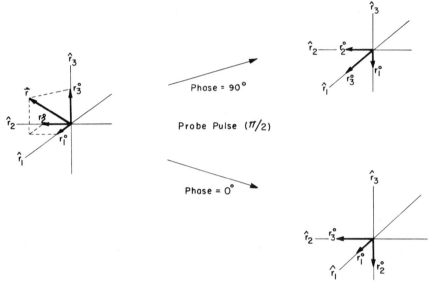

Figure 4-7. Optical detection of the coherent components and the use of the probe pulse. When the $\pi/2$ is applied at a $0°$ phase angle, the coherent components r_2 and r_3 are rotated such that the original r_2 now appears along the $\hat{r}_3$ frame coordinate. When the $\pi/2$ pulse is applied with a $90°$ phase, the r_1 and r_3 components are rotated such that the initial r_1 component now appears along the $\hat{r}_3$ frame coordinate. In both cases the coherent component is converted into a population difference dependent upon the phase of the applied field.

and after the probe pulse it is given by

$$\langle \hat{\mu}_n \rangle = \rho_{aa}(k_{an} + k_{bn}) + \frac{r_2(t)}{2}(k_{an} - k_{bn}) \tag{4-45}$$

so the difference is simply

$$\Delta\langle \hat{\mu}_n \rangle = \frac{r_2(t)}{2}(k_{an} - k_{bn}) \tag{4-46}$$

The decay will be observable for any situation for which $k_{an} \neq k_{bn}$ and even includes the rare case in which $|k_{an}| = |k_{bn}|$, but the polarization directions are different for emission from $|a\rangle$ and $|b\rangle$. There is only one situation in which it will not work, $\mu_{an} \equiv \mu_{bn}$, including polarization direction. In this case the quantum beat is present and may be monitored without a probe pulse by looking at emission from the state represented by Eq. (4-42), yielding, from Eq. (4-41),

$$\langle \hat{\mu}_n \rangle = (\rho_{aa} + \rho_{bb})k_n + k_n(r_1 \cos \omega t + r_2 \sin \omega t) \tag{4-47}$$

which could in the most favorable case yield a 100% modulation of the emission with a frequency $\omega = \omega_0$.

4.2.4. Experiments Utilizing Optically Detected Coherence

4.2.4.1. Introduction

The probe pulse method outlined in the previous section demonstrates that optical detection of coherence is not only possible but that it is simple enough to apply easily to the large number of coherence experiments that have been developed over the last 15 years. The facts to bear in mind are that most conventional pulse sequences need only be altered with the probe pulse, and that the method of detection is now like boxcar integration, in which the signal resulting from the probe pulse is time-averaged, the probe pulse delay time-changed and the signal time-averaged again. The decay shape is thus mapped out slowly in real time but can be accurately related to the actual rapid decay.

This section describes several canonical pulse experiments utilizing optical detection. For the most part, detailed descriptions of the pulse sequences will not be made here, since they constitute a volume in themselves and are amply documented in books of Farrar and Becker (1971) and Abragam (1961) as well as in Chapter 3 of this volume. The analytical expressions for the sequences will come from Eq. (4-20) or from (4-15) where applicable, since the general solution, Eq. (4-32), is too cumbersome to handle easily.

The pulse sequences are easily represented as a nested series of transformations on the $t = 0$ density matrix. For example, the spin echo sequence involves a sequence of $\pi/2$, dephasing time, π, rephasing, $\pi/2$ probe, that is represented in the evolution operators as $Q(\pi/2)$, $Q(\tau)$, $Q(\pi)$, $Q(\tau')$, and $Q(\pi/2)$, respectively.

If the driving field phase is shifted by an amount ϕ from the previous pulse, it is denoted by $Q(\pi, \phi)$, for example. From an operational point of view it is simpler to indicate a phase change in the applied field by rotating the initial values of the in-plane components, r_1 and r_2, an amount ϕ, rather than to include it in the expression for Q. Thus, for a phase-shifted pulse of 90° $r_1(0)$ becomes $r_2(0)$ and $r_2(0) \rightarrow -r_1(0)$. In this fashion Q will always appear to cause the vector to precess about an ω_{eff} direction that lies within the xz plane. Dephasing times, $Q(\tau)$, imply that ω_1 is set to zero, and Q is determined for a "pulse length" of τ.

Actual experimental examples will also be given for each sequence in which zero-field optically detected phosphorescence microwave double-resonance experiments have been performed on organic excited states. The details of these experiments are outlined in Section 4.3 but are sufficient to realize at this point that the experiments represent only a few examples of double-resonance detected coherence.

4.2.4.2. Transient Nutation

Before any coherence experiments can be performed, one must first ensure that the entire ensemble of systems can be coupled coherently. If this can be done, the ensemble will behave as a single entity, allowing coherence to be established. The on-resonance ($\Delta\omega = 0$) transient nutation (Torrey, 1949) represents the simplest form of coherent coupling, in which the populations undergo sinusoidal oscillations. This experiment is necessary for establishing the times required for π and $\pi/2$ pulses in addition to verifying coherent coupling.

Analytically only one Q-matrix calculation is required for a pulse of length t. The initial values of the density matrix elements correspond to the incoherent state $\rho_{aa} = \rho^0_{aa}$, $\rho_{bb} = \rho^0_{bb}$, $r_1 = r_2 = 0$ [Eq. (4-13)]. With these considerations the components are described by

$$r_1(t) = 0$$

$$r_2(t) = \frac{\exp(-k_{av}t)}{\alpha^2}[r^{ss}_2(k_D^2 - \omega_1^2 \cos \alpha t) - (r_3^0 - r^{ss}_3)\alpha\omega_1 \sin \alpha t$$

$$+ (\rho^0_{aa} - \rho^{ss}_{aa} + \rho^0_{bb} - \rho^{ss}_{bb})\omega_1 k_D(1 - \cos \alpha t)] + r^{ss}_2$$

$$\rho_{aa}(t) = \frac{\exp(-k_{av}t)}{\alpha^2}\{(\rho^0_{aa} - \rho^{ss}_{aa})[\alpha \cos(\alpha t/2) - k_D \sin(\alpha t/2)]^2$$

$$+ (\rho^0_{bb} - \rho^{ss}_{bb})\omega_1^2 \sin^2(\alpha t/2) \qquad\qquad (4\text{-}48)$$

$$- r^{ss}_2\omega_1 \sin(\alpha t/2)[\alpha \cos(\alpha t/2) - k_D \sin(\alpha t/2)]\} + \rho^{ss}_{aa}$$

$$\rho_{bb}(t) = \frac{\exp(-k_{av}t)}{\alpha^2}\{(\rho^0_{bb} - \rho^{ss}_{bb})[\alpha \cos(\alpha t/2) + k_D \sin(\alpha t/2)]^2$$

$$+ (\rho^0_{aa} - \rho^{ss}_{aa})\omega_1^2 \sin^2(\alpha t/2)$$

$$+ r^{ss}_2\omega_1 \sin(\alpha t/2)[\alpha \cos(\alpha t/2) + k_D \sin(\alpha t/2)\} + \rho^{ss}_{bb}$$

where the notation has been defined in Eqs. (4-13), (4-14), and (4-22). The term α is a special case of Ω in Eq. (4-22) for which $\Delta\omega = 0$; then α is given by

$$\alpha = (\omega_1^2 - k_D^2)^{1/2}$$

Although Eq. (4-48) appears quite complicated, its basic shape is that of a damped harmonic wave, which is what would be expected intuitively. Note that sine terms effectively shift the phase from a pure cosine, and k_D shifts the nutation frequency. These effects are not readily observable, however, for the following reasons. If ω_1 is small or comparable to k_D, the exponential k_{av} term is appreciable and dominates the waveform, resulting in a highly damped or overdamped oscillation in which the nutation frequency is not well defined. In order to obtain a meaningful nutation frequency, the driving field power must be increased so that the exponential decay is sufficiently surpressed, $\omega_1 \gg k_{av}$, but this also necessarily implies that $\omega_1 \gg k_D$, and the waveforms approach simpler expressions that are similar to the Bloch equation transient nutation forms for high driving fields:

$$r_1(t) = 0$$

$$r_2(t) \cong -r_3{}^0 \exp(-k_{av}t) \sin \omega_1 t - r_3{}^0 k_a k_b / \omega_1 k_{av}$$

$$\rho_{aa}(t) \cong [\exp(-k_{av}t)/2][(N^0 - N^{ss}) - r_3{}^0 \cos \omega_1 t] + N^{ss}/2$$

$$\rho_{bb}(t) \cong [\exp(-k_{av}t)/2][(N^0 - N^{ss}) + r_3{}^0 \cos \omega_1 t] + N^{ss}/2$$

$$(4\text{-}49)$$

The new terms

$$N^0 \equiv \rho_{aa}^0 + \rho_{bb}^0$$

$$N^{ss} \equiv \rho_{aa}^{ss} + \rho_{bb}^{ss} = \frac{F_a + F_b}{k_{av}}$$

$$(4\text{-}50)$$

have been introduced to emphasize that the total population in the excited-state system can have different values between conditions of no driving field and a strong driving field, despite the fact that the two levels are being fed at a constant rate which is itself independent of the driving field. In fact, the time evolution of the total population may be obtained by combining ρ_{aa} and ρ_{bb} from Eq. (4-49) to give

$$\rho_{aa} + \rho_{bb} = N(t) = e^{-k_{av}t}(N^0 - N^{ss}) + N^{ss} \qquad (4\text{-}51)$$

from which it is seen that the total population builds to N^{ss} with a characteristic time constant given by k_{av}.

Figure 4-8 depicts an optically detected transient nutation that reflects the coherent coupling on two zero-field spin sublevels in an excited triplet state. The oscillations clearly establish the nutation frequency and provide pulse times for $\pi/2$, π, etc., rotations of the r vector.

4.2.4.3. Free Induction Decay and Spin Echo

The spin echo (Hahn, 1950) is perhaps the simplest coherence experiment that yields information on intermolecular interactions that cannot be obtained easily from the absorption spectra or by other methods. The sequence $Q(\pi/2)$, $Q(\tau)$, $Q(\pi)$, $Q(\tau')$, $Q(\pi/2)$ yields an optically detected

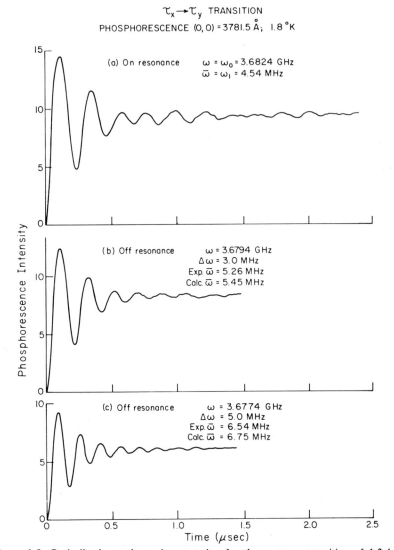

Figure 4-8. Optically detected transient nutation for the τ_x to τ_y transition of 1,2,4,5–tetrachlorobenzene in h_{14}-durene. The observed microwave change in phosphorescence is plotted versus the duration of the microwave pulse (zero denotes the zero change in phosphorescence).

electron spin echo. The shape, when averaged over the inhomogeneous line-shape equation (4-33), is proportional to

$$r_3 = r_3^{\,0} \exp\left[-(\tau + \tau')/T_2\right] \exp\left[(\tau - \tau')/T_2^*\right] \cos \overline{\Delta\omega}(\tau - \tau') \qquad (4\text{-}52)$$

which is identical in form to the conventionally detected echo shape. The

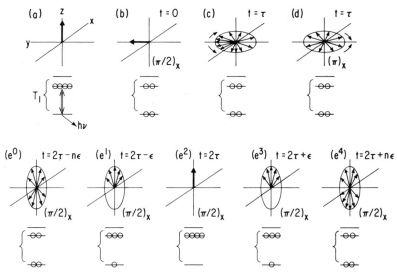

Figure 4-9. Optically detected Hahn echoes in excited triplet states. The relationship between the laboratory frame and the interaction representation or the rotating frame. In the interaction representation (a) the spins are prepared coherently with a $\pi/2$ pulse applied along x; (b) the ensemble dephases; (c) the ensemble is rephased with a π pulse applied along x; (d) the spins rephase. Coherence can be monitored by an appropriate $\pi/2$ pulse applied along x as the echo is rephasing. This is illustrated in the figures marked e^0, e^1, e^2, e^3, and e^4. The probe pulse converts the coherent state which is forming an echo in the xy plane to a population distribution along the z axis, and hence the echo can be monitored point by point in time.

relationship between the probe pulse, phosphorescence intensity, and the state of coherence is illustrated in Figure 4-9. An example of such an optically detected electron spin echo is given in Figure 4-10 for both on- and off-resonance applied driving fields. The shape consists of back-to-back free induction decays. A plot of the echo maximum as a function of the dephasing parameter yields, in the traditional manner, the homogeneous linewidth, which may be directly linked to models that incorporate fluctuating interaction Hamiltonians.

The ability to perform an optically detected electron spin echo allows one to assume that practically all of the coherence experiments that have been developed over the years for ground-state magnetic resonance may be immediately applied to the optically detected method for excited-state systems. This is true in principle, but of course certain technological differences may prohibit the application of some of these methods.

4.2.4.4. Echo Trains and Coherent Averaging

From the point of view of spectroscopy and its applications, the simple echo sequence described in the previous section suffers from the fact that the

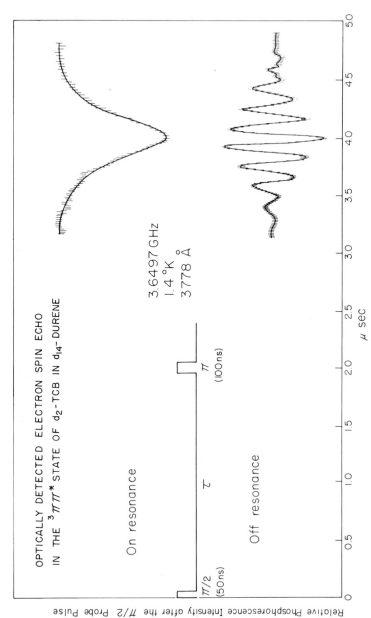

Figure 4-10. Optically detected zero-field electron spin echoes on- and off-resonance. The echoes at the right are traced out point by point in time as the probe pulse is applied at the time given on the abscissa. The vertical lines are the noise at each point in time in the experiment. Oscillations in the lower echo are a result of the off-resonance frequency.

echo maximum decay may result from a number of irreversible interactions that are occurring simultaneously, and as a consequence one has no way of sorting out these contributions. Fortunately, a number of pulse train sequences have been developed that allow one to separate the interactions in certain cases.

One example is a Carr–Purcell train (Carr and Purcell, 1954), which was originally designed to mitigate the effects of physical diffusion of a liquid sample within the inhomogeneous magnetic field in proton magnetic resonance. Similar diffusion-like processes occur with electron spins within solids; however, this phenomenon is caused by randomly fluctuating hyperfine fields resulting from mutual spin flips of nuclear species surrounding the unpaired electrons at the excited center. Since the hyperfine Hamiltonian acts as a small perturbation in determining the zero-field splitting $\hbar\omega_0$, fluctuating nuclear spin states will alter this splitting in much the same way that the proton Zeeman splitting fluctuates as the spin-$\frac{1}{2}$ species drifts randomly through an inhomogeneous external static field. If the nuclear spin diffusion process is an important contribution to the T_2 measured in the simple spin echo sequence, one would expect a dramatic increase in the relaxation time obtained from a Carr–Purcell pulse sequence, as illustrated at the top of Figure 4-11. The sequence has the effect of constantly refocusing the inhomogeneous contributions allowing a spin echo to occur every time τ between the π pulses spaced 2τ apart. As the dephasing–refocusing time τ approaches the time scale of the random spin flips, irreversible dephasing will be greatly suppressed since the local environment does not fluctuate appreciably between echoes. Figure 4-12 illustrates this rather dramatic increase in relaxation time as the parameter τ is shortened, resulting in a hundredfold difference between $\tau = 3$ and $\tau = 0.5$ s.

The Carr–Purcell echo sequence serves as one example of a "coherent averaging process," a concept first pioneered by Waugh and coworkers in 1968 (Haeberlen and Waugh, 1968; Waugh *et al.*, 1968; Haeberlen and Waugh, 1969). It was recognized that if the relaxing spin system was viewed at only a single point in time of a repetitive pulse sequence, the sequence would have the effect of removing or averaging the true relaxation Hamiltonian, and the system would appear to behave as if it were responding to this new average Hamiltonian. As an example, the system may be viewed only at the maximum of the Carr–Purcell echoes, and it can be seen that this in effect "removes" both the inhomogeneous relaxation and the relaxation due to diffusion. With this powerful concept, one may then design a specific pulse sequence that will selectively remove a specific relaxation Hamiltonian and thereby unravel a number of fluctuating interactions that arise from different sources but are occurring simultaneously.

A rather straightforward argument serves to demonstrate how the hyperfine interaction is averaged out by the Carr–Purcell sequence.

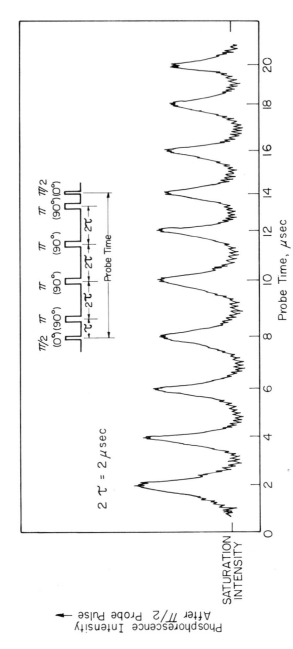

Figure 4-11. Optically detected Carr–Purcell–Meiboom–Gill spin echo train in the $^3\pi\pi^*$ state of h_2-TCB (Y-trap). The probe pulse terminates the pulse train and is swept continuously in time. Only the first few echoes are shown. Echoes were detectable to 1 ms.

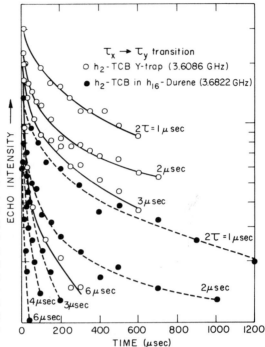

Figure 4-12. Optically detected echo maxima in the optically detected Carr–Purcell–Meiboom–Gill spin echo train for various times 2τ between the rephasing pulses. The solid circles illustrate the relaxation profile of h_2-TCB in h_{14}-durene, and the open circles represent the dephasing relaxation profiles of h_2-TCB (trap) in h_2-TCB. The differences are due to the fact that the nuclei in the host crystals affect nuclear-spin diffusion differently in the two systems.

Following Haeberlen and Waugh (1968), the Hamiltonian is partitioned into three contributions

$$\mathscr{H} = \mathscr{H}_0 + V(t) + \mathscr{H}_R(t) \qquad (4.53)$$

where $\mathscr{H}_0$ and V have been defined previously, and $\mathscr{H}_R(t)$ is a time-dependent relaxation Hamiltonian. The crucial step is to choose a repetitive cycle time determined by the pulse sequence. If the evolution operators associated with the sequence are grouped such that their product yields the unit matrix,

$$S(1)S(2)S(3)\cdots S(n) = \mathit{1} \qquad (4.54)$$

then the cycle time is determined by the sum of times associated with each pulse, plus the waiting times between pulses:

$$\tau_c = t_1 + \tau_{12} + t_2 + \tau_{23} + \cdots + \tau_n \qquad (4-55)$$

Provided that (1) the pulses affect the system so strongly that relaxation may be ignored during the pulse, (2) τ_c is short compared to the effective time for irreversible dephasing, and (3) the system is sampled periodically at the same point in each cycle, the density matrix will have an equation of motion

$$i\hbar \frac{\partial \rho}{\partial t} = [\mathscr{H}, \rho] \qquad (4-56)$$

where the average Hamiltonian, $\bar{\mathcal{H}}$, is given by

$$\bar{\mathcal{H}} = \sum_{k=1}^{n} \left[\left(\prod_{i=1}^{k} S_i^{-1} \right) \hat{\mathcal{H}}_R \left(\prod_{i=1}^{k} S_i \right) \right] \frac{\tau_k}{\tau_c} \tag{4-57}$$

and $\hat{\mathcal{H}}_R$ is the interaction representation form of $\mathcal{H}_R$,

$$\hat{\mathcal{H}}_R = U^{-1} \mathcal{H}_R U \tag{4-58}$$

For the Carr–Purcell sequence

$$S(\pi)S(\pi) = 1 \tag{4-59}$$

and for triplet spins in zero field (Breiland *et al.*, 1975), the cycle time is 4τ, provided that the time for a π pulse is sufficiently short. This results in an average Hamiltonian given by

$$\bar{\mathcal{H}} = [\hat{\mathcal{H}}_R + S^{-1}(\pi)\mathcal{H}_R S(\pi)] 2\tau / \tau_c \tag{4-60}$$

The relaxation Hamiltonian has the general form

$$\mathcal{H}_R = \begin{bmatrix} \mathcal{H}_{aa} & \mathcal{H}_{ab} \\ \mathcal{H}_{ba} & \mathcal{H}_{bb} \end{bmatrix} \tag{4-61}$$

where all the elements can be complicated functions of time. The off-diagonal elements $\mathcal{H}_{ab}$ contribute to spin–lattice relaxation, which, for the experiment described above, is negligible. Thus, we may assume that the relaxation Hamiltonian is diagonal, and the terms $\mathcal{H}_{aa}$ and $\mathcal{H}_{bb}$ represent the random fluctuations in the two-level spacing described earlier. Since $\tau_c = 4\tau$, we have, from Eq. (4-60),

$$\bar{\mathcal{H}} = \frac{1}{2} \left\{ \begin{bmatrix} \mathcal{H}_{aa} & 0 \\ 0 & \mathcal{H}_{bb} \end{bmatrix} + \begin{bmatrix} 0 & 0 \\ 1 & 0 \end{bmatrix} \begin{bmatrix} \mathcal{H}_{aa} & 0 \\ 0 & \mathcal{H}_{bb} \end{bmatrix} \begin{bmatrix} 0 & 1 \\ 1 & 0 \end{bmatrix} \right\} = \frac{1}{2}(\mathcal{H}_{aa} + \mathcal{H}_{bb}) \tag{4-62}$$

Thus, $\bar{\mathcal{H}}$ is a scalar that will commute with ρ and the density matrix will be unchanged.

The Carr–Purcell sequence will therefore completely remove the effects of T_2 relaxation provided that (1) the pulses used are sufficiently strong to overcome the T_2 process (i.e., the pulse "spans the homogeneous linewidth"), and (2) 4τ must be short compared to the correlation time associated with the relaxation fluctuations. Of course, as τ is lengthened, the Carr–Purcell sequence will become less effective and eventually the total unaveraged relaxation will be manifest, which is verified by Figure 4-12.

4.2.4.5. Spin Locking and Coherent Averaging

From Figure 4-12 it is evident that even at $\tau = 0.5 \ \mu$s the relaxation has not been completely averaged. As $\tau \to 0$ the Carr–Purcell sequence becomes

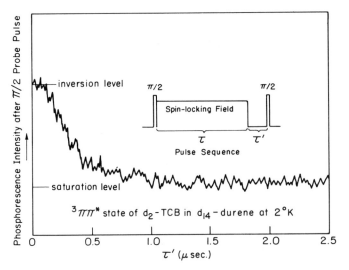

Figure 4-13. Optically detected free induction decay after a spin-locking time of 15 μs.

the classical spin-locking experiment (Solomon, 1959b) described in Section 4.2.2.6. The pulse sequence for optical detection (Harris *et al.*, 1973) would be $Q(\pi/2, 0^0)$, $Q(\alpha\pi/2, 90^0)$, $Q(\pi/2, 0^0)$, where $\alpha\pi/2$ simply indicates a high-power pulse applied for an arbitrary length of time. In terms of the vector components we have, for large ω_1,

$$r_3 = -r_3{}^0 \exp(-k_{av}t) \tag{4-63}$$

This demonstrates analytically what was mentioned earlier, that feeding subsequent to the establishment of the spin-locked vector has no effect on the spin-locked ensemble. In order to verify that the spin-lock sequence is indeed preserving a coherent component, an optically detected free induction decay can be performed by delaying the probe pulse. An experiment of this kind is depicted in Figure 4-13, and a plot of the decay of the spin-lock signal itself is given in Figure 4-14. Notice that in the absence of relaxation, the equation above states that the spin-lock signal will decay with the average of the two-level rate constants. For the situation in Figure 4-14, $1/k_{av} = 63$ ms, which is only slightly longer than the observed 42 ms decay, indicating that the spin-locking experiment is quite effective as a means of prolonging the coherence time in the two-level system. This feature is extremely important for applications that are designed to measure relaxation phenomena that compete with radiative or nonradiative decay. The effects of coherent averaging in spin locking are particularly easy to formulate, since spin-locking represents the limiting case of the Carr–Purcell train when $2\pi \to 0$. The common procedure for performing Carr–Purcell trains is

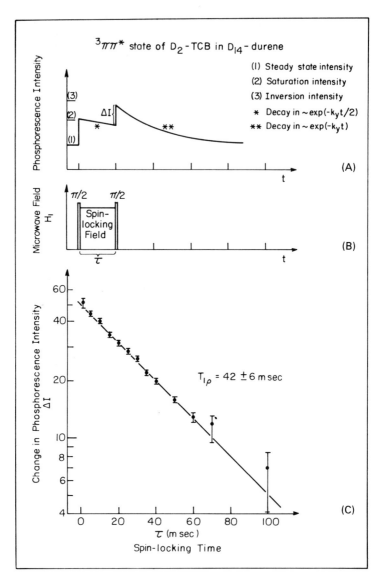

Figure 4-14. (A) Relationship of the phosphorescence intensity to the pulse sequence using spin locking. (B) Microwave pulse sequence for spin locking. (C) Relaxation time, $T_{1\rho}$, for the $^3\pi\pi^*$ state of d_2-TCB in d_{14}-durene at 2°K.

to phase-shift the π pulses by 90° relative to the $\pi/2$ pulses. This modification, introduced by Meiboom and Gill in 1958, effectively cancels out cumulative errors due to incorrect π pulse length. If one now reduces 2τ in the Carr–Purcell–Meiboom–Gill method to zero, the ensemble r vector is now spin-locked along H_1 in the rotating frame. We have assumed that H_1 is so strong that it completely determines the evolution of the spin system during a pulse; this implies that $\mathcal{H}_R(t)$ would be completely dominated by H_1 in spin locking. Since the cycle time is now simply the length of two π pulses (or four, in the general case where communication among all three sublevels may be important), the average Hamiltonian $\bar{\mathcal{H}}$ for zero waiting time is zero. Coherence should be very long-lived.

Specifically, in the limit of large $\gamma\mathcal{H}_1$ fields,

$$\bar{\mathcal{H}} = \frac{2\tau}{t_c} \int_0^{4\pi} (\hat{\mathcal{H}}_R + S^{-1}(\theta, \phi)\mathcal{H}_R S(\theta, \phi) \to 0 \qquad (4\text{-}64)$$

when $2\tau \to 0$. The decay from spin locking is given by $T_{1\rho}$, which we have seen is immune to any decay due to $\mathcal{H}_R$. The importance, then, of spin locking for measuring kinetic phenomena whose relaxation time competes with the radiative decay is that $T_{1\rho}$ measured from a spin-locking experiment is in essence related to the Fourier transform of the life-time-limited linewidth when kinetic processes result in a new state whose Larmor frequency, ω_0', differs from that of the spin-locked state's Larmor frequency, ω_0, by an amount large compared to γH_1. Therefore, when $|(\omega_0 - \omega_0')| > \gamma H_1$ and the lifetime in the primed state, τ', is longer than the γH_1, $T_{1\rho}$ measures the scattering rate from the initial spin-locked state. The latter condition ensures that spin memory is lost in the primed state. In cases where $|(\omega_0 - \omega_0')| < \gamma H_1$, such as in modulation of the hyperfine tensor via nuclear spin diffusion, all effects should be averaged to zero and $T_{1\rho}$ should approach the radiative or nonradiative lifetime. This is illustrated in Figures 4-15 and 4-16 for the triplet state of cyclopentanone (Harris and Tarrasch, 1977).

A word of caution is in order with respect to the spin-locking experiment. If the driving field is not applied exactly on resonance, the system is no longer isolated from incoming population. In fact, an off-resonance high-power pulse can produce very substantial coherent components (Breiland *et al.*, 1976). If one attempts to create a spin-lock signal with an *off-resonance* driving field, a spurious coherent component will add to the spin-lock component, building up to a steady-state value with a time constant k_{av}. This component will not decay as long as the driving field is on, and will interfere with the apparent decay of the "true" spin-locked signal.

As an additional comment, the off-resonance long pulse will produce the largest steady-state coherent component, owing to the fact that high powers can be used. The signal will not saturate as it would on resonance, and this allows one to exploit the power-dependent T_e relaxation term to obtain substantially larger steady-state components than could be obtained with an on-resonance signal. The extremely long (by ESR standards) coherence times on and off resonance open up the possibility of cross-

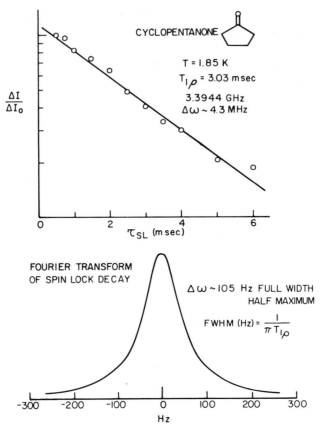

Figure 4-15. Decay of coherence in spin-locking in the triplet state of cyclopentanone and the Fourier transform of $T_{1\rho}$. Before coherent averaging the linewidth was 4.3 MHz, while during coherent averaging in the spin locking, the linewidth was reduced to 105 Hz.

relaxation as performed by Hartmann and Hahn (1962). Indeed, in a series of preliminary experiments, it was demonstrated by Schuch and Harris (1975) that nuclear spins in the surroundings of the excited triplet states in this lattice could be optically detected by a cross-relaxation of the electron spins by the nuclear spins with high sensitivity.

4.2.4.6. Rotary Echoes and Driving-Field Inhomogeneities

If spin-locking times $T_{1\rho}$ become very long for high-power driving fields, then the transient nutation should also persist for very long times, contrary to what is observed in Figure 4-8. The experimentally observed decay, however, is due almost exclusively to driving-field inhomogeneities which can be quite significant. If different parts of the crystal experience different $\mathcal{H}_1$ fields, then a range of nutation frequencies are produced in different regions of the sample. This leads to an apparent decay of the oscillations and to an ambiguity in the times for $\pi/2$ pulses, as well as rendering the transient nutation somewhat useless as a tool for measuring T_1 and T_2 processes.

The dephasing due to driving-field inhomogeneities can be reversed by reversing the phase of the driving field and forming a rotary echo (Solomon, 1959a). The fan of vectors produced in the yz plane then refocuses, forms an echo, and dephases again, as illustrated in Figure 4-17. By continually shifting the phase from 0° to 180°, a series of transient nutation echoes form whose heights correspond to points on the transient nutation envelope that would occur in the absence of driving field in homogeneities. This is illustrated in Figure 4-18. One should note that in this instance, however, the driving field is on continuously during the course of the experiment, and hence the decay time of the rotary echo train is subject to average Hamiltonian theory. The net result is a significant lengthening over the homogeneous decay time obtained by Hahn echo experiments. The decay of rotary echoes can also be used to measure kinetic phenomena in the same sense as the decay during spin locking. The average Hamiltonians, however, are not

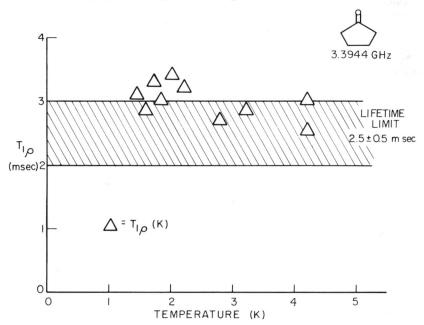

Figure 4-16. $T_{1\rho}$ versus temperature for the triplet state of cyclopentanone. Over the region measured, the relaxation time was essentially temperature-independent and limited to the lifetime for radiative decay.

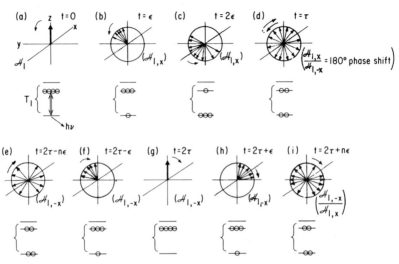

Figure 4-17. Relationship of the laboratory-frame populations in the triplet spin sublevels and the rotating-frame magnetization in a rotary echo. The field is applied along x, causing the pseudo-magnetization to precess in the zy plane with subsequent dephasing. Rephasing is accomplished by a 180° phase shift in the applied field and an echo forms along $+z$ in the rotating frame. The rotary echo is optically detected as a modulation of the phosphorescence.

identical. Rotary echoes are useful when the decay times of kinetic phenomena are on similar time scales as nuclear spin diffusion. The advantages of rotary echoes over spin locking are only instrumental insofar as the change in the phosphorescence intensity, as shown in Figure 4-19. This results from the fact that as τ becomes very short, the probe pulse in the spin locking becomes difficult to distinguish from the initial change in intensity due to the first $\pi/2$ pulse. The rotary echoes, however, form along z in the rotating frame and can be monitored with ease down to times as low as 100 ns. Several recent examples of their use will be presented in Section 4.4.

4.2.4.7. Adiabatic Demagnetization and Rapid Passage

One of the earliest experiments that was performed on excited triplet-state systems was adiabatic rapid passage (Harris and Hoover, 1972). In this experiment the microwave field is rapidly swept through resonance, but slowly enough that many nutations may occur during the process. In this way the vector components "follow" the effective field direction and a population inversion results. The rapid-passage technique has the advantage of being able to invert very broad esr lines, since the inhomogeneous distribution does not matter for a complete passage, but the method is restricted to inversion or saturation only, and in that respect is somewhat limited.

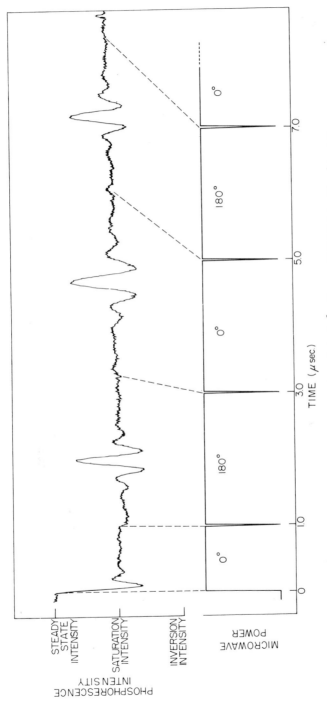

Figure 4-18. Optically detected rotary echo train in the $^3\pi\pi^*$ state of 1,2,4,5-tetrachlorobenzene.

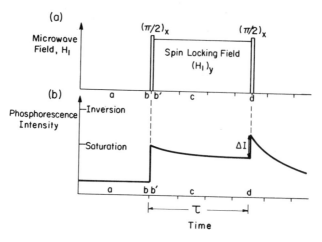

Figure 4-19. Change in the phosphorescence intensity during spin lock. (a) the microwave pulse sequence. (b) Schematic of the corresponding change in the phosphorescence intensity as a function of the spin-locking time τ.

The adiabatic condition may be satisfied in another way by adiabatic demagnetization in the rotating frame, or ADRF (Slichter and Holton, 1961; Anderson and Hartmann, 1962), in which the driving field is reduced adiabatically from a spin-locked state, and the individual vector contributions within the inhomogeneous line "follow" the effective field direction as in the rapid passage. However, with this method a unique spin distribution is created, since those components whose frequencies are positive with respect to ω_0 will lie along the $+z$ axis, while those whose frequencies are less than ω_0 will be distributed along the $-z$ axis after demagnetization (see Figure 4-20). In effect, the inhomogeneous line is split into two parts; one side undergoes a population inversion and the other is unaffected. For excited triplet states the inhomogeneous line distribution has been "ordered" in a very special way, insofar as the process is isentropic, and the order of the spin-locked field is conserved, being converted to order in the local field. For excited triplet states, the initial entropy is that associated with the spin alignment of the magnetic sublevels produced by selective intersystem crossing. This is illustrated in Figure 4-21, where the spin sublevel distributions are depicted in a molecule containing a quadrupolar nucleus such as ^{35}Cl or ^{37}Cl.

Because of hyperfine interactions, the zero-field transition is split into two transitions, ω_s and ω_f. These correspond to frequencies "slower" and "faster" than the center frequency ω_0. It is clear that in this instance the local field is determined by the orientation of the nuclear spin via the hyperfine interaction. In general, this will be true for triplet states, regardless of whether or not the hyperfine interactions are resolved. After ADRF (cf.

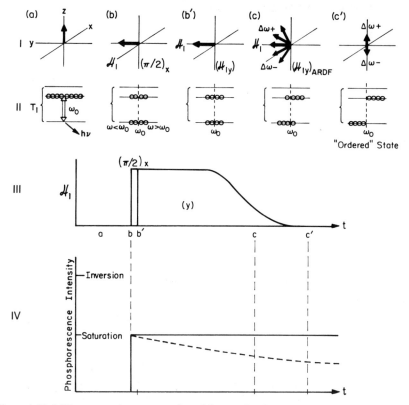

Figure 4-20. I. Diagrammatic representation of the rotating-frame pseudo-magnetization for a system in which intersystem crossing occurs only to one level. (a) r_3 before the application of microwaves. (b) Orientation of r_3 after an initial $\pi/2$ pulse is applied along the rotating frame x axis; (b') the H_1 field is phase-shifted by 90° to the y axis, spin locking the initial spin alignment in the rotating frame. (c) The H_1 field is reduced adiabatically, with the result that each spin isochromat remains parallel to the vector sum of the z component of its local field, $\Delta\omega/\gamma$, and the H_1 field. (c) When H_1 is reduced to zero, each isochromat is aligned along its own local field; thus isochromats with a positive $\Delta\omega$ point toward positive z, while those with negative $\Delta\omega$ point in the negative direction, $\Delta\omega$ being the difference between the isochromat's resonant frequency and the applied microwave frequency, ω_0. II. Illustration of the spin sublevel population in the laboratory frame under the application of the adiabatic demagnetization pulse sequence. The dashed vertical line diagrammatically represents the isochromats according to their resonance frequency ω relative to ω_0, the applied microwave frequency. When H_1 is zero, the population difference per unit frequency interval for $\omega < \omega_0$ is inverted, while for $\omega > \omega_0$ it is restored now to its initial value. The triplet ensemble has been "ordered," redistributing the spin alignment along local fields. III. Representation of the microwave pulse sequence. IV. Illustration of the change in phosphorescence intensity as a result of the adiabatic demagnetization pulse sequence. The curved dashed line represents the approach of the phosphorescence to a new saturation value, determined by averaging of populating and depopulating rate constants by the applied H_1 field. This change would be seen if the length of the spin lock were on the order of triplet sublevel lifetimes; we restrict our attention to short spin-lock times. Although the adiabatic demagnetization itself does not change the phosphorescence from its saturated value, the order remaining in the ensemble at a later time can be converted to a proportional change in the phosphorescence intensity by a spin–echo pulse sequence.

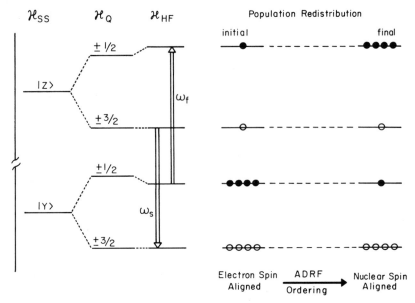

Figure 4-21. Preparation of an ordered state by adiabatic demagnetization in excited triplet states having quadrupolar nuclei such as Cl. Ordering with the applied field at the center of resonance results in the conversion of the electron-spin alignment between the triplet spin sublevels to nuclear-spin alignment within the triplet spin sublevels. Note that the $\pm\frac{1}{2}$-spin distribution in the τ_y spin sublevel is transferred to the τ_z spin sublevel, with the result that the τ_y spin sublevel is left with $\pm\frac{3}{2}$ nuclear spins while the τ_z spin sublevel contains $\pm\frac{1}{2}$ spins.

Figure 4-21) the nuclear spin distribution is changed. Specifically, for this example the $\pm\frac{1}{2}$ nuclear spins in the τ_y electron spin sublevel are inverted with the $\pm$spins in the τ_z manifold, with the result that the electron spin alignment between τ_y and τ_z has been transferred to the nuclear spins in each spin sublevel. In τ_y the nuclei are aligned in the $\pm\frac{3}{2}$ level, while in τ_z the nuclei are aligned in the $\pm\frac{1}{2}$ state. The unique feature of the ordered state is that the electron spins are aligned along their local hyperfine fields and the population in this ordered state is distributed along the Z axis in the rotating frame, and hence the order is *not* sensitive to fluctuations in the Larmor frequency, as is the case in a spin-locked ensemble. Stated another way, the population is selectively distributed along z in such a way that $r_1 = r_2 = 0$ and a change in the local field (hyperfine field) is necessary to cause relaxation in the ordered state.

The ordered state is sensitive to two kinds of relaxation, as illustrated in Figure 4-22. The first is nuclear or electron spin lattice relaxation. It may destroy the order by causing transitions between the two levels, with the net result that the inverted part of the line will return to equilibrium. This mechanism can be considered to be an interfering term, since the ordered

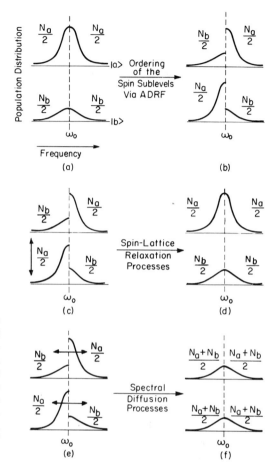

Figure 4-22. Relaxation processes in the ordered state (a) and (b) represent the population distributions between spin sublevels as a function of frequency. (c) and (d) illustrate diagrammatically spin–lattice relaxation processes while (e) and (f) represent the final population distributions that result from spectral diffusion in the absence of spin–lattice relaxation.

state is not required to measure this type of relaxation unless a special situation exists in which one part of the line experiences a different relaxation time than another. The second kind of relaxation is of greater interest since it represents a unique type of transfer between homogeneous components within the inhomogeneous line. To understand this feature it is important to note that spectral and spin diffusion are two entirely different processes for triplet systems. In spin diffusion a particular line in the inhomogeneous spectrum experiences fluctuating energy-level shifts, giving relaxation times on the order of 5 μs, which corresponds to a homogeneous linewidth of only 0.06 MHz. However, typical *inhomogeneous* linewidths are on the order of 2 MHz, indicating that there is a distribution of molecules in different sites that accounts for a static distribution of level splittings. Communication among these sites in the form of spectral diffusion usually

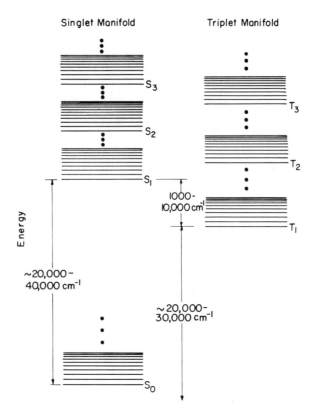

Figure 4-23. Schematic representation of the density of electronic and vibrational states in the singlet and triplet manifold.

involves some kind of energy migration within the crystal. Since the ordered state is insensitive to fluctuations in the Larmor frequency of intermediate states, it proves to be a unique probe for energy migration in molecular crystals. This will be dealt with in more detail in the last section.

4.3. Experimental Methods

4.3.1. Excited Triplet States and Phosphorescence Spectroscopy

To put things into perspective, an energy-level diagram for a typical aromatic hydrocarbon is pictured in Figure 4-23. The thin lines represent energies corresponding to normal modes of vibration. When one considers that for a molecule with N atoms, there are $3N-6$ normal modes, each

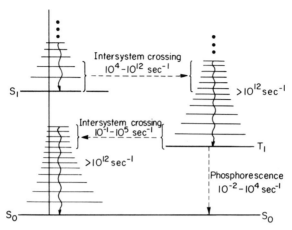

Figure 4-24. Radiative and nonradiative processes and representative relaxation times in aromatic hydrocarbons.

mode having an anharmonic-oscillator potential, it is obvious that a quasi-continuum of states rapidly develops in the vibronic structure, indicated in Figure 4-23 by the series of dots. Each excited-state level can either be singlet or triplet, the triplet states having the lower energy, owing to the exchange energy.

From Figure 4-23 it would appear that spectroscopy of such a molecule would be hopelessly complicated. Radiationless processes simplify the spectroscopy problem considerably, as indicated in Figure 4-24. Excitation of states above S_1 generally results in a rapid internal conversion (10^{-12} s) to the lowest vibrational modes in S_1, where the excitation remains for about 10^{-8} s, allowing fluorescence to the vibronic manifold in S_0. Strictly speaking, transitions between the singlet and triplet states would be forbidden. Fortunately, however, small spin-orbit coupling terms (McClure, 1952) mix these two states slightly, with the result that intersystem crossing to the triplet vibronic continuum occurs. Again, rapid internal conversion in the triplet manifold to T_1 takes place, on a time scale of 10^{-12} s. Relatively speaking, the lowest excited triplet lives for extremely long times (10^{-4}–10^2 s), while phosphorescence to the S_0 vibronic manifold takes place. If the intersystem crossing process is rapid enough, it can compete with fluorescence, and substantial population is thus channeled into the lowest triplet. For condensed phase systems, phosphorescence is usually not observed at room temperature, but sharp line spectra can usually be obtained from single crystals at liquid-helium temperatures. Figure 4-25 is an example of such a spectrum. Single crystals are usually necessary for sharp line spectra, to avoid multiple sites that would occur if a rigid glass were used. An

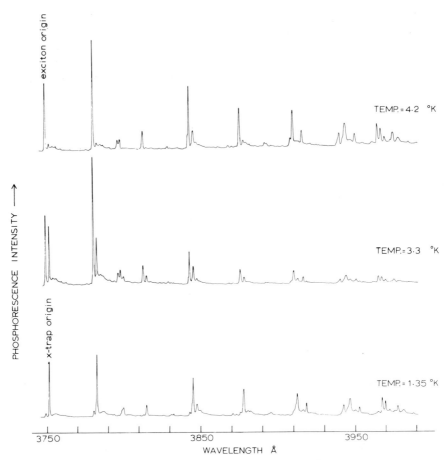

Figure 4-25. Typical phosphorescence spectrum from the $^{3}\pi\pi^{*}$ state of 1,2,4,5-tetrachlorobenzene as a function of temperature. The spectrum shows exciton emission at 4.2 and predominantly trap emission at 1.3°K. The linewidths are instrumentally limited.

exception to this is when Shpolskii-type glasses (Shpolskii, 1960, 1962, 1963) are used as the host system (Chan *et al.*, 1971).

By choosing substances with similar van der Waals' radii, a guest molecule may be doped substitutionally in a host crystal for study. Typical doping ratios are 0.01 mol % or less to reduce the possibility of dimer and higher aggregate formation of the guest species. It is important to choose a host whose excited-state energies are higher than the guest's, since the guest can act as an excited-state energy trap, eliminating the possibility of radiationless transfer to the host. If high excited-state populations in the guest molecule are desired, it is possible to excite the host singlet states. The highly abundant host states will act as energy collectors, and the excitation eventually gets funneled into the guest trap states.

Because of the desirability of a highly spin aligned triplet state, liquid-helium temperatures are normally employed. If the crystals are placed in a liquid-helium cryostat equipped with a large-capacity pump, temperatures down to 1.1°K can be routinely achieved. The cryostat is also equipped with quartz windows for viewing with a medium-resolution optical spectrometer ($\frac{3}{4}$ m is usually sufficient). Light from a high-pressure mercury arc lamp is filtered through 3 cm of water to remove infrared radiation and is passed through an interference filter or low-resolution monochromator and focused on the crystal. This is necessary in order to excite only the lower singlet states of the guest molecule. In some cases it will also prevent ionization and photocomposition. Alternatively, extremely selective and intense excitation may be achieved by using a tunable dye laser. The phosphorescence is usually monitored at 90° from the exciting radiation beam to reduce scattered light in the viewing monochromator and a cooled high-sensitivity photomultiplier tube is used to record the phosphorescence emission.

4.3.2. Conventional Techniques: Optically Detected Magnetic Resonance

The planar nature of aromatic molecules causes an anisotropic distribution of electron spin density in the excited triplet state, leading to a splitting of the triplet state even in zero external field. This is caused primarily by electron dipole–dipole interaction in organic molecules, although spin–orbit interactions can also contribute. The Hamiltonian for the spin–dipolar interaction is given by

$$\mathcal{H} = -XS_x^2 - YS_y^2 - ZS_z^2 \tag{4-65}$$

where X, Y, and Z are the principal values of the dipolar tensor, and S_x, S_y, and S_z are the triplet-state spin operators whose eigenfunctions $|\tau_x\rangle$, $|\tau_y\rangle$, and $|\tau_z\rangle$ are linear combinations of the high-field Zeeman states corresponding to $M_z = -1, 0, 1$. The dipolar tensor is traceless, so the zero-field triplet state may be expressed in terms of two parameters, D and E, which are related to the principal values by

$$
\begin{array}{ll}
D = \frac{1}{2}(X + Y) - Z & X = D/3 - E \\
E = -\frac{1}{2}(X - Y) & Y = D/3 + E \\
& Z = -2D/3
\end{array} \tag{4-66}
$$

The symmetry representations of the excited triplet sublevels are quite important, for they will indicate not only the polarization of phosphorescence emission to a vibronic level in the ground state, but they will also indicate the possible routes for spin–orbit coupling and intersystem crossing (El-Sayed, 1968). One of the most valuable features of the zero-field triplet state lies in the fact that intersystem crossing to the individual sublevels, which is determined largely by these symmetry considerations, is highly anisotropic (de Groot *et al.*, 1967). The differences manifest themselves at sufficiently low temperatures as highly non-Boltzmann population distributions within the triplet-state sublevels. This increases the sensitivity of optically detected magnetic resonance by orders of magnitude, as well as providing information on the intersystem crossing process itself. The review articles cited earlier present a thorough discussion of these features.

In optically detected magnetic resonance (ODMR) any two of the zero-field spin sublevels may be connected by a resonant field via a magnetic dipole transition. The time-dependent field is given by

$$V(t) = \gamma H_1 S_{x,y,z} \cos(\omega t + \phi) \qquad (4\text{-}67)$$

where γ is the gyromagnetic ratio and H_1 is the amplitude of the oscillating magnetic field. Typical zero-field transitions lie in the range 0.5–9 GHz. Since no static external field is being used to sweep through resonance, the microwave coupling device is usually sufficiently broad-banded as to allow one to sweep through a wide range of frequencies. A slow-wave helix meets all these demands reasonably well and has the added advantage of being quite simple to construct (Webb, 1962).

The crystal is mounted inside a slow-wave helix connected by coaxial lines to a microwave sweeper. It is immersed in a liquid-helium cryostat, and a particular vibronic phosphorescence transition is monitored while the microwaves are swept through resonance. The microwaves are usually amplitude-modulated, and changes in the phosphorescence are detected via lock-in-detection. When resonance occurs, the change in population between the two levels is detected as a modulation of the phosphorescence intensity. Figure 4-26 is an example of such an optically detected zero-field spin resonance spectrum. The multiplet structure is caused by hyperfine and quadrupole interactions and is well understood. For a more detailed discussion on the interpretation of the fine structure, see Harris and Buckley (1975).

4.3.3. Pulse Techniques in Optically Detected Magnetic Resonance

The conversion of an optically detected esr spectrometer to a high-power pulse spectrometer is enormously simplified by the fact that optical detection is being used, which completely eliminates detector overload problems. One only needs to add the pulse generators and high-power microwave amplifiers in order to produce an ODMR spectrometer that is capable of both pulse and standard slow-passage zero-field optically

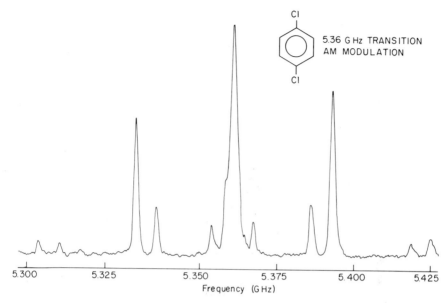

Figure 4-26. Optically detected magnetic resonance spectrum of the τ_x, τ_y transition in paradichlorobenzene. The center line is the electron-spin-only transition while the satellite results from forbidden transitions associated with chlorine nuclear quadrupole interactions in the excited state.

detected magnetic resonance. High-power microwave fields are conveniently generated by traveling-wave-tube amplifiers that have relatively large bandwidths, in addition to high gain; typically 20 W can be produced from a 30-mW input. Rapid switching is achieved by fast PIN-diode devices that achieve 10-ns rise times when driven properly. An important feature of the switching circuit must be that the difference in the microwave power between the "on" and the "off" state be high; 100 db is usually sufficient. This arises from the fact that it only takes microwatts or less to saturate a zero-field transition, so one must ensure that when the PIN diode is in the "off" position, there is not enough leakage to cause a significant change in the sublevel populations. Isolation factors of 100 db or more can be achieved by placing two PIN diodes in series a distance $\lambda/4$ apart.

The pulse trains are conveniently generated by constructing devices from transistor–transistor logic-integrated circuits. The devices are easy to design and build and are extremely reliable. Most circuits consist of straightforward combinations of NAND gates, monostable multivibrators, and edge-triggered flip-flops. For Carr–Purcell trains, an accurate time base is required that produces two pulses separated in time by several hundred microseconds, yet is accurate to within 10 ns. This is accomplished by the circuit shown in Figure 4-27, which counts pulses from a 1-MHz crystal

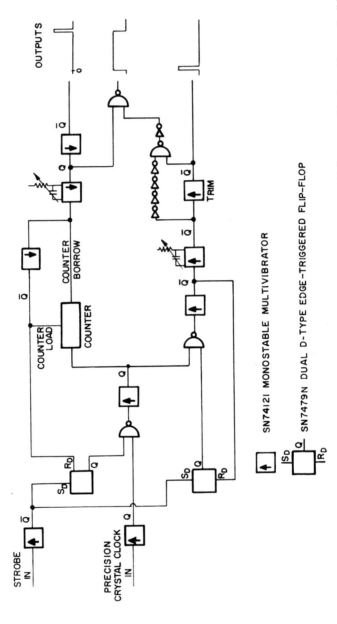

Figure 4-27. Schematic diagram of a digital plus variable delay circuit capable of producing two pulses widely separated in time with reproducibility on the order of 10 ns.

clock that is accurate to 1 part in 10^9. The probe pulse is generally swept slowly in time after being triggered by a predetermined number of pulses from the clock. This is easily accomplished by using as a delay a very accurate monostable multivibrator in conjunction with a precision 10-turn potentiometer that can be turned by a synchronous clock motor.

Phase-shifted pulses important for spin locking, Carr–Purcell trains, and other coherent averaging sequences can be accomplished by the

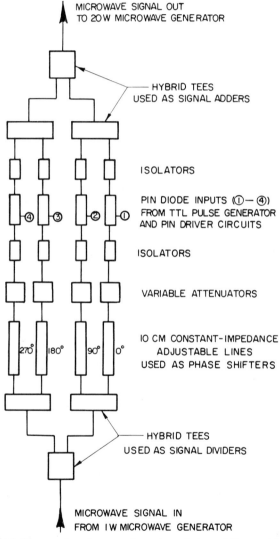

Figure 4-28. Block diagram of a microwave four-channel phase shifter used in the multiple-pulse experiments.

arrangement pictured in Figure 4-28. Accurate setting of phases is accomplished by the following procedure. Two phase-shifted signals are monitored by a crystal detector, and the pulses produced by each are added together. A 0° and a 180° pulse exactly cancel each other and produce no pulse when added together. A 90° phase shift is set after first setting the amplitudes and phases until the two pulses cancel. This can occur only if the amplitudes are both equal and 180° out of phase. The phase line is adjusted so that a maximum sum is achieved at 0° phase. Since the crystal detector is sensitive to the amplitude squared, a 90° phase shift will produce a sum of

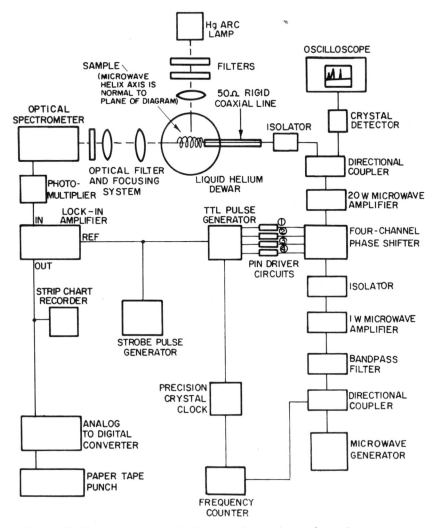

Figure 4-29. Block diagram of an optically detected magnetic resonance pulse spectrometer.

one half the maximum, which should also correspond to one half the total distance traveled by the adjustable delay line in going from maximum sum to null. In some situations the actual optically detected coherence signal will be phase-sensitive and can be maximized. One method of ensuring sharp, square pulses of equal amplitude is to drive the TWT close to saturation. If the resulting power is too high, attenuators may be placed at the output. The block diagram shown in Figure 4-29 incorporates the equipment discussed here with the ODMR spectrometer shown in Figure 4-30. A more complete discussion can be found in Breiland *et al.* (1975).

4.3.3.1. Transient Nutation and Pulse Timing

The transient nutation, first detected optically by Schmidt *et al.* (1971), is an essential feature of any coherence experiment, for it defines the times necessary to produce $\pi/2$, π, etc., pulses for a given microwave power in addition to forming the basis for rotary echoes. A probe pulse is not needed here, for the phosphorescence directly reflects the oscillating populations in the triplet sublevels. A direct method for measuring the transient nutation would be to apply a long, high-power pulse and record the oscillations directly with a device that could resolve the ~5-MHz oscillations that occur. An equivalent method (Schmidt *et al.*, 1971) is to start with a short pulse of known length. The pulse produces a partial oscillation on microsecond time scales, but when turned off leaves the sublevel populations fixed at one point in the transient nutation, as given by Eq. (4-48). The phosphorescence intensity decays back to steady-state on millisecond time scales, and the pulse may then be repeated. This repetitive signal is then fed into a lock-in amplifier producing a dc signal that is proportional to the population difference at one point in the transient nutation. The microwave pulse is then lengthened slowly enough to be compatible with the time constant on the lock-in amplifier, and in this fashion the transient nutation is traced out point by point in time. This process is illustrated schematically in Figure 4-31. Points (b) and (c) in Figure 4-31 would yield the pulse times necessary for $\pi/2$ and π pulses, respectively.

An important aid in obtaining maximum power, and accordingly highest nutation frequencies, is a "trombone" impedance matching device placed between the slow-wave helix and the TWT amplifier. By using a pulse that is less than π, any increase in the nutation rate will cause an increase in the phosphorescence, and in this fashion the impedance matching can be made optimal, since maximum power to the helix is monitored by a maximum in the phosphorescence intensity at this pulse setting. Note that the pulse length is not arbitrary, since either an increase or decrease in nutation frequency will result in a decrease in phosphorescence intensity if the pulse width corresponded originally to a π pulse.

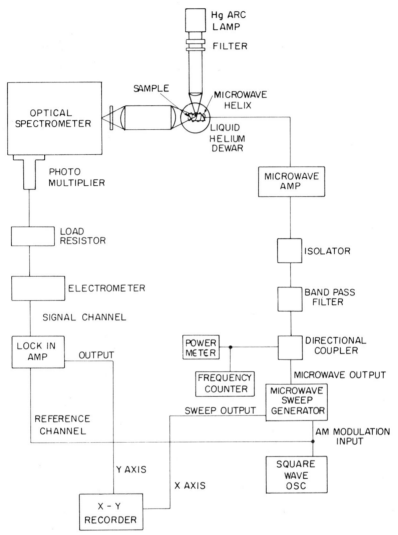

Figure 4-30. Experimental arrangement for a conventional optically detected magnetic resonance experiment in a zero field.

4.3.3.2. Short Coherence Sequences

A short sequence is defined to be one that is so fast that the entire sequence is over before appreciable feeding or decay can occur. In this situation the method described for the transient nutation is suitable, and the experiment proceeds as follows. A Strobe pulse generator defines the repetition time for the experiment. It acts both as a reference for the lock-in

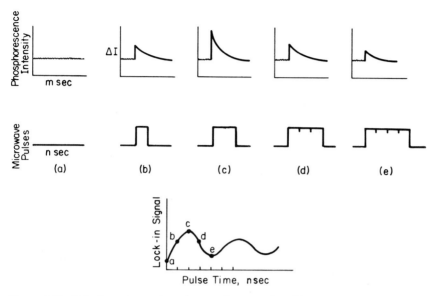

Figure 4-31. Point-by-point tracing of a transient nutation. The top trace represents the observed phosphorescence intensity on a millisecond time scale after being subjected to nanosecond-length microwave pulses. After the short pulse is over, the phosphorescence decays back to a steady state. As the short pulse is lengthened, the height of the initial phosphorescence change varies in direct proportion to the transient nutation signal. A plot of the initial height versus the pulse time yields the same curve that would be recorded by a fast transient analyzer capable of resolving the MHz transient nutation oscillation frequency. A lock-in amplifier allows one to make a continuous tracing of the nutation as a function of the pulse time.

and as a command pulse to start the microwave pulse sequence. The reference frequency is largely determined by the triplet-state lifetimes, the period being roughly determined as three or four times the average lifetimes of the states involved. Since the pulse sequence lasts only on the order of microseconds, it is reasonable to neglect changes in the phosphorescence during the sequence. The probe pulse will produce a phosphorescence spike that is proportional to the coherence remaining in the ensemble at the time the probe pulse was applied, and the same lock-in detection technique discussed earlier may be used to trace out the decay of the coherence. As an example, Figure 4-32 gives an oscilloscope trace of the actual spikes encountered in the echo experiment.

4.3.3.3. Long Coherence Sequences

These are experiments that last for times that approach or even exceed the triplet lifetimes. Under these conditions the lock-in method becomes

OPTICALLY DETECTED ELECTRON SPIN ECHO
IN THE $^3\pi\pi^*$ STATE OF h_2-TCB (y-trap)

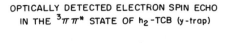

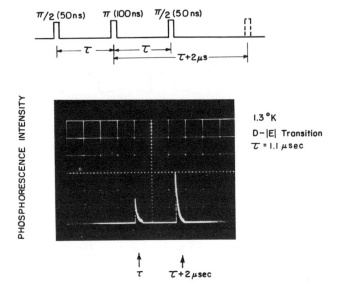

1.3 °K
D−|E| Transition
τ = 1.1 μsec

Figure 4-32. Phosphorescence intensity when spin coherence is rephased in the echo (τ) and the spin ensemble is completely dephased. ($\tau + 2$ microseconds.)

unsuitable and it becomes advantageous to time-average the transient signals, including the probe pulse "spike" on a digital CAT. The spin-locking sequence is performed this way. The only disadvantage of the method lies in the fact that a continuous display of the coherence loss is no longer possible.

4.3.3.4. Triplet-State Multiplets and Orientation Factors

A useful property of optical detection is that saturated signals do not disappear as they would with conventional detection, but approach a maximum. This allows one to display a complete multiplet structure by applying intense fields to bring out the nearly forbidden lines, and these will appear alongside the allowed transitions with a nearly equal signal strength, as seen in Figure 4-26. In this particular spectrum some of the lines in the multiplet are essentially forbidden, being only 10^{-4} that of the central peak. The allowedness of the transition is manifested directly in the nutation frequency, and one should not be led to think that all the lines seen in the saturated spectrum can be coherently coupled. To do this, one would need a 100-fold increase in the microwave power to produce the same nutation frequency for the forbidden line as one would need for the central allowed line.

Although zero-field magnetic resonance has an advantage in that the molecules within the crystal essentially select the proper polarization direction of the oscillating electromagnetic field, this feature is usually detrimental in coherence experiments. In randomly oriented samples such as microcrystalline powders or dilute guest species in a rigid glass matrix, the random orientations produce a range of nutation frequencies corresponding to the projection of the magnetic field on the dipole transition direction, and will not produce a meaningful transient nutation. Therefore, single crystals are usually desirable for most coherence experiments. Even in a single crystal the molecules are often arranged in several translationally inequivalent positions within the unit cell. This leads to "beats" within the transient nutation that result from a sum of sinusoidal oscillations with different frequencies that are determined by the projection of the driving-field magnetic vector on the transition dipole direction for each molecule in the unit cell. For three molecules per unit cell or less, a driving-field polarization can always be found that will produce a single nutation frequency. In some cases, of course, the disadvantages of orientational effects can be turned into advantages for studying physical rearrangements or for isolating one set of molecules within the unit cell.

4.4. Applications

4.4.1. Preliminaries

The information one can obtain from conventional molecular spectroscopy is generally limited in at least one important way. Observations are usually made on an inhomogeneous line; hence, the time-dependent correlation function describing physical phenomena that are responsible for energy fluctuations of states cannot be obtained directly from the line shape or linewidth. The net result is that the important interrelationships between the time-dependent molecular Hamiltonian and line-shape transform theories cannot be exploited (Kubo and Tomita, 1954; Kubo, 1954; Anderson, 1954). The power of coherence techniques lies in their ability to remove certain unwanted interactions from consideration, such as the inhomogeneous contributions to the linewidth or, in the case of excited triplet states, effects from nuclear-spin diffusion, laying bare the effective linewidth that results from the relaxation time of phenomena of more immediate interest. For example, if a kinetic process such as energy transfer or vibrational relaxation affects the electron-spin coherence in a time fast compared to other dephasing phenomena in a particular coherence experiment, then the

Fourier transform of the relaxation spectrum yields a homogeneously narrowed line that is centered at the zero-field transition frequency and that has a linewidth that is related to rate parameters for energy transfer. In this sense the time-dependent field radiation removes unwanted contributions to the linewidth and narrows the esr transition to a point where its residual width is due only to the kinetic process itself.

We see, however, that a prerequisite for studying kinetic phenomena with coherent spectroscopy is that the effective Hamiltonian responsible for the decay of the electron-spin coherence also be responsible for the phenomena one is interested in investigating Since hyperfine interactions and nuclear-spin diffusion dephase the triplet spins on a time scale of about 10 μs, either the relaxation time of the kinetic events must be shorter than this, or one must use one of the methods we have described for removing the nuclear interactions.

The coherence method one chooses to use depends somewhat on the range of relaxation times one expects from the kinetic phenomena. In principle, optically detected spin locking (Harris *et al.*, 1973) and optically detected ADRF (Brenner *et al.*, 1974) would suffice for most problems. Spin locking is not restricted to short times and can be performed down to several hundred nanoseconds. ADRF is practically limited, however, from experimental considerations from $\sim$220 μs to 1 s. Optically detected rotary echoes (Harris *et al.*, 1973) are also useful, from about 100 ns to 500 μs. In all methods the correlation time of the technique can be controlled by the cycle time [cf. Eqs. (4-55) and (4-57)], which in turn is controlled by the strength of the applied field $\gamma \mathcal{H}_1$. Since nuclear-spin diffusion affects the electron-spin coherence in a characteristic time of 10 μs, one needs an $\gamma \mathcal{H}_1$ field of about 0.5 MHz to remove fluctuating hyperfine fields from consideration. Under these circumstances kinetic phenomena that dephase the spins in the range 100–500 μs can be investigated. Relaxation times longer than 500 μs can be measured if $\gamma \mathcal{H}_1$ is increased to about 5 MHz to ensure complete removal of nuclear-spin effects.

4.4.2. Addition of Energy Exchange to the Equations of Motion

In order to quantitatively interpret the decaying spin coherence in any particular experiment, it is important to include the effects of spin memory that may result when excited triplet states are exchanging their energy between two closely lying levels. The equations we have developed for optically detected spin coherence [Eqs. (4-26)–(4-30)] can easily be modified to include exchange in the same manner as chemical exchange was included in the Bloch equations in nmr (McConnell, 1958).

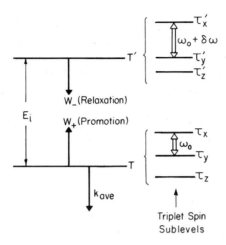

Figure 4-33. Spin exchange resulting from scattering between excited triplet states each having a different Larmor frequency. The rate W_+ represents the promotion rate out of the lowest excited triplet T; W_- is the relaxation rate from the excited state T. E_i is their energy separation. The triplet spin sublevels associated with these electronic states are diagrammatically illustrated on an energy scale which is of the order of a few tenths of a wavenumber, whereas E_i can vary from a few to a thousand wavenumbers.

Consider the coupled systems illustrated in Figure 4-33. For simplicity we shall assume that the excited state is at an energy E_i above the lowest triplet state T and that it has the same orientation of its zero-field tensor, although the magnitude of the zero-field splittings differ. Furthermore, we assume that the matrix elements for energy exchange are spin-independent so that the spin states exchange, conserving spin angular momentum, with rate W_+ from the lowest state, T, to the upper state, T', and W_- from T' to T. If the event leading to the exchange of energy between T and T' is fast compared to any time-dependent term in the density matrix, then all components of $\rho(t)$ are conserved during exchange. This is essentially the same assumption that is made in Dexter's exchange (Dexter, 1953). Under these conditions energy is being exchanged between two time-dependent density matrices, $\rho(t)$ for T and $\rho'(t)$ for the state T'; hence, Eq. (4-26) must be coupled via exchange to an equivalent equation for $\rho'(t)$. Following McConnell (1958) one can write the elements of $\rho(t)$ in exchange as $\rho(t)_e$, where the r vector components $\rho'(t)$. Following McConnell (1958) one can write the elements of $\rho(t)$ given by

$$\left(\frac{dr_1}{dt}\right)_e = \frac{dr_1}{dt} + \frac{r_1'}{\tau} - W_+ r_1 \tag{4-68a}$$

$$\left(\frac{dr_2}{dt}\right)_e = \frac{dr_2}{dt} + \frac{r_2'}{\tau} - W_+ r_2 \tag{4-68b}$$

$$\left(\frac{d\rho_{aa}}{dt}\right)_e = \frac{d\rho_{aa}}{dt} + \frac{\rho'_{aa}}{\tau} - W_+\rho_{aa} \tag{4-68c}$$

$$\left(\frac{d\rho_{bb}}{dt}\right)_e = \frac{d\rho_{bb}}{dt} + \frac{\rho'_{bb}}{\tau} - W_+\rho_{bb} \tag{4-68d}$$

The leading terms dr_1/dt, dr_2/dt, $d\rho_{aa}/dt$, and $d\rho_{bb}/dt$ are given explicitly by Eq. (4-26). The elements of $\rho'(t)$ which are coupled to the equations above are given by

$$\left(\frac{dr'_1}{dt}\right)_e = \frac{dr'_1}{dt} + W_+r_1 - r'_1/\tau \tag{4-69a}$$

$$\left(\frac{dr'_2}{dt}\right)_e = \frac{dr'_2}{dt} + W_+r_2 - r'_2/\tau \tag{4-69b}$$

$$\left(\frac{d\rho'_{aa}}{dt}\right)_e = \frac{d\rho'_{aa}}{dt} + W_+\rho_{aa} - \rho'_{aa}/\tau \tag{4-69c}$$

$$\left(\frac{d\rho'_{bb}}{dt}\right)_e = \frac{d\rho'_{bb}}{dt} + W_+\rho_{bb} - \rho'_{bb}/\tau \tag{4-69d}$$

Following van't Hof and Schmidt (1975a), if we make the assumption that the frequency difference between the upper and lower states, $\delta\omega$, is large compared to the strength of the driving field, ω_1, then for all practical purposes the upper state is unaffected by the applied microwave field. Under these conditions the rotating frame in the upper state is off-resonance by an amount $\Delta\omega'$, where

$$\Delta\omega' = (\omega_0 - \omega) + \delta\omega \tag{4-70a}$$

or

$$\Delta\omega' = \Delta\omega + \delta\omega \tag{4-70b}$$

We see that if the experiment is applied on-resonance in the lower state, then $(\omega_0 - \omega) = 0$ and $\Delta\omega' = \delta\omega$. Furthermore, if the upper state's relaxation is determined solely by W_- and it is created from T at a rate given by W_+, then

$$\frac{dr'_1}{dt} = -\Delta\omega'r'_2 \tag{4-71a}$$

$$\frac{dr'_2}{dt} = \Delta\omega'r'_1 \tag{4-71b}$$

$$\frac{d\rho'_{aa}}{dt} = 0 \tag{4-71c}$$

$$\frac{d\rho'_{bb}}{dt} = 0 \qquad (4\text{-}71d)$$

and one arrives at a set of simplified density matrix elements for $\rho'(t)_e$ in exchange with $\rho(t)_e$. The elements are given as

$$\left(\frac{dr'_1}{dt}\right)_e = -\Delta\omega' r'_2 + W_+ r_1 - r'_1/\tau \qquad (4\text{-}72a)$$

$$\left(\frac{dr'_2}{dt}\right)_e = \Delta\omega' r'_1 + W_+ r_2 - r'_2/\tau \qquad (4\text{-}72b)$$

$$\left(\frac{d\rho'_{aa}}{dt}\right)_e = W_+ \rho_{aa} - \rho'_{aa}/\tau \qquad (4\text{-}72c)$$

$$\left(\frac{d\rho'_{bb}}{dt}\right)_e = W_+ \rho_{bb} - \rho'_{bb}/\tau \qquad (4\text{-}72d)$$

Equations similar to these coupled equations are common to a variety of problems in magnetic resonance and have been solved by a number of authors, including Barrat and Cohen-Tannoudji (1961), for exchange in excited atomic states: Swift and Connick (1962), for ligand exchange; and van't Hof and Schmidt (1975a), for the effects of vibrational relaxation in excited triplet states.

   Under conditions in which the upper state is considered to be completely off-resonance, $\delta\omega \gg \omega_1$, the eight coupled equations become uncoupled. The driving field does not affect the populations of ρ' and the coherent components r_1 and r_2 in Eqs. (4-72a) and (4-72b) appear approximately constant to the rapidly fluctuating r'_1 and r'_2 components. On the other hand, r'_1 and r'_2 fluctuate so rapidly that their contribution to Eqs. (4-69a) and (4-69b) will be an effective time-averaged quantity obtained from the steady-state solutions to Eqs. (4–72), assuming r_1 and r_2 to be constant. We thus set the time derivatives in Eqs. (4-72) to zero and solve for r'_1 and r'_2.

$$r'_1 = \frac{\tau W_+ (r_1 - \Delta\omega' \tau r_2)}{1 + (\Delta\omega')^2 \tau^2} \qquad (4\text{-}73a)$$

$$r'_2 = \frac{\tau W_+ (r_2 + \Delta\omega' \tau r_1)}{1 + (\Delta\omega')^2 \tau^2} \qquad (4\text{-}73b)$$

$$\rho'_{aa} = W_+ \tau \rho_{aa} \qquad (4\text{-}73c)$$

$$\rho'_{bb} = W_+ \tau \rho_{bb} \qquad (4\text{-}73d)$$

Substitution of these values into Eqs. (4-68) yields the final time-dependent equation for the lower state:

$$\left(\frac{dr_1}{dt}\right)_e = -\left[\Delta\omega + \frac{\Delta\omega' W_+\tau}{1+(\Delta\omega')^2\tau^2}\right]r_2 - \left[\frac{1}{T_e} + \frac{W_+(\Delta\omega')^2\tau^2}{1+(\Delta\omega')^2\tau^2}\right]r_1$$

$$\left(\frac{dr_2}{dt}\right)_e = \left[\Delta\omega + \frac{\Delta\omega' W_+\tau}{1+(\Delta\omega')^2\tau^2}\right]r_1 - \left[\frac{1}{T} + \frac{W_+(\Delta\omega')^2\tau^2}{1+(\Delta\omega')^2\tau^2}\right]r_2 - \omega_1(\rho_{aa} - \rho_{bb})$$

$$(4\text{-}74)$$

$$\left(\frac{d\rho_{aa}}{dt}\right)_e = \frac{d\rho_{aa}}{dt}$$

$$\left(\frac{d\rho_{bb}}{dt}\right)_e = \frac{d\rho_{bb}}{dt}$$

One should note that these equations are identical in form to Eqs. (4-26)–(4-30), which we derived earlier. One needs only to replace the off-resonance frequency $\Delta\omega$ in Eqs. (4-26) by an *effective* value, including exchange given by

$$(\Delta\omega)_e = \Delta\omega + \frac{\Delta\omega' W_+\tau}{1+(\Delta\omega')^2\tau^2} \qquad (4\text{-}75)$$

and the in-plane relaxation rate $1/T$ or $1/T_2$ in Eqs. (4-26) by an *effective* relaxation rate which includes exchange given by

$$\left(\frac{1}{T}\right)_e = \frac{1}{T} + \frac{W_+(\Delta\omega')^2\tau^2}{1+(\Delta\omega')^2\tau^2} \qquad (4\text{-}76)$$

and all subsequent solutions are also solutions for coherent coupling, including energy exchange.

Physically, exchange introduces two additional parameters into Eqs. (4-26). The first is a frequency-shift term which reflects the time-averaged contribution that the upper state makes to the lower state's Larmor frequency via spin memory. It is simply the time-weighted probability that a spin will be in the upper state, resulting in a time-weighted average of $\Delta\omega$ and $\Delta\omega'$ to give an effective $\Delta\omega$. The second contribution is a weighted kinetic term related to the rate at which the lower state scatters to the upper state, causing a partial relaxation of the spin coherence.

Two limits of Eqs. (4-75) and (4-76) are interesting to investigate because they represent (1) the limits when no spin memory is reflected in the scattering process, and (2) the case when the maximum spin memory is retained.

4.4.2.1. Loss of Spin Memory in the Slow Exchange Limit

We can see from Eqs. (4-75) and (4-76) that the value of $\Delta\omega'$ relative to τ determines the extent to which phase coherence or spin memory is retained in the scattering between the lowest state, T, and the upper state, T'. When the condition

$$\Delta\omega'\tau \gg 1 \qquad (4\text{-}77)$$

is met, all spin memory is lost in a single scattering event from T to T' and the states are said to be in slow exchange. For practical considerations only a 1% shift is introduced when

$$\Delta\omega'\tau \sim 10^2 \qquad (4\text{-}78)$$

The slow exchange limit requires that (1) the lifetime in T' be very long; (2) $\Delta\omega'$, the off-resonance shift in the upper state, be large; or (3) both $\Delta\omega'$ and τ be large. When the experiment is performed on resonance in the lower state, T, then $\Delta\omega' = \delta\omega$ and the condition in Eq. (4-77) becomes dependent on the product of the difference in Larmor frequencies between states T and T' times the lifetime in the upper state, i.e.,

$$\delta\omega\tau \gg 1 \qquad (4\text{-}79)$$

We see immediately for the slow exchange limit that $(\Delta\omega)_e$ given by Eq. (4-75) becomes equal to $\Delta\omega$ and, for on-resonance experiments, goes to zero.

Physically this corresponds to a scattering from T to T' with all loss of phase memory before returning to T. Such would be the case when a localized triplet state is promoted to an exciton band and is propagated away from the site so that τ becomes long in the upper state, T'.

In a similar manner, $(1/T)_e$ approaches a limiting value in the slow exchange required. We note that Eq. (4-76) approaches a value given by

$$(1/T)_e \rightarrow 1/T + W_+ \qquad (4\text{-}80)$$

in the limit $\Delta\omega'\tau \gg 1$. This is a very important result, for it allows one to extract the absolute rate, W_+, for promotion to T' from the decay of the coherent spin state in T. Since $(T)_e$ appears as a T_2-type relaxation, it can be easily extracted from most coherence experiments and W_+ can be determined using Eq. (4-28), since the radiative or nonradiative rates k_a and k_b can be obtained independently. W_+ is thus displayed as a term which competes with the radiative (or nonradiative) decay channels as a mechanism for dephasing the spins. Moreover, if one knows that the states T and T' are in Boltzmann equilibrium, then from microscopic reversibility, W_+ and τ are related by

$$W_+\tau = \exp\left(-E_i/kT\right) \qquad (4\text{-}81)$$

One should note that when the two levels undergoing exchange are in Boltzmann equilibrium, the slow exchange condition in Eq. (4-77) is also met by

$$\frac{\Delta\omega'}{W_+} \exp\left(-E_i/kT\right) \gg 1 \qquad (4\text{-}82)$$

4.4.2.2. Retention of Spin Memory in Scattering in the Fast Exchange Limit

When

$$\Delta\omega'\tau \ll 1 \qquad (4\text{-}83)$$

the system will retain the maximum amount of spin memory as it scatters from T to T' and back and can be said to be in the fast exchange limit. The condition that leads to this is a short lifetime in T'. As in the slow exchange limit, Eq. (4-75) approaches a limiting value in fast exchange. This is given by

$$(\Delta\omega)_e = (\Delta\omega + W_+ \Delta\omega'\tau) \qquad (4\text{-}84)$$

For on-resonance conditions in T, Eq. (4-75) becomes

$$(\Delta\omega)_e = W_+ \delta\omega\tau \qquad (4\text{-}85)$$

If a system is in this limit, one expects a frequency shift in the zero-field transition of T, proportional to the difference in the Larmor frequency in the state T' weighted by the kinetic terms (W_+) responsible for the steady-state population in T'. If the two levels T and T' are in Boltzmann equilibrium, then according to microscopic reversibility [i.e., Eq. (4-81)],

$$(\Delta\omega)_e = \delta\omega \exp\left(-E_i/kT\right) \qquad (4\text{-}86)$$

This is an important prediction of the theoretical development and has been observed by van't Hof and Schmidt in several instances (van't Hof and Schmidt, 1975a, 1976a).

Physically, $(\Delta\omega)_e$ relates the fact that it takes many scattering events between T and T' to lose spin memory and that such scattering processes tend to average the Larmor frequencies of the two scattering states, hence the $\delta\omega$-dependent frequency shift. The maintenance of phase or spin memory in this limit is further reflected by the fact that it becomes progressively more difficult to extract W_+ from the relaxation parameter $(1/T)_e$ since as $\tau \to 0$, then $(1/T)_e \to 1/T$, according to Eq. (4-76). Stated more colloquially, if there is little dephasing in T', then the spin has difficulty knowing that it has been promoted to T' from T via W_+ processes; hence, an absolute value of W_+ cannot be extracted easily from the experiment. One should note in passing that in such cases one expects very little exponential behavior of the effective relaxation time with temperature. One should observe only the frequency-shift term. In practice, however, most systems are intermediate between fast and slow exchange and both a frequency shift and change in relaxation time should be resolvable experimentally. Nature is kind to us in these circumstances; we are able to determine $\delta\omega$, W_+, τ, and E_i simultaneously if we know that the system is in Boltzmann equilibrium. If the system is not in Boltzmann equilibrium, then τ and W_+ are not related by Eq. (4-81) but are kinetically determined (Fayer and Harris, 1974a) and the problem becomes more complicated. In these instances, τ, $\delta\omega$, W_+, and E_i are interrelated and not as clearly revealed.

In the remainder of this chapter we will confine ourselves to several problems or applications of the techniques we have described. These will be drawn from only two areas of triplet spectroscopy: energy transfer dynamics and vibrational relaxation in triplets. We should note, however, that these techniques can be applied to many other areas, including problems of more immediate interest in magnetic resonance.

4.4.3. Energy Transfer Studies Using Coherent Spectroscopy Techniques

It is not our purpose to go into great detail on the myriad of interesting questions that arise in the area of energy transfer dynamics in solids but rather to illustrate how coherent spectroscopy can provide qualitative insight and quantitative information on the routes and mechanisms for the processes. The salient features of *one* problem (Fayer and Harris, 1974a,b; Brenner *et al.*, 1974, 1975; Lewellyn *et al.*, 1975), that of measuring the rate at which a localized state is promoted to an exciton-band state where it becomes a mobile wave packet, is illustrated in Figure 4-34. It is clear that two limits are expected depending on the energy of separation from the

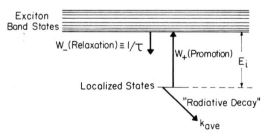

Figure 4-34. Scattering of triplet excitation between a localized state beneath an exciton band and delocalized band states. W_+ is the promotion rate and W_- is the relaxation rate out of the exciton state. E_i is the energy separation and k_{av} is the total radiative and nonradiative decay to the ground state in the presence of an applied field.

band, E_i. If E_i is large enough relative to kT, then it becomes highly unlikely that phonons can promote the localized state within the lifetime of the excitation to the band. In such cases, W_+, the promotion rate, becomes less than the average radiative rates given by k_{av} in Eq. (4-28), and one expects in a coherence experiment which removes contributions from a fluctuating hyperfine field that the dephasing mechanism for the electron spins is due simply to radiative or radiationless decay to the ground state.

In another extreme, phonons can promote the excitation to the band at a reasonable rate W_+. In such instances a mobile exciton is formed. Often in these cases the recurrence time τ from the band to the localized site is long and satisfies the condition

$$\Delta\omega'\tau \gg \tag{4-87}$$

No spin memory is maintained in these instances in the promotion and retrapping process. The general expression for the relaxation time in a coherence experiment, including relaxation effects from fluctuating hyperfine fields, can be seen from Eqs. (4-28) and (4-75).

$$\left(\frac{1}{T_e}\right)_{eff} = k_{av} + \frac{1}{T_{2e}} + \frac{W_a + W_b}{2} + \frac{W_+(\Delta\omega')^2\tau^2}{1+(\Delta\omega')^2\tau^2}$$

In spin locking, the average Hamiltonian for interactions leading to $1/T_{2e}$ vanishes, as seen by Eq. (4-64); hence, $1/T_{2e} = 0$ and the decaying coherence, as measured by spin locking (usually termed $T_{1\rho}$), is related only to decay channels to the ground state or to the band.

In the case where little promotion occurs to the band, $W_+ \ll k_{av}$; hence

$$(T_e)^{-1} \equiv (T_{1\rho})^{-1} = k_{av} \tag{4-88}$$

This is illustrated in Figure 4-35 for an excited triplet state of 1,2,4,5-tetrachlorobenzene doped in durene. In this example (Harris et al., 1973) the band state is ~ 1500 cm^{-1} above the localized state. As was expected, $T_{1\rho}$ approached the average lifetime of the triplet-spin sublevels, k_{av}; furthermore, no temperature dependence in $T_{1\rho}$ was discerned from the data, consistent with a temperature-independent radiative decay of the excited state to the ground state. In addition, no temperature-dependent frequency shift was observed, indicating that the system is in the slow exchange limit. In this experiment the Fourier transform of the relaxation spectrum as given by $T_{1\rho}$ is a Lorentzian absorption line having a width of 24 Hz, which is due almost entirely to uncertainty broadening resulting from the 63-ms average lifetime of the excited triplet state. The linewidth of the transition without coherent averaging is 2×10^6 Hz.

In another example, h_2-1,2,4,5-tetrachlorobenzene (TCB) is doped as a dilute isotopic substituent into a d_2-1,2,4,5-tetrachlorobenzene host

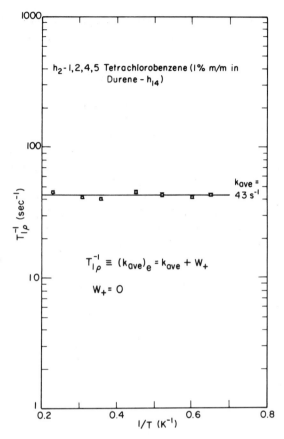

Figure 4-35. Decay of coherence due to radiative processes when the energy separation E_i is very large. In this instance it is approximately 1400 wavenumbers. Note that the decay is temperature-independent and that $T_{1\rho}$ in effect measures the average radiative and nonradiative rate to the ground state.

crystal. In this instance, the localized state of h_2-TCB is only ~ 20 cm^{-1} below the d_2-TCB exciton band, with the result that phonons can scatter the excitation from the localized state to the band at a rate given by W_+ (Fayer and Harris, 1974b). In this instance the promotion rate W_+ is much faster than the radiative decay rate k_{av} and W_+ is highly temperature-dependent. Furthermore, no frequency shifts are observable with temperature. The loss of spin coherence in this experiment is given by Eq. (4-88), where $(T_{2e})^{-1}$ is equal to zero. Since k_{av} is temperature-independent, one expects a constant term at the lowest temperature which is equal to the decay channel lifetimes. Since there is no frequency shift with temperature, the system is in the slow exchange limit, so to a first, but good, approximation,

$$(T_e)^{-1} \equiv T_{1\rho}^{-1} = k_{av} + W_+ \tag{4-89}$$

If the exciton band and the localized state were in Boltzmann equilibrium, one could extract the lifetime, τ, of the exciton state before retrapping, since W_+ and τ are related by

$$W_+ = \rho(E)(\tau)^{-1} \exp{(E_i/kT)} \qquad (4\text{-}90)$$

where $\rho(E)$ is the exciton density-of-states function.

The experimental results are illustrated in Figure 4-36. At the lowest temperatures the decay of spin coherence as measured by spin locking approaches k_{av} and shows a pronounced exponential temperature dependence whose exponential factor is the energy separation between the localized states of the band. The results clearly show that the localized state is decaying nonradiatively into the exciton band and that the loss of electron-spin coherence at the higher temperature reflects quantitatively the absolute promotion rate to the band. The fact that the system is not in Boltzmann equilibrium, however, makes Eq. (4-90) inappropriate in this

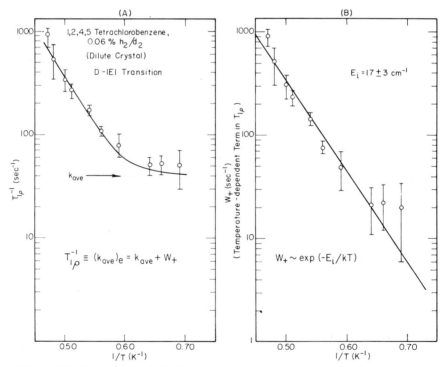

Figure 4-36. Determination of the decay rate of a localized state into a localized wave packet from an optically detected spin-locking experiment. (A) $T_{1\rho}$ versus $1/T$ yields a temperature-independent term, k_{av}, and a temperature-dependent term, W_+, respectively. (B) The temperature-dependent promotion rate versus $1/T$ shows an exponential dependence, the exponent being the energy separation E_i between the localized state and the band.

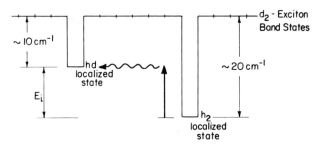

Figure 4-37. Schematic representation of phonon-assisted tunneling between localized states at different energies. At high concentration the separation between an h_2-localized state and an hd-localized isotope state can become short enough that a phonon of energy E_i can promote the h_2-localized state and allow tunneling to the hd-localized state without going via the exciton band.

instance. As a result, τ cannot be obtained independently. It has been shown, however (Fayer and Harris, 1974b), that an exponential temperature dependence is expected because the low-temperature phonon distribution responsible for scattering the localized state to the exciton band is given by a Planck distribution function, $\langle N(E)\rangle$, where

$$\langle N(E)\rangle = \rho(E)[\exp(E/kT) - 1]^{-1} \qquad (4\text{-}91)$$

which at low temperatures is given by

$$\langle N(E)\rangle = \rho(E)\exp(-E/kT) \qquad (4\text{-}92)$$

where $\rho(E)$ is the phonon density-of-states function. Hence, W_+ exhibits an exponential temperature dependence, since

$$W_+ \sim \langle N(E)\rangle \qquad (4\text{-}93)$$

This is expected in either the Boltzmann or non-Boltzmann regime and is an important finding, since in either temperature region the path for energy transfer can be discerned from the energy gap E_i between the state that is coherently coupled and the decay channel responsible for relaxation.

In another example (Lewellyn *et al.*, 1975) the concentration of h_2-TCB was increased to 5%. In these crystals coherent spectroscopic techniques could be used to establish the phonon-assisted tunneling rate between h_2-TCB and an hd-TCB separated by d_2-TCB barriers. This is illustrated in Figure 4-37. Crystals having h_2-TCB molecules in low concentration, as we have seen, decay via promotion to the exciton band with a 17-cm^{-1} energy gap. When the concentration is increased to 5%, however, the decaying electron-spin coherence, as measured in a spin-locking experiment, changes. The temperature-dependent term reflects a much lower energy separation of 10 ± 3 cm^{-1} in the concentrated crystals (cf. Figure 4-38). This is equal to the separation between the h_2-TCB and hd-TCB

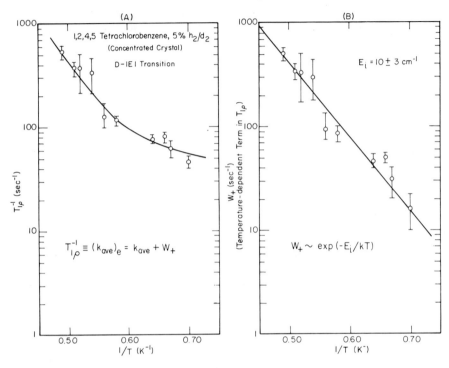

Figure 4-38. Determination of the promotion rate for phonon-assisted tunneling between delocalized states as illustrated in Figure 4-37. (A) $T_{1\rho}$ versus $1/T$ yields a temperature-independent and a temperature-dependent term. (B) The determination of the temperature-dependent term reflects the energy gap, which is equal to the separation between the isotopic localized state energies.

triplet states as established independently through the emission spectra. Thus, the decay of the spin coherence unequivocally reveals this new channel for energy transfer in a straightforward manner.

4.4.4. Vibrational Relaxation Studies Using Coherence Techniques

In the examples above, the states were not in Boltzmann equilibrium; hence, the lifetime in the upper states could not be established from Eq. (4-90). Recently, van't Hof and Schmidt (1975b, 1976a) have applied these techniques to the measurement of vibrational relaxation in the triplet state of *p*-benzoquinone. This is an ideal system since the two levels T and T' are the lowest triplet state, $^3B_{1g}$, and a vibronic level, ^3A_u, 21 cm^{-1} higher in energy. Moreover, they can be considered isolated in the sense that the decay channel for the loss of coherence in the lower state is restricted to a single state, the ^3A_u, and the two states can be shown to be in Boltzmann equilibrium. The relaxation times and differences in Larmor frequencies between the two states were such as to allow the authors to measure both the temperature-dependent loss of coherence via a rotary echo experiment and to also resolve the temperature-dependent frequency shift from a conventional epr experiment.

We note from Eqs. (4-75) and (4-76) and the Boltzmann relationship between W_+ and τ from microscopic reversibility [Eq. (4-81)] that the temperature-dependent terms for on-resonance condition are given by

$$\delta\omega(T) = \frac{\delta\omega}{1+(\delta\omega)^2\tau^2} \exp(-E_i/kT) \qquad (4\text{-}94)$$

and

$$\left(\frac{1}{T}\right)_e = \frac{(\delta\omega)^2\tau \exp(-E_i/kT)}{1+(\delta\omega)^2\tau^2} \qquad (4\text{-}95)$$

The preexponentials of Eqs. (4-94) and (4-95) contain only $\delta\omega$ and τ; hence, they are simultaneously obtained. Moreover, the exponential factor gives the energy gap in two independent experiments. Finally, knowledge of τ and E_i gives W_+ from the Boltzmann condition. In short, all parameters are established by the combination of coherence experiments which yield $(1/T)_e$ and conventional experiments which yield $\Delta\omega(T)$.

Van't Hof and Schmidt (1975a) used optically detected rotary echoes (Harris *et al.*, 1973) to measure the decay of the electron-spin coherence in these experiments. This has the effect of removing the inhomogeneous contributions of the driving field and some nuclear-spin diffusion terms, leaving the kinetic parameters revealed in a manner similar to spin-locking.

4.4.5. Energy Transfer Studies Using an Ordered State

Thus far we have described coherence experiments that rely on a dephasing mechanism for the electron-spin coherence due to fluctuations in the Larmor frequency that the triplet excitation experiences in different electronic states. As has been demonstrated, this provides a direct and unique means for investigating a variety of phenomena. In some cases (Brenner *et al.*, 1975), however, this dephasing of the ensemble can be a limitation. For example, in our previous example, when the triplet state is promoted to an exciton band and later returns to a localized state, we have no way of knowing a priori whether or not the excitation returns to the same site from which it came or whether the exciton is subsequently localized on another site in a different region of the crystal. The decay of coherence in spin locking (with or without exchange) only gives rates for thermalization to the band. It does not distinguish an individual site from the whole ensemble. The uniqueness of the ADRF experiments developed in Section 4.2.4.7 is that in an "ordered" state, the triplet spins are aligned along the local field of a particular site and the order is established as a unique population distribution. The result is that the "order" is not necessarily lost irreversibly by fluctuations in the Larmor frequency as in the case in a spin-locked ensemble. Instead, the order is lost by a time-dependent redistribution of local fields, which in the present case requires that the excitation be propagated to a different lattice site after promotion from the localized state. Simply retrapping on the same site will not alter the local field of the triplet's spin. The time-dependent loss of order, therefore, measures the rate of energy migration between *different* localized states and is a rate for the composite processes of promotion, migration, and retrapping at a different site. The time-dependent loss of coherence in spin locking measures only the rate of promotion to the exciton band. The ratio of these two experiments,

$$\frac{(k_{av})_{ADRF}}{(k_{av})_{spin\ lock}}$$

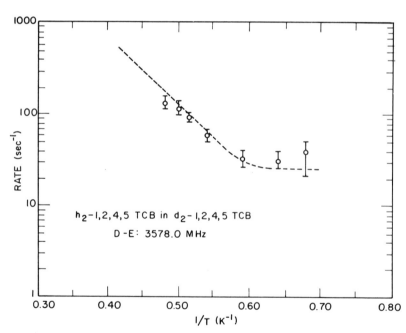

Figure 4-39. Decay of "order" in a demagnetized triplet state. The ordered state decays with temperature-dependent terms and a temperature-independent term. The temperature-independent term is given by k_{av}; the dependent-term is related to the migration rate constants from one site in the lattice to a different site via promotion to the band.

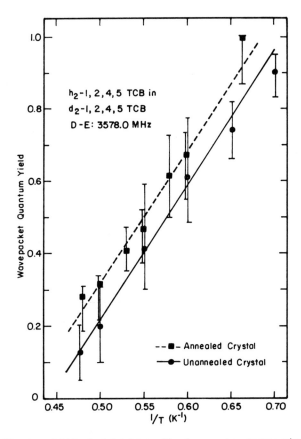

Figure 4-40. Quantum yield for h_2-1,2,4,5-tetrachlorobenzene trap-to-trap migration and d_2-1,2,4,5-tetrachlorobenzene as determined by an optically detected adiabatic demagnetization experiment. The quantum yield is related to the ratio of the migration rate constant to the promotion rate constant and represents the fraction of excited states which migrate to different sites after promotion to the exciton band.

at a given temperature gives the fraction, α, of spins which migrate to other localized sites upon promotion and is more-or-less related to a quantum yield for the creation of an exciton wave packet. The dependence on $(k_{av})_{ADRF}$ with temperature (Brenner *et al.*, 1975) is given in Figure 4-39, and the dependence of a on temperature is given in Figure 4-40.

It is clear from the experiments that ADRF of triplet spins provide a quite different and complementary coherence technique to spin locking or rotary echoes. The most effective use of coherence techniques is in combination with each other and with conventional methods.

Coherence techniques are like a surgeon's tools. Each has an advantage in a particular circumstance, but the dissection of a problem requires the careful interplay of the techniques at the hands of a skilled experimentalist. No doubt in the coming few years many new applications will emerge as the field of coherent spectroscopy, in its various forms, develops. A few of these areas have been outlined in this volume. Although the experimental systems are different, the principles of the methods are the same, in large measure, and thus they form a powerful approach to solving varied problems in chemical dynamics in both ground and excited states.

ACKNOWLEDGMENTS

This work was supported in part by the Energy Resources and Development Agency under the auspices of the Lawrence Berkeley Laboratory, the Division of Material and Molecular Science, and in part by the National Science Foundation and in part by NATO. Our appreciation goes to Vijaya Narasimhan, for her invaluable assistance in the preparation of this manuscript.

References

Abragam, A., 1961, *The Principles of Nuclear Magnetism*, Oxford University Press, Inc., New York.
Anderson, A. G., and Hartmann, S. R., 1962, *Phys. Rev.* **128**:2023.
Anderson, P. W., 1954, *J. Phys. Soc. Japan* **9**:316.
Barrat, J. P., and Cohen-Tannoudji, C., 1961, *J. Phys. Radium* **22**:329–443.
Bloch, F., 1946, *Phys. Rev.* **70**:460.
Breiland, W. G., Harris, C. B., and Pines, A., 1973, *Phys. Rev. Lett.* **30**:158.
Breiland, W. G., Brenner, H. C., and Harris, C. B., 1975, *J. Chem. Phys.* **62**:3458.
Breiland, W. G., Fayer, M. D., and Harris, C. B., 1976, *Phys. Rev.* **A13**:383.
Brenner, H. C., Brock, J. C., and Harris, C. B., 1974, *J. Chem. Phys.* **60**:4448.
Brenner, H. C., Brock, J. C., Fayer, M. D., and Harris, C. B., 1975, *Chem. Phys. Lett.* **33**:471.
Brock, J. C., Brenner, H. C., and Harris, C. B., 1977, unpublished results.
Carr, H. Y., and Purcell, E. M., 1954, *Phys. Rev.* **94**:630.
Chan, I. Y., Schmidt, J., and van der Waals, J. H., 1969, *Chem. Phys. Lett.* **4**:269.
Chan, I. Y., van Dorp, W. G., Schaafsma, T. J., and van der Waals, J. H., 1971, *Mol. Phys.* **22**:741.
de Groot, M. S., Hesselmann, I. A. M., and van der Waals, J. H., 1967, *Mol. Phys.* **12**:259.
Dennison, A. B., 1973, *Magn. Resonance* **2**:1.
Dexter, D. L., 1953, *J. Chem. Phys.* **21**:836.
El-Sayed, M. A., 1968, *Acc. Chem. Res.* **1**:8.
El-Sayed, M. A., 1971, *Acc. Chem. Res.* **4**:23.
El-Sayed, M. A., 1974, in: *Excited States* (E. Lim, ed.), Vol. 1, p. 35, Academic Press, Inc., New York.
El-Sayed, M. A., 1972, in: *MTP International Review of Science*, Vol. 3, "Spectroscopy" (A. D. Buckingham and D. A. Ramsey, eds.), Butterworths, London.
El-Sayed, M. A., and Olmsted, J., III, 1971, *Chem. Phys. Lett.* **11**:568.

Farrar, T. C., and Becker, E. D., 1971, *Pulse and Fourier Transform NMR*, Academic Press, Inc., New York.

Fayer, M. D., and Harris, C. B., 1974a, *Phys. Rev.* **B9**:748.

Fayer, M. D., and Harris, C. B., 1974b, *Chem. Phys. Lett.* **24**:149.

Fayer, M. D., Harris, C. B., and Yuen, D. A., 1970, *J. Chem. Phys.* **53**:4719.

Feynman, R. P., Vernon, F. L., and Hellwarth, R. W., 1957, *J. Appl. Phys.* **28**:49.

Francis, A. H., and Harris, C. B., 1971a, *Chem. Phys. Lett.* **9**:181–188.

Francis, A. H., and Harris, C. B., 1971b, *J. Chem. Phys.* **55**:3595.

Geschwind, S., Devlin, G. E., Cohen, R. L., and Chinn, S. R., 1965, *Phys. Rev.* **A137**:1097.

Haeberlen, U., and Waugh, J. S., 1968, *Phys. Rev.* **175**:453.

Haeberlen, U., and Waugh, J. S., 1969, *Phys. Rev.* **185**:420.

Hahn, E. L., 1950, *Phys. Rev.* **80**:580.

Harris, C. B., 1971, *J. Chem. Phys.* **54**:972.

Harris, C. B., and Buckley, M. J., 1975, *Advan. Nucl. Quadrupole Resonance* **2**:15.

Harris, C. B., and Hoover, R. J., 1972, *J. Chem. Phys.* **56**:2199.

Harris, C. B., and Tarrasch, M., 1977, unpublished results.

Harris, C. B., Tinti, D. S., El-Sayed, M. A., and Maki, A. H., 1969, *Chem. Phys. Lett.* **4**:409.

Harris, C. B., Schlupp, R. L., and Schuch, H., 1973, *Phys. Rev. Lett.* **30**:1019.

Hartmann, S. R., and Hahn, E. L., 1962, *Phys. Rev.* **128**:2042.

Hutchison, C. A., Jr., and Mangum, B. W., 1958, *J. Chem. Phys.* **29**:952.

Hutchison, C. A., Jr., and Mangum, B. W., 1961, *J. Chem. Phys.* **34**:908.

Kuan, T. S., Tinti, D. S., and El-Sayed, M. A., 1970, *Chem. Phys. Lett.* **4**:341.

Kubo, R., 1954, *J. Phys. Soc. Japan* **9**:935.

Kubo, R., and Tomita, K., 1954, *J. Phys. Soc. Japan* **9**:888.

Kwiram, A. L., 1967, *Chem. Phys. Lett.* **1**:272.

Kwiram, A., 1972, *Int. Rev. Sci., Ser. I*, **4**:271.

Lewellyn, M. T., Zewail, A. H., and Harris, C. B., 1975, *J. Chem. Phys.* **63**:3687.

Lewis, G. N., and Kasha, M. J., 1944, *J. Amer. Chem. Soc.* **66**:2100.

Lewis, G. N., and Kasha, M. J., 1945, *J. Amer. Chem. Soc.* **67**:994.

McClure, D. S., 1952, *J. Chem. Phys.* **20**:682.

McConnell, H. M., 1958, *J. Chem. Phys.* **28**:430.

Meiboom, S., and Gill, D., 1958, *Rev. Sci. Instr.* **29**:688.

Putzer, E. J., 1966, *Amer. Math. Monthly* **73**:2.

Redfield, A. G., 1955, *Phys. Rev.* **98**:1787.

Schmidt, J., 1972, *Chem. Phys. Lett.* **14**:411.

Schmidt, J., and van der Waals, J. H., 1968, *Chem. Phys. Lett.* **2**:460.

Schmidt, J., Veeman, W. S., and van der Waals, J. H., 1969, *Chem. Phys. Lett.* **4**:341.

Schmidt, J., van Dorp, W. G., and van der Waals, J. H., 1971, *Chem. Phys. Lett.* **8**:345.

Schuch, H., and Harris, C. B., *Z. Naturforsch.* **30a**:361.

Sharnoff, M., 1967, *J. Chem. Phys.* **46**:3263.

Shpolskii, E. V., 1960, *Soviet Phys. Usp.* **3**:373.

Shpolskii, E. V., 1962, *Soviet Phys. Usp.* **5**:522.

Shpolskii, E. V., 1963, *Soviet Phys. Usp.* **6**:411.

Slichter, C. P., and Holton, W. C., 1961, *Phys. Rev.* **122**:1701.

Solomon, I., 1959a, *C. R. Acad. Sci. Paris* **248**:92.

Solomon, I., 1959b, *Phys. Rev. Lett.* **2**:301.

Swift, T. J., and Connick, R. E., 1962, *J. Chem. Phys.* **37**:307.

Tinti, D. S., El-Sayed, M. A., Maki, A. H., and Harris, C. B., 1969, *Chem. Phys. Lett.* **4**:409.

Torrey, H. C., 1949, Phys. Rev. **76**:1059.

van't Hof, C. A., and Schmidt, J., 1975a, *Chem. Phys. Lett.* **36**:457.

van't Hof, C. A., and Schmidt, J., 1975b, *Chem. Phys. Lett.* **36**:460.

van't Hof, C. A., and Schmidt, J., 1976a, *Chem. Phys. Lett.* **42**:73.

van't Hof, C. A., and Schmidt, J., 1976b, unpublished results.

van't Hof, C. A., Schmidt, J., Verbeek, P. J. F., and van der Waals, J. H., 1973, *Chem. Phys. Lett.* **21**:437.

Veeman, W. S., and van der Waals, J. H., 1970, *Chem. Phys. Lett.* **7**:65.

Waugh, J. S., Wang, C. H., Huber, L. M., and Vold, R. L., Jr., 1968, *J. Chem. Phys.* **48**:662.

Webb, R. H., 1962, *Rev. Sci. Instr.* **33**:732.

Winscom, C. J., and Maki, A. H., 1971, *Chem. Phys. Lett.* **12**:264.

5

Resonant Scattering of Light by Molecules: Time-Dependent and Coherent Effects

F. A. Novak, J. M. Friedman, and R. M. Hochstrasser

5.1. Elementary Time-Dependent Theory Related to Luminescence

5.1.1. Introduction

The object of this article is to provide a quantum-mechanical description of the interaction of light and molecules and to describe the time dependence of the absorption and reemission of the light. It will be shown that the incident and scattered photons can be coherent under special circumstances, and also that popular light sources that produce many photons manifest varying degrees of coherence. In many of the experiments described here, it is the light source that cannot be easily described by a definite wavefunction and that needs the density matrix description, whereas single molecules serve to probe this coherence through their light-scattering and stimulated effects. However, in many instances the molecular and photon ensembles have features in common that are useful in the design of experiments. Thus,

---

F. A. Novak (Henry P. Busch Fellow), *J. M. Friedman, and R. M. Hochstrasser* • Department of Chemistry and Laboratory for Research on the Structure of Matter, The University of Pennsylvania, Philadelphia, Pennsylvania; J. M. Friedman's present address is Bell Telephone Laboratories, Murray Hill, New Jersey. Research for this article was supported by grants from the U.S. Department of Health (NIH GM1252), the National Science Foundation, and by the NSF/MRL at Pennsylvania.

one object of the present article will be to distinguish between various interpretations of coherence, in light, in scattering, and in ensembles.

Spontaneous emission cannot be readily inferred from the semiclassical theory that describes the interaction of light characterized by classical fields, with matter characterized by quantum equations. It is therefore essential that the theoretical presentation of this topic be entirely quantum-mechanical. On the other hand, semiclassical arguments incorporating Wigner–Weisskopf constructions (Weisskopf and Wigner, 1930, 1931; Weisskopf, 1931, 1933) can be justified and spontaneous emission can be shown, in a correspondence principle argument, to have a rate proportional to the appropriate transition multipole moment matrix element between the two states (Condon and Shortley, 1957; Fowler, 1962). Thus, there always exists a semiclassical description of the scattering phenomena herein, and we will provide that description also in a number of instances.

The phenomenon of light scattering is of course a well-studied process and has been a useful tool in studying many properties of matter. For example, Raman scattering has long been used to obtain molecular structural information as well as information about the motion of molecules in liquids (Gordon, 1965). The theories that are usually used in discussing the scattering of light are either classical or are based on the use of the golden rule in quantum mechanics. Both of these approaches look at what is essentially the steady behavior of the scattered light when a nonresonant monochromatic wave is incident on the sample. The approach used in this article stresses the time-dependent aspects of the process of light scattering, and therefore explicit account is taken of the more detailed aspects of the exciting light source.

It is important when considering resonant or near-resonant processes to incorporate explicitly the effects of level widths without resort to phenomenological devices; thus, the approach used here is to provide exact quantum-mechanical solutions for many of the problems. The reader is referred to the recent book by Berne and Pecora (1976) for discussions of more classical effects of light scattering.

Recent years have seen the appearance of many papers concerned with Raman scattering, resonance Raman scattering, fluorescence, and resonance fluorescence. The relationships between these effects is an old topic that was discussed in the early work of Weisskopf (Weisskopf and Wigner, 1930, 1931; Weisskopf, 1931, 1933) and later brought out by Heitler (1954). Much later experiments with lasers, including the resonance Raman effect, and time-resolved studies of scattering from iodine vapor (Williams *et al.*, 1974), helped to stimulate a resurgence of theoretical work on the time dependence of light scattering processes (Friedman and Hochstrasser, 1974b; Berg *et al.*, 1974; Mukamel and Jortner, 1975; Jortner and Mukamel, 1974; Mukamel, 1975; Mukamel *et al.*, 1976; Ben-Reuven and

Mukamel, 1975). A central question in all of this work has been whether resonance Raman scattering and resonance fluorescence should be distinguished from one another conceptually. In many instances this is a question of semantics and it would be clearer to ask whether there are different ways in which light nearly in resonance with a molecular transition can be scattered, and leave aside the question of which process is called resonance fluorescence, resonance scattering, resonance Raman, etc. The term "resonance fluorescence" was used by Heitler (1954) in relation to the excitation of resonances that were wider spectrally than the light field. Nowadays the term "fluorescence" is used mainly to refer to a molecular property both in relation to its frequency (i.e., whether or not it is resonant with a molecular transition), its linewidth, and its transform-limited characteristic decay time. As we shall see more rigorously later, only the peak wavelength of the Heitler limit scattering matches the characteristics of what is now termed fluorescence in the literature of photochemistry and solid-state physics. Since the linewidth of preresonance Raman scattering is essentially independent of the homogeneous width of the near-resonant state—when the light frequency is distant from resonance, the spectral width of the light and the damping in the final state mainly determine the linewidth of the scattered light— it seemed reasonable (Friedman and Hochstrasser, 1974b) to label the Heitler limit the resonance scattering limit (or resonance Raman limit). The fluorescence limit would then occur when the light field, either stationary or pulsed, is spectrally broad compared with the homogeneous width of the resonance, the scattered light then being determined by the spontaneous decay time of the molecular state.

In addition to the resonance fluorescence versus Raman question, there is the difficulty of relating the scattered waves to the incident light in a real experiment with an ensemble of molecules. Obviously, two extreme situations exist, in which the dephasing time for the two-level system of ground and resonant states is either very large or small compared with the effective scattering time. Once again it is not unreasonable to classify scattering into Raman or fluorescence, depending on which of these situations is operative (Shen, 1974; Kushida, 1976).

It is necessary to consider the selection rules for spectroscopic processes in the various limits discussed above, especially whether the various transitions that can occur in fluorescence or Raman are dependent on properties of the field. It will be seen below that while first-order processes—that is, processes that involve the linear response of a system to the light field—are dependent on the spectral width of the light, the second-order optical effects such as two-photon absorption, or induced Raman processes, also depend on statistical properties of the light. This chapter will contain a theoretical discussion of these various effects, especially where they relate to spectroscopic phenomena. In order to make the presentation more self-contained, background has been included that relates to the nature of the electromagnetic field and its observables.

5.1.2. Scattering Theory

In the usual time-dependent perturbation treatment of the interaction of a molecule with an electromagnetic field, the time dependence of the interaction is basically limited to either a sudden switch-on approximation or an adiabatic switch-on approximation for the perturbation. The use of scattering theory allows the removal of the time dependence of the interaction from the perturbation Hamiltonian and a focus on the temporal properties of the interacting particles. By describing the interacting system in terms of temporal overlay of wavepackets (molecule and photons), a continuous gradation of interaction times is feasible. What will be assumed is that by specifying an initial quantum state the system will evolve in time following the dynamical laws of quantum mechanics. For convenience in what follows, some standard quantum-mechanical definitions will be provided.

The linear nature of the Schrödinger equation in frequency units,

$$i\dot{\Psi} = H\Psi \tag{5-1}$$

means that symbolically speaking, the evolution of a state $\Psi(t)$ from an initial state $\Psi(t_0)$ can be accomplished by the use of a linear operator $U(t, t_0)$:

$$\Psi(t) = U(t, t_0)\Psi(t_0) \tag{5-2}$$

The operator $U(t, t_0)$ therefore obeys the relation

$$i\frac{\partial U(t, t_0)}{\partial t} = HU(t, t_0) \tag{5-3}$$

which can be readily integrated for Hamiltonians that do not depend explicitly on time. For the systems of interest the explicit time dependence will be carried within the wave packet of the scattering particles so that the Hamiltonian will indeed be time-independent. Integrating, Eq. (5-3) yields the operator equation,

$$U(t, t_0) = e^{-iH(t-t_0)} \tag{5-4}$$

which has meaning in that if $\Psi(t_0)$ is expanded in the basis set of eigenfunctions of H, $\{\varphi_i\}$, Eq. (5-2) would be written

$$\Psi(t) = \sum_i e^{-i\omega_i(t-t_0)} C_i(t_0)\varphi_i \tag{5-5}$$

ω_i is the eigenvalue of H for φ_i. For systems described by a Hamiltonian H that is readily partitioned into a sum of potentials V and a zeroth-order Hamiltonian H_0 so that

$$H = H_0 + V \qquad (5\text{-}6)$$

it becomes very convenient to utilize the interaction representation in finding a workable expression for the time evolution operator. State functions in the interaction representation $\Psi_I(t)$ do not display the oscillatory time dependence exhibited by $\Psi(t)$ in the Schrödinger representation. The two state functions are related by the following unitary transformation:

$$\Psi_I(t) = e^{iH_0 t}\Psi(t) \qquad (5\text{-}7)$$

which, with Eq. (5-1), readily yields the time-dependent equation for the wavefunction in the interaction representation in which the time dependence follows directly from the perturbation $V_I = e^{iH_0 t}V e^{-iH_0 t}$:

$$i\dot{\Psi}_I(t) = V_I\Psi_I(t) \qquad (5\text{-}8)$$

The expression for $U(t, t_0)$ in the interaction representation is obtained with the use of Eq. (5-4):

$$U_I(t, t_0) = e^{iH_0 t}e^{-iH(t-t_0)}e^{-iH_0 t_0} \qquad (5\text{-}9)$$

In the following theoretical treatment the interaction representation will be used almost exclusively; consequently, the subscript I will be dropped from the state functions and $U_I(t, t_0)$ except for ambiguous situations. The linear operator $U(t, t_0)$ smoothly transforms an eigenfunction of H_0 at time t_0 under the influence of H into a state that properly describes the system at time t. It therefore becomes possible to follow the time evolution of a quantum state $\Psi(t_0)$ by expressing $\Psi(t_0)$ in terms of the eigenfunctions of H_0. The result of operating on this decomposed $\Psi(t_0)$ with $U(t, t_0)$ is a linear combination of eigenfunctions of H_0 that can now be recomposed to yield $\Psi(t)$:

$$\Psi(t_0) = \sum_i C_i|\varphi_i\rangle, \quad H_0|\varphi_i\rangle = \omega_i|\varphi_i\rangle \qquad (5\text{-}10a)$$

$$\Psi(t) = U(t, t_0)\Psi(t_0) = \sum_{i,j} C_i|\varphi_j\rangle\langle\varphi_j|U(t, t_0)|\varphi_i\rangle \qquad (5\text{-}10b)$$

In one of the cases to be treated, the initial system will be taken to be a system consisting of a molecule and photon that are not interacting. To accommodate photons of variable extents in time, including the temporally indeterminate monochromatic photon, t_0 is set at $-\infty$ if $t = 0$ is to be the instant of the onset of observable manifestations of the scattering event. This step accommodates the fact that for a true monochromatic photon, the system is interacting over all time, so that it is only in the limit of $t_0 = -\infty$ that the system can be expressed as a simple product of a molecule (m) and photon (p) state, i.e.,

$$\Psi(t_0 = -\infty) = |\varphi_m\rangle|\varphi_p\rangle \qquad (5\text{-}11)$$

By first differentiating Eq. (5-9) with respect to t_0 and then integrating between the limits $t_0 = -\infty$ to t the following explicit form for $U(t, -\infty)$ is obtained:

$$U(t, -\infty) = 1 + \lim_{\eta \to 0} + \frac{e^{iH_0t} 1 V e^{-iH_0t}}{H_0 - H + i\eta} \qquad (5\text{-}12)$$

The η arises out of the necessity to introduce an infinitesimal energy uncertainty, i.e., a *nearly* infinite lifetime for the eigenstates of H_0 in order to ensure a well-defined integration limit at $t_0 = -\infty$. The limit of $\eta \to 0$ is to be taken after the integrations involving the operator in Eq. (5-12). The H_0 in the denominator operates on the initial state so that if $\Psi(-\infty)$ is expressed as a superposition of eigenfunctions of H_0 with frequency eigenvalues ω, that factor becomes the Green's operator $(\omega - H + i\eta)^{-1} = G$.

Equation (5-12) is essentially equivalent to the result obtained by Roman (1965), who uses a different limiting procedure. His results and Eq. (5-12) are both equivalent to expanding the time-dependent state in terms of the Ψ^+ or "in" stationary scattering states for the photon–molecule system. Some of these different but equivalent representations for the time evolution operator are more useful than others in verifying the important mathematical properties of that operator.

5.1.3. Approximate Model for the Photon States

To illustrate the theoretical methods, consider an explicit form for the photon state:

$$|\varphi_p\rangle = (\Gamma_p/2\pi)^{1/2} \int_{-\infty}^{\infty} d\omega \frac{1}{\omega - \omega_p + (i\Gamma_p/2)} |l\omega\rangle \qquad (5\text{-}13)$$

This representation of a light field has certain important theoretical properties:

$$\langle\varphi_p|\varphi_p\rangle = 1; \qquad \langle N\rangle^2 = \langle N^2\rangle = 1$$

The first expresses normalization, the second indicates that there is no variance in the photon number N; i.e., one photon is represented. Calculations based on the interaction of the field with a molecule will yield probabilities for which the usual statistical interpretation in quantum mechanics is valid. So (Eq. (5-13)) might be regarded as the form of a light pulse. On the other hand, although a localized single photon is analytically useful, it is hardly realizable in the laboratory.

The term "single photon" is used here within the context of a quantized radiation field that has a low density of excited photon states. A radiation field consisting of separate photons could arise from the radiative decay of a sufficiently dilute population of excited, noninteracting atoms or molecules so that the possibility of stimulated emission is eliminated. Radiation fields arising from laser emission in which there is a reasonably well defined phase can no longer be adequately described as an accumulation of distinct photons. The state of the field is then appropriately described as a type of coherent state. These coherent states have an average fixed number of photons that is not altered by molecular interactions with the field as in the first case. Coherent states and other types of light will be discussed in Section 3.1.

Equation (5-13) describes a Lorentzian electromagnetic field for which Γ_p is the frequency bandwidth and Γ_p^{-1} describes the exponential decay of the field envelope at a point in space. Equation (5-13) is a quantum analog of the widely used classical light pulse having spectral width Γ_p and electric field for $t > 0$, $E(t)$, given by

$$\mathbf{E}(t) = \mathbf{E}_0(e^{-i\omega_p t} + e^{+i\omega_p t})e^{-\Gamma_p t/2} \tag{5-14}$$

where the absolute square of the Fourier transform of $\mathbf{E}(t)$ is the same Lorenzian frequency distribution that is obtained from the absolute square of the coefficients in Eq. (5-13). The pulses (5-13) and (5-14) exert no fields prior to $t = 0$ because they have sharp leading edges followed by exponential decay.

5.1.4. Molecular States

The nature of the operator $U(t, -\infty)$ is such that $\Psi(-\infty)$ must be expressed in terms of the basis of eigenfunctions of H_0. The partitioning of H into an H_0 and V will be such that H_0 is the sum of a molecular Hamiltonian, H_0^m, and a radiative Hamiltonian, H_0^r, with V containing the molecular–photon interaction. Depending on the choice for H_0^m, V may also contain intramolecular interaction terms such as the nuclear kinetic energy operator. If we assume for the moment that we have partitioned H such that $|\Psi_g\rangle$, the initial state of the molecule at $t = -\infty$ is an eigenfunction of H_0^m (i.e., a stationary state) with the arbitrary eigenvalue of zero, then $\Psi(-\infty) = |\Psi_g\rangle|\varphi_p\rangle$. The operation of $U(t, -\infty)$ [Eq. (5-12)] on $\Psi(-\infty)$ chosen to incorporate Eq. (5-13) yields the state at time t:

$$\Psi(t) = \Psi(-\infty) + \lim_{\eta \to 0^+} \int_{-\infty}^{\infty} d\omega \frac{e^{iH_0 t} G(\omega) V e^{-i\omega t}}{\omega - \omega_p + (i/2)\Gamma_p} |\Psi_g\rangle |l_\omega\rangle \qquad (5\text{-}15)$$

where $G(\omega)$ is the Green's function operator, given in this case by $(\omega - H + i\eta)^{-1}$. Introduction of sums over complete sets $\{|i\rangle\}$ in the basis of eigenfunctions of H_0 yields

$$\Psi(t) = \Psi(-\infty) + \lim_{\eta \to 0^+} \sum_{j,i} \int_{-\infty}^{\infty} d\omega \frac{|j\rangle G_{ji}^{(\omega)} \langle i|V|g, l\omega\rangle e^{-i(\omega-\omega_j)t}}{\omega - \omega_p + (i/2)\Gamma_p} \qquad (5\text{-}16)$$

$\Psi(t)$ is the complete wavefunction of the interacting system expressed in the basis of eigenfunctions of H_0. Equation (5-16) is the master formula from which the time dependence of the molecule–photon interaction can be calculated for an exponentially decaying light pulse. Assuming the system to be in a definite state at $t = -\infty$, the probability that at time t the system is in state $|f\rangle$ is just $|\langle f|\Psi(t)\rangle|^2$, which gives an explicit formula for the time dependence. Two additional steps are needed to complete the analysis: the evaluation of the matrix elements of G, and the completion of the integration over the frequency. It is, however, important to consider further the issue of basis functions, which can be confusing in applications of the results to real systems.

As mentioned above, H can be partitioned in any of a number of ways. The initial state is a product of molecule and photon states so there is freedom in the choice of V in that various basis sets can be used to describe the molecular state. A convenient choice is suggested by the time scale of the interaction between the photon and the molecule. For example, if the nuclei did not move during this time, the basis set of nuclear coordinate fixed functions ("crude adiabatic") would not be an inreasonable choice, because the initially prepared state had effectively stationary nuclei during the interaction. The characteristic relaxation of the system will be in the femtosecond regime, and Γ_p^{-1} must also be in the femtosecond regime in order to prepare such a state. The same conclusion was arrived at by Lefebvre (1971), who considered the mechanism of exciting molecular levels with light pulses by expanding the crude adiabatic states in terms of the adiabatic states. With subsequent nuclear motion, large numbers of crude adiabatic states contribute to the evolution of the system under the influence of the constant part of V. A more conventional pulse would have an interaction time such that the state projected out [i in Eq. (5-16)] will present an average over the molecular nuclear oscillations that occur during the interaction. The most suitable choice of basis (partitioning of H) is dictated by the nature of the experiment, including the choice of light source and the coupling of the resonant state to both molecular and radiative continua through V. Even if $H_0{}^m$ describes Born–Oppenheimer states, the state of H_0 reached by virtue of V is *always* a superposition of the states of $H_0{}^m$; i.e., it is never a Born–Oppenheimer state. In order to prepare mainly one particular resonant state, it is necessary to excite the molecule with photons that have the appropriate resonant frequency, and the excitation must occur during the uncertainty time defined by the interaction between the resonant state (as chosen by partitioning H) and the molecular and radiative continua into which the resonant state decays. These points arise from the form of Eq. (5-16), which is solved below exactly for certain model situations. When considering molecular relaxation and decay processes, an appropriate form for the molecular wavefunction is in terms of states that satisfy the adiabatic approximation, and in such a case the nuclear kinetic energy can be included in V. The reasonableness of this partitioning is based on the frequency bandwidth of the light source, being usually larger than or comparable with rates caused by nonadiabatic interactions in molecules. Said in another way, the emission decay times of molecular states are mostly much longer than the inverse bandwidth of the exciting field. The states of the system, molecule plus light, that occur because of V may be stationary or nonstationary, but they are always superpositions of eigenstates of H_0. We will investigate the time-dependent probability that a molecule originally in one adiabatic state undergoes a transition into a near-resonant adiabatic state due to the interaction of the molecule with a photon, i.e., the process $|\Psi_g\rangle|\varphi_p\rangle \to |\Psi_i\rangle|0\rangle$. In addition, we will obtain the time-dependent amplitude for this system reaching the state $|\Psi_j\rangle|1_{\mathbf{k}}\rangle$, where $|\Psi_j\rangle$ is a molecular state that does not directly couple to $|\gamma_g\rangle$ by radiation, and a new photon, $|1_{\mathbf{k}}\rangle$, is emitted. It is worth noting that even for nearly monochromatic photons ($\Gamma_p \approx 0$), where the prepared resonant state is nearly a time-independent superposition (i.e., a stationary state) of adiabatic states of the appropriate energies and radiative strengths, only a very small number of these adiabatic states overtly contribute to the photon scattering process. The majority of these act to nonradiatively damp or dilute the scattering process or state, respectively.

5.1.5. Matrix Elements of $G(\omega)$

For a number of idealized systems, some of which closely correspond to experimentally encountered situations, it is possible to easily arrive at simple analytical expressions for the relevant matrix elements of G. The procedure used here for the evaluation of these matrix elements closely parallels the treatment given by Mower (1966, 1968). This approach can be shown to be equivalent to the employment of the Dyson equation:

$$G = G_0 + G_0 V G \qquad (5\text{-}17)$$

where

$$G_0 = \frac{1}{\omega - H_0}$$

used by Nitzan *et al.* (1972) in treating sequential and parallel decay problems.

Consider first a single excited molecular adiabatic state labeled $|i\rangle$ to which the ground state is coupled via the field (5-13). The adiabatic state $|i\rangle$ is also coupled through the molecular part of V (nuclear kinetic energy, spin–orbit coupling) to a quasi-continuum of adiabatic states labeled n, and through the radiative part of V to a continuum of states labeled r. The r states are product states consisting of a lower-lying adiabatic state (with respect to ω_i) and monochromatic photon states of wavevector $\mathbf{k}'$, i.e., $|l_{\mathbf{k}'}\rangle$. The continuous range of frequencies $\omega' = c|\mathbf{k}'|$ provides the continuum of states into which $|i\rangle$ can radiatively decay. In Figure 5-1, which is a pictorial representation of this coupling scheme, the dashed lines connect the various levels that will be focused upon in a typical scattering experiment. The final state $|j\rangle$, which is a member of the set labeled r, can typically be the product state composed of an excited vibrational level of the ground electronic state and a monochromatic photon state $-|j\rangle|l_{\mathbf{k}'}\rangle$. There is associated with every $|i\rangle$ a continuous distribution of states $|l_{\mathbf{k}}\rangle$. The matrix elements of G that we wish to evaluate for this scheme are G_{ii}, G_{ji}, and G_{ni}.

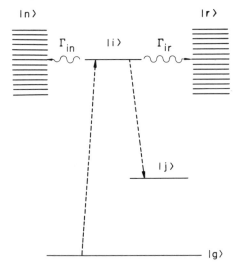

Figure 5-1. Model three-level system with interacting radiative and nonradiative continua.

Following Mower (1966, 1968) we start with the identity

$$(\omega - H_0 - V + i\eta)G = 1 \tag{5-18}$$

A series of coupled equations can then be obtained by introducing the projection operator for the complete set of eigenfunctions of H_0, $\sum_l |l\rangle\langle l| = 1$, and by then taking the appropriate matrix elements:

$$(\omega - H_0 - V + i\eta) \sum_l |l\rangle\langle l| G = 1 \tag{5-19a}$$

$$\langle i|(\omega - H_0 - V + i\eta) \sum_l |l\rangle\langle l| G|i\rangle = 1 \tag{5-19b}$$

$$\langle j|(\omega - H_0 - V + i\eta) \sum_l |l\rangle\langle l| G|i\rangle = 0 \tag{5-19c}$$

$$\langle \eta'|(\omega - H_0 - V + i\eta) \sum_l |l\rangle\langle l| G|i\rangle = 0 \tag{5-19d}$$

By partitioning the projection operator into the n and r sets that couple to $|i\rangle$ through V, Eqs. (5-19b) become

$$(\omega - \omega_i + \eta)G_{ii} - \sum_n V_{in}^{(n)} G_{ni} - \sum_r V_{ir}^{(r)} G_{ri} = 1 \tag{5-20a}$$

$$-V_{ji}^{(r)} G_{ii} + (\omega - \omega_j + i\eta)G_{ji} = 0 \tag{5-20b}$$

$$-V_{n'i}^{(n)} G_{ii} + (\omega - \omega_{n'} + i\eta)G_{n'i} = 0 \tag{5-20c}$$

where we have assumed that n does not couple to r and that higher-order terms such as $V_{im}V_{mr}$ are negligible. Solving for G_{ji} and $G_{n'i}$ in terms of G_{ii} and substituting into (5-20a), it is found that

$$G_{ii} = 1 \Big/ \left[\omega - \omega_i + i\eta - \sum_n \frac{V_{in}^{(n)} V_{ni}^{(n)}}{\omega - \omega_n + i\eta} - \sum_r \frac{V_{ir}^{(r)} V_{ri}^{(r)}}{\omega - \omega_r + i\eta} \right] \tag{5-21}$$

In a statistical limit both $\{n\}$ and $\{r\}$ form a continuum of states so that the sums may be replaced with integrations over energies:

$$\sum_n \to \int d\omega_n \rho(\omega_n), \qquad \sum_r \to \int d\omega_r \rho(\omega_r) \tag{5-22}$$

where ρ is the appropriate density of states function. The integrals I_n and I_r contribute a real part, which is a photon source-dependent energy shift that will be incorporated into ω_i, and an imaginary part, which will be evaluated as follows, assuming that $V_{in}^{(n)}$ is a slowly varying function of energy:

$$\text{Im}\,(I_n) = \int d\omega_n \frac{\rho(\omega_n)|V_{in}^{(n)}|^2 \eta}{(\omega - \omega_n)^2 + \eta^2} \tag{5-23}$$

By completing a contour in the negative imaginary quadrants of complex frequency space, the residue theorem may be applied to Eq. (5-23), and it is found that

$$\text{Im}\,(I_n) = -\pi\rho_n(\omega)|V_{in}^{(n)}|^2 \tag{5-24}$$

and similarly for Im (I_r). G_{ii} now becomes

$$G_{ii} = \frac{1}{\omega - \omega_i + (i/2)\Gamma_{in} + (i/2)\Gamma_{ir}} = \frac{1}{\omega - \omega_i + (i/2)\Gamma_i} \tag{5-25}$$

where

$$\Gamma_{in} = 2\pi\rho_n(\omega)|V_{in}^{(n)}|^2, \qquad \Gamma_{ir} = 2\pi\rho_r(\omega)|V_{ir}^{(r)}|^2$$

and

$$\Gamma_i = \Gamma_{ir} + \Gamma_{in}$$

Γ_{ix} is a frequency width or dilution of $|i\rangle$ arising from the coupling of $|i\rangle$ to the continuum of states labeled $\{x\}$. Substituting Eq. (5-25) into (5-20b) and (5-20c), we have

$$G_{ji} = V_{ji}^{(r)}\left[(\omega - \omega_j + i\eta)\left(\omega - \omega_i + \frac{i\Gamma_i}{2}\right)\right]^{-1} \tag{5-26a}$$

$$G_{n'i} = V_{n'i}^{(n)}\left[(\omega - \omega_n + i\eta)\left(\omega - \omega_i + \frac{i\Gamma_i}{2}\right)\right]^{-1} \tag{5-26b}$$

The trivial extension of the treatment above to include the coupling of $|j\rangle$ and $|n'\rangle$ to other continua results in $\Gamma_j/2$ and $\Gamma_{n'}/2$, respectively, replacing η in Eq. (5-26a). The added possibility of coupling among the various continua through V_{nr}-type terms has been treated by Nitzan *et al.* (1972), and indeed there is a vast hierarchy of model problems in dynamics that can be tackled in this fashion.

If there is an additional discrete state $|k\rangle$ to which $|\Psi_i\rangle$ couples, the set of coupled equations involving G_{ii} and G_{ki} reduces to

$$\begin{bmatrix} \omega - \omega_i + i/2\Gamma_i & -V_{ik} \\ -V_{ki} & \omega - \omega_k + i/2\Gamma_k \end{bmatrix} \begin{bmatrix} G_{ii} \\ G_{ki} \end{bmatrix} = \begin{bmatrix} 1 \\ 0 \end{bmatrix} \tag{5-27}$$

These equations can be solved for G_{ii} and G_{ki} by the standard use of determinants. We find that

$$G_{ii} = \left(\omega - \omega_k + \frac{i\Gamma_k}{2}\right)\left[\left(\omega - \omega_k + \frac{i}{2}\Gamma_k\right)\left(\omega - \omega_i + \frac{i\Gamma_i}{2}\right) - |V_{ik}|^2\right]^{-1} \tag{5-28a}$$

$$G_{ki} = V_{ki}\left[\left(\omega - \omega_k + \frac{i}{2}\Gamma_k\right)\left(\omega - \omega_i + \frac{i}{2}\Gamma_i\right) - |V_{ik}|^2\right]^{-1} \tag{5-28b}$$

5.1.6. Excitation of an Isolated Resonant State: A Two-Level System

First we consider the excitation of the state $|i\rangle$, where this implies a molecular state achieved by the removal of a photon from the initial state. We will assume that $|i\rangle$ is generated exclusively from the interaction of the ground state with photons so that the amplitude $\langle i|\Psi(t)\rangle$ is obtained from Eqs. (5-16) and (5-25), since all terms in the sums vanish except for those involving $|i\rangle$. Integration over the energy yields for the probability that the system is in state $|i\rangle$:

$$P_i(t) = |\langle i|\Psi(t)\rangle|^2 = \frac{2\pi\Gamma_p|V_{ig}^{(r)}|^2}{(\omega_i-\omega_p)^2+(\Gamma_i-\Gamma_p)^2/4}$$

$$\times[e^{-\Gamma_i t}+e^{-\Gamma_p t}-2e^{-(\Gamma_p+\Gamma_i)t/2}\cos(\omega_i-\omega_p)t],$$

$$t\geq 0; \qquad P_i(t)=0, \qquad t<0 \qquad (5\text{-}29)$$

The time restrictions in Eq. (5-29), are specified by the contour integrations in the complex ω plane necessary for the evaluation of the integral in (5-15), and are expressions of causality.

In the most usual case $V_{ig}^{(r)}$ is the electric-dipole transition moment dotted into the electric field. Figures 5-2 and 5-3 show the form of $P_i(t)$ for some choices of parameters Γ_i, Γ_p and ω_{ip}. An important point is that both the photon width Γ_p (spectral width) and the width of the molecular state due to radiative and nonradiative couplings, Γ_i, play a role in the dynamics of the excitation of the state $|i\rangle$. The time that it takes the state $|i\rangle$ to build up probability is determined by Γ_p or Γ_i, depending on which is the largest.

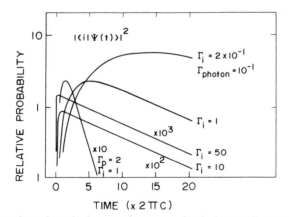

Figure 5-2. Time dependence in the case of resonance for the intermediate-state probability. The level widths are given in cm^{-1}. Note that of the four curves with $\Gamma_p = 0.1$ cm^{-1}, the buildup is determined by the magnitude of Γ_i and the decay by Γ_p.

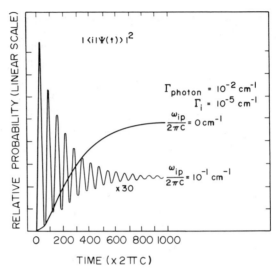

Figure 5-3. Time dependence in the off-resonance case for the intermediate-state probability. Note that the situation $\Gamma_p \gg \Gamma_i$ pertains.

5.1.7. Semiclassical Analogy

If in a semiclassical model we consider a molecule to have stationary states Ψ_g (at zero frequency and having no homogeneous width) and Ψ_i (at ω_i), then in the presence of a light pulse having center frequency ω and having the form (5-13), the system is driven into a nonstationary state $\Psi(t) = a_g(t)\Psi_g + a_i e^{-\Gamma_i t/2} e^{-i\omega_i t} \Psi_i$. The probability that the system will be in state Ψ_i at time t is $|a_i(t)|^2 e^{-\Gamma_i t}$. The coupled equations for the coefficients are

$$\dot{a}_g(t) = i\gamma_{gi}[e^{-i(\omega_i - \omega_p)t} + e^{-i(\omega_i + \omega_p)t}]^{-(\Gamma_i + \Gamma_p)t/2} a_i(t) \qquad (5\text{-}30)$$

$$\dot{a}_i(t) = i\gamma_{ig}[e^{i(\omega_i - \omega_p)t} + e^{i(\omega_i + \omega_p)t}]e^{i(\Gamma_i - \Gamma_p)t/2} a_g(t) \qquad (5\text{-}31)$$

where γ_{gi} is $(\mathbf{E}_0 \cdot \boldsymbol{\mu}_{gi}/\hbar)$ and $\omega_i = E_i/\hbar$. In the weak-signal limit, where $\gamma_{gi}^2 \ll [(\omega_i - \omega_p)^2 + (\Gamma_i - \Gamma_p)^2/4]$ and with the rotating-wave approximation (consider terms with $\omega_i + \omega$ to be comparatively unimportant for physically important times), we can put $a_g(t) \simeq 1$ for all t, and the first-order result for $a_i(t)$ is

$$a_i(t) = \frac{\gamma_{ig}}{(\omega_i - \omega_p) - i(\Gamma_i - \Gamma_p)/2}[e^{i(\omega_i - \omega_p)t + (\Gamma_i - \Gamma_p)t/2} - 1] \qquad (5\text{-}32)$$

so the probability for $t \geq 0$ is found to be the same as in (1-29):

$$P_i(t) = \frac{\gamma_{ig}^2}{(\omega_i - \omega_p)^2 + (\Gamma_i - \Gamma_p)^2/4}[e^{-\Gamma_p t} + e^{-\Gamma_i t} - 2e^{-(\Gamma_i + \Gamma_p)t/2} \cos(\omega_i - \omega_p)t] \qquad (5\text{-}33)$$

By analogy with the more traditional semiclassical approach, it is readily seen what is involved in reaching (5-29), a weak-signal limit where the pulse intensity is low enough to keep $|a_i(t)|^2$ very small at all times, and a rotating-wave approximation (RWA). The RWA is not a necessary approximation, and it could readily be excluded in cases when it is needed, such as in the conventional Raman scattering when $(\omega_i - \omega_p)$ is large.

It can be seen from Figures 5-2 and 5-3 that at resonance there is simply a smooth build-up and decay. At long times the decay is characteristic of the smaller of Γ_p and Γ_i. Off-resonance there is an oscillation in $P_i(t)$ that eventually gets damped out by an $\exp\{-[\frac{1}{2}(\Gamma_p + \Gamma_i)t]\}$ term.

The foregoing results apply to a homogeneous ensemble of molecules, such as in a perfect beam. In the event that the system is inhomogeneous, such as a velocity distribution in a gas or a distribution of sites in a solid, it is necessary to average $P_i(t)$ over the distribution in order to obtain the time response of the system. Clearly such averaging will cause the rapid oscillations in $P_i(t)$ to be removed from the observed time evolution. For example if the distribution of frequencies is Gaussian spanning a width β^{-1} about ω_i, Eq. (5-33) would become:

$$\bar{P}_i(t) = \int_{-\infty}^{\infty} d\omega P_i(t) e^{-\beta^2(\omega_i - \omega)^2} \tag{5-33a}$$

5.2. Applications of Scattering Theory to Model Systems

5.2.1. Three-Level System

The next step in the calculation is to introduce a third molecular level $|j\rangle$ that can be reached by dipole transitions from the $|i\rangle$ of Section 5.1. On the other hand, $|j\rangle$ is considered not to couple with the ground state. To provide explicit results, the three-level model will be first solved for the exponentially decaying pulses (5-13) and (5-14) and again a comparison of the quantum-mechanical and semiclassical theories will be brought out. The three-level system is an appropriate model for resonance Raman scattering in cases where these are well-separated states or where the state represented by $|i\rangle$ dominates other excited states in its strength of radiative coupling to the ground state. Many of the essential features of the time dependence and spectral properties of the light scattered by systems containing many levels can be understood by studying the three-level system.

5.2.2. Scattering of an Exponentially Decaying Pulse

The time dependence of the light scattering for an exponentially decaying pulse of the form (5-13) follows immediately from (5-16) in association with the matrix element for the off-diagonal matrix element of the Green's operator (5-26):

$$P_{j\mathbf{k}}(t) = |\langle j; 1_\mathbf{k} | \Psi(t) \rangle|^2$$

$$= \frac{|V_{ji}^{(r)} V_{ig}^{(r)}|^2 A(t)}{(\omega_{ji}^2 + \tfrac{1}{4}\Gamma_i^2/4)(\omega_{jp}^2 + \Gamma_p^2/4)(\omega_{ip}^2 + \Gamma_{ip}^2/4)} \qquad (5\text{-}34)$$

$$A(t) = (\Gamma_p/2\pi) \sum_{a \neq b \neq c} \{ e^{-\Gamma_a t}(\omega_{bc}^2 + \Gamma_{bc}^2/4)$$

$$- 2e^{-(\Gamma_a + \Gamma_b)t/2}[\sin \omega_{ab}t[(\omega_{ac}\Gamma_{cb}/2) - (\omega_{bc}\Gamma_{ca}/2)]$$

$$+ \cos \omega_{ab}t(\omega_{bc}\omega_{ac} - \Gamma_{bc}\Gamma_{ac}/2)\}$$

where $\omega_{ab} = \omega_a - \omega_b$, $\Gamma_{ab} = \Gamma_a - \Gamma_b$, and a, b, and c symbolize j, p, or i, and $\Gamma_j = 0$. Equation (5-34) contains a great deal of information, and in order to experimentally verify all its ramifications, time-gated measurements of the spectral distribution are needed. However, two aspects are simple both in theory and experiment. First, consider an experiment which measures the lifetime of the scattered radiation in the usual manner of detecting the emitted light with a bandwidth for frequency detection that is large compared with the width of the resonant state. In this case the measured intensity signal is proportional to

$$I(t) = \frac{d}{dt} L^3 \int \frac{d^3\mathbf{k}}{(2\pi)^3} P_{j\mathbf{k}}(t) \qquad (5\text{-}36)$$

which is given explicitly, and quite accurately, by

$$I(t) = \Gamma_{ir} P_i(t) \qquad (5\text{-}37)$$

[Γ_{ir} is the radiative width of state r as in (5-34)]. Equation (5-36) is a very simple result which says that it is the intermediate-state probability which governs the time dependence of the scattered light. Figure 5-4 shows plots of $P_i(t)$ for various amounts of resonance mismatch. When the exciting light is exactly on resonance, the decay has a long-lived component containing information about the lifetime of the molecular state. On shifting the light frequency further off-resonance, the scattered light is seen to begin to follow the time character of the excitation pulse. Experimental observation of this type of effect is due to Williams et al. (1974), whose work provided an important stimulus for the reexamination of the theory of light scattering.

The other aspect of Eq. (5-34) that is straightforward is the spectral distribution of the scattered light, which is obtained by allowing $t \to \infty$ in Eq. (5-34). In order that the proper limit be found as $t \to \infty$ from Eq. (5-34) it is necessary to exclude any instability associated with the state $|j\rangle$. Thus, if $\Gamma_j \neq 0$, the probability will tend to zero as $t \to \infty$. Γ_j is therefore set equal to

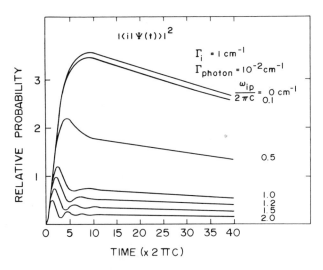

Figure 5-4. Time dependence in the off-resonance case for the intermediate-state probability in the three-level system. Note that the situation $\Gamma_i \gg \Gamma_p$ pertains (cf. Figure 5-3). As the resonance mismatch ω_{ip} becomes larger, the probability becomes localized in the early time regime (note that $\Gamma_p = 0.01$ cm^{-1} corresponds to a time of 10 in units of $2\pi c$).

zero in order to obtain the spectrum. The distribution of scattered photons (the line shape) is found to be

$$P_{jk}(\infty) = \frac{\Gamma_p/2\pi \left| V_{ji}^{(r)} V_{ig}^{(r)} \right|^2}{(\omega_{ji}^2 + \Gamma_i^2/4)(\omega_{jp}^2 + \Gamma_p^2/4)} \tag{5-38}$$

where ω_p and Γ_p are the field variables. Note that ω_{jp} is given by

$$\hbar\omega_{jp} = E_j - \hbar\omega_p + \hbar\omega_s \tag{5-39}$$

where E_j is the energy of the molecular state, ω_s is the frequency of the scattered light, and $E_j + \hbar\omega_s$ is the energy of the final molecule–photon state $|j\rangle$. Thus, it is seen that for ω_p chosen exactly on resonance with the state i, we get a line shape that is the product of two Lorentzians:

$$P_{jk}^{(res)}(\infty) = \Gamma_p/2\pi \frac{\left| V_{ji}^{(r)} V_{ig}^{(r)} \right|^2}{[(\omega_s - \omega_{ij})^2 + \Gamma_i^2/4][(\omega_s - \omega_{ij})^2 + \Gamma_p^2/4]} \tag{5-40}$$

centered about the molecular energy $(E_i - E_j)$. Two interesting limits arise: $\Gamma_i \gg \Gamma_p$ and $\Gamma_i \ll \Gamma_p$. In the first of these limits, the radiation bandwidth is small compared with the linewidth of the resonant state, and the linewidth becomes a Lorentzian whose width is determined by Γ_p. In the other, the radiation bandwidth is large and the linewidth of the scattered light is essentially that of the molecular state $|i\rangle$.

The lineshape for the case that the state $|j\rangle$ is damped can be calculated by summing over *all* final states containing the photon of interest. The result for the probability of signal in the direction $\hat{k}$ of the light scattered at ω_s is

$$P_s(\infty) = \Gamma_p/2\pi \frac{|V_{gi}^{(r)} V_{ij}^{(r)} V_{jg}^{(r)}|^2}{[\omega_{ij}^2 + \frac{1}{4}(\Gamma_i + \Gamma_j)^2]}(F_1 + F_2) \tag{5-41}$$

$$F_1 = \{\frac{1}{2}\omega_{ji}[\omega_{jp}(\Gamma_i + \Gamma_j) + \omega_{ji}(\Gamma_p + \Gamma_i)$$
$$+ \frac{1}{4}(\Gamma_j - \Gamma_i)[2\omega_{ji}\omega_{jp} - (\Gamma_i + \Gamma_j)(\Gamma_p + \Gamma_i)]\}$$
$$\times \{\Gamma_j[\omega_{ij}^2 + \frac{1}{4}(\Gamma_j - \Gamma_i)^2][\omega_{jp}^2 + \frac{1}{4}(\Gamma_p + \Gamma_i)^2]\}^{-1} \tag{5-42}$$

$$F_2 = \{\frac{1}{4}(\Gamma_j - \Gamma_i)[2\omega_{ji}\omega_{ip} + (\Gamma_j + \Gamma_i)(\Gamma_p + \Gamma_i)]$$
$$- \frac{1}{2}\omega_{ji}[\Gamma_j + \Gamma_i)(\Gamma_p + \Gamma_i)]\}$$
$$\times \{\Gamma_i[\omega_{ij}^2 + \frac{1}{4}(\Gamma_j - \Gamma_i)^2][\omega_{ip}^2 + \frac{1}{4}(\Gamma_p + \Gamma_i)^2]\}^{-1} \tag{5-43}$$

5.2.3. Semiclassical Treatment of the Three-Level System

The results of the previous section can be obtained by assuming that the amplitude for spontaneous emission from the nonstationary state $\Psi(t)$, which has coefficients given by Eq. (5-32), to the stationary state Ψ_j is determined by

$$2\,\mathrm{Re}\,[\langle\Psi(t)|\mu|\Psi_j\rangle] = M_j(t) \tag{5-44}$$

where μ is the dipole moment operator. A justification for this can be found in the correspondence principle. The intensity and polarization of spontaneous emission from an upper to a lower state is the same as that given by the classical radiation theory for an oscillating dipole, which is described by the appropriate matrix element of the dipole moment operator. In the semiclassical theory the induced processes present no problem; thus, $|\langle\Psi(t)|\Psi_j\rangle|^2$ is the probability that by time t the system will be in the quasi-stationary state Ψ_j, and this would correspond to two-photon absorption or stimulated Raman. Assuming that $\Psi(t)$ arises in a three-level system where the incident field only couples the ground and ith states, but matrix elements of the dipole operator exist between Ψ_i and Ψ_j, we can use Eq. (5-32) and find that

$$M_j(t) = a_i^*(t)\mu_{ij}e^{-\Gamma_i t/2 - i\omega_{ji}t} + \text{c.c.} \tag{5-45}$$

From the fact that the intensity of radiation emitted into 4π rad by an oscillating dipole (d) is $(2/3c^2)\ddot{d}^2$, we find the scattered intensity in the three-level system to be

$$I_s(t) = \frac{2}{3c^3}|\ddot{M}_j(t)|^2 \tag{5-46}$$

This is the total scattered intensity at all frequencies. The energy distribution of scattered light at ω is desired, so we require a formula for the spectral resolution of the energy of dipole radiation. This is calculated using the classical formula (Jackson, 1975) for the energy distribution scattered into solid angle $d\Omega$:

$$\frac{d^2 W(\omega_s)}{d\omega_s \, d\Omega} = \sin^2 \Theta / 2\pi c^3 |\mathbf{M}_j(\omega_s)|^2$$

$$= \sin^2 \Theta / 4\pi^2 c^3 \left| \int_0^\infty dt \, e^{i\omega_s t} \ddot{\mathbf{M}}_j(t) \right|^2$$

$$= \omega_s^4 \sin^2 \Theta / 4\pi^2 c^3 \left| \int_0^\infty dt \, e^{i\omega_s t} \mathbf{M}_j(t) \right|^2 \qquad (5\text{-}47)$$

where Θ is the angle between $\ddot{\mathbf{M}}_j(t)$ and n, a unit vector in the direction of solid angle $d\Omega$. Substitution of Eqs. (5-32) and (5-45) into (5-47) yields the lineshape for scattering in the three-level system as

$$\frac{d^2 W(\omega_s)}{d\omega_s \, d\Omega} = \frac{\omega_s^4 \sin^2 \Theta \gamma_{gi}^2 |\mu_{ij}|^2}{4\pi^2 c^3}$$

$$\times \frac{1}{[(\omega_{pj} - \omega_s)^2 + \frac{1}{4}\Gamma_p^2][(\omega_{ij} - \omega_s)^2 + \frac{1}{4}\Gamma_i^2]} \qquad (5\text{-}48)$$

This result is identical with that obtained in Eq. (5-38), although care is needed with the definitions of the frequencies [see Eq. (5-39)]. In the quantized field model, $\hbar\omega_j$ is the energy of the jth eigenstate of the zero-order Hamiltonian that contains both molecular and photon parts. In the semiclassical model $\hbar\omega_j$ is just the energy of the jth molecular eigenstate. These examples show that the semiclassical approach, while involving familiar concepts and algebraic methods, yields the same results in these simple problems as the more advanced scattering theory.

The same dependence of the scattered light intensity can be obtained in a form identical with Eq. (5-35) from that part of $M_j(\omega_s)$ that is contributed in the time interval $t = 0$ to t. In other words, $M_j(\omega, t)$ is obtained by performing the integration in Eq. (5-47) up to the time t only. The absolute square of $\ddot{M}_j(\omega, t)$ is then proportional to the probability that the light of frequency ω is observed by time t, a quantity that is automatically obtained in the quantum treatment.

5.2.4. Resonance and Near-Resonance Raman Scattering

The three-level system provides a good model for near-resonance Raman scattering. For a molecule having many lower states $|j\rangle$ onto which

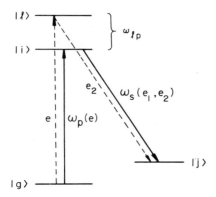

Figure 5-5. Model four-level system. Both $|i\rangle$ and $|l\rangle$ are assumed to couple radiatively to both $|g\rangle$ and $|j\rangle$.

spontaneous emission from $\Psi(t)$ can occur, the probability for each downward transition can be calculated separately, so if the resonant state is isolated, each Raman band can be considered in a three-level system. For molecules the condition that the resonant state be isolated, i.e., that $\Psi(t)$ contains only two components, is seldom rigorous because of the large number of closely spaced vibrational and rotational levels near any resonance. A more appropriate model involves a situation where the ground state couples strongly in the field to one excited state by virtue of near-resonance conditions and less strongly to other states, because they are further from resonance. A typical situation is sketched in Figure 5-5, where $|i\rangle$ is again the resonant state, and $|l\rangle$ is another state that may couple to the ground state, and again $|j\rangle$ is the final state of the scattering process. Consider the case where the molecule is excited by a pulse of light of the form (5-13), so that the system is described by

$$|\Psi(t)\rangle = |g\rangle\varphi_p(t) + a_i(t)|i; 0\rangle + a_l(t)|l; 0\rangle + \sum_{\mathbf{ke}} a_{j\mathbf{ke}}(t)|j; \mathbf{k, e}\rangle \quad (5\text{-}49)$$

Suppose further that the transition $i \to j$ is polarized $\mathbf{e}_1$ and $l \to j$ is polarized $\mathbf{e}_2$ with $\mathbf{e}_1 \cdot \mathbf{e}_2 = 0$, and that the exciting light can interact with both $|i\rangle$ and $|l\rangle$. When the spectrum of scattered light is measured about ω_{pj} using a polarizer set at $\mathbf{e}_2$ only $|l\rangle$ can be detected. If the scattering experiment is performed with polarizers in particular relative orientations, the interference terms, which complicate the present calculations (but make the experiment more interesting!), can be removed. The measurement of a spectrum of the light emitted between two times defined by the interval $t_2 - t_1$ is of practical importance. In that case the gated spectrum is proportional to

$$\int_{t_1}^{t_2} \frac{d}{dt}|a_{j\mathbf{ke}}(t)|^2 \, dt \quad (5.50)$$

The amplitude $a_{jke}(t)$ may be explicitly displayed as a sum of two time integrals [cf. Eq. (5-92)], one which convolutes the exciting pulse with the molecular decay characteristics and another which convolutes the exciting pulse with the scattered light frequency. When the polarizer is set at $\mathbf{e}_2$ so that it can pick out the off-resonant state $|l\rangle$, we see that the probability varies as

$$P_{jke}(t) \sim \omega_{lP}^{-2} \left| \frac{f_1(t)}{\omega_{pj} + i\Gamma_p/2} + \frac{f_2(t)}{\omega_{lP}} \right|^2 \tag{5-51}$$

where f_1 is a function having appreciable value only during the light pulse and f_2 is a function having value during the molecular decay time as well as the light pulse duration, so that as long as the detuning frequency is large, a scattered light response characteristic of the exciting light pulse is observed. If one then gates at $t_1 > \Gamma_p^{-1}$, the $\mathbf{e}_2$ spectrum is very small. On the other hand, the $\mathbf{e}_1$ spectrum follows both the light pulse and the molecular decay, so that even if $t_1 > \Gamma_p^{-1}$, one may still see a strong spectrum if $t_1 < \Gamma_i^{-1}$. This is, of course, just a statement of the well-known fact that a time delay suffered by a particle is resonant scattering is longer than in nonresonant scattering (Goldberger and Watson, 1964). The situation modeled by this example is commonly encountered in molecular spectroscopy. It would arise in a resonance Raman experiment if the laser were tuned to near resonance with one electronic state when the vibrational shift selected by the mono-chromator could appear in the Raman process involving a higher electronic state as well.

5.3. Nature of the Electromagnetic Field

5.3.1. Definition of the Field Variables

The scattering theory presented in Section 5.1 is essentially that used in describing material particle interactions (Goldberger and Watson, 1964). Since a photon is a particle of a relativistic quantum field, further discussion is needed on the use of this mathematical framework to describe light scattering. The assumption of an exciting light field containing only one photon also requires more detailed evaluation in relation to real experiments, where it is usual for many photons to be present in the field, especially when the experiments are done with lasers. We will therefore discuss the quantum mechanics of the radiation field when no charges or currents are present. The object of this chapter is to develop the quantum theory of coherence sufficiently that a fully quantum-mechanical description of light scattering will follow automatically.

Since we are interested in the interaction of molecules and radiation in free space and not in some specific type of cavity resonator, it is convenient to decompose the field into modes corresponding to running waves normalized with periodic boundary conditions corresponding to some large volume L^3. The vector potential operator may then be written as (Loudon, 1973)

$$\mathbf{A}(\mathbf{r}) = \mathbf{A}^{(+)}(\mathbf{r}) + \mathbf{A}^{(-)}(\mathbf{r})$$
$$\mathbf{A}^{(-)}(\mathbf{r}) = [\mathbf{A}^{(+)}(\mathbf{r})]^{\dagger} \tag{5-52}$$

$$\mathbf{A}^{(+)}(\mathbf{r}) = \sum_{k,\lambda} (\hbar/2\varepsilon_0 L^3 \omega_k)^{1/2} e_{k\lambda} e^{i\mathbf{k}\cdot\mathbf{r}} a_{k\lambda}$$

The Coulomb gauge has been chosen (i.e., $\nabla \cdot \mathbf{A} = 0$) and the annihilation and creation operators (wave vector $\mathbf{k}$, polarization λ) satisfy the usual commutation rules:

$$[a_{k\lambda}, a_{k'\lambda}^{\dagger}] = \delta_{kk'}\delta_{\lambda\lambda'} \tag{5-53}$$

In regions of space far from any charges and currents, the scalar potential can be chosen as $\varphi = 0$, and the electric and magnetic fields are obtained as follows:

$$\mathbf{E}(\mathbf{r}) = -\dot{\mathbf{A}}(\mathbf{r}) = -\frac{i}{\hbar}[H_r, \mathbf{A}(\mathbf{r})]$$

$$\mathbf{B}(\mathbf{r}) = \nabla \times \mathbf{A}(\mathbf{r}) \tag{5-54}$$

so that the Hamiltonian for the free field is

$$H_r = \tfrac{1}{2} \int d^3\mathbf{r}(\varepsilon_0 \mathbf{E}^2 + \mu_0 \mathbf{B}^2)$$

$$= \sum_{k\lambda} \hbar\omega_k(a_{k\lambda}^{\dagger} a_{k\lambda} + \tfrac{1}{2}) \tag{5-55}$$

Equations (5-53) and (5-55) imply that the field is described by a set of simple harmonic oscillators.

5.3.2. Radiation–Matter Interaction

Before discussing states of the radiation field, it is useful to consider the interaction of a field with charged particles simply because it is this interaction which causes non-vacuum-state fields to come into existence. It is also useful for the interpretation of experiments to understand the states of the field in relation to their method of preparation.

By following the usual procedures, the quantum-electrodynamical Hamiltonian is found to be

$$H = \sum_j \frac{1}{2m_j}[\mathbf{P}_j - e_j\mathbf{A}(\mathbf{r}_j)]^2 + H_r + V \qquad (5\text{-}56)$$

where $\mathbf{P}_j$ is the canonical momentum operator for particle j and V is the Coulombic potential energy of the charges in the molecule. A unitary transformation of Eq. (5-56) using the generator (Power and Zienau, 1959; Atkins and Woolley, 1970; Woolley, 1971; Babiker *et al.*, 1974)

$$\frac{1}{\hbar}\int d^3\mathbf{r}\mathbf{P}(\mathbf{r}) \cdot \mathbf{A}(\mathbf{r}) \qquad (5\text{-}57)$$

yields a multiple Hamiltonian, which may be written

$$H = H_m + H_r - \int d^3\mathbf{r}\mathbf{P}(\mathbf{r}) \cdot \mathbf{E}(\mathbf{r}) - \int d^3\mathbf{r}\mathbf{M}(\mathbf{r}) \cdot \mathbf{B}(\mathbf{r}) + (\text{diamagnetic term}) \qquad (5\text{-}58)$$

We shall be interested in the first term of the expansion of $\mathbf{P}(\mathbf{r})$ about the molecular origin, which may be written as

$$\mathbf{P}(\mathbf{r}) \approx \mu\delta(\mathbf{r} - \mathbf{r}_0) \qquad (5\text{-}59)$$

μ being the dipole-moment operator for the molecule. Keeping the spatial integral in Eq. (5-58) helps to elucidate the coordinate dependence of the scattering process. Keeping only the electric-field terms in Eq. (5-58) yields, along with (5-59), the usual electric-dipole approximation for the radiation–matter interaction.

5.3.3. States of the Radiation Field

Since the Hamiltonian (5-55) is for a set of harmonic oscillators, a natural choice for a radiation-field basis is the set where the $\mathbf{k}$th oscillator is in the state $n_\mathbf{k}$ of the oscillator. Then there are $n_\mathbf{k}$ quanta or photons in mode $\mathbf{k}$. This Fock-space basis is denoted by

$$|\{n_\mathbf{k}\}\rangle = \prod_\mathbf{k} |n_k\rangle \qquad (5\text{-}60)$$

This state represents a beam of light consisting of a *definite* (i.e., sharp) number of photons having a particular momentum distribution and is a convenient basis set for the description of chaotic light, which is fully described by specifying a definite number of photons for each field mode. The Fock-space representation is also useful when a small number of quanta are present in a particular mode, such as in processes occurring at high energies or in processes involving black-body radiation. For the purposes of visualizing experiments, particularly those involving pulses of light, the Fock space presents some problems, since it treats photons of definite momentum and hence particles having total positional uncertainty. One might expect that a Fourier transform of these momentum states would lead to photon position states. Unfortunately, a photon of definite position is not consistent with the local nature of the electromagnetic field and has no meaning (Berestetskii *et al.*, 1971). Photons can be localized only in volumes having minimum dimensions on the order of the wavelength of the light. A superposition of states of definite momentum provides a coarse localization which can have a precise meaning on a laboratory scale since it represents a localization of the field. The simplest superposition is that of one-photon states of the form

$$|\varphi\rangle = \int d^3k f(\mathbf{k}) a_\mathbf{k}^\dagger |\text{Vac}\rangle \tag{5-61}$$

which is the form of the Lorentzian one-photon wavepacket used in Section 5.1. States of the form (5-61) provide a convenient mathematical form for the solution of problems but have the disadvantage that one does not really know how to make them in the laboratory.

A state of the radiation field that can be manufactured in the laboratory is one that comes about when a classical current density drives the field. A classical current density is one that is not affected by its own radiation. The mathematics of this interaction is fully discussed in a number of references (Glauber, 1963; Merzbacher, 1970), and it is found that the state is characterized by one complex number for every mode of the field. For a particular mode the state $|\alpha\rangle$ is given by

$$|\alpha\rangle = e^{-|\alpha|^2/2} \sum_n \frac{\alpha^n}{n!} |n\rangle \tag{5-62}$$

$|\alpha\rangle$ represents a state where the probability of finding a particular number of quanta is given by that for a Poisson distribution with the mean number of quanta in the mode given by $|\alpha|^2$, as may be seen by calculating the average of $a^\dagger a$ in Eq. (5-62).

The "coherent" state (5-62) possesses several important mathematical properties. First, the set defined by all complex numbers $\alpha = \text{Re}\,\alpha + i\,\text{Im}\,\alpha$ is overcomplete, so states labeled by different complex numbers are not orthogonal, i.e.,

$$\langle\beta|\alpha\rangle = e^{-|\alpha|^2/2 - |\beta|^2/2 + \alpha\beta^*} \neq 0 \qquad (\text{if } \alpha \neq \beta) \tag{5-63}$$

An important closure relation does exist, however:

$$\frac{1}{\pi} \int d^2\alpha\, |\alpha\rangle\langle\alpha| = 1 \qquad [d^2\alpha = d(\text{Re}\,\alpha)\, d(\text{Im}\,\alpha)] \tag{5-64}$$

Relations (5-63) and (5-64) are included here primarily to caution the reader used to dealing with the usual type of complete orthonormal set of states encountered in quantum mechanics. A property of particular importance is that a coherent state is an eigenstate of the annihilation operator,

$$a|\alpha\rangle = \alpha|\alpha\rangle \qquad (5\text{-}65)$$

Because of this property the averages of field operators in coherent states are solutions to the classical Maxwell equations. Coherent states thus provide a quantum-mechanical representation for classical fields. They are also the states closely approximated by a laser operating above threshold. The name "coherent state" is applied to Eq. (5-62) because of its use in describing optical interference experiments using quantum mechanics, a topic discussed in the next section. The coherent state for a multimode field is given by

$$|\{\alpha_k\}\rangle = \prod_k |\alpha_k\rangle \qquad (5\text{-}66)$$

in analogy with the Fock-space state (5-60).

5.3.4. Measurables of the Field and Photon Experiments

The basis states (5-60) and (5-66) provide useful theoretical representations for light beams, and it is important now to be able to understand simple experimental properties of light beams in terms of these theoretical constructions. A real light beam is a statistical entity in that it contains fluctuations other than those due to quantum-mechanical effects, and therefore it should be represented by an ensemble. The quantities that we measure in the laboratory are then quantum ensemble averages of the operators corresponding to the measurement.

The most common type of measurement is that of the intensity of a light beam without regard to its spectral content. Since essentially all intensity-measuring devices depend upon the photoelectric effect, this process is taken to define the intensity of the light in terms of the rate of production of electrons and a simple analysis using time-dependent perturbation theory shows that the counting rate of electrons $w^{(1)}(\mathbf{r}_0, t)$ is given by (Loudon, 1973; Glauber, 1963, 1970)

$$w^{(1)}(\mathbf{r}_0, t) = (\text{const.})\langle E^{(-)}(\mathbf{r}_0 t)E^{(+)}(\mathbf{r}_0 t)\rangle \qquad (5\text{-}67)$$

with $E^{(+)}(\mathbf{r}t)$ the positive frequency part of the electric field operator for the light in the Heisenberg picture:

$$\mathbf{E}^{(+)}(\mathbf{r}t) = i \sum_k (\hbar\omega_k/2\varepsilon_0 L^3)^{1/2} e_k e^{i(\mathbf{k}\cdot\mathbf{r}-\omega_k t)} a_k \qquad (5\text{-}68)$$

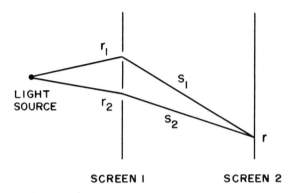

Figure 5-6. Schematic of the double-slit interference experiment.

where $E^{(-)}$ is its Hermitian conjugate. The average in Eq. (5-67) is over the initial state of the light field, and in general it is given by

$$\text{Tr}\left[\rho E^{(-)}(\mathbf{r}_0 t)E^{(+)}(\mathbf{r}_0 t)\right] \tag{5-69}$$

where ρ is the density operator for light beam. The counting rate is easily evaluated for a coherent state, since these are eigenstates of the annihilation operator [see Eq. (5-65)] and the result is just proportional to the mean number of photons in the mode. A more interesting experiment to describe is the measurement of the distribution in the number of photons counted over a fixed time interval. This can be done by counting for a fixed time, then waiting for a time long enough to ensure the absence of correlations before counting again. The distribution of counts around the mean $|\alpha|^2$ is found to have Poissonian form. Quantum-mechanically, the probability of counting N photons for a coherent state is

$$P_N(\alpha) = |\langle N|\alpha\rangle|^2$$

which gives immediately [using Eq. (5-62)] the appropriate Poisson distribution.

Experiments in classical optics are concerned essentially with the measurement of Eq. (5-67), so they are averages of normally ordered products of two field operators. To illustrate this point, consider the Young double-slit experiment pictured in Figure 5-6. The intensity pattern on the screen 2 arising from the presence of the two pinhole sources in screen 1 is measured. By Huygen's principle, one may take the electric field at $\mathbf{r}$ (the position of the detector) as

$$\mathbf{E}(\mathbf{r}t) = u_1\mathbf{E}(\mathbf{r}_1 t_1) + u_1 E(\mathbf{r}_2 t_2)$$
$$t_1 = t - s_1/c, \quad t_2 = t - s_2/c \tag{5-70}$$

The coefficients u_1 and u_2 are taken as constants for simplicity. From Eq. (5-63) the response of a photodetector at $\mathbf{r}$ is given up to a constant by

$$\mathrm{Tr}\,[\rho E^{(-)}(\mathbf{r}t)E^{(+)}(\mathbf{r}t)] = |u_1|^2\,\mathrm{Tr}\,[\rho E^{(-)}(\mathbf{r}_1 t_1)E^{(+)}(\mathbf{r}_1 t_1)]$$
$$+\,|u_2|^2\,\mathrm{Tr}\,[\rho E^{(-)}(\mathbf{r}_2 t_2)E^{(+)}(\mathbf{r}_2 t_2)]$$
$$+\,2\,\mathrm{Re}\,u_2^* u_1\,\mathrm{Tr}\,[\rho E^{(-)}(\mathbf{r}_2 t_2)E^{(+)}(\mathbf{r}_2 t_2)] \qquad (5\text{-}71)$$

The last term in Eq. (5-65) determines the visibility of the fringe on screen 2. The term causing the interference contains the autocorrelation function:

$$G^{(1)}(\mathbf{r}_2 t_2;\mathbf{r} t_1) = \mathrm{Tr}\,[\rho E^{(-)}(\mathbf{r}_2 t_2)E^{(+)}(\mathbf{r}_1 t_1)] \qquad (5\text{-}72)$$

which can be shown (for the case $\mathbf{r}_1 = \mathbf{r}_2$) to be related to the spectrum of a stationary light source by a generalization of the Wiener–Khintchine theorem (Glauber, 1963) as

$$S(\omega) = \frac{1}{\pi}\int_{-\infty}^{\infty} G^{(1)}(\mathbf{r}_0;\mathbf{r} t)e^{i\omega t}\,dt \qquad (5\text{-}73)$$

where $S(\omega)$ is the energy per unit angular frequency in the light beam. So it is seen that the double-slit experiment measures $G^{(1)}$. Experiments depending essentially on the $G^{(1)}$ function with $\mathbf{r}_1 = \mathbf{r}_2 = \mathbf{r}$ depend only on the spectrum of the exciting source. The linear response of a molecule at point r to a stationary light source is an example of this dependence. The quantum-mechanical description of the point-source double-split experiment for a coherent state is again straightforward. By the properties of coherent states given above, it is found that

$$G_\alpha^{(1)}(\mathbf{r}_2 t_2;\mathbf{r}_1 t_1) = G_\alpha^{(1)}(\mathbf{r}_0;\mathbf{r} t) \propto |\alpha|^2 e^{-i\omega_k t}$$

The Fourier transform of $G_\alpha^{(1)}$ is a delta function [use Eq. (5-73)]. This result is again bringing out the correspondence between the coherent state and the classical monochromatic coherent wave. The coherent state is the quantum-mechanical representation of the classical field. Note that the double-slit experiment is explained by $G_\alpha^{(1)}(\mathbf{r}_0;\mathbf{r} t)$ and therefore measures only the temporal coherence in the beam (for a point source), a property that for this example is described entirely by the spectral width of the light. The coherent state $|\alpha\rangle$ is monochromatic, but the product state $|\{\alpha_\mathbf{k}\}_i\rangle = \prod_\mathbf{k} |\alpha_\mathbf{k}\rangle$ has a first-order correlation function that corresponds to a superposition of monochromatic waves with the occupation numbers (the α's) determining the weights of each of the Fourier components. The spectral width in this case depends on the choice of the distribution of α's.

It is also possible to perform experiments which measure other properties of the field besides the intensity, and which therefore lead to a more

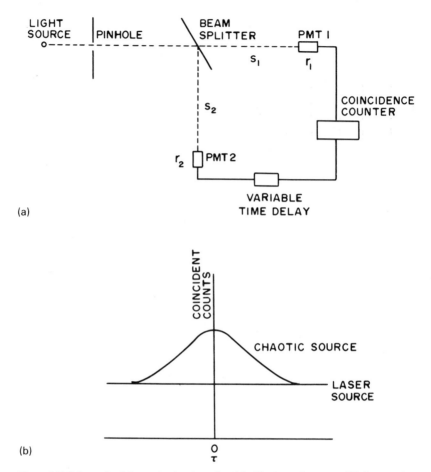

(a)

(b)

Figure 5-7. Schematic of the method and results of the Hanbury-Brown and Twiss experiment.

refined understanding of light beams. The classic experiment of this type pictured in Figure 5-7a is the Hanbury-Brown and Twiss (1975) experiment, and it measures the rate of coincidence counts registered in two photomultipliers [PMT(1) and PMT(2)] as a function of delay time. The rate of coincident counts depends on the second-order correlation function for the field:

$$w^{(2)}(\tau) = (\text{const.})^2 G^{(2)}(\mathbf{r}_1 t_1, \mathbf{r}_2 t_2; \mathbf{r}_2 t_2, \mathbf{r}_1 t_1) \qquad (5\text{-}74)$$

with $\tau = t_1 - t_2 - (1/c)(S_1 - S_2) =$ delay time. The experimental results are sketched in Figure 5-7b for two different light sources. The interpretation of this experiment leads to a definition of coherence due to Glauber (1970): a light source is said to be nth-order-coherent if its set of autocorrelation functions, up to an including n, factor as

$$G^{(n)}(x_1 \cdots x_n; y_1 \cdots y_n) = \prod_{j=1}^{n} G^{(1)}(x_j; y_{n-j+1}) \qquad (5\text{-}75)$$

This definition indicates that for coherent light, Eq. (5-69) becomes just the product of the intensities of the two beams from the beam splitter and is a constant with respect to τ. The peak in the coincidence signal at $\tau = 0$ is a manifestation of the fact that for a totally random stochastic process, one may write

$$G^{(n)}(x_1 \cdots x_n; y_1 \cdots g_n) = \sum_{p} \prod_{j=1}^{n} G^{(1)}(x_j; y_{pj}) \qquad (5\text{-}76)$$

where p designates the $n!$ permutations of the numbers 1 through n. Applying Eq. (5-76), Glauber (1970) shows that

$$G^{(2)}(x_1 x_2; x_2 x_1) = G^{(1)}(x_1; x_1) G^{(1)}(x_2; x_2) + |G^{(1)}(x_1, x_2)|^2 \qquad (5\text{-}77)$$

The last term in Eq. (5-71) is responsible for the peaking effect. This result has one very obvious but important consequence in laser spectroscopy, and that is that orders of coherence must be distinguished in experiments. A chaotic light source may yield a perfectly coherent beam inasfar as the first-order effects discussed above are concerned, yet its second-order coherence is quite different from that of a coherent state. The higher-order coherence measures the spatial bunching of photons, and it is not difficult to see why this bunching must be absent in a field made up from classical monochromatic waves.

The fundamental optical experiments briefly described here should provide a solid framework on which to build understanding of more complex processes involving molecules as the photodetectors. Clearly, it will be extremely useful to describe light scattering and other spectroscopic processes in terms of $G^{(1)}$ and $G^{(2)}$, but of course it will only be possible to cast the scattering formulas in this form in some kind of perturbation approach, and the next section describes one scheme for accomplishing this aim.

5.4. Theory of Light Scattering with Well-Defined Light Sources

5.4.1. More General Approach to Light Scattering

Rather than trying to adjust the parameters of a model that contains a qualitative representation of the light field, it seems better to set up the problem in such a manner that the real measurable characteristics of the light field discussed above are included in the formalism for light scattering (Hochstrasser and Novak, 1976). In what follows we present a method of solution for the near-resonant scattering of light having well-defined coherence properties.

A crucial part of the method of solution is the manner in which the Hamiltonian is partitioned. The total Hamiltonian for a system consisting of a molecule and a radiation field is, in the dipole approximation,

$$H = H_m + H_r + \boldsymbol{\mu} \cdot \mathbf{E}(\mathbf{r}_0) \tag{5-78}$$

where the electric field operator is given by Eq. (5-68) and its Hermitian conjugate. In a Raman or fluorescence experiment, there is light present mainly in two frequency regions, corresponding to the incident and the scattered frequencies. The Hamiltonian can be written in a form (5-79) that displays the interactions of the system with these two fields separately:

$$H = H_m + H_r^{(e)} + H_r^{(s)} - \boldsymbol{\mu} \cdot \mathbf{E}_e(\mathbf{r}_0) - \boldsymbol{\mu} \cdot \mathbf{E}_s(\mathbf{r}_0)$$
$$\equiv H_0 - \boldsymbol{\mu} \cdot \mathbf{E}_e(\mathbf{r}_0) \tag{5-79}$$

where s and e signify scattered and exciting fields. The time evolution operator in the interaction representation, $U_I(t)$, satisfies

$$i\frac{d}{dt}U_I(t) = V_I(t)U_I(t) \tag{5-80}$$

with

$$V_I(t) = \begin{cases} -e^{iH_0t}\boldsymbol{\mu} \cdot \mathbf{E}_e(\mathbf{r}_0)e^{-iH_0t} & (5\text{-}81) \\ -\boldsymbol{\mu}_I(t) \cdot \mathbf{E}_e(\mathbf{r}_0t) & (5\text{-}82) \end{cases}$$

where $\mathbf{E}_e(\mathbf{r}_0t)$ is the Heisenberg operator for the electric field of the incident light. In order to calculate the spectrum, the number of photons in a mode $\mathbf{k}$,

$$a_{\mathbf{k}}^{\dagger}(t)a_{\mathbf{k}}(t) = e^{iH_1t}a_{\mathbf{k}}^{\dagger}a_{\mathbf{k}}e^{-iH_1t} \tag{5-83}$$

must be averaged in the state $|\psi_I(t)\rangle = U_I(t)|\psi_I(0)\rangle$ for photons of wavevector $\mathbf{k}$ not in the exciting field and where $H_1 = H_0 - H_r^{(e)}$.

5.4.2. Spectral Content of a Scattered Coherent Pulse

Before proceeding to calculate the spectrum of radiation scattered by a molecule interacting with an arbitrary light beam, it is useful to investigate again the case closest to that described by a one-photon wavepacket—the interaction of a molecule with a coherent pulse of light. Characterize the pulse quantum-mechanically by the coherent state

$$|\{\alpha_\mathbf{k}\}_e\rangle = \prod_{\mathbf{k}\in e} |\alpha_\mathbf{k}\rangle \qquad (5\text{-}84)$$

so that if the molecule was initially in its ground state $|g\rangle$, the wavefunction becomes, in first order,

$$|\Psi_I(t)\rangle \approx |g; \text{Vac}_s; \{\alpha_\mathbf{k}\}_e\rangle$$
$$+ i \int_0^t dt' e^{iH_1,t'} \mathbf{\mu} e^{-iH_1,t'} \cdot \mathbf{E}_e(\mathbf{r}_0 t') |g; \text{Vac}_s; \{\alpha_\mathbf{k}\}_e\rangle \qquad (5\text{-}85)$$

Keeping only the $E_e^{(+)}$ part of the incident field, which is equivalent to making the rotating-wave approximation, the wavefunction becomes

$$|\Psi_I(t)\rangle \approx |g; \text{Vac}_s; \{\alpha_\mathbf{k}\}_e\rangle$$
$$+ i \int_0^t dt' e^{iH_1,t'} \mathbf{\mu} e^{-iH_1,t'} |g; \text{Vac}_s; \{\alpha_\mathbf{k}\}_e\rangle \cdot \mathbf{\mathcal{E}}_\alpha(\mathbf{r}_0 t') \qquad (5\text{-}86)$$

where use has been made of the fact that a coherent state is an eigenstate of the annihilation operator $\mathbf{\mathcal{E}}_\alpha(\mathbf{r}t')$ is the eigenvalue of the operator $\mathbf{E}_e^{(+)}$ for the coherent state.

Suppose that the molecule is characterized by three levels, $|g\rangle$, $|i\rangle$, and $|j\rangle$, and that the incident field contains only frequencies near ω_{ig}. Furthermore, assume for simplicity (but not necessity) that state $|j\rangle$ is undamped. Under these circumstances, it is possible to calculate the probability that after a long time compared with all lifetimes, a photon is scattered from the incident beam into a photon of frequency $\approx \omega_{ij}$. Note first that the term

$$e^{-iH_1 t'} |g; \text{Vac}_s; \{\alpha_\mathbf{k}\}_e\rangle \qquad (5\text{-}87)$$

represents no real-time evolution of the system, since there are no scattered photons in the initial state. It can only cause an energy shift of the ground state due to the interaction of the ground-state molecule with virtual photons in the scattered fields. Calling the corrected ground molecule state energy zero, the amplitude for scattering one photon into a frequency near ω_{ij} in s, leaving the excited field in some other state, is given by

$$i\langle\{n'_\mathbf{k}\}_e|\{\alpha_\mathbf{k}(t)\}_e\rangle \int_0^t dt' \langle \mathbf{k}; j| e^{-iH_1(t-t')} |i; \text{Vac}_s\rangle \mathbf{\mu}_{ig} \cdot \mathbf{\mathcal{E}}_\alpha(\mathbf{r}_0 t') \qquad (5\text{-}88)$$

where $\{n_\mathbf{k}\}_e$ represents a final state of the incident light pulse. Note that Eq. (5-88) represents the convolution of the exciting pulse with the internal propagator, an often-quoted result. The matrix element of the propagator

$\exp[-iH_1(t-t')]$ may be calculated using Green's operator techniques, defining

$$G(z) = (z - H_1)^{-1} \tag{5-89}$$

so that

$$e^{-iH_1(t-t')} = \frac{1}{2\pi i} \int_c dz\, e^{-iz(t-t')} G(z); \qquad t \geq t' \tag{5-90}$$

The matrix element of interest here is

$$\langle \mathbf{k}; j | G(z) | i; \text{Vac}_s \rangle = \frac{-\boldsymbol{\mu}_{ij} \cdot \langle \mathbf{k} | E_s^{(-)} | \text{Vac}_s \rangle}{(z - \omega_j - \omega_{\mathbf{k}})(z - \omega_i + \frac{1}{2} i \Gamma_i)} \tag{5-91}$$

where the width Γ_i is calculated by considering only modes in s. Using Eqs. (5-90) and (5-91), Eq. (5-88) becomes

$$-i \langle \{n_{\mathbf{k}}\}_e | \{\alpha_{\mathbf{k}}(t)\}_e \rangle \boldsymbol{\mu}_{ji} \cdot \langle \mathbf{k} | E_s^{(-)} | \text{Vac}_s \rangle [\omega_{\mathbf{k}} + \omega_j - \omega_i + \tfrac{1}{2} i \Gamma_i]^{-1}$$

$$\times \int_0^t dt'\, \boldsymbol{\mu}_i \cdot \boldsymbol{\mathscr{E}}_\alpha(\mathbf{r}_0 t') [e^{-i(\omega \mathbf{k} + \omega_j)(t-t')} - e^{-i[\omega_i - (1/2)i\Gamma_i](t-t')}] \tag{5-92}$$

Choosing a form for $\boldsymbol{\mathscr{E}}_\alpha(\mathbf{r}_0 t')$,

$$\boldsymbol{\mathscr{E}}_\alpha(\mathbf{r}_0 t') = \tfrac{1}{2} \boldsymbol{\mathscr{E}}_0 e^{-i(\omega_p - (1/2)i\Gamma_p)t'} \tag{5-93}$$

which represents a pulse with an initial sharp rise followed by an exponential decay at $\mathbf{r}_0$ the square modulus of the amplitude (5-92) is given at long times by

$$\tfrac{1}{4} |\langle \{n_{\mathbf{k}}\}_e | \{\alpha_{\mathbf{k}}(t)\}_e \rangle|^2 |\boldsymbol{\mu}_{ig} \cdot \boldsymbol{\mathscr{E}}_0|^2 |\boldsymbol{\mu}_{ji} \cdot \langle \mathbf{k} | E_s^{(-)} | \text{Vac}_s \rangle|^2$$

$$\times [(\omega_{\mathbf{k}} + \omega_j - \omega_i)^2 + \tfrac{1}{4}\Gamma_i^2]^{-1} [(\omega_{\mathbf{k}} + \omega_j - \omega_p)^2 + \tfrac{1}{4}\Gamma_p^2]^{-1} \tag{5-94}$$

In Eq. (5-94) the final state of the field is of no interest, and summation over all $\{n_{\mathbf{k}}\}_e$ gives a factor of unity for the coherent-state part,

$$\sum_{\{n_{\mathbf{k}}\}_e} |\langle \{n_{\mathbf{k}}\}_e | \{\alpha_{\mathbf{k}}(t)\}_e \rangle|^2 = \langle \{\alpha_{\mathbf{k}}(t)\}_e | \{\alpha_{\mathbf{k}}(t)\}_e \rangle = 1 \tag{5-95}$$

since Eq. (5-94) is the same as that given earlier (5-37) using a one-photon Lorentzian wavepacket; however, the physics is now made clear. There is no assumption of only one photon in the pulse, and the assumption that the exciting pulse is weak means

$$p = \frac{|\boldsymbol{\mu}_{ig} \cdot \boldsymbol{\mathscr{E}}_0|}{\hbar \Gamma_i} \ll 1 \tag{5-96}$$

since the perturbation treatment is a power-series expansion in p when all terms are resonant.

5.4.3. Scattering from a Gaussian Pulse

We have shown in Section 4.3 that the time dependence and spectrum of light scattering arises from Eq. (5-88), which contains the classical form $\mathscr{E}_\alpha(\mathbf{r}_0 t)$ of the light pulse used in the experiment. The implication is that any achieved time profile $\mathscr{E}_\alpha(\mathbf{r}_0 t)$ may be used in (5-88) and the light scattering calculated exactly. In addition to the exponentially decaying pulse (Friedman and Hochstrasser, 1974b), the literature already contains examples of explicit light-scattering calculations using squared Lorentzian pulses (Berg *et al.*, 1974) and flat-top pulses having exponential rises and decay (Mukamel and Jortner, 1975). A type of pulse that is often achievable in the laboratory is one that has a Gaussian time profile. For example, the output pulses of a transform-limited solid-state mode-locked laser are very nearly Gaussian. It is useful to understand light scattering using such pulses, so instead of Eq. (5-93), consider the positive frequency part of the electric field to correspond to the Gaussian pulse:

$$\mathscr{E}_\alpha(\mathbf{r}_0 t) = \mathscr{E}_0 e^{-i\omega_p t} e^{-(1/8)\Gamma_p^2[(x_0/c)-t]^2} \tag{5-97}$$

In Eq. (5-97) the propagation is specified along the x direction in a medium with no dispersion and Γ_p is the full width at $1/e$ of the maximum in the frequency spectrum. Substituting Eq. (5-97) into (5-88) and taking the square modulus of the only term remaining as $t \to \infty$,

$$P_{\omega_k}(\infty) = |\mathscr{E}_0 \cdot \boldsymbol{\mu}_{ig}|^2 |\boldsymbol{\mu}_{ji} \cdot \langle \mathbf{k}|E_s^{(-)}|\text{Vac}_s\rangle|^2$$
$$\times [(\omega_j + \omega_k - \omega_i)^2 + \tfrac{1}{4}\Gamma_i^2]^{-1} \left| \int_0^\infty dt' e^{i(\omega_k + \omega_j - \omega_p)t'} e^{-(1/8)\Gamma_p^2(x_0/c)-t']^2} \right|^2 \tag{5-98}$$

Now if x_0 is chosen so large that at $t' = 0$ there is essentially no field, at the molecule the lower limit may be extended to $-\infty$ in the integral in Eq. (5-98) and the integration analytically performed. The line shape is

$$P_{\omega_k}(\infty) = (8\pi/\Gamma_p^2)|\boldsymbol{\varepsilon}_0 \cdot \boldsymbol{\mu}_{ig}|^2 |\boldsymbol{\mu}_{ij} \cdot \langle \mathbf{k}|E_s^{(-)}|\text{Vac}_s\rangle|^2$$
$$\times e^{-(\omega_p - \omega_j - \omega_k)^2/\Gamma_p^2}[(\omega_j + \omega_k - \omega_i)^2 + \tfrac{1}{4}\Gamma_i^2]^{-1} \tag{5-99}$$

which is the product of a Lorentzian centered on the molecular resonance and a Gaussian centered on the exciting light frequency.

5.4.4. Scattering from a Weak Stationary Light Beam

Spectral lines are usually measured under steady-state excitation, and for this reason the line shape for radiation scattered at $\approx \omega_{ij}$ when a molecule is put in a stationary field is included here. Two types of commonly used light sources both have the same first-order correlation function. These are a pressure-broadened chaotic source (Loudon, 1973) and a random-phase-modulated coherent-state model for a laser (Glauber, 1964). For either of these sources,

$$G^{(1)}(r_0 t_2; r_0 t_1) = G_0 e^{i\omega_0(t_2 - t_1) - \zeta|t_2 - t_1|} \tag{5-100}$$

ζ is expected to be on the order of the collision frequency in a pressure-broadened chaotic source. For the laser output field case, ζ represents a phase diffusion constant related to the frequency fluctuations of the laser oscillator. The number of photons scattered by time t into a particular mode $\mathbf{k}$ is readily found to be

$$\langle a_k^\dagger(t) a_k(t) \rangle \approx \hbar^{-2} \int_0^t dt_1 \int_0^t dt_2 |\mu_{ig}|^2 G^{(1)}(r_0 t_2; r_0 t_1)$$

$$\times \langle \mathbf{k}; j | e^{-iH_1(t-t_1)} | i \text{ Vac} \rangle \langle j \text{ Vac} | e^{iH_1(t-t_2)} | \mathbf{k}; j \rangle \tag{5-101}$$

This expression yields the scattering in terms of the first-order correlation function for the field. Thus, it is seen that only those properties of the light source that determine $G^{(1)}(r_0 t_2; r_0 t_1)$ can influence the nature of the resonant or near-resonant Raman scattering. Equation (5-101) is quite general, and it is readily integrated for light of the form (5-100). At times long after the beam (5-100) has been turned on, the mean number of photons of wave vector $\mathbf{k}$ in the field is

$$\langle a_k^\dagger(t) a_k(t) \rangle \approx \frac{\omega_k}{2\varepsilon_0 L^3} |\boldsymbol{\mu}_{ji} \cdot \hat{e}_k|^2 |\mu_{i0}|^2 G_0 \zeta$$

$$\times [(\omega_k + \omega_j - \omega_i)^2 + \tfrac{1}{4}\Gamma_i^2]^{-1}$$

$$\times [(\omega_k + \omega_j - \omega_0)^2 + \zeta^2]^{-1} t + \text{const.}; t \text{ long} \tag{5-102}$$

Equation (5-102) indicates that the average number of photons increases linearly in time when the time is long. The quantity that is measured with a monochromator and a photomultiplier is, however,

$$\frac{d}{dt} \langle a_k^\dagger(t) a_k(t) \rangle = \text{const.} [(\omega_k + \omega_j - \omega_i)^2 + \tfrac{1}{4}\Gamma_i^2]^{-1}$$

$$\times [(\omega_k + \omega_j - \omega_p)^2 + \zeta^2]^{-1} \frac{\omega_k}{L^3} \tag{5-103}$$

and therefore apart from constants, the spectral distribution of the scattered light is the same as would be obtained were the scattering caused by a coherent pulse having spectral width ζ.

The results of previous sections have demonstrated that the formal theory of scattering is not needed to understand the nature of light scattering by molecules in weak fields. In fact, there is a danger in the approach described in Section 5.1, since it requires wavepackets, designed for non-relativistic particles, to describe the ultrarelativistic photons (Berestetskii *et al.*, 1971). In this section such subtle points were avoided, and improper questions were circumvented by using representations for the light field that have definite physical meaning in quantum optics experiments. The commonly used one-photon wavepacket concept is thereby seen to actually restrict the physical situation to light pulses which are first-order-coherent (Titulaer and Glauber, 1966). Nevertheless, the form of the scattering equations is found to be the same in both cases. It is just that measurable properties of the light field appear in the expressions developed in this section.

The relations for scattering deduced in this and previous sections give the result for a single three-level system, such as would be encountered for a large number of molecules in a molecular beam. In many of the more common experimental situations, there is a fixed distribution of excitation energies for the whole duration of the experiment. Some examples are a distribution of Doppler shifts in a low-pressure gas and a distribution of environments in a rigid medium. We will consider in Section 5.5 effects resulting from fluctuations in the transition energies, but the effect of static inhomogeneous line broadening is readily incorporated to the theory developed here. If the distribution of transition frequencies about ω is $J(\omega_{ig} - \omega)$, the probability of scattering onto the state $|j\rangle$, which has a sharply defined energy, is, from Eq. (5-88),

$$\int d\omega_{ig} J(\omega_{ig} - \omega) \left| \int_0^t dt' \langle \mathbf{k}; j | e^{-iH_1(t-t')} | i; \text{Vac}_s \rangle \boldsymbol{\mu}_{ig} \cdot \boldsymbol{\mathscr{E}}_\alpha(r_0 t) \right|^2 \qquad (5\text{-}104)$$

For example, if the effect is from Doppler broadening, $J(\omega_{ig} - \omega)$ is a Gaussian. In that case there are two consequences of Eq. (5-104) that are qualitatively different from the molecular beam experiment. First, when $\mathscr{E}_\alpha$ is a broad coherent pulse, the scattering width is no longer restricted to the maximum width Γ_i. Second, the coherent development of the scattering (see Figure 5-2) is different for each different transition energy, and Eq. (5-104) implies a smoothing out of the time evolution of the scattering from that shown in Figure 5-2.

In summary, it may be concluded that previous calculations of resonance light scattering that used a hypothetical one-photon packet having frequency width yield results that are correct in form for any light source. The restriction on the validity of the result is the well-known weak-signal limit where the interaction between the molecule and the incident photons is less than the resonant linewidth. The real experiment that most accurately mimics the imagined one-photon packet concept is one that uses a coherent pulse such as obtained from a mode-locked laser, the appropriate frequency-width parameter being the inverse time of the pulse. The first-order correlation properties of such a pulse depend only on its spectral width. Such is the case for any light source, and the scattering depends only on the spectral-width parameter for stationary sources as well. Thus, the physical interpretation for the parameter describing the light field in the denominator of all scattering formulas is simply the spectral width of the light.

5.5. Effects of Intermolecular Interactions on Luminescence

5.5.1. Resonance Scattering (Raman Fluorescence) in the Presence of Fluctuations

The discussion of light scattering has so far been restricted to the electrodynamics of isolated molecules subject to fluctuations only in the electromagnetic field. In practice many experiments are done in dense media and at nonzero temperatures. Therefore, fluctuations are expected to occur because of uncorrelated motions in the environment of a molecule which will influence the scattering of light. This is no great surprise, since it has been known since the early 1900s that fluctuations in density, for example, are important in understanding Rayleigh scattering. Light scattering experiments have been frequently used to study the statistical processes occurring in condensed phases. A complete understanding of the resonant scattering process requires consideration of the dynamics of the statistical perturbations experienced by a molecule during the scattering process.

A fundamental difference between resonant and nonresonant scattering is that in the resonant case the interaction of the resonant state and the environment needs to be considered, in contrast to the off-resonance case, in which a large number of intermediate states participate equivalently and the interactions between the final states and the medium are usually the most important. With these ideas in mind, a simple example will be considered below, where it is assumed that only the intermediate state experiences the random perturbations of the environment.

In most systems of practical interest, the energy levels are not rigorously fixed but are changing due to the occurrence of interactions between the molecules or between the molecules and their environment and external fields. At any instant the system consists of a distribution of transition energies. In addition a molecule in the ensemble at finite temperature will not remain in a given molecular eigenstate for appreciable times, owing to the exchange of thermal excitations between molecules. In this manner the transition frequency undergoes a fluctuation that depends on the anharmonicities in the thermally populated modes (Harris *et al.*, 1977). Because of the nature of the near-resonance scattering formulas described above, it is evident that if the energy levels were allowed to fluctuate or to undergo exchange, the spectra would be changed. When a system is irradiated with a light pulse, there is a characteristic time during which a molecule may be considered to be interacting with the light pulse. Clearly, the response of the system in regard to its resonance of near-resonance light scattering will be dependent on whether there can be fluctuations in the transition energy during the interaction with the light. In optical spectra the situation is qualitatively different from that normally occurring in magnetic resonance, since the time-varying fields from the medium are normally very slow compared with the optical frequency, whereas these time-varying fields will cause magnetic resonance transitions. Nevertheless, allowed transitions have lifetimes in the range 10^{-6}–10^{-9} s, and both laboratory light pulses and fluctuations can be encountered that incorporate this range of times. An extreme case to envisage is where the fluctuations are very rapid compared with the characteristic interaction times leading to scattering. In this case the overall effect of the fluctuations is to cause an essentially homogeneous sample to be exposed to the light field, since all the molecules in the system can be brought into equivalent near-resonant situations many times during the application of light pulse. The scattering must therefore be expected to be similar to the resonance Raman limit mentioned above except that the fluctuations, as well as the radiative and nonradiative decays, contribute to the homogeneous level width in the ensemble. The scattered light linewidth should not contain molecular excited-state dynamical information in this case, but is determined by the incident light field and fluctuations or decay in the final state of the two-step process. This is why we prefer the terminology "Raman scattering" for this limit. However, light may also be emitted at frequencies other than that of the incident light in the presence of these fluctuations because of the formation of an excited-state population. A certain amount of light can now be emitted at the molecular fluorescence frequency having a spectral width determined not only by the excited-state decay rates but also the transition energy fluctuations. The occurrence of this light is not dependent on the light pulse being present and may occur long after the exciting pulse has gone by if the decay time of the excited state is slow enough. It is natural to call this radiation "fluorescence." According to the present qualitative discussion, the fluorescence in the weak-signal limit is distinguished from Raman by virtue of its occurrence both during and after the light pulse, its spectral width being determined by the decay of the excited-state population and the transition energy fluctuations, and the peak frequency being independent of the excitation frequency.

In the limit where the fluctuations are very slow, it is as if for the whole duration of the interaction with the light, the sample consists of a distribution of fixed energies. In this case illumination with nearly monochromatic light should produce a single narrow Raman scattering, with the contribution of each molecule in the system being determined by how close the incident frequency is to each molecular transition. Such a situation could arise in a relatively low pressure gas because of the velocity distribution or in a matrix at low temperatures. In the latter situation the nonradiative relaxation of the levels would most likely cause the incoherent population of other levels, resulting in emission at many other frequencies besides the Raman frequency under discussion here.

In the treatment of the two-level system given above, the solutions to the coupled equations (5-30) and (5-31) were readily obtained in the weak-signal limit by neglecting diagonal elements of the time-dependent perturbation in the semiclassical model. These diagonal elements are contributions to the energy shifts of the levels and are needed in a first-principles discussion of the effect of fluctuations. The time-dependent interaction can be considered to consist of two parts, one the dipole interaction that oscillates near the transition frequency, and another a function representing a much slower time-varying potential. Only the latter is considered to have diagonal elements in the basis of molecular wavefunctions.

A theory that illustrates the important features of light scattering from a molecule interacting in a random manner with its environment may be based on solving the quantum mechanics of both the molecule and invironment or heat bath, for example by calculating the interaction of a scattering center in a crystal with phonons in that crystal or by using master equation techniques to enable a treatment based on the principles of quantum statistical mechanics. A second approach is more phenomenological and consists of assuming that the interaction between the molecule and its environment may be described by a random function of time—in other words, by using a stochastic Hamiltonian. The second approach will be outlined because it can be developed in a straightforward manner from the previous considerations.

The effect of fluctuations on the absorption-line shape in a magnetic resonance experiment was considered some time ago (Anderson, 1954; Kubo, 1954). The exchange and motional narrowing theories of Anderson and Kubo are likely to be applicable to optical absorption as well. However, for the case of fluorescence or Raman, the fluctuations occur in the intermediate state of the scattering process and the consequences have not yet been fully explored.

5.5.2. Random Modulation in Resonance Raman Scattering

The theory based on the quantized electromagnetic field does not lend itself simply to a calculation of light scattering from a system described by a stochastic Hamiltonian because of the explicit time dependence. The semiclassical treatment will therefore be used in this discussion. As a further simplification, a completely monochromatic exciting light source is assumed so as to have stationary excitation. Since the fluctuations are occurring because of the bath interaction, it is assumed that there are no frequency components near the optical frequencies such that no transitions are induced by this radiation source. The statistical interaction contributes to the dynamical equations matrix elements diagonal in $|i\rangle$ and $|g\rangle$. Under these conditions the coupled equations (5-30) and (5-31) become

$$\dot{a}_g(t) = i\gamma_{gi} e^{-i(\omega_i - \omega)t} e^{-\Gamma_i t/2} a_i(t) + i f_g(t) a_g(t) \tag{5-105}$$

$$\dot{a}_i(t) = i\gamma_{ig} e^{-(\omega_i - \omega)t} e^{\Gamma_i t/2} a_g(t) + i f_i(t) a_i(t) \tag{5-106}$$

where $f(t)$ is the random potential-energy fluctuation. In order to calculate the correspondence-principle dipole moment (5-45) in the weak-signal limit, achieved by setting $a_g(t) = 1$ for all time in (5-106), it is required to solve the first-order inhomogeneous differential equation:

$$\dot{a}_i(t) - i f(t) a_i(t) = i\gamma_{ig} e^{i(\omega_i - \omega)t} e^{\Gamma_i t/2} \tag{5-107}$$

where only excited-state fluctuations are explicitly considered. Equation (5-107) may be solved in a form well suited for the subsequent use of statistics by the method of Green's functions, which will be outlined here for the sake of completeness. The Green's function corresponding to Eq. (5-107) is a solution of

$$\frac{dG(t, t')}{dt} - i f(t) G(t, t') = \delta(t - t') \tag{5-108}$$

so that a particular solution of (5-107) is

$$a_i^P(t) = i \int_{-\infty}^{\infty} \gamma_{ig} G(t, t') e^{i(\omega_i - \omega)t} e^{\Gamma_i t/2} \, dt' \tag{5-109}$$

For $t < 0$, $\gamma_{ig} = 0$ and the general solution of Eq. (5-107), if c is a constant of integration, is

$$a_i(t) = c \exp\left[-i \int_0^t f(\tau)\right] + i\gamma_{ig} \int_0^{\infty} dt' G(t, t') e^{i(\omega_i - \omega_p)t'} e^{\Gamma_i t'/2} \tag{5-110}$$

An appropriate Green's function is found by the usual methods to be (Merzbacher, 1970)

$$G(t, t') = \theta(t - t') \exp\left\{-i \int_{t'}^t f(\tau) \, d\tau\right\} \tag{5-111}$$

where θ is the unit step function. Since $a_i(t) = 0$ for $t < 0$, c must be taken as zero in Eq. (5-110), and therefore the amplitude for the excited state is given by

$$a_i(t) = i\gamma_{ig} \int_0^t dt' \exp\left[-i \int_{t'}^t f(\tau) \, d\tau\right] e^{i(\omega_i - \omega_p)t'} e^{\Gamma_i t'/2} \tag{5-112}$$

The time-dependent dipole moment is then obtained from Eq. (5-45):

$$\mathbf{M}_j(t) = i\gamma_{ig}\,\boldsymbol{\mu}_{ije}^{-\Gamma_i t/2}e^{-i\omega_{ji}t}\int_0^t dt'\exp\left[-i\int_{t'}^t f(\tau)\,d\tau\right]e^{i(\omega_i-\omega_p)t'}e^{\Gamma_i t'/2}+\text{c.c.}$$

$$(5\text{-}113)$$

Equation (5-113) can be used to calculate the spectrum of the scattered light. The classical electric field associated with the oscillating dipole (5-113) is given by

$$\mathbf{E}_{cl}(\mathbf{r}t)=\text{const. (geometry term)}\cdot\ddot{\mathbf{M}}_j(t) \qquad (5\text{-}114)$$

where $\mathbf{r}$ is the position of the detector. To a good approximation the field may be written

$$\mathbf{E}_{cl}(\mathbf{r}t)\approx-\text{const. (geometry term)}\cdot\omega^2\mathbf{M}_j(t) \qquad (5\text{-}115)$$

Instead of using Eq. (5-47) to calculate the energy distribution of the scattered light, it is more convenient in this case to calculate the autocorrelation function

$$\langle\mathbf{E}_{cl}(\mathbf{r}_0,t+\tau)\mathbf{E}_{cl}(\mathbf{r}_0,t)\rangle \qquad (5\text{-}116)$$

whose Fourier transform with respect to τ represents the spectrum of the scattered light. The brackets in Eq. (5-116) denote ensemble average over the f's in (5-113). The prediction of the spectrum is then accomplished by calculating the two-time dipole correlation function

$$\langle M_j(t+\tau)M_j(t)\rangle \qquad (5\text{-}117)$$

This correlation function is calculated by substituting Eq. (5-113) for $M_j(t)$, after which the average over f is taken. It is found that the random perturbation enters the integrand as the quantity

$$\left\langle\exp\left[-i\int_{t''}^{t+\tau}d\tau f(\tau)+i\int_{t'}^t d\tau f(\tau)\right]\right\rangle \qquad (5\text{-}118)$$

This is a characteristic functional for the process described by $f(t)$ so that $f(t)$ must be specified in order to make further progress. One of the simplest choices for $f(t)$ is a Gaussian Markov process, so that at any instant of time f has a Gaussian probability density and the second-order joint probability density specifies the process completely. Under these conditions only the autocorrelation function for $f(t)$ is required: for stationary $f(t)$,

$$\langle f(t_2)f(t_1)\rangle=\varphi(t_2-t_1) \qquad (5\text{-}119)$$

As a further simplification it will be assumed that

$$\varphi(t_2-t_1)=\Delta^2\tau_c(t_2-t_1) \qquad (5\text{-}120)$$

which is the motional narrowing limit of magnetic resonance, with Δ the rms average of f, and τ_c the correlation time for f. A criterion for the validity of Eq. (5-120) is

$$\tau_c \Delta \ll 1 \tag{5-121}$$

With these assumptions (5-118) becomes simply

$$e^{-\Delta^2 \tau_c |t' - t''|} e^{-\Delta^2 \tau_c |\tau|} \tag{5-122}$$

The dipole correlation function (5-117) may now be obtained explicitly after some fairly lengthy algebraic manipulation, and it is found that at large t the function depends only upon τ. In order that the integrals converge, the following requirement must be met:

$$\Delta^2 \tau_c < \Gamma_i / 2 \tag{5-123}$$

The final result for the spectrum, when the exciting light is exactly on resonance, is

$$\int_{-\infty}^{\infty} e^{-i\omega_s \tau} \langle M_j(t + \tau) M_j(t) \rangle \, d\tau$$

$$\times (\text{const.}) \left[2\pi \frac{\frac{1}{2}\Gamma_i - \Delta^2 \tau_c}{\frac{1}{2}\Gamma_i + \Delta^2 \tau_c} \delta(\omega_s - \omega_i + \omega_j) \right.$$

$$\left. + \frac{\Delta^2 \tau_c}{(\omega_i - \omega_s - \omega_j)^2 + (\frac{1}{2}\Gamma_i + \Delta^2 \tau_c)^2} \right] \tag{5-124}$$

The first term in Eq. (5-124) is the Raman scattering. The spectral width of this term is represented by a δ function, because we assumed a stationary monochromatic source. The Raman term therefore describes scattering at ω_{ji}, the case of resonance, with a spectral width determined by the light source. The emission in this case is necessarily narrower than the homogeneous linewidth and corresponds to the Heitler effect limit $\Gamma_p \ll \Gamma_i$ in Eq. (5-40). The second term depends for its existence on a finite fluctuation amplitude. The spectral width of the part caused by the fluctuations is given by $(\Gamma_i/2 + \Delta^2 \tau_c)$ and may therefore become as large as the homogeneous width. This result, although for a simplified model, demonstrates clearly the existence of two types of resonance emission, which, although at the same center frequency, have quite different physical source terms and are readily distinguishable experimentally. The effect described here is motional narrowing of resonance scattering in the presence of fluctuations. The fraction of the total emission that is Raman scattering is readily obtained in terms of the dimensionless parameter $\alpha = \Gamma/2\Delta^2 \tau_c$ as

$$\frac{I_R}{I_S} = \frac{2(\alpha - 1)}{2\alpha - 1} \tag{5-125}$$

Theoretical developments concerned with the principles of the calculation outlined above represent a lively on-going research area of chemical physics. One of the first studies (Huber, 1968) showed how to cast the Kramers–Heisenberg formula for the photon counting rate in the form of the Fourier transforms of the polarizability correlation function. From this starting point it is possible to use the same standard stochastic methods that are used in magnetic resonance to generate results similar to (5-124) (the narrowing limit). The model leading to (5-124) contains no irreversible decay other than spontaneous emission, but Huber's (1968) approach can include phenomenological T_2 processes and a range of modulation cases to yield expressions that contain an explicit temperature dependence to account for the transitions caused by the heat bath. Collisional effects in gases have also been investigated by Huber (1969). The spectrum of the light scattered by such a collisionally perturbed gas is found to consist of two terms similar to those in Eq. (5-124). In that case, when Poisson rather than Gaussian processes are assumed, the result is similar to Eq. (5-124) with $\Delta^2\tau_c$ replaced by a collision width parameter, and an extra frequency shift in the denominator of the second term.

The fundamental assumptions involved in the applications of stochastic models to resonance light scattering in real physical situations have been discussed by Toyozawa (1976). Using a density matrix formalism, he has verified the validity of the Kramers–Heisenberg formula as used by Huber (1968). Toyozawa (1975) also points out the fact that shifts in molecular energy levels due to interaction with the heat bath cannot be taken into account in a stochastic model. The problem of resonance light scattering by an impurity center interacting with phonons through the quadratic electron–phonon coupling was treated by Kotani and Toyozawa (1976). Using diagram techniques, they calculate the lineshape of scattered light and find a sharp Raman-like emission and a broader emission corresponding to a luminescence. In recent work (Kubo *et al.*, 1975) methods of dealing with transitions in the three-level system when the intermediate state in the scattering is allowed to fluctuate have been considered. The theory can describe the change from fluorescence to Raman scattering as a change in the structure of the transition rate for scattered photons when there is a finite correlation time for the fluctuations. Results similar to those discussed here have been presented recently by Kushida (1976), who has treated the resonant scattering in a three-level system, including dephasing relaxation in the intermediate state using a transition-rate approach (Feld and Javan, 1969). This work stresses the different natures of the incoherently produced population n_i (say) and the average population N_i for the state $|i\rangle$ in a two-level system. In the weak-signal limit, the average population of the intermediate state for monochromatic light is obtained from Eq. (5-32) as

$$N_i = n_g|a_i(\infty)|^2 = \gamma_{gi}^2 n_g/[(\omega - \omega_{ig})^2 + \Gamma_i^2/4] \qquad (5\text{-}126)$$

where n_g is the ground-state population. Under the circumstances that

(5-126) pertains, the value of n_i may be zero. The radiation emitted by the system, on the other hand, is determined by n_g and n_i. The scattering involves the whole coherent process $|g\rangle \rightarrow |i\rangle \rightarrow |j\rangle$ and depends on n_g. The fluorescence is described by the process $|i\rangle \rightarrow |j\rangle$ and depends on the incoherently produced population n_i. The existence of n_i *depends on the excitation of* $|i\rangle$ by incoherent processes. Energy fluctuations or exchanges such as described above or incoherent spontaneous processes in the medium such as occur in vibrational relaxation of a multilevel system would serve to create n_i. These points are discussed further in the next section, after some practical aspects of luminescence have been surveyed.

5.5.3. Classical Character of Fluorescence

The term "fluorescence" has been used in many different ways in the past, but at least in spectroscopy and solid-state physics there is some general agreement about its phenomenological meaning. The vast majority of luminescence experiments have been carried out with rather indiscriminate excitation methods using spectrally wide excitation sources—that is to say, sources having low first-order coherence. In the event that luminescence experiments were carried out with relatively sharp line sources, the excitation was most often into levels that could be rapidly vibrationally or electronically relaxed into incoherent excited-state populations. The very low pressure mercury discharge lamp generates a line at 2537 Å that is Doppler-broadened to a width of ca. 0.06 cm^{-1}, assuming that the pressure is low enough for collisions to be avoided during the excited-state lifetime of the atoms in the lamp. While the Hg/2537 source matches many lasers, such as N_2 pumped dye lasers in coherence properties, the available energy density is many orders of magnitude less than with the laser. The spectral width of the low-pressure mercury lamp is considerably larger than homogeneous widths of spectral lines due to radiative effects, but considerably less than pressure-broadened or fluctuation-broadened spectra in the condensed phase at normal temperatures. However, in the absence of tunability to optimize the strength of scattering signals in relation to background radiation arising from light absorption, there has been little chance of exposing by means of conventional light sources many of the important consequences of the near-resonance light scattering theories presented here. Thus, the contradictions in the usage of terms such as absorption, fluorescence, scattering, resonance Raman, and resonance fluorescence have really only become a significant problem in the last few years. Rebane and Saari (1976) have considered numerous theoretical and experimental aspects of the relationship between fluorescence, Raman, and hot-luminescence in solids. Although Heitler (1954) discussed the problem of scattering of photons from homogeneously broadened transitions, these early theoretical ideas do not seem to lead to obvious or unique descriptions or language to describe radiative processes in moderately-high-pressure gases or condensed phases. Experimentalists, on the other hand, have used the terms "fluorescence" or "phosphorescence" (luminescence from a metastable state) to describe the following general behavior:

1. Fluorescence involves a characteristic decay, usually exponential if only a single molecular state is excited, that is determined by molecular parameters such as radiationless transitions and a vibrational average of the transition dipole moment squared. The decay results when a system is excited with a light pulse whose time width τ_p is short compared with the observed decay time Γ_i^{-1}. Alternatively, the steady excitation may be interrupted in a time τ_p.

2. Fluorescence may appear to originate in excited levels that are different from those responsible for the interaction of the system with the incident light. This is understood to occur by absorption and incoherent relaxation.

3. Fluorescence from a system of effectively homogeneously broadened levels displays a linewidth characteristic of the interactions causing the homogeneous broadening. If there is inhomogeneous broadening that persists for times longer than Γ_i^{-1}, fluorescence will have all the characteristics described here except that they will be displayed by only a part of the distribution if the light source is sufficiently discriminating.

4. Fluorescence occurs between two levels when the higher-energy one is populated, and the decay of the fluorescence corresponds to the decay of this population. Thus, the fluorescence is assumed to be described by the diagonal elements of a density matrix, and normal fluorescent systems are treated using ordinary kinetics with the rate of luminescence of the system calculable from a rate constant $k = \Gamma_i$.

5. Fluorescent light normally is not considered to be correlated with the excitation radiation, nor is the phase relationship between the two states coupled by the incident field maintained during the light emission. In other words, an incoherent population n_i can be used to calculate the fluorescence rate as $\Gamma_i n_i$.

5.5.4. Absorption and Scattering

The operational definitions of fluorescence described above are clearly distinguished from Raman scattering and this is perhaps most easily understood by considering more carefully the process of light absorption. Light incident on a sample of three-level systems having the properties described previously (i.e., $\mu_{gi} \neq 0$, $\mu_{gj} = 0$ and the light frequency far from ω_{ij}) establishes in the medium a macroscopic polarization $P(t)$ given by

$$P(t) = \mu_{gi}(\rho_{gi} + \rho_{ig})$$

where ρ_{0i} and ρ_{i0} are elements of a 2×2 density matrix. Consider first the case where the light frequency is far from resonance (but still not near ω_{if}). The polarization at a point in the medium oscillates at the light frequency and there is a well-defined (phase) relationship between the amplitudes at every point in the medium. The polarization in this instance is the source of an electromagnetic wave traveling through the sample with an appropriately reduced velocity and longer wavelength than would obtain in vacuum. At the same time, the average population in the excited state is close to zero. A small amount of Raman scattering will attenuate the beam to some extent. The Raman is spontaneous emission that diminishes the coherent polarization, and the scattering is said to be coherent. The situation may become different when the light frequency becomes close to ω_{0i}. In a dilute gas or beam of molecules, where the collision periods are very long compared with the radiative lifetime of the excited state, the system may have a significant *average* population of the excited state even though the polarization is at all times maximally coherent. There will again be coherent scattering (in the three-level system) of some of the radiation into directions other than the direction of propagation of the light, thereby more substantially diminishing the light transmitted by the system. There is no n_i in this case, in the sense that there was no mechanism of producing an incoherent population. Nevertheless, the coherent average excited-state population decays away at the radiative rate. Increasing the pressure will result in an increase in n_i because the collisions will irreversibly destroy the coherence in the system. In a solid or high-pressure gas, the linewidth will often be mainly determined by dephasing collisions (i.e., T_2 processes). The polarization in the system depends on the existence of the off-diagonal elements of the density matrix, so by introducing energy fluctuations the perfect spatial coherence of the polarization may be destroyed even as the incident light field is allowed to vary over a range considerably broader than T_1^{-1}. The absorbed energy resulting from dephasing collisions or interactions is ultimately scattered into 4π rad in a process that we term fluorescence. In this case, in contrast to the case where T_2 is infinite, it is reasonable to assert that a change in population of states has occurred as a result of the dephasing and that the fluorescence intensity is proportional to the number of incoherently produced excited molecules. In the same experiment, there may also be some coherent polarization, in which case coherent scattering, or Raman scattering, will occur as well. Obviously, in the general case these two processes are not independent. It is tempting to assert that light is not absorbed unless there is some T_2 process occurring. On the other hand, the light beam is attenuated because of scattering even in the absence of dephasing interactions. It seems least confusing to retain the term "absorption" to indicate any attenuation of the light, regardless of whether or not an incoherent excited-state population is produced. According to Eq. (5-37), the scattered light intensity at any time is proportional to the probability that the intermediate state is reached by that time. Similarly, the total rate of nonradiative spontaneous processes originating from $|i\rangle$ is given by

$$I_{nr}(t) = \Gamma_{in}P_i(t) \qquad (5\text{-}127)$$

Although they were not considered explicitly in the present discussion, nonradiative processes—such as internal conversion, intersystem crossing, chemical reactions—also destroy the polarization and cause attenuation of the incident beam.

The foregoing concepts will be rather easily studied using tunable lasers. In one idealized approach, a boxcar integrator could be incorporated into a conventional Raman experiment such that the spectrum of the scattered light can be examined for various gated time intervals. Figure 5-8 shows one essential character of the results expected for a case where the line broadening due to dephasing is very large compared with the natural radiative width and the laser spectral width. If we suppose that the radiative lifetime (τ) is 10^{-6} s, the spectral linewidth 10 cm^{-1}, and that the laser consists of a 2 ns pulse of linewidth 0.001 cm^{-1} at ω_L, the example would refer to a type of impurity state in a solid or low-temperature liquid in which relaxation processes in the final state are negligible. The fraction of the total fluorescence (time-integrated) that is observed during 2 ns is 2×10^{-3}, so if the boxcar gate is set at 0–2 ns, we would expect to observe this small fraction of the fluorescence having linewidth 10 cm^{-1} and a large fraction of the Raman scattering, which in the absence of final-state relaxation would have a spectral width determined by the convolution of the light and radiative bandwidths. If the boxcar is gated at t to $t+2$ ns, where $\tau \gg t > 2$ ns, we should observe a fraction $2 \times 10^{-3} e^{-t/\tau}$ of the fluorescence. However, with the gate open at the later time there should be no Raman scattering, and all the emission will have a characteristic width of ca. 10 cm^{-1} and spectral positions ω_{0i} and ω_{ij} regardless of the laser frequency. On the contrary, for the early time gate, the spectrum will display, in addition to the wide line, a sharp band at ω_{pj}.

It is easy to see that in a steady-state experiment the relative amounts of Raman scattering and fluorescence integrated over all time, such as is accomplished in a conventional Raman spectrometer, is determined by the competition between dephasing and the natural radiative rate. The rate of dephasing $2T_2^{-1}$ determines the rate of production of excited-state incoherent population, which then must ultimately fluoresce, assuming that other

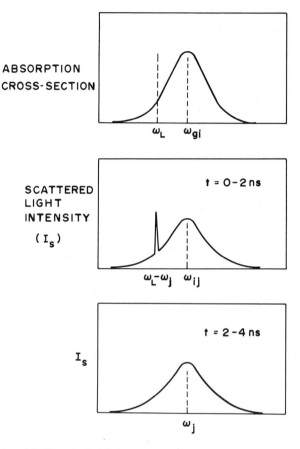

Figure 5-8. Figure 5-8. Hypothetical light-scattering experiment. The upper box shows the laser (ω_L) in relation to the absorption (at ω_{gi}), whose width is partly determined by dephasing effects. The middle and lower boxes show the emission at ω_{ij} gated during the pulse and after the pulse.

dissipative processes cannot occur (although to include them is straightforward). The rate of Raman scattering is determined by Γ_i (τ^{-1} in the example above). Thus, the ratio of the fluorescence to the Raman quantum yield is $2/T_2\Gamma_i$. This describes the fact that the Raman process will only occur to the extent that the interaction between the molecule and the light field is allowed to persist in an uninterrupted fashion for a sufficiently long time that natural radiative decay can occur. In the example of the previous subsection the quantum yield ratio (fluorescence/Raman) in the continuous-wave (cw) experiment would be 1.8×10^6. The ratio of incoherently produced to the average population also provides a measure of how much fluorescence will occur.

5.5.5. Spectroscopic Selection Rules for Resonance Raman, Fluorescence, and Phosphorescence

The previous discussion should make it clear that there are no differences in the spectroscopic selection rules or angular distributions of scattered radiation for fluorescence and resonance Raman in the model three-level system. There are, however, substantial differences in the time dependence and in the spectral distribution of the scattered light for the two kinds of scattering in the presence of collisions or a heat bath. If for a real system in which relaxation can occur there is incoherent feeding of excited states resulting from relaxation processes during excitation of the system, then of course the fluorescence may have quite different selection rules and angular distribution from the Raman. The fluorescence is in this case *relaxed fluorescence*, so called because it does not originate from one of the levels directly involved in the coupling of the molecule and the radiation field. The relaxed fluorescence will not display a coherent component (such as shown in Figure 5-8) during any time interval. In a real system containing many levels that can be fed incoherently by spontaneous decay of the intermediate state of the scattering process, relaxed fluorescence will arise in principle from all these levels. The spectral width of the relaxed fluorescence is determined by contributions from the fluctuations and their T_1 decay into either other levels or into fluorescence.

As was discussed in Section 5.2.4, there are certain differences in spectroscopic selection rules occurring during and after pulse excitation in a model five-level system. These are caused by time-dependent interferences among the components of the nonstationary state and have to do with the different time dependences of resonance versus off-resonance scattering. Studies of Raman excitation profiles should provide an easy way to expose such effects, which nevertheless show up differences between fluorescence and Raman. Contributions to the scattering from mainly distant resonances will often display strong interference effects in the Raman profile of a particular resonant state. Another possibility is that certain Raman lines will simply not be enhanced at all by passing the excitation frequency through a particular resonance.

All the same conclusions drawn in this section in relation to fluorescence are valid in principle for phosphorescence from triplet states. The additional feature is that when the resonant intermediate state is a triplet state, the three-level model may not always be useful. In actuality there are three nearly degenerate intermediate states, and if more than one of them can be coupled with the ground state by a radiation field, the resulting scattering phenomena may be more complex than previously described.

Three typical situations are depicted in Figure 5-9 for the case of two nearby radiative levels that can couple to the ground state. The states could

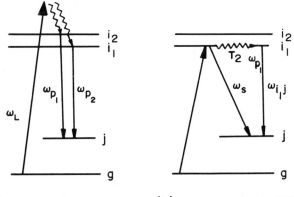

(a) INCOHERENT FEEDING;
PHOSPHORESCENCE

(b) COHERENT SCATTERING
AND PHOSPHORESCENCE

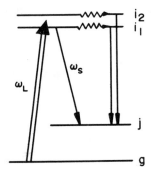

(c) COHERENT PUMPING ·
BEAT SCATTERING

Figure 5-9. Some possible schemes for scattering from multiplets in the condensed phase. In each case two states are shown that can be selectively coupled to the ground state with light (e.g., two components of a molecular triplet state).

be two triplet sublevels of a sufficiently low symmetry molecule, or two fine-structure levels of an atom or molecule in a dilute gas. The various manifestations of coherent pumping of nearby levels of atoms in dilute gases have often been reviewed (Bernheim, 1965; Series, 1970), so only some extra features of a condensed phase example will be dealt with here. It is quite possible to perform experiments at sufficiently low temperature that spin–lattice relaxation between the levels i_1 and i_2 is negligibly slow (van der Waals and de Groot, 1967; Harris and Breiland, Chapter 4). However, fluctuations in the transition energies ω_{gi_1} and ω_{gi_2} occur at finite temperatures and result in the creation of (incoherent) populations of molecules in particular spin sublevels. Figure 5-9a depicts the conventional technique of creating phosphorescent states in the condensed phase, which is excitation of certain levels followed by their incoherent feeding of i_1 and i_2. There is no scattering in the region ω_{ij} in this case. The light may be chosen to be resonant with ω_{gi}, as in Figure 5-9b, while the coupling to i_2 may be avoided by using a particular incident field polarization. Scattering occurs in this case, and the triplet level i_2 is also incoherently populated by the T_2 process, resulting from fluctuations in ω_{gi_2}. Figure 5-9c is the conventional quantum beat arrangement where both triplets are pumped coherently. If there is an inhomogeneous broadening of the levels that exceeds the fine-structure splitting, then coherent pumping can be readily achieved, but a light pulse of width Γ_p larger than the separation of the levels is clearly an appropriate source. The decay of the system in this case should display quantum beats. However, the coherent scattering will be quenched, owing to the fluctuations in ω_{gi_1} and ω_{gi_2}, which will again result in incoherent populations of i_1 and i_2 being formed. Figure 5-9a–c, with the addition of an electromagnetic field at $(\omega_{i_2} - \omega_{i_1})$, will be recognized as the schemes for optically detected magnetic resonance dealt with by Harris and Breiland in Chapter 4. An additional complexity of having nearby levels that each are coupled to spontaneous emitting sources, such as the incoherent feeding case (Figure 5-9a), is that if the overall relaxation process is fast enough, $T_1^{-1} \gg (\omega_{i_2} - \omega_{i_1})$, the levels may be excited coherently (Hochstrasser, 1970). However, such possibilities and their ramifications are outside the scope of this article. It is expected that laser systems will soon be sufficiently advanced that many of the optical experiments relating to coherence loss in the condensed phase scattering will become possible.

5.6. Two-Photon Induced Light Scattering

5.6.1. Two-Photon Processes

In the foregoing part of this chapter we have considered only spontaneous emission for the second step in the molecule–photon interaction. Spontaneous scattering also occurs when $2\omega_p$ is close to a molecular resonance. In that case the system is driven into a nonstationary state by two photons, but otherwise the theory for the resulting spontaneous emission is similar to that given previously. An additional feature in the dynamics of two-photon absorption, or two-photon scattering, arises when the molecular system has a transition close to the frequency of one of the photons. Interesting questions now arise regarding the coherence of the overall two-photon process, and whether there are differences between the resonance-enhanced and off-resonance cases. From the previous sections it will be obvious that answers to these questions will depend on whether the excitation is cw, pulsed, coherent, or chaotic; whether the system is subject to phase interruptions; and whether these interruptions are frequent or infrequent compared with the overall process leading ultimately to spontaneous emission. A brief account of some of these phenomena will be given in this section, starting with non-resonantly enhanced two-photon-induced scattering, followed by a qualitative account of coherent light effects in the resonantly enhanced case.

Some of the common situations are depicted in Figure 5-10 starting with (a), which shows the conventional scheme for two-photon induced fluorescence or scattering. In this case the scattering rate depends on the

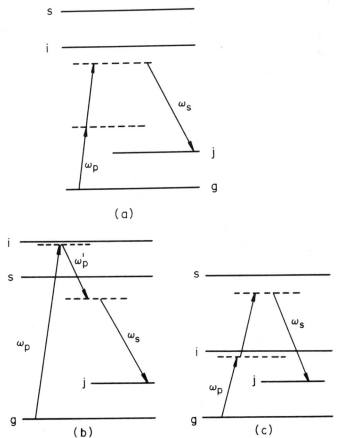

Figure 5-10. Mode four-level systems picturing some typical situations for two-photon induced scattering. (a) Conventional two-photon excited scattering. (b) Resonance-enhanced two-photon excited scattering. (c) Resonance-enhanced two-photon excited scattering.

second-order coherence of the light source. The light frequency is far from resonance, but twice this frequency is close to resonance. Figure 5-10b depicts the resonance-enhanced two-photon scattering. Obviously, the rate of scattering in this case is dependent on the relaxation of the state i. Finally, in Figure 5-10c is depicted the more common type of resonance-enhanced two-photon excited fluorescence. The first step in Figure 5-10c is the usual resonance-enhanced two-photon absorption. This process could also be termed conventional excited-state absorption followed by fluorescence in cases where the dephasing effects are sufficiently large. In a system of isolated noninteracting identical molecules, there is no difference in principle between consecutive and simultaneous absorption of the two photons. The coherence is maintained throughout the whole process, including the spontaneous emission from the nonstationary state near i, even when light pulses are used. It is only in the presence of heterogeneities in the medium that the probability of scattering tends toward being the uncorrelated probability of the separate events absorption (i.e., creation of a population) and Raman scattering in the excited state i'. In many cases of practical interest, it should be possible to describe multiphoton absorption using kinetic equations involving only populations.

5.6.2. Scattering Induced by Two-Photon Excitation: Hyper Raman Scattering

The method of calculation used in Section 5.4 to understand the usual type of resonant scattering process can be readily extended to the case of two-photon induced scattering at least under pulsed conditions. This, then, permits the calculation of the dynamics of the process in a manner free from phenomenological decay constants and complex energy denominators. Resonant or nearly resonant intermediate states then can be incorporated naturally into the theory. Very interesting points having to do with the second-order coherence properties of the exciting light are brought out at the end of this section, but first the case of coherent pulse excitation will be considered, to make clear the dynamics of the two-photon excitation.

Equation (5-86) can be extended to include second-order effects by the addition of the term

$$-\int_0^t dt_1 \int_0^{t_1} dt_2 e^{-iH_1(t-t_1)} \mathscr{E}_\alpha(\mathbf{r}_0 t_1) \cdot \mathbf{\mu} e^{-iH_1(t_1-t_2)} \mathbf{\mu} \cdot \mathscr{E}_\alpha(\mathbf{r}_0 t_2) | g; \text{Vac}_s; \{\alpha_k\}\rangle$$

$$(5\text{-}128)$$

A light pulse such as (5-93) may now be specified, a set of intermediate states $\{l\}$ introduced, and the amplitude for scattering from a final state S onto a state j and a photon obtained:

$$\frac{1}{4}\sum_{l}(\boldsymbol{\epsilon}_0\cdot\boldsymbol{\mu}_{sl})(\boldsymbol{\epsilon}_0\cdot\boldsymbol{\mu}_{lg})\int_0^t dt_1\int_0^{t_1}dt_2\langle j\mathbf{k}|e^{-iH_1(t-t_1)}|s;\text{Vac}_s\rangle$$
$$\times\langle l;\text{Vac}_s|e^{-iH_1(t_1-t_2)}|l;\text{Vac}_s\rangle e^{-i[\omega_p-(1/2)i\gamma](t_1+t_2)}\qquad(5\text{-}129)$$

The matrix elements of the exponential operators in Eq. (5-129) may be calculated using the Green's operator or resolvent (5-89), and the results are essentially the same as those given at the end of Section 5.1.

A convenient way of proceeding, especially when the interest is in the long-time behavior of the amplitude (5-129), is to complete the integral with respect to t_2 and use the convolution theorem from the theory of Laplace transforms; then after a Laplace-transform inversion the amplitude is obtained in the form

$$\frac{1}{8}\sum_{l}(\boldsymbol{\mu}_{sl}\cdot\boldsymbol{\mathscr{E}}_0)(\boldsymbol{\mu}_{lg}\cdot\boldsymbol{\mathscr{E}}_0)\int_c dz\,\frac{\langle j\mathbf{k}|G(z)|s;\text{Vac}\rangle e^{-izt}}{(z-2\omega_p-i\gamma)[z-\omega_p-\omega_l-\frac{1}{2}i(\gamma+\Gamma_l)]}\quad(5\text{-}130)$$

where of course G is defined by Eq. (5-89) and c is the usual contour of Goldberger and Watson (1964). The long-time behavior of Eq. (5-130) may then be extracted by replacing the contour c by c', where c' encircles only poles on the real axis. This amplitude is now very similar in form to those in Section 5.2, which are valid for one-photon excitation. When the two photons have different central frequencies, the formulas are more complicated, but Eq. (5-128) still provides a general starting point for any calculation. A useful yet straightforward application of (5-130) is the calculation of the lineshape of scattered light at frequency $\approx\omega_{sj}$. If only one intermediate state, $|l\rangle$, is responsible for the two-photon process, only one term exists in the summation in (5-130). The matrix element of G appearing in (5-130) has already been calculated and is given by (5-91) with the appropriate change in indices. Since the spectral shape is calculated from the long-time behavior of (5-130), only the pole on the real axis at $z=\omega_j+\omega_k$ need be considered in the amplitude. Performing the integral using the residue theorem and taking the absolute square in order to obtain the probability yields the line shape

$$\frac{1}{8}\frac{|\boldsymbol{\mu}_{sl}\cdot\boldsymbol{\mathscr{E}}_0|^2|\boldsymbol{\mu}_{lg}\cdot\boldsymbol{\mathscr{E}}_0|^2|\boldsymbol{\mu}_{sj}\cdot\langle\mathbf{k}|\mathbf{E}_s^{(-)}|\text{Vac}_s\rangle|^2}{[(\omega_j+\omega_k-\omega_i)^2+\frac{1}{4}\Gamma_i^2][(\omega_j+\omega_k-2\omega_p)^2+\gamma^2][(\omega_j+\omega_k-\omega_p-\omega_l)^2+\frac{1}{4}(\gamma+\Gamma_l)^2]}$$
$$(5\text{-}131)$$

The first two Lorentzian denominators in (5-131) are similar to the result for a linear scattering process except that the Lorentzian containing the exciting light frequency is peaked at $2\omega_p$ instead of ω_p and the width is now twice the exciting light spectral width. The third Lorentzian in (5-131) is seen to contain the effect of the intermediate state, both in its frequency mismatch with the exciting light and in its homogeneous width. The line shape (5-131) is for resonance two-photon Raman scattering in which the development of the final-state amplitude proceeds uninterrupted by dephasing interaction. In contrast, the two-photon-induced fluorescence, in which incoherent effects actually produce a population in $|s\rangle$ and will also contain a linewidth contribution from the state s, is discussed in Section 5.5.

The calculation (5-131) is only for coherent pulses of light and does not expose the effects of optical coherence on the process. It is unfortunate that the method of calculation does not lend itself readily to understanding the two-photon excitation from a steady beam as well as it does one-photon excitation. However, the origin of the light coherence effect is seen in a qualitative manner from a Golden Rule argument. The two-photon transition rate in the second order of perturbation theory is given by (Loudon, 1973):

$$\frac{1}{\tau} = \frac{2\pi e^4}{\hbar^4} \left| \sum_i \frac{(\hat{e}_i \cdot \mathbf{\mu}_{fi})(\hat{e}_2 \cdot \mathbf{\mu}_{ig})}{\omega_p - \omega_i} \right|^2 \delta(2\omega_p - \omega_f)\langle \mathbf{E}^{(-)}(\mathbf{r}_0)\mathbf{E}^{(-)}(\mathbf{r}_0)\mathbf{E}^{(+)}(\mathbf{r}_0)\mathbf{E}^{(+)}(\mathbf{r}_0)\rangle$$

$$(5\text{-}132)$$

so it is the second-order autocorrelation function which determines the rate. It was shown in Section 5.3 that chaotic and coherent light sources, although they may have the same degree of first-order coherence, differ in their degree of second-order coherence. The same photon bunching effects observed by Hanbury-Brown and Twiss (1957) makes the two-photon absorption rate twice as fast in chaotic light as in coherent light. There have been a number of attempts to measure this effect in solids (Shiga and Imanura, 1967; Carusotto *et al.*, 1969), but so far it has not been observed for a dilute gas. The effect of light coherence on the two-photon induced fluorescence of a dilute gas is important because the photodetector of the incident light and its statistical characteristics is a molecule which only then produces fluorescence that is observed with a conventional photomultiplier.

5.7. Recent Resonance Fluorescence Concepts and Experiments

The recent improvements in laser technology have resulted in a significant upsurge in the popularity of resonance Raman scattering as a probe of molecular structure. In spite of the large number of papers involving one or another facet of resonance Raman, only a small fraction of these are experiments directed at understanding the pertinent phenomena associated with resonance scattering. In most instances the technique is used as an extension of conventional ground-state vibrational spectroscopy. In this section we discuss some of those experiments that focus on the nature of the scattering process and consequently upon the excited-electronic-state properties that generate and modify the inherent character of the resonance scattering.

Experimentally, Raman scattering has been characterized in several different ways. Perhaps the most widely accepted characterization of Raman scattering is through the frequency relationship between the incident and scattered radiation. The scattered radiation retains a fixed-frequency separation from the incident radiation, regardless of the absolute energy of the latter, and when this fixed-frequency relationship also holds for a resonant excitation between the incident and scattered radiation, the scattered radiation is often classified as resonance Raman scattering. Usually (although recall the theory in Section 5.2.4), as the frequency of the incident radiation nears a resonance, the Raman scattering cross section will increase. The enhancement of the Raman cross section with approaching resonance has been observed and studied in numerous systems (Suzuka *et al.*, 1972) and has been termed "preresonance enhancement." This enhancement is influenced by interference in systems with more than three levels. However, when resonance is reached, it often happens that the Raman emission, seen off-resonance, is replaced by a strong emission that remains at fixed absolute energy. This emission is identical in character to the relaxed fluorescence which is typically generated by the common spectroscopic practice of irradiating fluorescent samples with a broad-spectrum light source well within the absorption manifold. At room temperature, even with a spectrally sharp excitation, this relaxed fluorescence is significantly broader in energy than either the conventional Raman and preresonance Raman emission or the incident light. In these instances it is often not clear whether the Raman is replaced by the more intense broad emission or merely obscured because of either inherent or experimental problems associated with observing a weak sharp line on an intense broad background containing many photons at the frequency of the sharp line.

On the other hand, there are molecular systems where sharp resonant excitation within an absorption manifold generates sharp Raman-like emission that tracks with excitation frequency over the full spectral range of the absorption. This sharp emission may or may not be associated with a broad emission background that is inherent to the resonant absorber. In some instances the sharp Raman-like reemission is simply the fluorescence line narrowing (Szabo, 1970, 1971; Riseberg, 1972a,b, 1973; Personov *et al.*, 1972; Motegi and Shionoya, 1974; Watts and Holton, 1974; Delsart *et al.*, 1974; Marchetti *et al.*, 1975). This effect occurs when the absorption band of a fluorescent molecule is actually an envelope for a continuous distribution of discrete spectral lines that originate from molecules that are in different noncommunicative environments (inhomogeneous broadening). For example, in frozen solutions at liquid-helium temperatures, molecules are trapped in their particular environment. Excitation with radiation that is spectrally sharp with respect to the absorption envelope permits only a narrow population of molecules to enter into the scattering process and only Raman can occur. As the exciting frequency is shifted, different resonant populations become involved and the subsequent emission will be at a new frequency—essentially following the excitation frequency like a Raman process. However according to the discussion in Section 5.5 it is easily seen that the identification of this emission as scattering or fluorescence is not established by observing the frequency alone. By raising the temperature, thermal effects can lead to the exchange of environmental sites by the molecules. Line broadening attributed to this exchange has been monitored in systems of even large molecules such as perylene in undecane (Personov *et al.*, 1972). At room temperature the site-exchange rate is usually much faster than typical nanosecond fluorescence lifetimes, and the bulk of the emission generated even by resonant monochromatic radiation is going to be broad. Of course, the site exchange represents a dephasing process in the spirit of Section 5.5.2. Vibrationally relaxed fluorescence that is spectrally sharper than the corresponding absorption band is expected when an inhomogeneous system is pumped by a narrow-band source up to a vibrational level higher in energy than the one emitting. In this case the linewidth of the fluorescent light carries information about the vibrational relaxation rate or additional homogeneous contribution to the linewidth of the higher level (Marchetti *et al.*, 1975). These results are in contrast to expectations for a homogeneously broadened transition, in which case each molecule has the identical absorption line shape for the time scale of the probe process.

The commonly encountered practicalities that were just discussed, i.e., strong broad fluorescence obscuring the Raman process and line narrowing effects, are complications that arise in the study of resonance Raman as a result of environmental perturbations upon the relevant molecule or atom. It would therefore be desirable to gain full knowledge of homogeneous systems as a first step in the investigation of resonance scattering. Recently, several groups have examined the resonant emission originating from the excitation of dilute atomic beams with single-frequency cw dye lasers (Wu *et al.*, 1975; Gibbs and Venkatesan, 1975, 1976). In these experiments, the Doppler broadening of the absorption was very small and the incident radiation was spectrally sharper than the absorption. The absorption line shape was very nearly homogeneous, with a half-width corresponding to roughly the reciprocal of the lifetime. The lifetime in these isolated atomic systems is very nearly purely radiative in nature. The intense resonance emission displayed a spectral width that was sharper than the natural radiative-lifetime-determined width observed in the absorption spectrum. This result is, of course, predicted by theory (Weisskopf, 1931; Heitler, 1954; see also Section 5.2.5). Fundamentally, the narrowing arises because without coupling to energy sinks and reservoirs, the system defined by the radiation and atom must conserve energy. If an isolated system is irradiated either on or off resonance with monochromatic light, there is no source for a line-broadening mechanism (other than some very small quantum fluctuations) for the scattered radiation. This result emphasizes an important point—that for an isolated atom of molecule, there is one and only one resonant radiative process. The resonant state contributes only once to the second-order perturbation expression for the scattering. The transition from off-resonance scattering to resonance-enhanced scattering involves the increased contribution of the resonant level over and above the superposition of states that comprise the polarizability. Monochromatic light when interacting with an isolated atom or molecule generates a stationary state of well-defined energy that can be described as a superposition of many contributing nonstationary states of the total system. As resonance is approached, the contribution to this superposition of states from the resonant nonstationary atomic or molecular state increases to a degree determined by the relevant electric-dipole-moment coupling elements and intramolecular nonradiative damping. For resonant states that are highly radiative, the resonant contribution to the scattering can be many orders of magnitude more intense than the remaining off-resonance contributions from the effective polarizability. By nonradiatively damping the resonant contribution sufficiently, one can be left with only the off-resonance scattering. This effect was proposed for Br_2 (Friedman *et al.*, 1976a). In the dilute gas phase, Br_2 displays resonance emission from the single vibrational levels of the $B(^3\Pi_{0u}^+)$ state (Kiefer, 1974). However, when Br_2 is matrix-isolated in argon, although the corresponding discrete quasi-sharp absorption lines remain, there is little if any resonance enhancement of the Raman scattering when the cw dye laser is tuned on to these discrete absorption peaks $X \rightarrow {}^3\Pi_{0u}^+$. This loss of the resonance contribution is as attributed to a medium-induced short nonradiative lifetime due to vibrational relaxation which occurs on a time scale at least an order of magnitude faster than the several-microsecond radiative lifetime (Friedman *et al.*, 1976b).

The degree to which resonance Raman could be damped or quenched was for awhile the criterion for distinguishing between scattering and resonance fluorescence. It was even suggested that the latter could be quenched while the former could not. Several experiments (Holzer *et al.*, 1970; Fonch and Chang, 1972; Berjot *et al.*, 1972, St. Peters *et al.*, 1973) were performed to study the quenchability of resonance scattering, and a lively controversy arose centered on the alleged distinction between the two resonance processes. Williams *et al.* (1974) demonstrated experimentally in the time domain that there was a continuous transition from off-resonance to on-resonance scattering and that associated with this transition was a change in the time dependence of the scattering. Using excitations of nanosecond laser pulses, they saw the time response of the scattering vary from cotemporal with the laser pulse to free induction decay as the peak excitation frequency was tuned from off-resonance to on-resonance, respectively. This effect corresponds to the principles discussed in Section 5.2.2. Thus, it was demonstrated that for isolated molecules there is only one resonance scattering process. The dichotomy of types of resonance scattering processes arises from the relative time scales of the radiative process and the quenching process. That is, if inelastic collisions or nonradiative damping occur on a time scale that is short compared with the radiative lifetime of the free molecule, the resonance emission will be effectively quenched by these processes. The effect of homogeneous relaxation processes is readily seen from Eq. (5-38). Consider the time-integrated scattering—cw-induced scattering—from a single resonant state i that has a homogeneous width Γ_i. This width may be entirely radiative, Γ_{ir}, or be broadened by homogeneous processes such as vibrational relaxation with a characteristic decay time Γ_{VR}^{-1} (any number of homogeneous processes could be included in Γ_{VR}). Then for the resonance case $\omega_p = \omega_{gi}$, and detection at the peak frequency of the scattered light, $\omega_s = \omega_{ij}$, Eq. (5-38) yields the ratio of resonance Raman in the presence of vibrational relaxation to that in the absence of quenching, or

$$P(\infty, \Gamma_{VR})/P(\infty, \Gamma_{ir}) = (\Gamma_{ir}/\Gamma_{VR})^2 \qquad (5\text{-}133)$$

It is important that this result is quite independent of the spectral width of the light source and of the time width of an exciting pulse if the signal is integrated over all times. In typical cases involving impurity centers in solids, sharp absorption spectra are observed for only lower-lying excited states (Hochstrasser and Prasad, 1974). The vibrational relaxation times are in the range 10 ps except for the zero-point level, which usually displays close to a radiative lifetime, typically 10–500 ns. Thus, the ratio (5-133) is expected to be typically 10^{-6} to 4×10^{-10} for relaxable vibrational levels in solids. The scattering from these levels is therefore expected to be less than is found in dilute gases by these same factors in cases where only radiative decay can occur in the gas. On the other hand, for the 0–0 transition in solids, there may be only radiative decay, in which case the scattering will be much more efficient for resonance with 0–0 than with any other band in the spectrum. This discussion shows clearly that there should not generally be a similarity in the relative peak strengths within a profile for resonant scattering and absorption. One exception to this occurs when the electronic state from which the resonance occurs is not the lowest excited state of the system. The best-studied examples of this case are the hemeproteins (Spiro and Strekas, 1972). The visible and Soret $\Pi-\Pi^*$ transitions are rapidly relaxed into low-energy metal and d–d types of states (Hochstrasser, 1970). Thus, the resonance Raman intensities for excitation in the 0–0 region are no stronger than those in the vibrational levels (Spiro and Strekas, 1974; Friedman and

Hochstrasser, 1973). The effect of vibrational or other nonradiative relaxations on the resonance Raman profile are readily distinguished from the effects of interference (Rimai *et al.*, 1971; Friedman and Hochstrasser, 1973). Since when interference is contributing to the profile, the scattering will not only decrease with increasing frequency, but it will ultimately increase again. For relaxation, on the other hand, only the scattering intensity will usually continue to decrease or become constant as the frequency increases. In an isolated system the scattering can be either a long or a short process; consequently, different systems will exhibit different degrees of damping in various media. In addition, for systems that can be expected to require a long interaction time for scattering, the transition from off-resonance to on-resonance should be accompanied by a marked change in quenchability for the scattering. Another situation which results in an exceptionally large value for (5-133) arises when there is very efficient internal conversion into the ground state, for example S_1 of azulene (Friedman and Hochstrasser, 1974b).

In most resonance scattering experiments, the molecules of interest are not isolated from environmental perturbations that can act to quench or alter the resonant scattering. We therefore consider below those experiments that expose this interplay among resonance scattering, damping, and environmentally induced fluctuations.

For an isolated molecule irradiated with monochromatic radiation, only one type of resonance scattering process can be expected to occur, but additional features are expected in the presence of collisions. In both diatomic iodine and atomic Na it was observed (Rousseau *et al.*, 1975; Carlsten and Szoke, 1976) that in addition to the Raman-like emission which tracks the excitation in frequency, there was also a pressure-dependent component to the scattering that originated from the resonant state having a fixed peak frequency independent of excitation frequency. The two contributions to the scattering (the spectral properties of which indicate that the same resonant state is involved for both processes) have different time dependence. In pulsed laser experiments it was shown that the time dependence of the Raman-like near-resonant scattering was like that of the exciting pulse, while the resonant-fluorescence-like scattering decayed on a time scale longer than one independent of the exciting pulse. The observation of the long-lived excitation-frequency-independent scattering was explained as originating from pseudo-elastic collisions which destroy the phase relationship between the incident and scattered radiation. Because of the short time interval for the collision, these pseudo-elastic collisions redistribute the excitation over the full width of the same resonant state, giving rise to the excitation-frequency-independent emission.

In the language developed in Section 5.5 we would understand that the collisions can bring about the formation of a phase uncorrelated population of excited states which would not otherwise exist. For dilute gas-phase diatomic molecules and atoms, the discrete resonant levels are very nearly uncoupled from other discrete levels; however, when the molecule is in the condensed phase, the solvent or matrix can induce important coupling among the discrete levels. Resonant excitation of an excited vibronic level can therefore be expected to generate emission from both the resonant vibronic level and from relaxed or partially relaxed levels (Section 5.5.6). At low temperatures the degree to which the environment can take up the excess vibrational energy of the resonantly excited molecule determines in part the partitioning of excitation into resonant and relaxed emission. At sufficiently low temperatures, the perturbation of the molecular levels by the matrix results in a nonradiative contribution to the damping constant due to vibrational relaxation. The stronger the coupling, the faster vibrational relaxation and the greater the damping of the resonant scattering. The fraction of resonance Raman to relaxed emission is approximately proportional to the ratio of Γ_i to Γ_{VR} (Friedman and Hochstrasser, 1975), where Γ_i is the natural damping constant for the vibronic state i and Γ_{VR} is the linewidth contribution due to vibrational relaxation (the total damping constant is $\Gamma_i + \Gamma_{VR}$). For levels with long natural lifetimes, the resonance scattering may represent an insignificant fraction of the total quantum yield of emission.

In matrix-isolated Br_2 (Friedman *et al.*, 1976a), resonant excitation of the discrete low-lying vibrational levels of the $B^3\Pi_{0u}^+$ electronic state results in emission that is within signal to noise, all relaxed except for a weak non-resonance-enhanced Raman signal. With a natural lifetime in the microsecond regime for these levels and an upper limit to the vibrational relaxation time of nanoseconds for $n \geq 11$ (Friedman *et al.*, 1976a), the expected ratio of quantum yields is roughly four orders of magnitude in favor of relaxed emission. A dramatic example of this effect is seen in the Raman scattering of the perylene crystal (Hochstrasser and Nyi, 1976). There is a strong preresonance Raman and a strongly peaked Raman excitation profile corresponding to the 0–0 band of the spectrum at 1.6°K. However, no Raman scattering is observed when the resonances occur with levels at higher energy than 0–0. The perylene crystal presents a case where excimers are formed after excitation and all the conventional light emission is at ca. 4000 cm^{-1} to lower energy than the 0–0 band. Thus, the Raman in this case is providing a method of measuring the relative rates of the two very fast processes of excimer formation, starting from two different vibrational levels of a separated dimer.

The effects of vibrational relaxation discussed above occur only in relaxable level resonances, which at low temperatures encompass all except the zero-point level (zero-phonon level) of the emitting state, and this includes phonon sidebands. At higher temperatures even the zero-phonon band widens by virtue of vibrational relaxation resulting from induced absorption of phonons and other dephasing effects. Thus, the effect of increasing the temperature is expected to cause partial recovery of the uniform Raman profile found in the dilute gas, but the dominant result of excitation at any frequency will be the creation of an incoherent population in the manner described in Section 5.6.2.

By adding nonradiative contributions to the width of the vibrational levels of the resonant electronic state corresponding to channels other than vibrational relaxation, one can increase the fraction of resonance scattering to relaxed emission. This effect was demonstrated by Friedman and Hochstrasser (1975) by increasing the intersystem crossing rate in fluorescein in solution by the addition of KI.

As the actual lifetime of a state becomes shorter, resonance scattering becomes a progressively greater fraction of the total quantum yield of emission. It was shown that in azulene (Friedman and Hochstrasser, 1974a), where the levels of S_1 have lifetimes on the order of picoseconds, the emission originating from near-resonant excitation of low-lying vibrational levels of the first excited singlet state of azulene (in a naphthalene host crystal) consists of Raman scattering, nonthermalized relaxed emission, and completely relaxed emission from the zero-point level. As anticipated, the relative quantum yields of these three types of emission were comparable.

For levels with subpicosecond lifetimes, it is expected that resonance Raman scattering can compete effectively with both the formation of an incoherent population and the vibrational relaxation process, even at high temperatures in the condensed phase. In heme proteins the porphyrin Q electronic-state vibrational levels have been shown to have subpicosecond lifetimes on the basis of quantum yield measurements (Friedman and Hochstrasser, 1974a; Adar *et al.*, 1976), absorption line shape, and RRS excitation profiles (Friedman *et al.*, 1976b). For the heme proteins, with diffuse porphyrin absorption bands, resonant excitation generates emission that is nearly all resonance Raman even at room temperature. In these systems the protein acts as a matrix for the porphyrin chromophore, providing a continuum of lattice levels through which the excitation can relax (Friedman and Hochstrasser, 1973). Actually, the observation of laser sharp emission when these systems are irradiated in their homogeneously broadened absorption bands is perhaps the clearest demonstration of the so-called Heitler effect.

In recent experiments (Friedman *et al.*, 1976b) on ferrocytochromes b_5 and c at 6°K, the relative quantum yield of relaxed fluorescence to Raman scattering from the zero-point energy-level region was shown to increase from near zero to nearly 50% as the excitation was tuned onto the low-lying vibrationally excited levels. This system presents some characteristics typical of those ideas discussed in Section 5.6. At sufficiently low temperature

(6°K) excitation in the zero-point level yields only a laser sharp Raman-like scattering. The level width is about 100 cm^{-1}, and this result is indicative of the fact that broadening is homogeneous, but not by virtue of a T_2 process. If dephasing had been competitive, an excited-state population would have been produced and a broad emission would have resulted. On the other hand, excitation of the vibrational level at $0+700 \text{ cm}^{-1}$ results in a broad emission from the origin region, suggesting that the vibrational relaxation process is incoherently generating the appropriate excited-state population in the sense of Section 5.6.5. Experiments such as this one on extremely complex systems expose fundamental aspects of the scattering process. The key feature here is that the homogeneous widths of all levels are very large compared with laser linewidths and with readily achievable spectral resolution. Also, the extremely short lifetimes of the states make time-resolved experiments essentially trivial. This is because boxcar integrators and similar devices are not needed in the circumstance that rapid homogeneous non-radiative processes already provide a built-in time gating.

It was previously emphasized that for condensed phase systems, shorter nonradiative lifetimes (arising from a process other than vibrational relaxation) result in resonance Raman being a greater fraction of the emission quantum yield. However, as seen from Eq. (5-38), the shorter the nonradiative lifetime, the smaller is the fraction of the incident field converted to Raman. The conversion to Raman scattering in the steady-state experiment can be obtained from Eq. (5-102) by summing the contributions scattered into all wavevectors, and dividing by the incident light intensity. Thus, from Eq. (5-102) the total Raman photon counting rate per molecule is

$$\frac{R_R}{N} = \sum_{\text{pol.}} L^3 \int \frac{d^3k}{(2\pi)^3} \frac{d}{dt} \langle a_k^\dagger(t) a_k(t) \rangle = \gamma_{gi}^2 \Gamma_{ir}^{(j)} \frac{(\Gamma_i + \Gamma_j)/\Gamma_i}{\omega_{ip}^2 + (\Gamma_i + \Gamma_p)^2/4} \tag{5-134}$$

where γ_{gi} is the Rabi frequency and $\Gamma_{ir}^{(j)}$ is the radiative rate of the process $i \rightarrow j$. So the fraction α_R converted to scattering is given by R_R divided by the number of incident photons per second:

$$\alpha_R = L \cdot N_g \cdot \omega_p |\mu_{ig}|^2 \Gamma_{ir}^{(j)} (\Gamma_i + \Gamma_p)/3\hbar c \Gamma_i (\omega_{ip}^2 + (\Gamma_i + \Gamma_p)^2/4] \tag{5-135}$$

where N_g is the number of molecules per unit volume N/L^3). On resonance ($\omega_{ip} = 0$), α_R is given by

$$\alpha_R = (\text{const.}) \, \Gamma_{ir}^{(j)}/\Gamma_i(\Gamma_i + \Gamma_p) \tag{5-136}$$

Thus, when the incident light spectral width (Γ_p) is large compared with the full width of the resonant state, the conversion varies at Γ_i^{-1}, whereas for small Γ_p it varies as Γ_i^{-2}. For a given level, as the nonradiative lifetime is decreased, both the absorption and Raman cross sections decrease; however, at the absorption maximum, the absorption cross section decreases as

$1/\Gamma_i$, whereas the Raman scattering cross section for essentially mono-chromatic excitation decreases as $1/\Gamma_i^2$. Consequently, as the lifetime decreases, the peak intensity of the absorption decreases, but the resonance Raman intensity will decrease faster. Simultaneously, both the absorption line shape and Raman excitation profile will broaden and more closely resemble each other. A lifetime dependence of this type was observed (Adar *et al.*, 1976) in ferrocytochrome b_5 at 6°K, where the two equivalent components of a nearly degenerate state have different lifetimes. The high-resolution resonance Raman excitation profile over these levels had the same lineshape as the corresponding absorption, but the relative inten-sities of the two components were markedly different in the two spectra. The relative intensity of the Raman scattering from the shorter-lived component was much smaller than was observed in absorption, as predicted from the theory.

It is possible to approach a quantitative understanding of the Raman process from phenomenological considerations. By virtue of the extensive discussion of the three-level system presented here it is evident that the quantum yield of Raman scattering is given by

$$\varphi_R = \frac{R_R}{R_p} = \frac{\Gamma_{ir}^{(j)}}{\Gamma_i} \tag{5-137}$$

where R_R is the rate of Raman scattering and R_p is the rate at which photons are removed from the incident light beam. Note that the quantum yield φ_R is independent of frequency and is not necessarily a useful practical quantity. The large yield arises from the fact that Raman is seen as spontaneous emission following the interaction of the light field with the Lorentzian resonance. When the peak frequency of the light is far from resonance, both R_R and R_p are small, but φ_R may be near to 1. The quantity R_p can be calculated in terms of measurable parameters, then R_R (s^{-1}) can be obtained from Eq. (5-137). For R_p, one finds that

$$R_p = \frac{L^3}{\hbar\omega_p} \int k(\nu)I(\nu)\,d\nu \tag{5-138}$$

where L^3 is the volume of the active region, $k(\nu)$ is the absorption coefficient (cm^{-1}), and $I(\nu)$ is the light intensity (ergs cm^{-2} s^{-1} per unit frequency), which is hardly attenuated. The integral in Eq. (5-138) is the convolution of the incident light pulse with the absorption spectrum. A more useful practical parameter is the fraction α_R. Equations (5-137) and (5-138) yield

$$\alpha_R = \frac{\Gamma_{ir}^{(j)}}{\Gamma_i} \frac{L \int I(\nu)k(\nu)\,d\nu}{\int I(\nu)\,d\nu} \tag{5-139}$$

If Lorentzian frequency distributions are chosen for the light pulse and the absorption spectrum, then with the conventional relation between the Einstein coefficient for induced absorption and the integrated absorption coefficient, it is readily shown that α_R is as given in Eq. (5-135). The foregoing relationships refer to a three-level system in near-resonance excitation. In the conventional Raman effect or a three-level system excited very far off-resonance it is necessary to include the rapidly oscillating terms that were excluded in the rotating wave approximation.

The relationship between excited-state lifetime—radiative, nonradiative, dephasing—and resonance Raman cross section as an experimentally useful physical parameter is only beginning to be realized. Only a very limited number of experiments have been performed along these lines, and these have been quite preliminary. The contents of this research indicate that resonance Raman can, in fact, be a very powerful probe of excited-state dynamical processes and of otherwise difficult-to-expose phenomena that affect and perturb the excited-state lifetimes.

Appendix. Contour Integration

The majority of the mathematical techniques used in this chapter are familiar methods in quantum mechanics and elementary calculus. An extremely important exception is the use of complex analysis to evaluate the integrals containing matrix elements of the Green's operator (sometimes called the resolvent):

$$G(z) \equiv (z - H)^{-1} \tag{5-A1}$$

where H is some Hamiltonian operator and z is a complex frequency, as in Eq. (5-89). In (5-18) z is chosen as $\omega - i\eta$, for reasons justified below.

Since a Hamiltonian operator is by physical necessity Hermitian, $G(z)$ is analytic; i.e., its matrix elements are differentiable with respect to z in the entire complex plane except along the real axis, where H has eigenvalues. If the spectrum of H is entirely discrete and nondegenerate, the points where $G(z)$ is not analytic form a set of singularities called simple poles, since by definition a function $f(z)$ has a simple pole at z_0 if

$$\lim (z - z_0)f(z) = A_0 \neq 0 \tag{5-A2}$$

If the spectrum of H forms a continuum, as it does in the situations discussed in this chapter, $G(z)$ is a multivalued operator, and a branch cut is said to exist along the real axis where H has continuous eigenvalues. $G(z)$ tends to different limits as the branch cut is approached from above or below. The domain of integration for all the integrals considered in this chapter is a line running above the real axis from $-\infty + iz'$ to $\infty + iz'$, where $z' = 0$. The choice of $z = \omega + i\eta$ in (5-18) is valid because the limit $\eta \to 0$ is taken after the integration is performed, since the choice of z' is immaterial.

Consider the integration specified by (5-90) (where c runs from positive ∞ to negative ∞ above the real axis:

$$\langle a|e^{-iH(t-t')}|a\rangle = \frac{1}{2\pi i}\int_c dz\, e^{-iz(t-t')}\langle a|G(z)|a\rangle; \qquad t \geq t' \quad (5\text{-A3})$$

This relation is given ample justification in Goldberger and Watson (1964). The matrix element of $G(z)$ in (5-A3) has the form (5-25):

$$(z - \omega_a + \tfrac{1}{2}i\Gamma_a)^{-1} \qquad (5\text{-A4})$$

and the integral is then

$$\frac{1}{2\pi i}\int_c dz\, e^{-iz(t-t')}(z - \omega_a + \tfrac{1}{2}i\Gamma_a)^{-1} \qquad (5\text{-A5})$$

where in our situations Γ_a is taken independent of z. If any contribution from the branch cut is ignored, the contour c may be changed to an infinite semicircle γ_R in the lower complex plane, closed by a line above the real axis. The integral about the semicircle is zero by Jordan's lemma, since if $\alpha < 0$, $f(z)$ tends uniformly to zero as $|z| \to \infty$ and $\pi \leq \arg z \leq 2\pi$:

$$\lim_{R\to\infty}\int_{\gamma_R} dz\, e^{i\alpha z} f(z) = 0 \qquad (5\text{-A6})$$

The integral about the closed contour is now equal to the integral along c. The integral about the closed contour may be evaluated by the residue theorem,

$$\int_c dz\, g(z) = 2\pi i \sum_{\substack{j \\ \text{inside } c}} A_j \qquad (5\text{-A7})$$

where for simple poles the residues A_j are given by (5-A2). This is valid where $g(z)$ has only simple poles as singularities. The integral of (5-A5) has only one pole at $z = \omega_a - i\Gamma_a/2$ and application of (5-A2) and (5-A7) gives exponential decay,

$$|\langle a|e^{-iH(t-t')}|a\rangle|^2 = e^{-\Gamma_a(t-t')} \qquad (5\text{-A8})$$

Jordan's lemma and similar mathematical reasoning involving the residue theorem are the tools used in evaluating the integrals in the text. Some caution should be used in closing the contour in the lower half-plane as discussed above, because of the multivalued nature of $G(z)$. A more rigorous evaluation is discussed by Goldberger and Watson (1964, p. 445), who consider the analytic continuation of $G(z)$ into the second Riemann sheet where the poles actually exist. The classic text by Titchmarsh (1964) may be consulted for a complete discussion of complex analysis.

References

Adar, F., Gouterman, M., and Aronovitz, S., 1976, *J. Phys. Chem.* **80**:2184.
Anderson, P. W., 1954, *J. Phys. Soc. Japan* **9**:316.
Atkins, P. W., and Woolley, R. G., 1970, *Proc. Roy. Soc. (London)* **A321**:557.
Babiker, M., Power, E. A., and Thirunamachandran, T., 1974, *Proc. Roy. Soc. (London)* **A338**:235.
Ben-Reuven, A., and Mukamel, S., 1975, *J. Phys.* **A8**:1313.
Berestetskii, V. B., Lifshitz, E. M., and Pitaeveskii, L. P., 1971, *Relativistic Quantum Theory*, Part 1, pp. 1–4, in: *Course of Theoretical Physics* (L. D. Landau and E. M. Litshitz, eds.), Pergamon Press, New York; Addison-Wesley Publishing Company, Inc., Reading, Mass.
Berg, J. O., Langhoff, C. A., and Robinson, G. W., 1975, *Chem. Phys. Lett.* **29**:305.
Berjot, M., Jacon, M., and Bernard, I., 1972, *J. Can. Spectrosc.* **17**:60.
Berne, B. J., and Pecora, R., 1976, *Dynamic Light Scattering*, John Wiley & Sons, Inc., New York.
Bernheim, R., 1965, *Optical Pumping*, pp 1–86, W. A. Benjamin, Inc., Menlo Park, Calif.
Carlsten, J. L., and Szoke, A., 1976, *Phys. Rev. Lett.* **36**:1667.
Carusotto, S., Polacco, E., and Vaselli, M., 1969, *Lett. Nuovo Cimento* **11**:628.
Condon, E. V., and Shortley, G. H., 1957, *The Theory of Atomic Spectra*, p. 87, Cambridge University Press, New York.
Delsart, C., Pelletiei-Allard, N., and Pelletiei, R., 1974, *Opt. Commun.* **11**:84.
Feld, M. S., and Javan, A., 1969, *Phys. Rev.* **177**:540.
Fonch, D. G., and Chang, R. K., 1972, *Phys. Rev. Lett.* **29**:536.
Fowler, G. N., 1962, Semiclassical theory of radiation, in *Quantum Theory* (D. R. Bates, ed.), Vol. 3, pp. 48–51, Academic Press, Inc., New York.
Friedman, J. F., and Hochstrasser, R. M., 1974, *Chem. Phys.* **1**:457.
Friedman, J. F., and Hochstrasser, R. M., 1974a, *Chem. Phys.* **6**:145.
Friedman, J. F., and Hochstrasser, R. M., 1974b, *Chem. Phys.* **6**:155.
Friedman, J. F., and Hochstrasser, R. M., 1975, *Chem. Phys. Lett.* **33**:225.
Friedman, J. F., Rousseau, D. L., and Adar, F., 1976a, manuscript in preparation.
Friedman, J. F., Rousseau, D. L., and Bondybey, V., 1976b, manuscript in preparation.
Gibbs, H. M., and Venkatesan, T. N. C., 1975, *IEEE J. Quantum Electron.* **QE11**:91D.
Gibbs, H. M., and Venkatesan, T. N. C., 1976, *Opt. Commun.* **17**:87.
Glauber, R. J., 1963, *Phys. Rev.* **131**:2766.
Glauber, R. J., 1964, Optical coherence and photon statistics, in *Quantum Optics and Electronics* (C. De Witt, A. Blondin, and C. Cohen-Tannoudji, eds.), pp. 78–84, Gordon and Breach Science Publishers, Inc., New York.
Glauber, R. J., 1970, Quantum theory of coherence, in *Quantum Optics* (S. M. Kay and A. Maitland, eds.), pp. 53–125, Academic Press, Inc., New York.
Goldberger, M. L., and Watson, K. M., 1964, *Collision Theory*, Chaps. 3 and 8, John Wiley & Sons, Inc., New York.
Gordon, R. G., 1965, *J. Chem. Phys.* **42**:3658.
Hanbury-Brown, R., and Twiss, R. Q., 1957, *Proc. Roy. Soc. (London)* **A243**:291.
Harris, C. B., Shelby, R. M., and Cornelius, P. A., 1977, *Phys. Rev. Letts.* **38**:1415.
Heitler, W., 1954, *Quantum Theory of Radiation*, Oxford University Press, Inc., New York.
Hochstrasser, R. M., 1970, Structural sensitive aspects of the electronic spectrum, in: *Probes of Structure and Function of Macromolecules and Membranes* (B. Chance, C. Lee, and J. K. Blasie, eds.), pp. 57–64, Academic Press, Inc., New York.
Hochstrasser, R. M., and Novak, F. A., 1976, *Chem. Phys. Lett.* **41**:407.
Hochstrasser, R. M., and Nyi, C. A., 1977, in press.
Hochstrasser, R. M., and Prasad, P., 1974, Optical spectra and relaxation in molecular solids, in: *Excited States* (E. Lim, ed.), Vol. 1, pp. 79–128, Academic Press, Inc, New York.

Holzer, W., Murphy, W. F., and Bernstein, H. J., 1970, *J. Chem. Phys.* **52**:399.

Huber, D. L., 1968, *Phys. Rev.* **170**:418.

Huber, D. L., 1969, *Phys. Rev.* **178**:93.

Jackson, J. D., 1975, *Classical Electrodynamics*, 2nd ed., p. 668, John Wiley & Sons, Inc., New York.

Jortner, J., and Mukamel, S., 1974, Preparation and decay of excited molecular states, in: *Proceedings of the First International Congress on Quantum Chemistry* (R. Daudel and B. Pullman, eds.), pp. 145–209, D. Reidel Publishing Company, Dordrecht, The Netherlands.

Kiefer, W., 1974, *Appl. Spectry.* **28**:115.

Kotani, A., and Toyozawa, Y., 1976, *J. Phys. Soc. Japan* **41**:1699.

Kubo, R., 1954, *J. Phys. Soc. Japan* **9**:935.

Kubo, R. Takagawara, T., and Hanamura, E., 1975, presented at Oki Seminar on Physics of Highly Excited States in Solids (Tomakomai, Hokkaido, Sept. 10–13).

Kushida, T., 1976, *ISSP Preprint*, Ser. A, No. 773, University of Tokyo.

Lefebvre, R., 1971, *Chem. Phys. Lett.* **8**:306.

Loudon, R., 1973, *The Quantum Theory of Light*, Oxford University Press, Inc., New York.

Marchetti, A. P., McColgin, W. C., and Eberly, J. H., 1975, *Phys. Rev. Lett.* **35**:387.

Merzbacher, E., 1970, *Quantum Mechanics*, 2nd ed., pp. 362–369, John Wiley & Sons, Inc., New York.

Motegi, N., and Shionoya, S., 1974, *J. Luminescence* **8**:1.

Mower, L., 1966, *Phys. Rev.* **142**:799.

Mower, L., 1968, *Phys. Rev.* **165**:145.

Mukamel, S., 1975, Thesis, Tel-Aviv University.

Mukamel, S., and Jortner, J., 1975, *J. Chem. Phys.* **62**:3609.

Mukamel, S., Ben-Reuven, A., and Jortner, J., 1976, *Chem. Phys. Lett.* **38**:394.

Nitzan, A., Jortner, J., and Berne, B. J., 1972, *J. Chem. Phys.* **57**:2870.

Personov, R. I., Al' Shitz, E. I., and Bykovskaya, L. A., 1972, *Opt. Commun.* **6**:169.

Power, E. A., and Zienau, S., 1959, *Phil. Trans. Roy. Soc.* (*London*) **A319**:549.

Rebane, K. and Saari, P., 1976, *Hot Luminescence and Relaxation Processes in the Centers of Luminescence*, Report at the International Conference on Luminescence, September 1–5, 1975, Tokyo, Japan.

Rimai, L., Heyde, M. E., Heller, H. C., and Gill, D., 1971, *Chem. Phys. Lett.* **10**:207.

Riseberg, L. A., 1972a, *Phys. Rev. Lett.* **28**:789.

Riseberg, L. A., 1972b, *Solid State Commun.* **11**:469.

Riseberg, L. A., 1973, *Phys. Rev.* **A7**:671.

Roman, P., 1965, *Advanced Quantum Theory*, Addison-Wesley Publishing Company, Inc., Reading, Mass.

Rousseau, D. L., Patterson, G. D., and Williams, P. F., 1975, *Phys. Rev. Lett.* **34**:1306.

Series, G. W., Optical pumping and related topics, in: *Quantum Optics* (S. M. Kay and A. Maitland, eds.), pp. 395–482, Academic Press, Inc., New York.

Shen, Y. R., 1974, *Phys. Rev.* **B9**:622.

Shiga, F., and Imanura, S., 1967, *Phys. Lett.* **25A**:706.

Spiro, T. G., and Strekas, T. C., 1972, *Proc. Natl. Acad. Sci. U.S.* **69**:2622.

Spiro, T. G., and Strekas, T. C., 1974, *J. Amer. Chem. Soc.* **96**:338.

St. Peters, R. L., Silverstein, S. P., Lapp, M., and Penney, C. M., 1973, *Phys. Rev. Lett.* **30**:191.

Suzuka, I., Mikami, N., Udagawa, Y., Kaya, K., and Ito, M., 1972, *J. Chem. Phys.* **57**:4500.

Szabo, A., 1970, *Phys. Rev. Lett.* **25**:924.

Szabo, A., 1971, *Phys. Rev. Lett.* **27**:323.

Titchmarsh, E. C., 1964, *The Theory of Functions*, Oxford University Press, Inc., New York.

Titulaer, O. M., and Glauber, R. J., 1966, *Phys. Rev.* **145**:1041.

Toyozawa, Y., 1976, *J. Phys. Soc. Japan* **41**:400.

Van der Waals, J. H., and De Groot, M. S., 1967, Magnetic interactions related to phosphorescence, in: *The Triplet State* (A. B. Zahlen, ed.), pp. 101–132, Cambridge University Press, London.

Watts, R. K., and Holton, W. C., 1974, *J. Appl. Phys.* **45**:873.

Weisskopf, V., 1931, *Ann. Phys.* **9**:23.

Weisskopf, V., 1933, *Z. Physik* **85**:451.

Weisskopf, V., and Wigner, E., 1930, *Z. Physik* **63**:54.

Weisskopf, V., and Wigner, E., 1931, *Z. Physik* **65**:18.

Williams, P. F., Rousseau, D. L., and Dworetsky, S. H., 1974, *Phys. Rev. Lett.* **32**:196.

Woolley, R. G., 1971, *Proc. Roy. Soc. (London)* **A321**:557.

Wu, F. Y., Grove, R. E., and Ezekiel, S., 1975, *Phys. Rev. Lett.* **35**:1426.

Author Index

Subject Index